To Bryan Booker,

thanks for participating in our Users' conference.

With best regards

A. Alan B. Pritsker

SLAM II

NETWORK MODELS FOR DECISION SUPPORT

SLAM II

NETWORK MODELS FOR DECISION SUPPORT

A. Alan B. Pritsker, Ph.D., P.E.
Pritsker & Associates, Inc., and
Purdue University

C. Elliott Sigal, M.D., Ph.D.
Cardiovascular Research Institute, and
Department of Medicine
University of California, San Francisco

R. D. Jack Hammesfahr, Ph.D.
College of Business Administration
Memphis State University

Library of Congress Cataloging-in-Publication Data

PRITSKER, A. ALAN B.
Slam II network models for decision support / A. Alan B. Pritsker, C. Elliott Sigal, R. D. Jack Hammesfahr.
p. cm.
Includes bibliographies and index.
ISBN 0-13-812819-7
1. Decision-making—Computer simulation. I. Sigal, C. Elliott (Charles Elliott) II. Hammesfahr, R. D. Jack, III. Title.
IV. Title: Slam 2 network models for decision support. V. Title: Slam two network models for decision support.
T57.95.P75 1989
658.4'03'52—dc19 88-28121
CIP

Editorial/production supervision: Patrice Fraccio
Cover design: Lundgren Graphics, Ltd.
Manufacturing buyer: Mary Noonan

Published by Prentice-Hall, Inc.
A Division of Simon & Schuster
Englewood Cliffs, New Jersey 07632

Printed in the United States of America

10 9 8 7 6 5 4 3 2 1

ISBN 0-13-812819-7

Prentice-Hall International (UK) Limited, *London*
Prentice-Hall of Australia Pty. Limited, *Sydney*
Prentice-Hall Canada Inc., *Toronto*
Prentice-Hall Hispanoamericana, S.A., *Mexico*
Prentice-Hall of India Private Limited, *New Delhi*
Prentice-Hall of Japan, Inc., *Tokyo*
Simon & Schuster Asia Pte. Ltd., *Singapore*
Editora Prentice-Hall do Brasil, Ltda., *Rio de Janeiro*

DEDICATION

To Kendra, Zachary, Bryan, Lauren, and Mark
for whom modeling is an end, not a means.
Alan

To my sons, Jason and Samuel.
Elliott

To the memory of MACDUFF.
Jack

PREFACE

"Computer simulation, now far more than a testing tool, remakes the factory floor ... U.S. manufacturers who do not wring what they can out of this new technology may soon find themselves falling behind not only their more nimble domestic competitors but also their offshore rivals." (Business Week, August 17, 1987). This is but one of many calls for more modeling and simulation. This book responds to the demand for fundamental information on modeling and simulation practice. It describes the practical aspects of the modeling and simulation process and demonstrates, through case studies, how that process is being used to support decision making. It is based on modeling and simulation applications performed over the last 15 years by simulation specialists at Pritsker & Associates. The material has been organized to portray a unified approach to problem solving using network modeling and simulation.

The SLAM II network language is used throughout the book and provides a common basis for model presentation. It has the advantages of wide usage, full support, and advanced capabilities. It is self-documenting and supports an organized and structured approach to problem solving. This book includes SLAM II network models for the following decision areas: project planning, production scheduling and control, computer integrated manufacturing design, material handling equipment evaluation, risk analysis, logistics analysis, inventory policy setting, reliability assessment, quality control, and service systems design. Throughout the book, the case studies emphasize performance measurement, key decision elements, network models, and the use of computer simulation to support decision-making. Increased profitability has resulted from these modeling and simulation projects.

The text introduces modeling concepts and simulation assessments. This material can be taught at either the advanced undergraduate level or in an introductory graduate level course. Because the text provides an understanding of problem situations, performance measures, and models for a wide variety of industrial situations, it also serves as a basic reference book.

Book Organization

SLAM II Network Models for Decision Support is a presentation of modeling and simulation practice. The book is organized into 21 chapters. Chapter 1 introduces management decision-making and the role that can be played by network modeling in the decision-making process. Chapter 2 describes the steps of the modeling and simulation process. The sequential and iterative application of the steps is discussed. In Chapters 3 through 5, a summary of SLAM II® is provided, including definitions of network symbols, support routines, input procedures, and output reports.

In Chapter 6, the concepts and procedures associated with decision trees, decision networks, and risk assessment are presented. SLAM II network models for these areas are described, and the procedures for using such network models in management decision-making are explored.

Chapters 7 through 10 are devoted to project planning. Basic concepts, terminology, and performance measures are contained in Chapter 7. Procedures are provided for estimating project completion-time distributions, activity start times, activity criticality indexes, and project cost for PERT/CPM network models. Chapter 8 combines the concepts of risk and decision making into project planning by illustrating how SLAM II networks can be used to model activity and project failures. Chapter 9 uses network models to schedule the activities of a project. Procedures for computing the early and late start times of each activity are demonstrated, and estimates of an activity's slack time are obtained. In addition, networks for assessing project plans under the condition of limited resources are developed. Applications of network modeling are presented in Chapter 10 for decisions made in contract negotiations, R&D planning, and the long-term funding of nuclear fusion research.

Chapters 11 and 12 address the two central questions of inventory control: when to place an order and how much to order. SLAM II models for periodic review and transaction reporting inventory control procedures are developed. The modeling of backorders and lost sales is demonstrated. Diverse methods for characterizing demands are discussed, along with a model of a multi-commodity inventory system. Chapter 13 defines reliability and quality control concepts in terms of SLAM II network models. Strategies for when to replace equipment and how to safeguard facilities from theft or sabotage are evaluated using a technique for finding the shortest path in a SLAM II network.

The terminology and modeling requirements associated with logistics systems analysis are presented in Chapter 14. This chapter integrates networks representing reliability, maintainability, supply, transportation, personnel and

---

SLAM and SLAM II are registered trademarks of Pritsker & Associates, Inc.

training, and support equipment into a single system model. An example of decision making for a logistics system at an Air Force depot is presented.

Chapters 15 through 19 discuss decision-making and model-building procedures for production planning and manufacturing operations. Chapter 15 presents the basic concepts of production planning, including performance measures and their use for decision-making. Models of computer-integrated manufacturing (CIM) systems and just-in-time (JIT) procedures are presented. Special topics related to production planning are detailed in Chapter 16. These include labor-limited queueing situations, inspection and maintenance processes, and order routing. Chapter 17 describes how network models can be used to evaluate scheduling and sequencing procedures.

Models for evaluating material handling systems are described in Chapters 18 and 19. Procedures for analyzing conveyors, pipelines, and overhead cranes are given. In Chapter 19, the SLAM II Material Handling Extension is described. Models of an automated guided vehicle system (AGVS) and a flexible manufacturing system (FMS) are built.

Chapters 20 and 21 describe the modeling of service systems. In Chapter 20, the differences between service and manufacturing systems are described. The performance measures appropriate for service systems are presented, and suggested procedures for performing projects in a service system environment are discussed. Models are developed for scheduling operating rooms in a hosptial, analyzing check-processing capabilities at a bank, and evaluating claims flow in an insurance company. Procedures for modeling computer communication networks are presented in Chapter 21.

Acknowledgements

The material in this book covers a wide spectrum of applications and network model developments. We have drawn on the research and the applications of many people. Several of the applications and analysis projects of Vic Auterio, Ed Clayton, Albert Garcia, Larry Moore, Bob Mortensen, Chuck Taylor, and John Vanston have been used directly. At Pritsker & Associates, we have made use of material developed by Steve Duket, Mark Erdbruegger, Bob Gaskins, Hank Grant, Bill Lilegdon, Dan Murphy, Ken Musselman, Jean O'Reilly, Charles Standridge, Jim Whitford, and Dave Wortman. We acknowledge the work of these individuals. Our discussions with them made this a better book.

In addition to the above, we have had support from many individuals interested in the development of network models for decision support. A complete list of such individuals would be hard to compile. We do want to recognize the efforts of Dave Bartkus, Jerry Chubb, Steve Drezner, Salah Elmaghraby, Mary

Grant, Bill Happ, Joe Mize, Don Phillips, Jerry Sabuda, Stan Settles, Dick Smith, Ware Washam, Jim Wilson, Phil Wolfe, and Gary Whitehouse. Throughout the years, there have been many colleagues, associates, and students who have worked on network modeling and simulation, and we gratefully acknowledge their contributions, assistance, and encouragement.

Putting the vast amount of material that we developed over the years into this book has been a difficult challenge. We want to thank our editors at Prentice Hall, Valerie Ashton and Patrice Fraccio, for encouraging and supporting the presentation and design features included in the book. We want to thank the following individuals for helping us prepare the models and figures for the numerous examples: Rita Hipps, Dave Laval, Jean O'Reilly, Anne Pritsker, Jeff Pritsker, Bill Lilegdon, Kathy Shoemaker, Tim Simpson, Miriam Walters and Ruth Whitis. Throughout the writing of the manuscript, many of the above people reviewed and edited parts of the manuscript. In particular, we would like to thank Charles Standridge for his complete review, Dan Murphy for his minority opinions, and Jim Evans for recommending a rewrite of Chapters 2 through 5.

The preparation, integration, and typesetting for the book was performed at Pritsker & Associates. Kathy Shoemaker did an excellent job of preparing the final pages including much of their design and layout. The project and author coordinator was Miriam Walters who performed these functions expertly and always in a friendly way. Kathy and Miriam have our deep appreciation for putting the book in its final form.

A. Alan B. Pritsker
West Lafayette, Indiana

C. Elliott Sigal
San Francisco, California

R. D. Jack Hammesfahr
Memphis, Tennessee

TABLE OF CONTENTS

1 MANAGEMENT DECISION MAKING

1.1 THE DECISION PROCESS

A ***decision*** is a selection of a course of action. A decision ***process*** specifies the steps used to make such a selection. In this book, we are interested in the process leading to management decision making. We assume that managers and systems analysts working together can develop implicit or explicit objectives associated with decision problems, can formulate alternative choices, can define performance measures to evaluate and compare these choices, and can assign qualitative or quantitative values to these performance measures. Our view of the decision process follows the paradigm shown in Figure 1-1, which is similar to the one given in [26].

The starting point in the decision process is the recognition of a problem. For some problems, decision-making can proceed immediately without any intervening steps. Other problems need to be better defined in terms of a system description, performance measures, and objectives. After defining the problem, the decision-making process can be initiated. A line with a double arrow is shown between a problem and problem definition. This is meant to indicate that, based on the problem definition, the understanding of the problem situation could change. That is, iterations of problem definition could take place. Throughout the description of Figure 1-1, lines with double arrows are used to indicate such an iterative process.

There are two other paths from problem definition to decision making: direct experimentation and model building. Direct experimentation includes observations and measurements on the defined system, which, after analysis, could lead to decision making. The last route to making a decision involves model building, model simulations and computations, analysis, and then decision making. It is this route that the material in this book best supports.

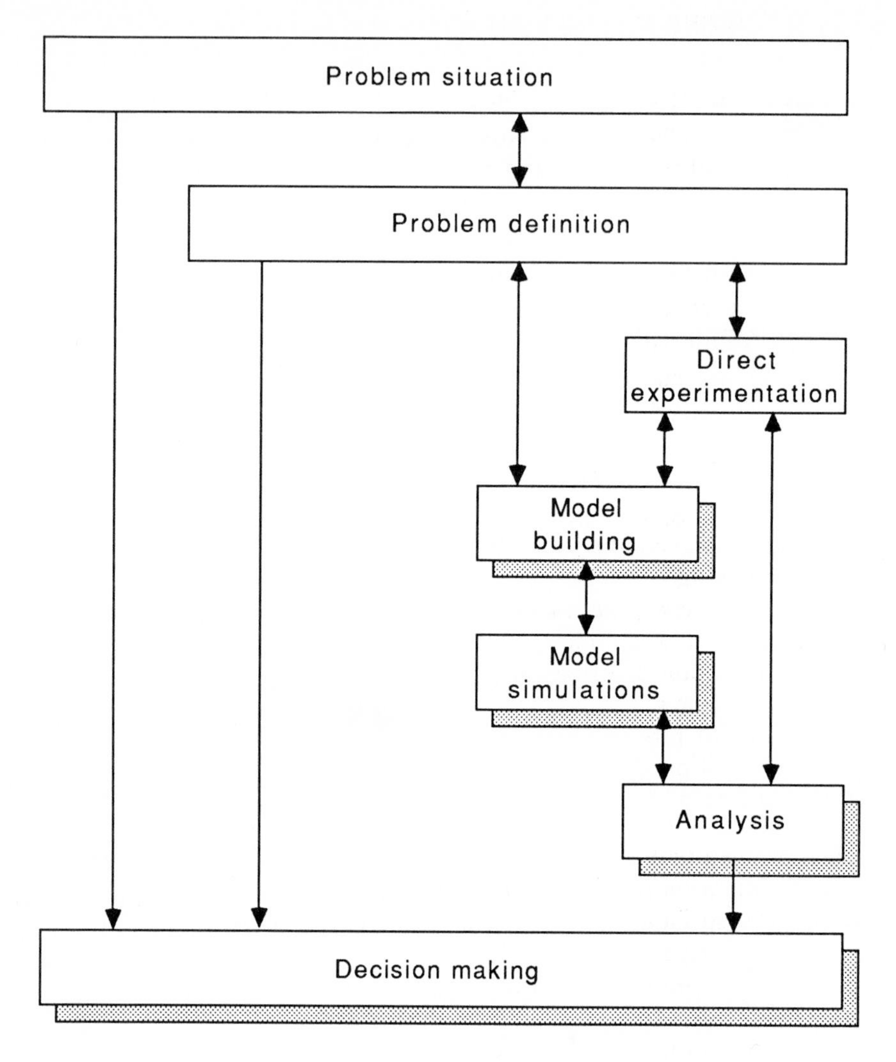

Figure 1-1 Paradigm for the decision process.

The advent of computers and advanced simulation software facilitates the performance of the steps in the decision process. The foundation of the approach presented herein emphasizes the building of a model on which simulations can be performed. The model should include a definition of the external factors that affect a decision and a specification of the measures to judge the worth of a decision. When using a simulation language, performance measures for each decision alternative are estimated by exercising the model on a computer. These performance measure estimates are then used in decision making.

1.2 CATEGORIZING MANAGEMENT DECISION MAKING

Decisions can be categorized into three levels: strategic planning, management control, and operational control [6]. ***Strategic planning*** is the process of deciding on the objectives of an organization, on changes in these objectives, on the resources needed to achieve these objectives, and on the policies that are to govern the acquisition, use, and disposition of resources. ***Management control*** is the process by which managers assure that the required resources are obtained and used effectively and efficiently in the accomplishment of the organization's objectives. ***Operational control*** is the process of assuring that specific tasks are carried out effectively and efficiently. A list of problem situations categorized by these three levels is presented in Table 1-1.

Table 1-1 Areas of Decision Making for Procedural Systems

Strategic planning

1. Design of new processes
2. Design of new policies
3. Determination of effect of different priorities
4. Design of new systems
5. Forecast of production levels
6. Determination of required resources
7. Estimation of cost of alternatives

Management control

8. Determination of how to improve throughput
9. Determination of effect of changes in resource capacities
10. Determination of effect of delays in raw materials
11. Determination of how to relieve bottlenecks
12. Determination of effect of change in demand
13. Determination of effect of equipment failures
14. Determination of system efficiency

Operational control

15. Determination of capacity
16. Determination of bottlenecks
17. Determination of operational requirements
18. Assessment of in-process inventories
19. Determination of utilizations
20. Determination of critical operation rates
21. Determination of best staffing configurations
22. Scheduling jobs
23. Scheduling resources

For each of the levels we can identify three types of decision: structured, semistructured, and unstructured [4, 6]. ***Structured decisions*** are those that are understood well enough to be automated. ***Semistructured decisions*** are those that involve some judgment and subjective analysis and are sufficiently well defined to enable the use of models. ***Unstructured decisions*** are those for which the alternatives, objectives, and consequences are ambiguous. Our main thrust is on decisions that are semistructured at each of the three decision-making areas. The models presented assist managers in making decisions about problems that are sufficiently complex to require the aid of a computer and that involve judgment and subjective analysis by the manager. Because of this, the decision maker must be drawn into the problem definition and model-building stages of the decision process.

To support the decision process, a database is required. The integration of the decision maker, decision models, and decision support database constitutes a decision support system (DSS). A pictorial view of the organization of these subsystems is shown in Figure 1-2, which is similar to the one given in [25].

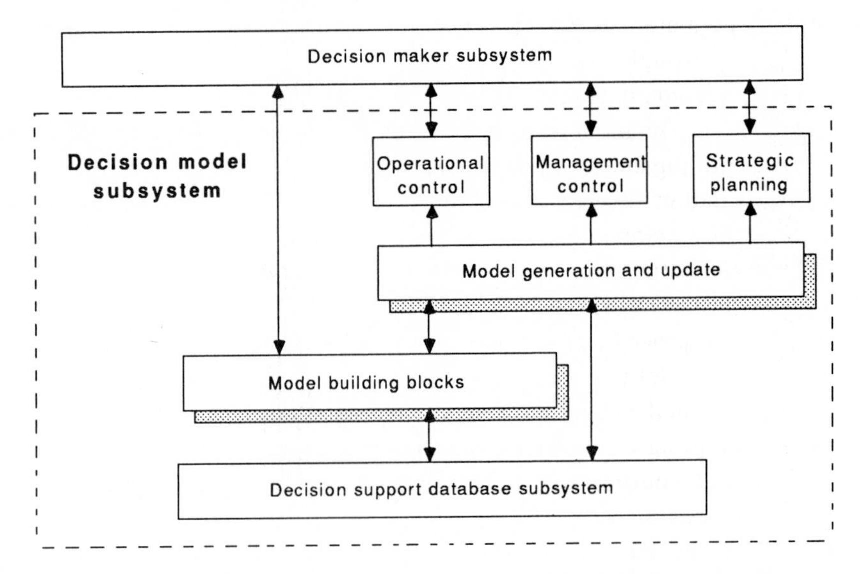

Figure 1-2 Conceptual framework of a model-based decision support system.

In this book, many problem situations are described and models are developed to show how to analyze semistructured decisions in strategic planning, management control, and operational control. The interfaces between the decision model, the decision support system, and the decision maker are described. The major emphasis of the book, however, is on the building of models for supporting the making of decisions.

1.3 MODELING FOR DECISION MAKING

Models are descriptions of systems. In the physical sciences, models are usually developed based on theoretical laws and principles. The models may be scaled physical objects, mathematical equations and relations, or graphical representations. The usefulness of models has been demonstrated in describing, designing, and analyzing systems. Many students are educated in their disciplines by learning how to build and use models. Model building is a complex process and in most fields is an art. The modeling of a system is made easier if (1) physical laws are available that pertain to the system, (2) a pictorial or graphical representation can be made of the system, and (3) the variability of system inputs, elements, and outputs is manageable.

Industrial engineers, managers and administrators, management scientists, and operations researchers deal primarily with procedural systems. These individuals and their respective fields are attempting to bring order out of chaos through the modeling and analysis of procedural systems. For our purposes, ***procedural systems*** are those that rely on information flow and decision making to implement stated or implied policy. Emphasis is placed on improving performance through procedural changes or through new designs regarding scheduling, sequencing, distribution, allocation, layout, and similar functions (see Table 1-1). The modeling of procedural systems is often more difficult than the modeling of physical systems for the following reasons: (1) few fundamental laws are available; (2) procedural elements are difficult to describe and represent; (3) policy statements are hard to quantify; (4) random components are significant elements; and (5) human decision making is an integral part of such systems.

In modeling procedural systems, the specification of the proper elements to include in the model is difficult. It is recommended that a dialogue between the model builder and the decision maker be established at the earliest possible time. The model-building process should be considered as an iterative one. ***First-cut*** models should be built, analyzed, and discussed. The time that elapses between the establishment of the purpose for building a model and the production of outputs of the first-cut model should be kept as small as possible. In many cases, this will require heroic assumptions and a willingness on the part of the modeler to expose his or her potential ignorance of the problem. In the long run, such a dialogue should pay off, as modeling inaccuracies will be discovered more quickly and corrected more efficiently than would be possible otherwise. Furthermore, discussions that include tentative projections tend to clarify issues and to promote suggestions.

A good modeler draws the decision maker into problem-solving activities for three reasons: (1) to ensure that a proper problem formulation has been developed, (2) to assist in determining the system details that can be omitted from the model, and (3) to set the stage for implementing the results of the analysis. In this light, an analyst should iteratively formulate the purpose for which the model was built and modify and rebuild the model as necessary. The modeling vehicle advocated in this book, SLAM II, has been designed to be used in this mode [21].

1.4 DECISIONS AND DECISION MAKERS

All decision makers are not alike nor should they be. Let us characterize decision makers with respect to how they might use a decision support system (DSS) if one were available. The ***hands-off*** user reads reports that are automatically generated by the DSS, but is not in direct contact with it either through requests or knowledge of the underpinnings of the DSS. This user's approach is to assume that, with the facts, a good decision can be made.

A second type of decision maker is the ***requester***. This type of user employs an intermediary to use the DSS. He or she frames the questions, interprets the results, and uses the answers to make decisions. The requester is not knowledgeable as to how the answers are obtained, but knows that they provide additional information that is helpful in making decisions.

The third type of decision maker is the ***hands-on*** variety. This type of user views the DSS input device as an extension of himself or herself and through predetermined interfaces performs experiments with models. Through direct interaction, he or she asks such questions as: "Why can't I get a particular type of information?" "Shouldn't a model be built that provides me with an estimate of the rate of return?" The hands-on user learns about the DSS by using it and provides inputs that can cause the DSS to adapt to his or her needs.

The fourth type of decision maker is the ***renaissance*** decision maker. This type of user functions as part of a team and does not feel uncomfortable talking in terms of database systems and modeling, but is at home when making decisions. The renaissance decision maker knows how to set requirements for information, can prescribe the types of reports he or she wants, and can ask questions concerning the details of a model. Although part of the team, the renaissance decision maker must not dominate the team and must be able to consider the decision process separate from the modeling and analysis process. This integrated team approach can avoid the expenditure of large amounts of the decision maker's time and the possible degradation of objectivity that is sometimes associated with a decision maker who builds and uses his or her own models. The hands-on and renaissance decision makers are the ones to whom this book is oriented.

1.5 NETWORK MODELING

A ***network model*** is a graphical picture of a system. In this book, we use network models to describe and analyze procedural systems. The network approach decomposes a complex problem situation into its elements by using network symbols to describe both the elements and their operational procedures. Network models provide a means for communicating system elements in the context of a problem definition. Such communication is an important part of the dynamics of any project group involved in the decision-making process. Furthermore, the network approach provides standardization, comprehension, and consistency in modeling activities.

Network models separate the model-building process from the analysis process. Thus, the building of a network does not require knowledge of analysis procedures for computing performance measures. This is analogous to modeling a system with a set of equations without being burdened with defining or understanding a solution technique for the set of equations. By segregating modeling and analysis, network languages encourage participation in modeling by all members of the project team, thus improving communication within the team. Complex analysis procedures are relegated to experts as a subsequent task.

The advantages of network modeling in the fields of electrical engineering, mechanical engineering, civil engineering, and industrial engineering have clearly been established. Examples of network models are electrical circuit diagrams that contain resistor, capacitor, inductance, and generator symbols; freebody diagrams showing the forces on structures; signal flowgraphs that present graphically simultaneous algebraic equations; block diagrams that illustrate system functions; flowcharts that describe logic operations; and activity networks that specify the order in which project tasks are to be performed.

In management science, network models have been employed in project planning, capacity flow, facility location, and other problem situations in which a mathematical programming formulation is applicable. A network language for modeling procedural systems has long been needed. This book demonstrates that SLAM II fills this need by illustrating SLAM II models for diverse problem situations of procedural systems.

1.6 SLAM II MODELING

SLAM II contains a network language that has been designed specifically for the modeling of procedural systems [20, 21]. A SLAM II network consists of a set of nodes and branches. The branches represent activities over which items, called entities, flow. The time for an entity to flow through a branch is the duration of the activity. The nodes of the SLAM II network represent events, storage points (queues), routing decisions, and resource allocation decisions. The complex procedures of a system are modeled through the selection and allocation rules associated with nodes. Unique types of nodes are used to model different aspects of procedural systems. By building a network of nodes connected by branches, complex problems can be modeled.

Entity is a generic term referring to objects, information, or any transaction type. Entities arrive at nodes, where they are stored or routed. When an entity leaves the node, it is routed over one or more of the branches emanating from the node representing the next activities to be engaged by the entity. The entity continues to move through the network until it encounters a terminate node. While the entity moves through the network model, it carries attribute values that can be changed as desired. Decisions regarding an entity's movement can be based on the entity's attributes.

As a part of the SLAM II modeling approach, statistical estimates of quantities to be used in decision making are collected. Reports summarizing the observed quantities are printed automatically as standard output from the analysis of the SLAM II network model.

SLAM II has been designed to allow an analyst and decision maker to model complex procedural systems within a logical systematic framework. SLAM II provides the following capabilities: (1) complex systems can be modeled by extending and integrating simple systems; (2) the SLAM II network model facilitates the discussion of how the model represents the operational system, that is, the network model is a communications mechanism; and (3) SLAM II network models can be developed from structural information concerning the system. Extensive data are not required to build the model and, in fact, the model serves as a means for specifying data requirements. SLAM II networks have been designed and developed over a 20-year period. Extensive applications have been made. Decision makers' needs have been identified during this period, and new features to support these needs have been included in the design of SLAM II. A history of network modeling developments is contained in [20].

1.7 DECISION MAKING USING SLAM II

A SLAM II network model is a graphical picture that characterizes the decision points and algorithms that affect the time-varying behavior of the system variables. These variables relate to entity flow and the time at which milestones (nodes) are reached. Information can be obtained directly from the network diagram relating to the structural aspects of a system's operation. This type of information provides a basis for a qualitative assessment and understanding of the mechanisms that are under the control of decision makers.

When data are available, a quantitative view of the actual flow of the entities through the SLAM II network model can be made and a further understanding of system operation obtained. Such a view indicates the quantity of flow, queue buildups, bottlenecks, and, to some extent, server and resource utilization. The standard outputs from the analysis program provide quantitative measures of these variables.

Decision making using SLAM II involves an assessment as to whether the system meets design requirements with respect to the types of questions that need to be answered to resolve the problem areas listed in Table 1-1. To evaluate alternatives, structural changes or parameter changes are made to the network and their impact determined. The sensitivity of the outputs to such changes is easily assessed, and a comparison between alternatives can be made until a satisfactory system operation is found. This evaluation of alternatives until a satisfactory alternative is found follows the "satisficing" approach to problem solving [14]. Throughout the decision-making process, objectives and requirements are continually reevaluated to determine if different goals are more appropriate for the problem situation under study. Specific details of the modeling and simulation process are presented in Chapter 2.

1.8 INFORMATION PROCESSING AND DATABASES

Database capabilities have grown extensively in the past ten years. Information-processing computers exist in most corporate settings. The availability of timely and accurate data and their conversion into knowledge is no longer a futuristic dream. The time has come for decision makers, analysts, and designers to employ such knowledge to resolve the problems facing them. Network modeling is one method for converting the data stored in databases into the knowledge required for management decision making. It is because of the current state of information processing and databases that the need for the material in this book becomes so important. Network models can now be used for business, manufacturing, and service systems in the same manner that models have always been employed for physical systems.

1.9 PROBABILITY CONCEPTS IN DECISION MAKING

There are many unknowns in decision making. The future cannot be forecast with certainty and all variables associated with a decision cannot be characterized explicitly. Because of this lack of complete information, decision making requires judgments and probabilistic assessments.

Consider the question of whether the selection of an alternative will result in a positive return. There is a chance that a negative outcome results from the selection. The recognition that there is a chance of a negative return is a first step in accepting probability concepts within the decision-making process. The chance can be characterized on a subjective or a relative-frequency basis. A subjective probability is based on a decision maker's feeling about a situation. Thus, if a decision maker believes that there is 1 chance in 10 that an adverse result will occur, the subjective probability estimate is 0.1. A probability based on a relative frequency is the number of times the adverse result occurred divided by the number of times it could have occurred. In this case, the decision maker considers all similar decisions and estimates the fraction of the decisions in which an adverse result occurred.

Potential reasons for the occurrence of a negative return could be an unanticipated environmental state or an incorrectly forecasted future state. A catastrophic event, low demand, or poor weather could also be the cause. Currently, weather prediction, catastrophic events, and economic forecasts are considered from a probabilistic viewpoint.

The recognition of the potential for a negative outcome should not interfere with the implementation process once the decision is made; that is, once an alternative is selected, the decision maker must stand behind it. Psychologically, the chances of a positive return are improved by the forcefulness of the decision maker in presenting the alternative selected. However, the implementation strategy needs to be sensitive to critical contingencies [18] and to be adaptive to meet the needs of the company and personnel affected.

When models are used to assist in the decision-making process, variables in the models can be defined in probabilistic terms. For example, the demand for a product could be characterized as a random variable, and the probability distribution associated with that random variable could be a part of the model. The underlying assumptions associated with different random variables have been described [13, 21] and depicted graphically. Computer programs have been developed to aid in selecting a distribution to use in a model [19]. Alternatively, a histogram of data collected for a situation can be employed.

By combining a probabilistic characterization of the variables included in a model with a probabilistic forecast of possible future states, estimates of the probability of making a correct or incorrect selection can be computed. This probability, coupled with the magnitude of the outcome associated with an alternative solution, provides the decision maker with a risk assessment for the alternative.

Network models provide the structure for including in the decision-making process each of the probabilistic concepts listed previously. The use of probabilistic concepts in the model is determined by the model builder. It is not always necessary to include advanced concepts in models nor is it necessary for the decision maker to be cognizant of all the details associated with each probabilistic concept. In fact, a network structure is ideal for representing probabilistic concepts in a form that is understandable to managers.

1.10 STATISTICAL CONCEPTS IN DECISION MAKING

There are three areas in which statistical concepts are important in decision making: (1) comparing the returns associated with alternatives, (2) determining the probability that an outcome may be beyond a prescribed limit, and (3) projecting the value of an outcome due to a change in one or more input values.

To compare alternatives on a statistical basis, hypotheses are established and statistical tests of hypotheses are performed. These statistical tests provide a means for specifying whether a return for one alternative is superior to a return for another alternative. If the hypothesis is that the returns are the same, a statistical test of the hypothesis estimates the probability of accepting that the values of the outcomes are different. This is referred to as a ***Type I error***. In some instances, it is possible to estimate the probability of accepting the alternatives as statistically the same when the hypothesis is false. This is a ***Type II error***. These types of errors relate to the probability of making a correct decision that was discussed in Section 1.9. When many alternatives are involved, the statistical procedures associated with the design of experiments, including analysis-of-variance (ANOVA) techniques, are used.

To determine the range for the value of an outcome, a confidence interval should be derived. A confidence interval is specified by a lower limit and an upper limit defined to contain a theoretical value with a prescribed probability. The establishment of a confidence interval is a standard procedure in statistical analysis.

Statistical concepts are also used to relate output values to input values through the methods of regression analysis [12, 21]. Regression analysis procedures are described in detail in most statistics books and courses [16].

In building models for management decision making, it is not necessary to have an extensive background in statistical concepts. However, it is important to know that such concepts exist and that additional information can be gleaned from the outputs of models. In this book, we do not dwell on statistical procedures. We advocate their use whenever appropriate to provide a better understanding of modeling and simulation results.

1.11 WILL MANAGEMENT ACCEPT MODELS?

It is often stated that an essential ingredient of successful management decision making is management know-how. Management know-how apparently relates to the ability of the manager to understand what is going on in his or her environment [8]. For management to accept the outputs of models as inputs to the decision-making process, the models must augment and complement the manager's understanding of the problem situation. Another way of stating this is that a manager will accept models if the models provide new knowledge. The knowledge that the manager requires is the know-how concerning the selection of the correct alternative event although conditions may change.

In the preceding sections of this chapter, we have described a decision process that includes a network modeling phase that leads to new knowledge. Our experience with SLAM II has demonstrated that management will accept models and that they will make decisions based on the information, analysis, and presentations from such models.

Semistructured problem situations are complex and require in-depth study. Managers recognize the need for an approach that includes modeling that can be useful in analyzing a broad spectrum of such problem situations. Network modeling and SLAM II modeling in particular provide such an analysis vehicle. This book documents our experiences in using SLAM II in this mode and lays a foundation for management's acceptance of models.

1.12 EXERCISES

1-1. Define the following terms used in Figure 1-1: problem definition; direct experimentation; model building; model simulations; analysis; and decision making. Specify a problem situation and relate the definitions to your approach in reaching a decision.

1-2. Specify potential performance measures for each of the areas of decision making listed in Table 1-1. Categorize the performance measures with respect to their use in strategic planning, management control, and operational control.

1-3. Compare accounting models with models used for decision making.

1-4. List the functions that are required for the model generation and updating blocks shown in Figure 1-2. Define the procedures for performing these functions and interrelationships between the functions.

1-5. Describe and characterize a procedural system. Give an example of a procedural system and discuss policies associated with the system. Define a sequence of steps that would facilitate the implementing of the policies.

1-6. Find five existing policy statements. Analyze the statements and identify elements that make them difficult to translate into operational guidelines.

1-7. In Section 1.4, decision makers are categorized by the following terms: hands-off; requester; hands-on; and renaissance. Relate the positions in a firm to these types of decision makers. Relate courses you have taken to these types of decision makers. Identify tools and techniques that could be used by each type of decision maker.

1-8. Describe network modeling procedures with which you are familiar and discuss how they separate modeling tasks from analysis tasks.

1-9. Discuss the information requirements for building a network model in comparison to the data requirements to analyze a network model.

1-10. Design output formats for presenting performance measures associated with one of the areas of decision making listed in Table 1-1.

1-11. Discuss how advances in database systems might have an impact on decision making and modeling for decision making.

1-12. Describe how you would react to a manager who indicates that he or she was not interested in the probability that an adverse outcome would be obtained, but was only interested in whether it would occur.

1-13. Discuss the ramifications of the analyst who makes the following recommendation to the plant general manager. "My model specifies that we should purchase two cranes, which will result in an increase in throughput of 30%. This increased throughput after deducting the cost of the cranes provides a 10% increase in net profit."

1-14. Discuss the use of regression analysis, analysis of variance, and tests of hypothesis with respect to a decision problem with which you are familiar.

1-15. Write a treatise on the subject "Will management accept models?"

1.13 REFERENCES

1. Banks, J., and J. S. Carson, II, *Discrete-Event System Simulation*, Prentice-Hall, Englewood Cliffs, NJ, 1984.
2. Bartee, E. M., *Engineering Experimental Design Fundamentals*, Prentice-Hall, Englewood Cliffs, NJ, 1968.
3. Blanchard, B. S., and W. J. Fabrycky, *Systems Engineering and Analysis*, Prentice-Hall, Englewood Cliffs, NJ, 1981.
4. Bonczek, R. H., C. W. Holsapple and A. B. Whinston, "The Evolving Roles of Models in Decision Support Systems," *Decision Sciences*, Vol. 11, 1980, pp. 337-356.
5. Bratley, P., B. L. Fox and L. E. Schrage, *A Guide to Simulation*, Springer-Verlag, New York, 1983.
6. Callahan, L. G., Jr., "Brief Survey of Operational Decision Support Systems," School of Industrial and Systems Engineering, Georgia Institute of Technology, February 1979.
7. Cellier, F. E., ed., *Progress in Modelling and Simulation*, Academic Press, New York, 1982.
8. Churchman, C. W., "Managerial Acceptance of Scientific Recommendations," in *Quantitative Disciplines in Management Decisions*, R. I. Levin and R. P. Lamore, eds., Dickenson Publishing Company, Belmont, CA, 1969.

9. Eisner, H., *Computer-Aided Systems Engineering*, Prentice-Hall, Englewood Cliffs, NJ, 1987.
10. Emshoff, J. R., and R. L. Sisson, *Design and Use of Computer Simulation Models*, Macmillan, New York, 1970.
11. Henriksen, J. O., "GPSS - Finding the Appropriate World-View," *Proceedings, Winter Simulation Conference*, 1981, pp. 505-516.
12. Kleijnen, J. P. C., van den Burg, A. J., and van der Ham, R. R., "Generalization of Simulation Results: Practicality of Statistical Methods," *European Journal of Operational Research*, Vol. 3, 1979, pp. 50-64.
13. Law, A. M., and W. D. Kelton, *Simulation Modeling and Analysis*, McGraw-Hill, New York, 1982.
14. March, J. C., and H. A. Simon, *Organizations*, Wiley, New York, 1958.
15. Mihram, G. A., *Simulation: Statistical Foundations and Methodology*, Academic Press, New York, 1972.
16. Mize, J. H., and J. G. Cox, *Essentials of Simulation*, Prentice-Hall, Englewood Cliffs, NJ, 1968.
17. Moore, L. J., and E. R. Clayton, *GERT Modeling and Simulation: Fundamentals and Applications*, Petrocelli/Charter, New York, 1976.
18. Morris, W. T., *Implementation Strategies for Industrial Engineers*, Grid Publishing, Columbus, OH, 1979.
19. Musselman, K. J., W. R. Penick and M. E. Grant, *AID: Fitting Distributions to Observations: A Graphical Approach*, Pritsker & Associates, West Lafayette, IN, 1981.
20. Pritsker, A. A. B., *Modeling and Analysis Using Q-GERT Networks*, 2nd ed., Wiley, New York, 1979.
21. Pritsker, A. A. B., *Introduction to Simulation and SLAM II*, 3rd ed., Wiley and Systems Publishing, New York and West Lafayette, IN, 1986.
22. Pritsker, A. A. B., "Compilation of Definitions of Simulation," *SIMULATION*, Vol. 33, 1979, pp. 61-63.
23. Pritsker, A. A. B., "Models Yield Keys to Productivity Problems, Solutions," *Industrial Engineering*, October 1983, pp. 83-87.
24. Rivett, P., *Model Building for Decision Analysis*, Wiley, New York, 1980.
25. Sprague, R. H., Jr., and H.J. Watson, "A Decision Support System for Banks," *OMEGA*, Vol. 4, No. 6, 1976, pp. 657-671.
26. Thrall, R. M., C. H. Coombes and R.L. Davis, eds., *Decision Processes*, Wiley, New York, 1954.
27. van Horn, R. L., "Validation of Simulation Results," *Management Science*, Vol. 17, 1971, pp. 247-258.
28. Zeigler, B. P., *Theory of Modelling and Simulation*, Wiley, New York, 1976.

2 MODELING AND SIMULATION PROCESS

2.1 INTRODUCTION

The modeling and simulation process focuses on formulating and solving a problem [2, 8, 10, 11, 14]. The modeling process is iterative because the act of modeling reveals important information piecemeal. This information supports actions that make the model and its output measures more relevant and accurate. The modeling process continues until additional detail or information is no longer necessary for problem resolution. During this iterative process, relationships between the system under study and the model are continually defined and redefined. The resulting correspondence between the model and the system establishes a tool that has value and relevance to the participating problem solvers.

Figure 2-1 presents the suggested steps in performing a project in which modeling and simulation are used. The six steps in the process are (1) formulate problem, (2) specify model, (3) build model, (4) simulate model, (5) use model, and (6) support decision making. The iterative nature of the process is indicated by the feedback branches in Figure 2-1. Within step 3, build model, the substeps of develop simulation model, collect data, and define experimental controls are performed concurrently. These substeps are dependent on the information generated by each other. This is also the case for the run model, verify model, and validate model substeps which are a part of step 4, simulate model.

Each of the six steps in the modeling and simulation process are described in this chapter. The approach to performing these steps relies heavily on the material presented in Chapter 1. In particular, evolutionary problem solving procedures are embedded in the process. To achieve problem resolution, an understanding is needed of corporate policy, the decision maker, and the formulated problem.

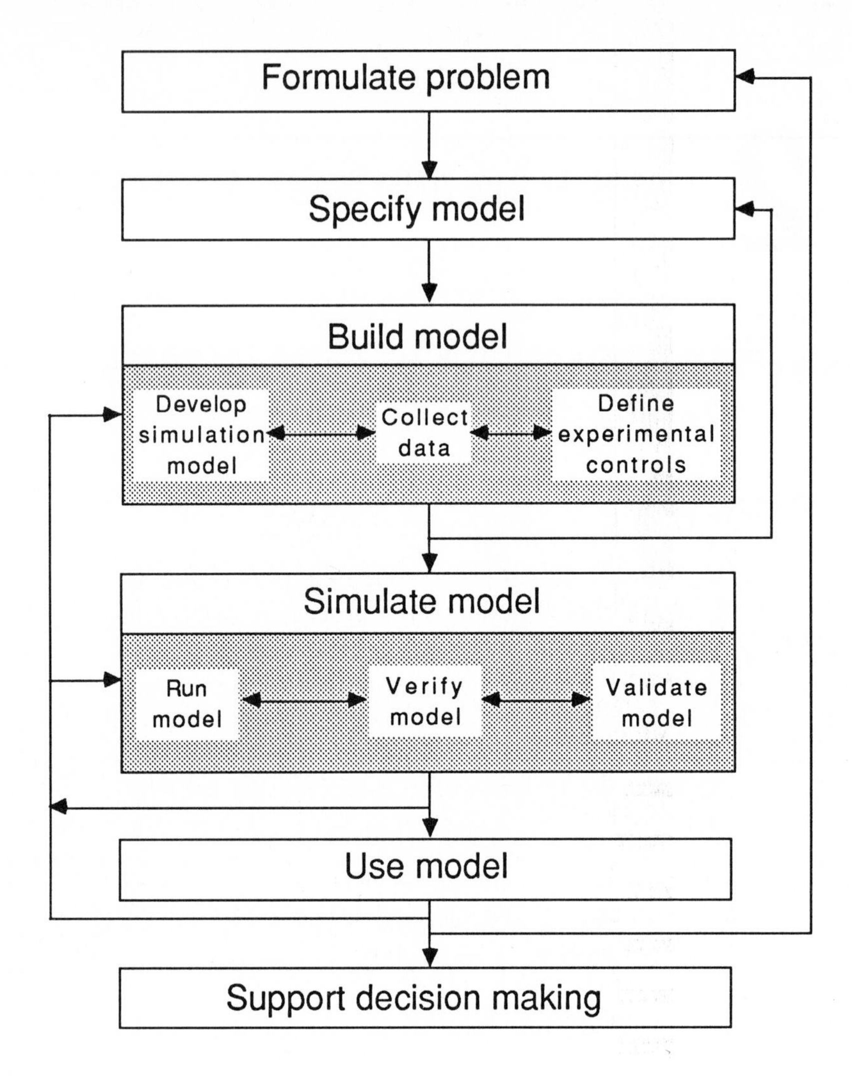

Figure 2-1 Modeling and simulation process.

2.2 FORMULATE PROBLEM

The first step in the problem solving process is to formulate the problem by understanding the problem context, identifying project goals, specifying system performance measures, setting specific modeling objectives, and defining the system to be modeled. These functions serve to guide and bound a project.

A project typically starts with a broad problem statement such as "the factory isn't meeting planned production quantities". Faced with this situation, basic questions should be asked to understand the problem context:

1. What operations and functions produce the system's output?
2. What procedural elements exist in the system's operation?
3. What interactions occur between functional units of the system?
4. What information is available to characterize the operations, functions, and procedures of the system?

The answers to these questions help to formulate a problem statement and to determine if modeling is an appropriate and efficient way to solve the problem. Modeling may not be appropriate because of data unavailability or because the resources required are more costly than the potential savings from solving the problem.

Measures of system performance are typically profits or costs which are a function of operational measures such as utilization, throughput, inventory levels, and quality. Project goals specify what is to be achieved in terms of these performance measures of the system. An example of a project goal is to lower in-process inventory in a factory without decreasing production rates.

Modeling objectives are statements of desired results in terms of performance measures. For example, if a goal is to cut operating costs through a reduction in inventory costs then an objective may be to introduce a control system to decrease inventory levels below 500 units without changing plant capacity. Objectives should be stated in such a way that an action can be taken or a decision can be made if the objective is met.

Once objectives are established, the system to be modeled can be defined. System boundaries and components (subsystems) should be determined based on the performance measures to be estimated. For example, if the throughput of an inspection station is to be estimated, then the inspection station must be included in the model. If the utilization of inspectors is a performance measure, then the inspectors and the inspection station must be represented.

After the project goal, critical performance measures, modeling objectives, and the system to be modeled have been defined, the use of a model should again be re-examined. In defining the problem, an obvious solution may become evident or the system definition may be too imprecise to warrant a modeling effort.

Problem formulation is extremely important and often is not given sufficient attention. When this step is performed correctly, modeling and analysis efforts are better focused. This leads to effective problem solving and large returns from the investment made in the project. Good problem formulation helps to avoid projects where the wrong problem is solved or where a model is developed when it is not needed.

2.3 SPECIFY MODEL

There is an art to conceptualizing a model. The modeler must understand the structure and operating rules of the system and be able to extract the essence of the system without including unneccessary detail. Good models tend to be easily understood, yet have sufficient detail to reflect realistically the important characteristics of the system. The crucial questions in model specification focus on what simplifying assumptions are reasonable to make, what components should be included in the model, and what interactions occur among the components. The amount of detail included in the model should be based on the modeling objectives established. Only those components that could cause significant differences in decision-making need be considered.

Close interaction among project personnel is required when formulating a problem and specifying a model. This interaction causes inaccuracies to be discovered quickly and corrected efficiently. Most important is that interactions induce confidence in both the modeler and the decision maker and help to achieve a successful implementation of results.

The second step, specify model, identifies the data requirements for the model. By conceptualizing the model in terms of the structural elements of the system and product flows through the system, a good understanding of the detailed data requirements can be projected. It establishes the schedules, algorithms, and controls required for the model. These decision components are typically the most difficult aspect of a modeling effort.

Models analyzed by simulation are easily changed, which facilitates making iterations between the problem formulation and the model specification steps. This is not the case for other model analysis techniques. Examples of the types of changes that can be made in simulation models which encourage flexibility in model specification are:

1. Setting arrival patterns and activity times to be constant, samples from a theoretical distribution, or using a history of values.
2. Setting due dates based on historical records, material requirements planning (MRP) procedures, or sales information.
3. Setting decision variables based on a heuristic procedure or calling a decision-making subprogram that uses an optimization technique.
4. Including fixed rules or expert-system setting rules directly in the model.

2.3.1 Model Specification Illustration

In a recent study of a flexible manufacturing system (FMS), a detailed model specification was prepared [4]. The project goal was to determine if the FMS could achieve the designed throughput rate. The modeling objectives were to estimate the utilization of work centers, carrousels, material handling equipment, and tools sets, and to provide a tool for analyzing alternative designs and graphically animating the operation of the FMS.

A conceptual model of the flow of parts through the FMS system is shown in Figure 2-2. For each component of the production process depicted in Figure 2-2, a specification was written from which a model could be developed. In addition, the flow of orders was described. The specification provided basic characteristics, assumptions, operation sequences, resource contention issues, and input data requirements for each component in the production process. To illustrate, the model specifications are given for the flow of orders in Figure 2-3 and the review station in Figure 2-4.

2.4 BUILD MODEL

The build model step consists of three substeps: develop simulation model, collect data, and define experimental controls. In the first substep, the simulation model is developed which contains the structural and procedural elements that represent a system. Experimental controls describe the procedures for performing a simulation and analysis of the model. The term ***scenario*** is used to refer to a particular combination of a simulation model, data, and experimental controls [15, 16].

2.4.1 Develop Simulation Model

The secret to being a good modeler is the ability to remodel. Model building should be interactive and graphical because a model is not only defined and developed but is continually refined, updated, modified, and extended. An up-to-date model provides the basis for future models. The following five model building themes support this approach and should be used where feasible:

1. Develop generalized input schemes.
2. Divide the model into relatively small logical elements.
3. Differentiate between physical movement and information flow in the model.
4. Develop and maintain clear documentation directly in the model.
5. Leave hooks in the model to insert extensions or more detail.

Throughout this book, SLAM II network models are used to illustrate these model building procedures.

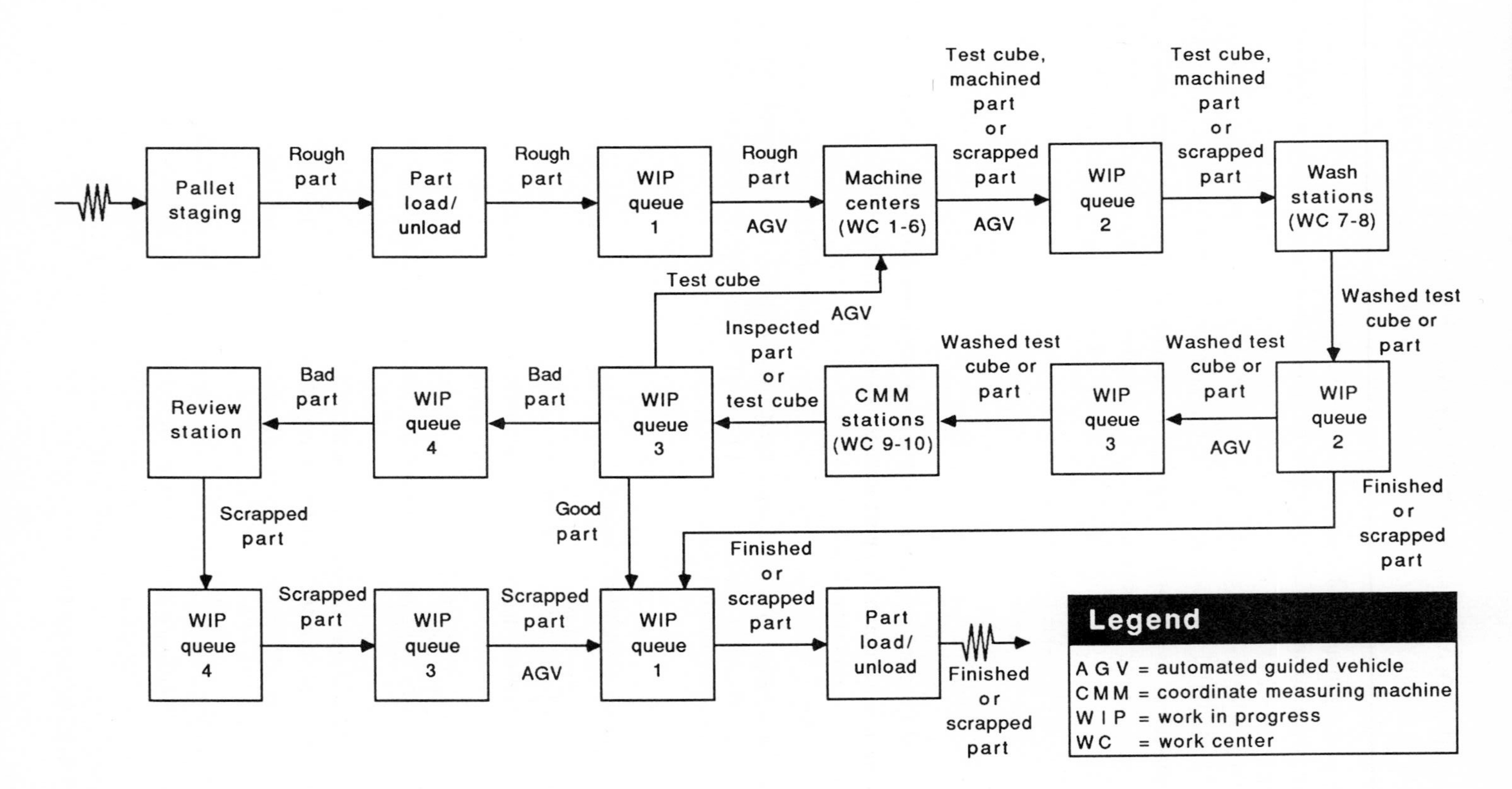

Figure 2-2 Flow of parts through a proposed flexible manufacturing system. Source [4]

FMS Model Specification: Order Processing

Basic characteristics

- There are 57 different part types.
- Each part type requires 1, 2, or 3 passes through the FMS.
- Parts are fixtured, machined, washed, and possibly inspected and reviewed.
- Part-operations are processed individually.
- Parts with multiple machining operations leave the FMS between operations.

Assumptions

- A scheduler module generates an order arrival sequence for the FMS based on current machining center backlogs, up/down states, and toolkit states.
- The sequence generated by the scheduler is an input to the simulation model.
- The scheduler is invoked at simulation start up time and also whenever required.
- The objectives of the scheduler are to balance the load across the machining centers, to be on time with respect to order due dates, limit toolkit changes at the machining centers, limit fixture changes on pallets, and limit resource contention blockages.
- The first parts on the schedule determine the toolkits initially installed at the machines.
- Parts are diverted from a down machining center toward one that is functional.
- Scrapped parts are monitored so that the scheduler can adjust the production schedule accordingly.

Operation Sequences Figure 2-2 presents the flow of parts through the system.

Resource contention issues None

Model inputs

- For each order
  - Part identifications
  - Scheduled start date
  - Scheduled due date

- For each part of the order
  - Part identification
  - Operation identification
  - Time to load part into fixture
  - Identification of the required toolkit
  - Time to perform the machining operation
  - Time to wash the part
  - Inspection requirement
  - Time to inspect the part
  - Probability of failing inspection
  - Time to review a part
  - Time to unload part from fixture

Figure 2-3 Model specification for FMS order processing. Source [4].

FMS Model Specification: Review Station

Basic characteristics

- There is 1 review station.
- Parts are reviewed at the review station to determine why they failed inspection.
- After review, parts exit the system through WIP queue 1 carrousel.

Assumptions

- The review station has a capacity of 1 part.
- The review station is blocked until the reviewed part is removed from the input/output shuttle and loaded onto the carrousel.
- A shuttle can have preventative maintenance and failures.

Operational sequences

- Load bad part onto review station.
- Free input/output shuttle of review station.
- Review part.
- Wait for/seize space in WIP queue 4 carrousel.
- Wait for/seize input/output shuttle of review station.
- Remove reviewed part from review station and place on shuttle.
- Wait for/seize control of carrousel.
- Free review station.
- Rotate shuttle 180 degrees.

Contention issues

- At input/output shuttle entities have the following priorities:

Priority	Contending system entity
1	Scrapped parts at the review station.
2	Bad parts in WIP queue 4 carrousel.

Model inputs

- Time to transfer a part from the shuttle to the review station.
- Time to review a part for each part-operation combination.
- Probability that a part passes the review for each part-operation combination.
- Time to transfer a part from the review station to the shuttle.

Figure 2-4 Model specification for review station. Source [4]

In developing a SLAM II network model, entities are defined that move through the network. An entity represents an object, part, person, or message. It can be thought of as an information packet with a set of attributes as shown in Figure 2-5 where the value of attribute 1 is 22, attribute 2 is 1 and so on. As an entity moves through a network model, its information content is changed. Some changes only modify the characteristics of the entity while other changes identify a new type of entity. For example, a raw material entity, after a processing step, is converted to an unfinished part entity; a demand entity is transformed into a sale entity; a sick patient entity is transformed into a hospitalized patient entity; and an information record entity is converted to a patient entity. Entities are created, cloned, and destroyed. These changes to entities are reflected in changes made to the attributes of the entity.

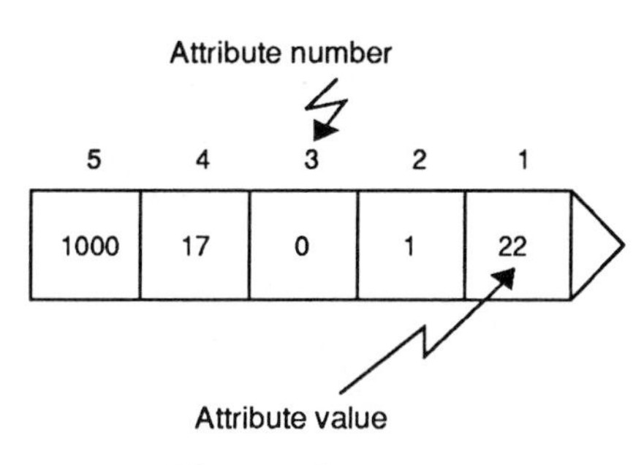

Figure 2-5 An entity.

As an entity moves through a network model, decisions are made at nodes to hold it at the node, to start an activity that explicitly models a delay in its movement through the network, to route it or its clone to specific nodes, to collect information based on its reaching a node, to allocate resources to it, or to determine if it should be combined with other entities. The nodes and activities defined to perform these operations on entities constitute the network modeling language. The symbols and syntax associated with the network language are the inputs to a simulation processor that is used to simulate the network model. The network portion of the SLAM II language is presented in Chapter 3. User-written procedures for augmenting the decision logic within a SLAM II network are presented in Chapter 5. SLAMSYSTEM [15] and TESS [16] are software systems which include SLAM II and support the modeling and simulation process described in this chapter on a wide variety of hardware platforms. These software systems also provide a means for producing a documentation trail regarding model development, analysis, and use.

2.4.2 Collect Data

The types of data which need to be collected to support the modeling and simulation process include data describing the system, data measuring the actual performance of the system, and data describing the alternatives to be evaluated. Data which describes the system is concerned with the system structure, the individual components of the system, component interactions, and system operations. The possible states of the system are established from this information.

Data collection may involve performing detailed time studies, getting information from equipment manufacturers , and talking to system operators. Actual system performance histories are collected whenever possible to help validate model outputs. Data describing proposed solutions are used to modify the basic model for each alternative to be evaluated. Each alternative is typically evaluated individually but performance data across alternatives are displayed together. The model specifications presented in Figures 2-3 and 2-4 provide a good list of the type of data needed for simulating the flow of parts in a manufacturing system.

2.4.3 Define Experimental Controls

A ***simulation run*** is an experiment which calculates and records model status from an initial state to a final state. In a network model, model status is reflected by the number of entities in the network and their location on activities and at nodes. For example, an entity on a machining activity indicates that the machine's status is busy performing an operation. The number of entities waiting at a node preceding this same activity provides the status of the number of jobs waiting for processing by the machine. Other indicators of model status are the number of resource units allocated to entities, the number of entities that have completed an activity or reached a node, and observations made on model variables and attributes of entities during the course of the run.

One of the key advantages of simulation is that model status changes are clearly identified with and related to system status changes. Evaluations made for model variables may be directly translated into evaluations for system variables. Statistical techniques used to analyze system variables can be applied directly to model variables.

In defining experimental controls, the following type of information is specified:

1. Project title and modeler identification
2. Beginning and ending times, number of runs, and other control information
3. Report types and frequency desired
4. Status variables on which performance is desired and the form in which outputs are to be presented, such as plots, tables, histograms, and charts
5. Initial values for status variables and initial location of entities in the model

6. Statistical estimation procedures including warm-up time periods, variance estimation procedures, confidence interval calculation methods, and variance reduction techniques
7. Files and databases where input data exists and output values are to be stored

2.4.4 Build Model Interactions

Figure 2-1 shows that the three substeps of develop simulation model, collect data, and define experimental controls are performed concurrently. Several examples are given to illustrate how these substeps are interdependent.

After a first-cut model is developed, it may be necessary to include weekend operations into the model. This modeling decision may require that the data collected on jobs be divided into those that can be performed on the weekend and those that must be performed during the weekday. The inclusion of weekend activity may also require that experimental controls be established to collect statistics to estimate utilization by weekday and by weekend. In collecting data on operators, it may be uncovered that different skill levels are used on weekends, and the time to perform a job must be based on the skill level of the personnel available. This could entail including different resource types in the model and different job processing times which are dependent on resource type. It might further require additional statistics collection to ascertain the number of jobs completed by resource type and day of the week. As can be seen, the three substeps of the build model step have a high degree of interdependence.

After completing the build model step, it is typically necessary to update the model specification. Although not always done, maintaining an up-to-date model specification makes reuse of the model possible and future extensions probable. In addition to updating the model specification, there are projects on which the model is not completely specified initially. For example, a weekly review of job priorities may have been omitted from the original specification. If this is the case, then it is necessary to establish how priorities are assigned to jobs and how jobs are to be reordered If an MRP system is used for establishing job priorities through due dates, then it is necessary to specify what aspects of the MRP system need to be included in the model. In some projects, it is determined that the level of detail of the model need not be as extensive as specified. In such an instance, it is necessary to question why the specification included the items being considered for deletion. If deletion is confirmed, then the specification should be shortened.

2.5 SIMULATE MODEL

The simulate model step requires that the build model step be completed at least once. The experimental controls establish an initial state of the model. A network simulation advances time in accordance with the movement of entities through the nodes and activities of the model. This is the run model substep. Before the model can be used to support decision making, it must be shown to run in accordance with model specifications. This is accomplished as a verification substep, which consists of determining that the simulation model behaves as intended. A validation substep seeks to establish that the simulation model is a reasonable representation of the system. These substeps are performed concurrently and usually require a return to earlier steps in the modeling and simulation process as described in Section 2.5.4.

2.5.1 Run Model

The running of a SLAM II model is facilitated by embedding SLAM II in a simulation support system that is programmed for specific computer platforms. SLAMSYSTEM provides procedures for simulating SLAM II models on personal computers. The input of SLAM II models is handled through graphical and forms input. The linking of SLAM II networks with user inserts is performed automatically. The storage of models for reuse as the basis of new models is provided. The capabilities for watching entity movement on a step-by-step basis or over an interval of time are available. Animations of model status changes on system diagrams can be presented to support model understanding. Outputs can be presented during a run or stored for display following a completed run. TESS provides these capabilities for larger computer systems. TESS and SLAMSYSTEM support the steps of a modeling and simulation project and simplify the process of using the computer as a simulation vehicle.

2.5.2 Verify Model

Verification is the process of determining that a simulation run is executing as intended [10, 12]. One method of verification is to check that each model element is described correctly and that modeling elements are interfaced as specified. Model verification can be a manual process of reviewing data inputs and outputs and insuring that no significant discrepancies exist between expected and observed model performance. SLAM II contains the Interactive Execution Environment which allows modelers to watch the running of a model as each status change occurs or at defined breakpoints established by the user [9]. At these breakpoints, the modeler can incrementally execute the simulation, examining the changes of the state of the model caused by simulation events.

The Interactive Execution Environment is also used to follow the logical flow described in the model. The modeler can simulate to a decision point, save the system status, and execute the simulation further from that breakpoint. The modeler can then examine the model's response to alternate inputs by restoring the simulation to the previously saved state, changing the appropriate inputs, and then restarting the simulation from the breakpoint. In this way, the verification process is supported by allowing the user to make runs interactively and to access model status variables during an interactive session. A brief description of the Interactive Execution Environment is presented in Section 5.13.

2.5.3 Validate Model

Validation is the process of determining that the simulation model is a useful or reasonable representation of the system [5, 11, 12, 13, 17]. Validation is normally performed in levels involving an examination of data inputs, model elements, subsystems, and interface points. Validation of simulation models is a significantly easier task than validating other types of models because there is a correspondence between model elements and system components. Testing for reasonableness involves a comparison of a model's organization with the system's structure as well as comparing the number of times elements in the model are exercised with data describing the number of times tasks in the system are performed.

Specific validation procedures evaluate reasonableness using all constant or expected values or assessing the sensitivity of outputs to parametric variation of data inputs. In making validation studies, the comparison yardstick should encompass both past system outputs and experiential knowledge of system behavior. A point to remember is that past system outputs are but one sample record of what could have happened.

For models of existing systems, validation is typically done in two ways. First, the structure and operation of the system is compared to the structure and operation of the model. Individual components as well as component interfaces are examined. Second, the values of the critical performance measures of the entire system and any identifiable subsystems are compared to the outputs of the entire model and the parts of the model representing the subsystems. If discrepancies exist, then the performance measure values for the system may have been incorrectly determined, or the data describing the system may contain errors. Both types of data should be checked before rebuilding the model.

For models of new system designs, the validation process is more difficult because model outputs cannot be compared to measures of actual system performance. The first means of validating a new system design is to compare the structure and expected operation of the system design to the structure and operation of the model. Each individual component and the interface between components are examined. The second means of validating new designs is to review model outputs for reasonableness considering the performance of similar existing systems. Individuals familiar with the system play a valuable role in validating a model. To support them, it is important to develop reports, traces, and graphical outputs that are expressed in the context of their system.

2.5.4 Simulate Model Interactions

The dependent nature of the run, verify, and validate model substeps is intuitive. To verify that the model is operating as intended, it is necessary to run the model. The same is true for the validation substep. In verification, runs of the model are made with all randomness removed. Counts are made on the number of times an event occurs and on the number of entities that are processed through various portions of the model. These counts are checked against the number expected as calculated from the model specification. Runs of the model are then made with randomness included and counts are compared to expected values. Verification also involves making parameter changes and ascertaining if performance measures change in an expected manner. For example, as arrival rates increase, utilizations of resources are expected to increase. The validation substep typically questions the occurrence of events and combinations of events in the simulation run. For example, why did a robot resource fail twice in the first 100 hours of the run? Validation is facilitated by watching animations of the model and comparing them to the operation of the real system. If system data is available , then it can be input into the animation system and the animation of the model can be compared to the animation of the system data. In the course of validating a model, verification questions arise concerning whether the model is operating in accordance with system operating rules.

Feedback from the simulate model step to the build model step is common. During verification, if an incorrect processing of jobs is uncovered, a new job processing algorithm may be required for the model. During validation, a shortfall on throughput for a given set of conditions may be observed. After establishing that the model was incorrectly developed, the model is rebuilt to eliminate the shortfall in throughput. A direct feedback from the simulate model step to the specify model step is not shown in Figure 2-1. Typically, a return from the simulate model step is made to the build model step from which a return to the specify model step can be made.

2.6 USE MODEL

The use of the model involves the making of runs and the interpretation and presentation of the outputs. When simulation results are used to draw inferences or to test hypotheses, statistical methods as described in Chapter 1 should be employed. Planning for the use of the model entails strategic and tactical considerations [1, 3, 6, 18].

A simulation model does not require a method for determining an optimal solution. Rather, the model specifies a means to obtain experimental data from which an alternative can be selected. For example, to determine the number of machines needed at a particular station to achieve a production rate of 150 parts per day, runs are made with 4 machines, then 5 machines, then 6 machines until the desired production rate is achieved. The key point here is that the simulation model does not determine the number of machines needed, it only evaluates the performance of systems with specified machine level alternatives.

In addition to evaluating alternatives, the procedure to be followed when evaluating alternatives is also defined. The order in which alternatives are evaluated is important, especially if the results of one evaluation are used to specify the next alternative. For example, if a bottleneck at a station is alleviated by adding additional equipment, another station may then become the bottleneck. Runs are then made to alleviate this second bottleneck. If additional alternatives arise, their impact on the entire analysis is considered. When using the model, the specification of the type and form of outputs is extremely important. When possible, model outputs should relate directly to system outputs. The types of outputs from SLAM II models is presented in Chapter 4.

2.6.1 Use Model Interactions

After the use model step is completed, there is a potential return to the simulate model step, build model step, or formulate problem step. After using the model and analyzing outputs, a better setting of a resource level may be discovered. This parameter change is easily made at the simulate model step where the model is rerun under the new resource level. In another situation, the outputs from the model may indicate that the scheduling system is not as robust as necessary. In this case, a return to the build model step is made where a new scheduling system is developed. It may be necessary to first determine if the model can accommodate a new scheduling procedure. If it cannot, then a return to the specify model step is made where the scheduling procedure must be specified before the model is embellished to contain it.

In some cases, after examining outputs from the model, it is determined that the scope of the problem is broader than anticipated and that the modeling objectives need revision. This should lead to a return to the formulate problem step. Examples of a broader type of scope are the need to purchase a new material handling system to alleviate a bottleneck or the incorporation of a production control system to provide for the efficient release of orders to the shop floor. When a return is made to the formulate problem step, the decision maker should be brought back into the process to provide information regarding the feasibility and desirability of any proposed change in scope or objectives. In many instances, the knowledge that a bottleneck will exist or a production control system is necessary provides sufficient information to take an action to meet the modeling objectives or the project goal.

2.7 SUPPORT DECISION MAKING

The final step in the modeling and simulation process is to support decision making. No simulation project should be considered complete until its results are used in the decision-making process. The success of this step is largely dependent upon the degree to which the modeler has successfully performed the other steps. If the model builder and model user have worked closely together and both understand the model, its outputs and its uses, then it is likely that the results of the project will be implemented with vigor. On the other hand, if the problem formulation and underlying assumptions are not effectively communicated, then it is more difficult to have recommendations implemented, regardless of the elegance and validity of the simulation model.

2.8 SUMMARY

Simulation is a technique that has been employed extensively to solve problems. Simulation models are abstractions of systems . They should be built efficiently, explained to all project personnel, and changed when necessary. The steps outlined in this chapter are rarely performed in a structured sequence beginning with problem formulation and ending with decision making. A project may involve false starts, erroneous assumptions, reformulation of the modeling objectives, and repeated redesign of the model. If properly performed, however, this iterative process yields new knowledge of the system and management policy. It fosters communication and leads to alternative designs for equipment and operations. It is the essence of the modeling and simulation process and results in properly assessed alternatives and improved systems.

2.9 EXERCISES

2-1. Discuss the types of data that are needed to build a structural model of a system. Discuss the type of data that is needed to estimate parameters of a model. Give examples of structural data and detailed data.

2-2. Discuss the different uses of statistics in the modeling and simulation process.

2-3. Models are used as explanatory vehicles, design accessors, and as a basis for control mechanisms. Give a definition of each model use and prepare an illustration to demonstrate each type of use.

2-4. Build an explanatory model of the university system from the point of view of an undergraduate student. Describe the activities that are required in each of the steps of the modeling and simulation process that estimates the utilization of the scarcest resource in your definition of the university system.

2-5. Develop the steps involved in applying a modeling and optimization process. Compare the steps to the process described in this chapter.

2-6. Compare the procedures used in problem formulation between seeking an optimal solution and using a satisficing approach to alternative selection.

2-7. A model of a system can be viewed as data describing the system. Develop a database schema that can be used to store models.

2-8. For each step in the modeling and simulation process described in this chapter, make a list of decisions that are required to accomplish each step. Describe the types of information necessary to make the decisions that you have listed.

2-9. Specify methods for obtaining time estimates for each data input specified in the order processing illustration given in Figure 2-3.

2.10 REFERENCES

1. Conway, R. W., B. M. Johnson, and W. L. Maxwell, "Some Problems of Digital Systems Simulation," *Management Science*, Vol. 6, 1959.
2. Emshoff, J. R., and R. L. Sisson, *Design and Use of Computer Simulation Models*, Macmillan, New York, 1970,.pp. 92-110
3. Fishman, G. S., *Principles of Discrete Event Simulation*, Wiley, New York, 1978.
4. Gaskins, R. J. , and J. P. Whitford, "Simulation Model Specification for a Flexible Manufacturing System for an Aerospace Company", Pritsker & Associates, West Lafayette, IN, 1988.
5. Gass, S. I., "Decision-aiding Models: Validation Assessment, and Related Issues for Policy Analysis," *Operations Research*, Vol. 31, 1983, pp. 601-631.
6. Kleijnen, J. P. C., *Statistical Tools for Simulation Practioners*, Dekker, New York, 1986.
7. Law, A. M., and W. D. Kelton, *Simulation Modeling and Analysis*, McGraw-Hill, New York, 1982.
8. Mihram, G. A., "The Modeling Process," *IEEE Transactions on Systems, Man and Cybernetics*, Vol. SMC-2, 1972, pp. 621-629.
9. O'Reilly, J. J., and W. R. Lilegdon, *SLAM II Quick Reference Manual*, Pritsker & Associates, West Lafayette, IN, 1987.

10. Pritsker, A. A. B., *Introduction to Simulation and SLAM II*, *3rd* ed., Wiley and Systems Publishing, New York and West Lafayette, IN, 1986.
11. Rolston, L. J., and R. J. Miner, *MAP/1 Manufacturing Analysis Program Using Simulation*, Pritsker & Associates, West Lafayette, IN, 1988.
12. Sargent, R. G., "An Expository on Verification and Validation of Simulation Models," *Proceedings,Winter Simulation Conference*, 1985, pp. 15-22.
13. Schlesinger, S., and others, "Terminology for Model Credibility," *Simulation*, Vol. 32, No. 3, 1979, pp. 103-104.
14. Shannon, R. E., *Systems Simulation: The Art and Science*, Prentice-Hall, Englewood Cliffs, NJ, 1975.
15. *SLAMSYSTEM User's Guide*, Pritsker & Associates, West Lafayette, IN, 1988.
16. Standridge, C. R., and A. A. B. Pritsker, *TESS: The Extended Simulation Support System*, Wiley and Pritsker & Associates, New York and West Lafayette, IN, 1987.
17. van Horn, R. L., "Validation of Simulation Results," *Management Science,* Vol. 17, 1971, pp. 247-258.
18. Welch, P. D., "The Statistical Analysis of Simulation Results," *Computer Performance Modeling Handbook*, S. S. Lavenberg, ed., Academic Press, Palisades, NY, 1983.
19. Wilson, J. R., "Statistical Aspects of Simulation," *Proceedings, 1984, IFORS*, pp. 825-841.

3 SLAM II NETWORKS

3.1 INTRODUCTION

A SLAM II network consists of nodes and branches. A branch represents an ***activity*** that involves a processing time or a delay. ***Nodes*** are placed before and after activities and are used to model milestones, decision points, and queues. Flowing through the network are ***entities*** (see Section 2.4.1). Entities can represent physical objects, information, or a combination of the two. A set of attributes is used to describe an entity and to distinguish one entity from another. For example, a part entity may have attributes describing its type, weight, and value. Entities are directed through the network from one node to another according to routing conditions placed on activities. The SLAM II network is a model of the operations and procedures of a system.

In a SLAM II network model, an entity originates at a CREATE node and is routed over the activities emanating from the CREATE node. For each activity, the entity is delayed in reaching the end node of the activity by the activity's duration. When reaching the end node, the disposition of the entity is determined by the node type, the attributes of the entity, and the status of the system as modeled by SLAM II variables. Typically, the entity is routed over one or more activities emanating from the node to which it arrived. Alternatively, the entity is held at the node until a server is available or a resource is allocated to it. The entity continues through the network moving from node to activity to node until a TERMINATE node is encountered or no further routing can be performed. When this occurs, the entity is terminated, that is, deleted from the network.

A simulation procedure is used to analyze a SLAM II network by generating entities and processing their movements through the network. As entities flow through the network model, observations are made of travel times, node release times and the status of servers, resources, and queues. Data collection nodes are inserted directly in the network model to collect observations on model variables. Statistics are presented for these nodes and for SLAM II variables.

This chapter provides a primer on SLAM II network constructs. Emphasis is placed on how network elements are used to model systems. No attempt is made to be complete nor to present all the syntax and semantics of SLAM II. With SLAMSYSTEM and TESS, SLAM II models can be built graphically. The knowledge of the syntax of SLAM II statements is only necessary to be able to read a developed model, not to produce it. In this text, both the graphical SLAM II model and its statement equivalent are presented. SLAM II input procedures and output reports are described in Chapter 4. Extensive documentation for SLAM II is available [4, 6].

3.2 SLAM II NETWORK ILLUSTRATION

To introduce SLAM II network concepts, a manufacturing system is modeled. The system consists of a single machine which produces a finished part from an unfinished part. The unfinished part is delivered to the machine area where an operation is performed by the machine. Following machining, the finished part leaves the machine area. Our interest is focused on the following three aspects of the system:

1. The arrival of unfinished parts to the machine area.
2. The buildup of unfinished parts awaiting machine processing.
3. The machining operation.

This is a single server queueing system. The parts to be processed are the entities which move through the system. The machine is modeled as a service activity. The time to perform the machining operation is the time the part entity spends in the service activity. The buildup of parts before the machine is modeled as a queue. In Figure 3-1, a pictorial diagram of this one-machine work

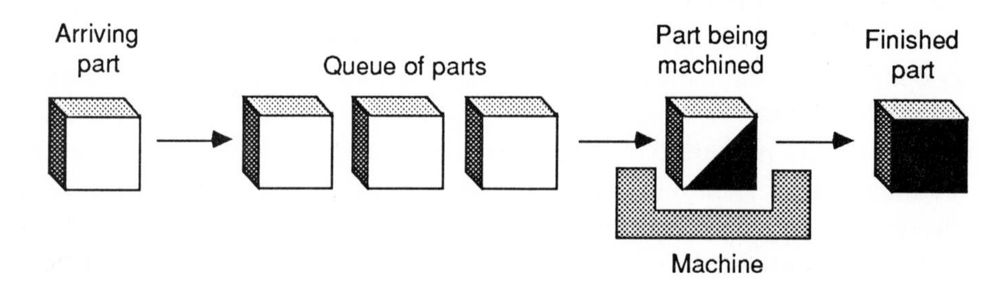

Figure 3-1 Diagram of a one-machine work center.

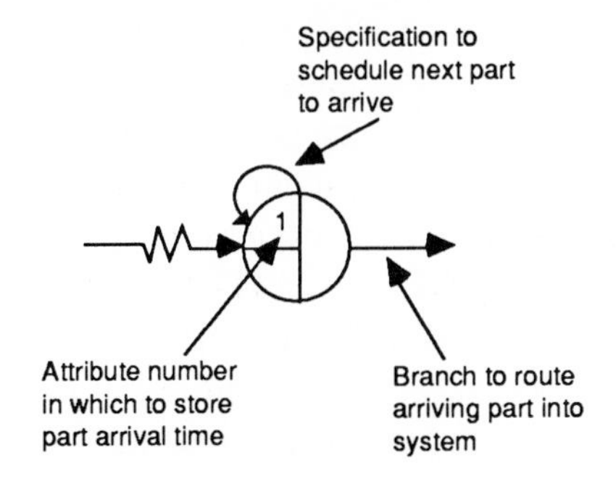

Figure 3-2 CREATE node for part entity.

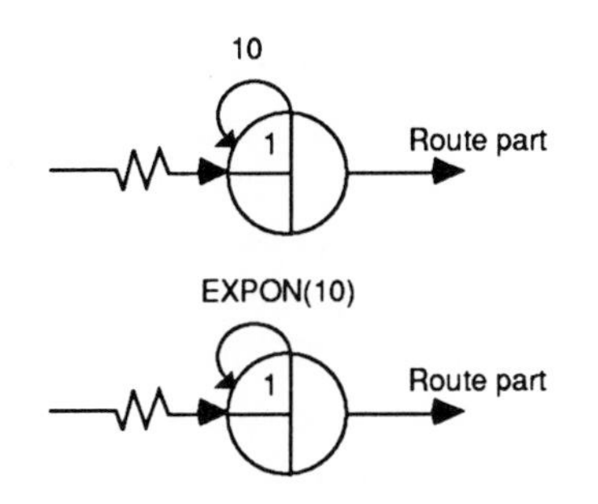

Figure 3-3 CREATE node examples.

center is shown. The part entity may have attributes to define characteristics such as part type, due date, part value, and arrival time. In this illustration only an arrival time attribute is used.

The arrival of parts is modeled in SLAM II by creating successive arrivals at a CREATE node as shown in Figure 3-2. The squiggly line before the CREATE node indicates the function of the node is to be initiated at the beginning of a simulation run. The time until the next creation of an arriving part is specified on the loop which is part of the CREATE node symbol. This could be every 10 time units or a sample from an exponential distribution (or some other distribution) whose average is 10 time units. EXPON is a SLAM II abbreviation for EXPONENTIAL. Examples of CREATE nodes with constant and exponential times between creation (TBC) are shown in Figure 3-3. The 1 in the upper left quadrant of the CREATE node specifies that attribute 1 of the part entity is to be set equal to the time of arrival (creation) of the part. An entity created at time 20 will have its first attribute set to 20. In SLAM II, attribute values are referenced as ATRIB(I) where I is the attribute number. In this example, a part arriving at time 20 has ATRIB(1) = 20. The activity following the CREATE node routes the arriving part to the machining area. Since only one activity follows the CREATE node, only one entity is routed from the CREATE node for each arrival.

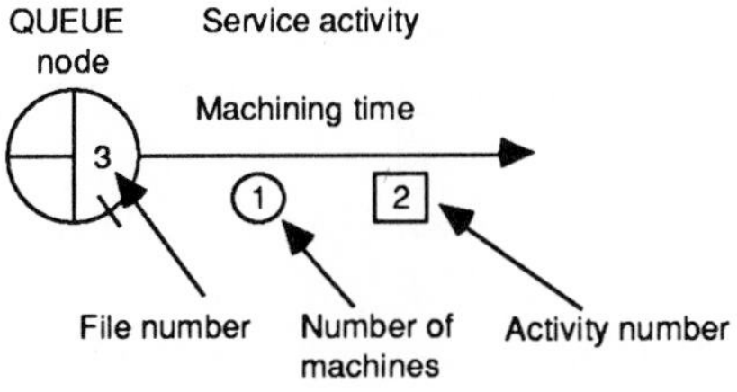

QUEUE node
Service activity
Machining time
3
1
2
File number
Number of machines
Activity number

Figure 3-4 QUEUE node-service activity combination.

The machining operation is a service activity. If the service activity is ongoing, that is, the machine is processing a part, then the arriving part entity waits at a QUEUE node. A QUEUE node must precede a service activity to provide a holding area for entities when servers are busy. In Figure 3-4, the machine area is modeled by a

QUEUE node-service activity combination. The QUEUE node performs the function of holding the entity if the machine is working on a part or letting the part pass through the waiting area to start its processing by the service activity (machine). The 3 on the right hand side of the QUEUE node identifies a file number where all parts waiting for this machine are held. The order in which these part entities are held is established by specifying a ranking rule for file 3. An example of a ranking rule is the entity that arrives first leaves first. This is a first in, first out (FIFO) ranking rule. Another ranking rule is to order entities according to their value of attribute 1, for example, the entity with the lowest value of attribute 1 is ranked first.

The activity emanating from the QUEUE node represents the service activity. A service activity has a duration and may represent 1 or more servers. The number in the circle below the branch in Figure 3-4 specifies the number of servers. The number in the square below the branch provides an activity number which is used to reference activity status information and to label statistical outputs relating to the fraction of time the activity spends processing entities. For sevice activities, this is the utilization of the servers.

Following machining, assume that the part is routed to another area which is outside the scope of the model being developed. To evaluate the operation of this one machine center, it is desired to estimate the time that a part entity spends at it. This time is the sum of the part's waiting time plus its machining time. It is the interval of time from when the part arrives, ATRIB(1), to the time at which machining is completed. Current simulation time in SLAM II is referenced by the variable TNOW. Following the activity representing machining, an ASSIGN node can be used to compute the time in the work center. SLAM II provides global variables for the user's convenience in maintaining information about model status. In Figure 3-5, the SLAM II global variable XX(1) is assigned the value TNOW - ATRIB(1) at an ASSIGN node. The value of XX(1) represents the time in the model for the entity that just arrived to the ASSIGN node, that is, the entity that just completed being machined.

Figure 3-5 ASSIGN node.

The value of XX(1) can be collected by routing the entity to a COLCT node. Each time an entity arrives to the COLCT node, a value of XX(1) is collected. The COLCT node, shown in Figure 3-6, combines these observations and SLAM II produces summary statistics on the combined observations. For example, estimates of the average time in the system, the standard deviation of the times in the system and the minimum and maximum time in the system are printed on

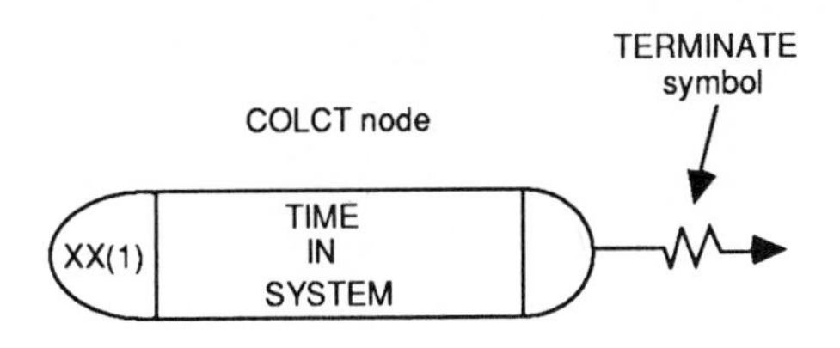

Figure 3-6 COLCT node.

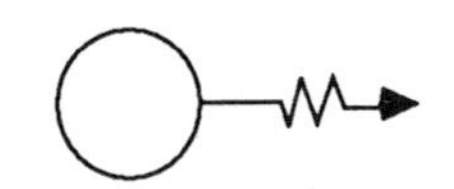

Figure 3-7 TERMINATE node.

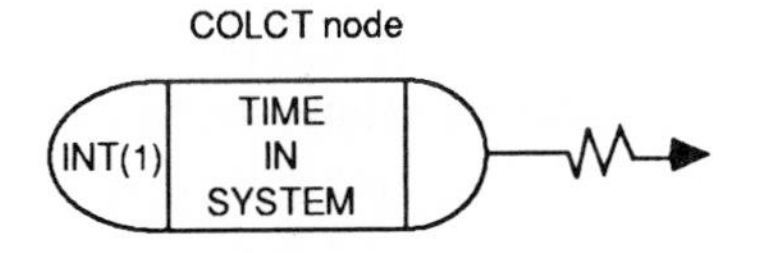

Figure 3-8 Time interval collector.

the SLAM II summary report. The squiggly line following the COLCT node is an indicator that the entity arriving to the COLCT node is to be terminated. The squiggly line is a separate symbol and can be used following any SLAM II node. A TERMINATE node is also included in SLAM II to terminate an entity following any activity. Its symbol is shown in Figure 3-7.

Because it is common to collect time intervals in a SLAM II network, a special syntax is available at a COLCT node to indicate that an interval of time is to be observed. The code word is INTERVAL(I) where I is the number of the attribute in which the reference time for the interval is stored. Three letter abbreviations are allowed throughout SLAM II and INTERVAL is commonly abbreviated as INT. Thus, in the above example, the time in the system for the entity could be obtained by the COLCT node shown in Figure 3-8 which replaces both the ASSIGN node of Figure 3-5 and the COLCT node of Figure 3-6.

A network to model the single server queueing system can now be drawn by combining Figures 3-2, 3-4, and 3-8 as shown in Figure 3-9. In this model, entities representing parts are created every ten time units and the time of creation is placed in ATRIB(1). The part entity is then routed to the QUEUE node in zero time, which is the default value for the duration of any activity for which a time is not specified. At the QUEUE node, a decision is made based on the status of activity 2 which models the machine. The **2** in the box underneath the activity identifies the service activity number. If the server is idle then the entity is placed into activity 2 and service is started. Service is specified to be completed 9 time units later. If the service activity is ongoing then the arriving part entity is placed in file 3 as specified by the 3 on the right hand side of the QUEUE node. There is one server indicated by the 1 in the circle underneath the service activity. After the service activity is completed, the part entity is routed to the

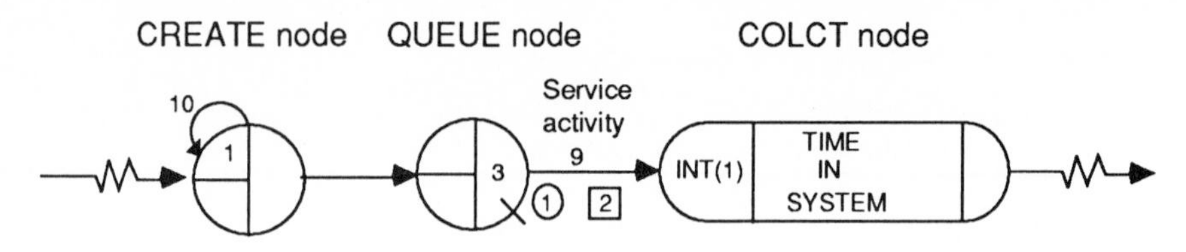

Figure 3-9 Network model of machining system.

COLCT node where INT(1) indicates that an observation of the interval of time, TNOW - ATRIB(1), is to be made. Statistics on these observations are printed on the SLAM II summary report under the label TIME IN SYSTEM. Following the COLCT node, the part entity is terminated.

When the service activity is completed, the file of the QUEUE node preceding the service activity is interrogated. If an entity resides in file 3, it is removed and it is placed in service activity 2. The attributes of the entity removed from file 3 are kept with the entity. Thus, when service is completed and the entity is routed to the COLCT node, the proper value of its arrival time in ATRIB(1) is available to compute the time-in-system interval at the COLCT node.

In this example, part entitites are generated into the network at the beginning of the simulation and every 10 time units thereafter. If the beginning of the simulation is at time 0, then part entities arrive to the network at time 0, 10, 20, and so on. There is an implied zero time for the part entity to travel to the QUEUE node, so that the entities arrive to it at the same time as they were created. Since the service time is 9 time units, the parts leave the system at time 9, 19, 29 and so on. The time spent in the system for each part is 9 time units. By changing the service activity time specification to be exponentially distributed with a mean of 9, parts will wait at the QUEUE node every time the service activity time exceeds 10. This change in the model is easily accomplished by changing service activity 2 in Figure 3-9 to EXPON(9). With this change, each service activity time is a sample from an exponential distribution with a mean 9.

3.2.1 Embellishments to the Network Model

The modeling procedure advocated in this book is to build a first cut model and then to embellish it. An example of this approach was given in the preceding section where a constant service time was changed to an exponentially distributed service time. Although this change is minor, it is a major analytical change. With a constant service time of 9 time units, there is no waiting for parts and an analysis of the model is direct. With an exponential service time, analysis is more complicated.

Consider the situation in which an inspection is performed following the service activity which classifies parts as good or bad. This can be modeled by adding a GOON node to the network following the service activity. A GOON node separates activities and is used to route entities arriving to the GOON node over the activities emanating from it. In Figure 3-10, a GOON node is used to randomly route 90% of the part entities to a COLCT node which collects observations on good parts and to route the other 10% of the entities to a COLCT node which estimates the time in system for bad parts. When an entity arrives at the GOON node, it is routed over one of the two branches. This is indicated by the 1 inside the GOON node and is referred to as the node's M-number (see Section 3.3.4). On the activities following the GOON node, a time duration of zero is specified and a probability that the activity is selected is prescribed. These activities have been assigned the activity numbers of 3 and 4. Since there is no QUEUE node to store entities prior to the activities, there is no limit on the number of entities that may flow over either of the activities. Such activities are referred to as ***regular activities***. No specification for the number of servers is ever given for a regular activity.

Two COLCT nodes are now included in the network and each has been given a node label which is displayed graphically as a rectangle beneath the node with an identifier placed in the rectangle. Node labels may be appended to any node in a network in the form shown in Figure 3-10 for the COLCT nodes. In this illustration, COLCT node COL1 records observations on the time in the model for good parts, and COLCT node COL2 collects observations on the time in the model for bad parts. The time between part arrivals has also been changed to be uniformly distributed between 8 and 12 time units at the CREATE node.

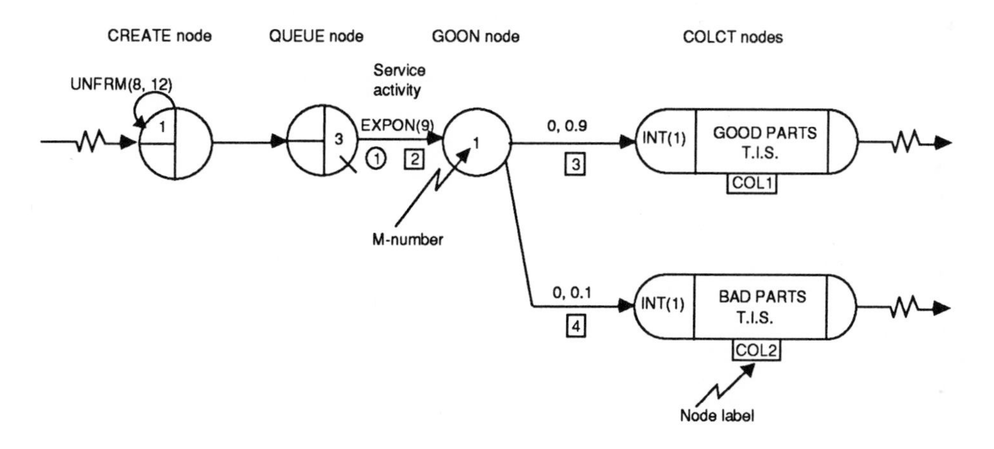

Figure 3-10 Network model with probabilistic branching.

3.3 ACTIVITIES

Each activity in a SLAM II network has a duration which specifies a time delay for each entity traversing the activity. Graphically, the duration is specified directly on the branch. The duration may be a constant, SLAM II variable, SLAM II random variable, or a *wait until* condition. SLAM II variables used to specify an activity duration may be taken as (1) an attribute of the entity starting the activity, ATRIB(I), (2) a global variable, XX(I) or ARRAY(I,J), or (3) a user written function USERF(N). Figure 3-11 illustrates activity duration specifications for each of these types of variables. The duration is specified as the third attribute of the entity traversing the activity in Figure 3-11a. In Figure 3-11b, the duration for each entity routed over the activity is specified as the global variable XX(2). The value of XX(2) may be changed at an ASSIGN node. The use of the two dimensional global variable ARRAY is shown in Figure 3-11c and 11d. In Figure 3-11c, the duration is taken from row 1, column 5 of ARRAY. In Figure 3-11d, the fifth attribute of the entity starting the activity is used as the column number specification and the value is taken from row 1, that is, ARRAY(1,ATRIB(5)). The duration of an activity may be computed in an externally written user function, as shown in Figure 3-11e. The specification USERF(6) invokes a call to function USERF(N) with N equal to 6. The function is written to return a value of USERF as the duration for the entity starting the activity.

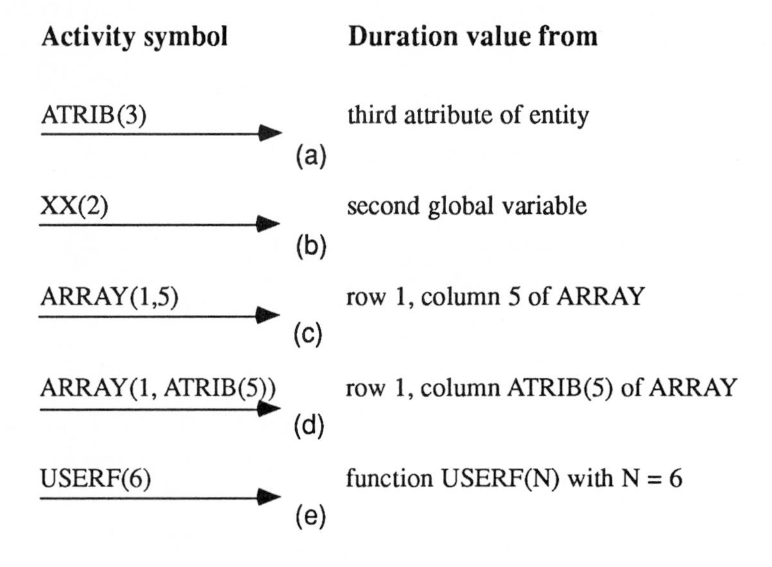

Figure 3-11 Activity duration specifications.

A duration specified as a random variable indicates that a sample value from the distribution of the random variable is to be used. The SLAM II random variable sampling functions are listed in Table 3-1.

Table 3-1 SLAM II Random Variables

Variable name[a]	Definition
DRAND(IS)	A pseudorandom number stream IS.
EXPON(XMN,IS)	A sample from an exponential distribution with mean XMN using stream IS.
UNFRM(ULO,UHI,IS)	A sample from a uniform distribution in the interval ULO to UHI using stream IS.
WEIBL(BETA,ALPHA,IS)	A sample from a Weibull distribution with scale parameter BETA and shape parameter ALPHA using stream IS.
TRIAG(XLO,XMODE, XHI,IS)	A sample from a triangular distribution in the interval XLO to XHI with mode XMODE using stream IS.
RNORM(XMN,STD,IS)	A sample from a normal distribution with mean XMN and standard deviation STD using stream IS.
RLOGN(XMN,STD,IS)	A sample from a lognormal distribution with mean XMN and standard deviation STD using stream IS.
ERLNG(EMN,XK,IS)	A sample from an Erlang distribution which is the sum of XK exponential samples each with mean EMN using stream IS.
GAMA(BETA,ALPHA,IS)	A sample from a gamma distribution with parameters BETA and ALPHA using stream IS.
BETA(THETA,PHI,IS)	A sample from a beta distribution with parameters THETA and PHI using stream IS.
NPSSN(XMN,IS)	A sample from a Poisson distribution with mean XMN using stream IS.
DPROBN(IRCUM,IRVAL,IS)	A sample from a probability mass function where the cumulative probabilities are in row IRCUM of ARRAY and the corresponding sample values are in row IRVAL of ARRAY using stream IS.

[a] Each parameter for a distribution can be specified as a constant, ATRIB(I), or XX(I).

The duration of an activity need not be determined when an entity starts, but may depend on the next release time of a node by using the specification REL(NLBL), where NLBL is a label of a node in the network. When the duration is specified in this manner, the entity is held in the activity until the next release of node NLBL. For many nodes, the next release occurs when an entity arrives at the node. However, there are ACCUMULATE, MATCH, BATCH and SELECT nodes that require more than one entity arrival for their release.

The activity duration can also be made to depend on an assignment. This is accomplished using the STOPA(NTC) specification, where NTC is an integer code to distinguish between entities in such activities. An activity whose duration is specified by STOPA(20) continues in operation, holding the entity in the activity, until an assignment is made which sets STOPA to 20. REL and STOPA specifications are methods for providing a *wait until* specification.

The routing of an entity from a node involves selecting one or more branches to process the entity (or identical copies of the entity). The selection of a single branch may be probabilistic, in which case, a probability is part of the activity description. The selection of one or more branches may be conditional, in which case a condition is part of the activity description. If no probability or condition is specified (a common situation), the activity is selected unless the M-number associated with the activity's start node has already been satisfied. When conditions are specified, the M-number of a node is defined as the maximum number of activities over which an entity may be routed from that node.

In previous discussions, service and regular activities were described. The symbol for a branch representing an activity is shown in Figure 3-12. Activities following QUEUE nodes are service activities. The number of parallel, identical servers is specified for service activities. The number of entities processed concurrently in the service activity cannot exceed the number of servers. Activities that are not preceded by a QUEUE node are regular activities. Any number of entities may be processed concurrently in a regular activity.

Activities are identified by optional activity numbers. If the number I is prescribed for an activity, then the variable NNACT(I) is maintained as the number of entities currently being processed through activity I. Also, the

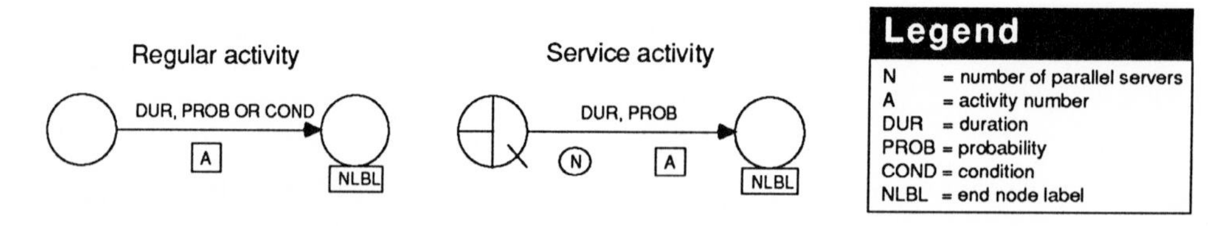

Figure 3-12 Activity symbol.

variable NNCNT(I) is available and is the number of entities that have completed activity I. For service activities and numbered regular activities, SLAM II automatically provides statistics on the values of NNACT and NNCNT.

3.3.1 Probabilistic Routing

Probabilistic routing of an entity from a node involves the selection of an activity from a set of activities based on a relative frequency. Probabilistic routing may be used to (1) categorize entity flow in accordance with the percentage of entities that flow over a given portion of a network, and (2) represent the fraction of time a specific type of activity duration is to be employed for entities departing from a node. Probabilistic routing is modeled by specifying a probability value for each activity emanating from a node. The probability can be a specified value, 0.3, or a value taken from the SLAM II variables ATRIB, XX, or ARRAY. Arithmetic calculations are allowed in the probability specification. Examples of probabilistic routing from a GOON node are shown in Figure 3-13. In Figure 3-13a, 30% of the entities arriving to node G1 are routed over activity 1 and 70% are routed over activity 2. Activity 1 has a duration of 7 time units and activity 2 has a duration of 12 time units. In Figure 3-13b, the probabilities are specified in attribute 5 of the entity to be routed over either activity 3 or activity 4. Attribute 5 of the entity must be established as a number between 0 and 1 at an ASSIGN node prior to the entity arriving to GOON node G2. The value of ATRIB(5) specifies the probability of routing the entity over activity 3 whose duration is exponentially distributed with a mean of 10. With probability 1-ATRIB(5), the entity is routed over activity 4 whose duration is normally distributed with a mean of 10 and a standard deviation of 2. This illustrates how a routing probability can be made a function of the specific entity being routed.

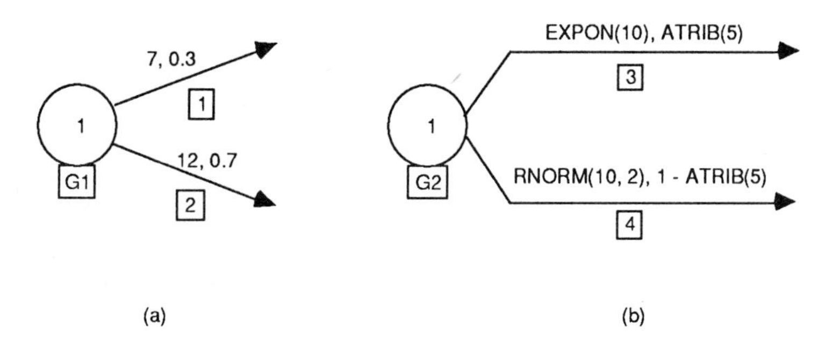

Figure 3-13 Examples of probabilistic routing.

3.3.2 Conditional Routing

Conditional routing selects activities emanating from a node based on a condition prescribed for an activity. A condition specification is only allowed for regular activities as service activities must be initiated if an entity is waiting to be served. Conditions are prescribed on activities in the following form

VALUE. OPERATOR. VALUE

where VALUE can be a constant, a SLAM II variable, or a SLAM II random variable. The OPERATOR may be one of the following standard relational codes:

Relational Code	Definition
LT	Less than
LE	Less than or equal to
EQ	Equal to
NE	Not equal to
GT	Greater than
GE	Greater than or equal to

The SLAM II variables that can be used in a condition are listed in Table 3-2.

Table 3-2 SLAM II Variables Used in Conditional Routing

Variable Name	Definition
TNOW	Current time.
ATRIB(I)	Attribute I of a current entity
XX(I)	Global vector.
NNACT(I)	Number of active entities in activity I at current time.
NNCNT(I)	The number of entities that have completed activity I
NNGAT(GATE)	Status of gate GATE at current time: 0 → open; 1 → closed.
NNRSC(RLBL)	Current number of units of resource type RLBL available.
NRUSE(RLBL)	Current number of units of resource type RLBL in use.
NNQ(I)	Number of entities in file I at current time.
USERF(N)	A value obtained from the N*th* user-written function USERF.
ARRAY(I,J)	Global array.

A list of conditions is given below that may be placed on a regular activity to route an entity arriving at the start node of the activity.

Activity condition	Arriving entity is routed over the activity
TNOW .GT. 100	if the current time is greater than 100
ATRIB(2) .LE. ATRIB(4)	if the value of attribute 2 of the entity is less than or equal to the value of attribute 4 of the entity
NNCNT(5) .EQ. 1000	if the number of entities that have completed activity 5 is equal to 1000
NNQ(4) .GE. UNFRM(2,6)	if the number of entities in file 4 is greater than or equal to a uniformly distributed random sample between 2 and 6.

The union and intersection of two or more conditions can be prescribed for an activity using .AND. and .OR. specifications. If more than two conditions are combined using the .AND. and .OR. specifications, then the conditions are tested sequentially from left to right. Complicated logic testing requiring parentheses is not permitted directly on the network.

3.3.4 Entity Routing

The maximum number of entities routed over activities emanating from a node is specified by the M-number of the node. If probabilities are assigned to activities emanating from the node then the M-number for the node should be 1. If conditions are specified for the activities emanating from the node then the M-number represents the maximum number of activities that can be selected. For each activity selected, an entity, with the attributes of the entity that arrived to the node, is routed over the activity.

As long as the function of the node is not to split or unbatch entities, no more than one entity will be routed over any activity emanating from the node. Thus, an M-number of 3 specifies that at most 3 entities be routed over the activities emanating from the node. If there are 10 activities leaving the node then the first three activities, whose condition is satisfied, are the only activities selected for entity routing. If no condition is specified for an activity, it is assumed that an entity is to be routed over the activity as long as the M-number of the activity's start node has not been satisfied.

3.4 SLAM II NODE TYPES

In SLAM II, there are 20 node types. A list of the node types is given in Table 3-3 with a brief statement regarding each node's function. The entity routing capabilities described in Section 3.3 may be performed from 17 of the 20 node types, with the only exceptions being the QUEUE, SELECT, and MATCH nodes. A brief description of each of the node types of SLAM II is given in the following subsections. Complete descriptions and examples of the use of the nodes and their integration into SLAM II networks are provided in [6].

Table 3-3 SLAM II Node Types and Functions

Node	Function
Basic node types	
CREATE	Generates and marks entities
QUEUE	Hold entities for available servers
TERMINATE	Destroys entities
ASSIGN	Sets values to SLAM II variables
GOON	Separates serial activities and routes entities
COLCT	Collects observations
Resources and gates	
AWAIT	Holds entities waiting for resources or a gate
FREE	Makes resource units available
PREEMPT	Reallocates a resource
ALTER	Changes the capacity of a resource
OPEN	Opens a gate
CLOSE	Closes a gate
Logic and decision nodes	
ACCUMULATE	Produces one entity from arriving entities
BATCH	Produces a batched entity to represent a batch
UNBATCH	Reestablishes entities of a batch
MATCH	Holds entities until a set of related entities arrive
DETECT	Creates an entity when a condition occurs
SELECT	Routes entities to and from queues and to servers
Interface nodes	
EVENT	Calls subroutine EVENT
ENTER	Accepts entities routed from user-written code

3.5 BASIC NODE TYPES

The basic network node types in SLAM II are CREATE, QUEUE, TERMINATE, ASSIGN, GOON, and COLCT. These node types were introduced through illustrations in Section 3.2. A brief formal discussion of each node is given in this section. With these six basic network elements, many diverse network models can be built.

3.5.1 CREATE Node

The CREATE node generates entities and routes them into the network over activities that emanate from the CREATE node. The symbol and statement for the CREATE node are shown in Figure 3-14. The time for the first entity to be created by the CREATE node is specified by the value of TF. The time between creations of entities after the first is specified by the variable TBC. TBC can be specified as a constant, a SLAM II variable, or a SLAM II random variable. Entities continue to be created until a limit defined as the maximum number of creations, MC, is reached. When MC entities have been input into the system, the CREATE node stops creating entities. The time at which the entity is created may be assigned to an attribute of the entity. This time is referred to as the mark time of the entity and it is placed in the MA*th* attribute of the entity. The variable ATRIB(MA) stores this value.

CREATE, TBC, TF, MA, MC, M;

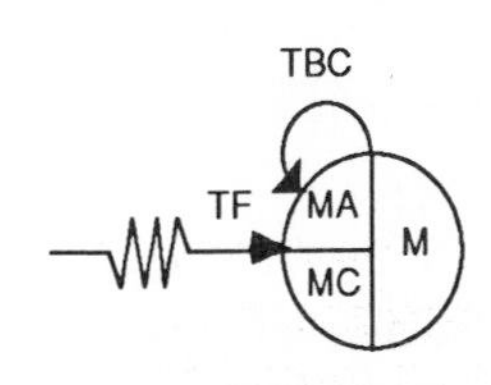

Figure 3-14 CREATE node

3.5.2 QUEUE Node

A QUEUE node is a location in the network where entities wait for service. When an entity arrives at a QUEUE node, its disposition depends on the status of the service activity that follows the QUEUE node. If the server is idle, the entity passes through the QUEUE node and goes immediately into the service activity. If all servers are busy, the entity waits in a file at the QUEUE node until a server becomes available. When a server becomes available, the entity is automatically taken out of the file and service is initiated. SLAM II assumes that no delay is involved from the time a server becomes available and the time service is started on an entity that was waiting at the QUEUE node.

When an entity waits at at QUEUE node, it is stored in file IFL which maintains the entity's attributes and the relative posistion of the entity with respect to other entities waiting in file IFL. The order in which the entities wait is specified outside the network on a PRIORITY statement that defines the ranking rule for file IFL. Files can be ranked in the following manner: first-in, first-out (FIFO); last-in, first-out (LIFO); low-value first based on attribute K (LVF(K)); and high-value first based on attribute K (HVF(K)). FIFO is the default priority for files.

Entities can initially reside at queues. The initial number of entities at a QUEUE node, IQ, is part of a QUEUE node's description. These initial entities all have their attribute values equal to zero. When IQ > 0, all service activities emanating from the QUEUE node are assumed to be busy working on initial entities. QUEUE nodes can have a capacity that limits the number of entities that can reside in the file of the QUEUE node at a given time. The symbol and statement for the QUEUE node are shown in Figure 3-15.

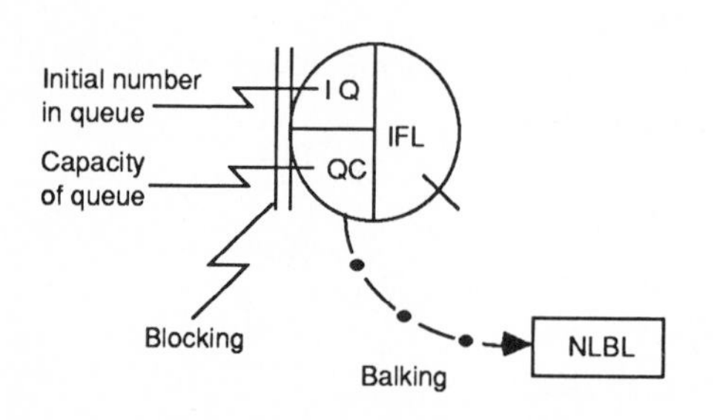

Figure 3-15 QUEUE node.

When an entity arrives at a QUEUE node that is at its capacity, its disposition is either that the entity should balk or be blocked. In the case of balking, the entity can be routed to another node of the network. This node is specified by providing the label of the node. If no balking node label is specified, the entity is deleted from the system.

When an entity is blocked by a QUEUE node, it waits until a space in the queue is available. The service activity that just served the entity that is blocked is also considered blocked. When a blocked entity joins the queue because a space is available, the blocked activity becomes free to process other entities waiting for it. *QUEUE nodes may only block preceding service activities.* No time delay is associated with unblocking operations.

A file number for a QUEUE node may be specified as an attribute of an arriving entity. When this is done, a range of file numbers must be given in the form ATRIB(I)=J,K where I is an attribute number and J through K are the allowable file numbers specified for IFL. When more than one service activity follows a queue and the service activities are not identical, a selection of the server to process an entity must be made. This selection is not made at the QUEUE node but at a SELECT nodes which follow it. SELECT nodes are described in Section 3.9.5.

3.5.3 TERMINATE Node

The TERMINATE node is used to delete entities from the network. It may be used to specify the number of entities to be processed on a simulation run. This number of entities is referred to as a termination count or TC value. When multiple TERMINATE nodes are employed, the first termination count reached ends the simulation run. If a TERMINATE node does not have a termination count, the entity is deleted from the network and no further action is taken. The symbols and statement for the TERMINATE node are shown in Figure 3-16. Every entity routed to a TERMINATE node is deleted. The squiggly line symbol allows an entity to be terminated after it is processed through any node of the network.

TERMINATE, TC;

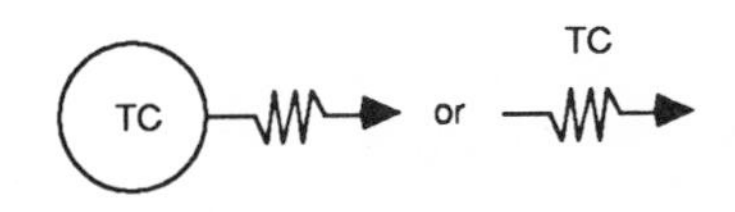

Figure 3-16 TERMINATE node

3.5.4 ASSIGN Node

The ASSIGN node is used to prescribe a value for an attribute of an entity passing through it or to set a value for a SLAM II global variable. Typically, assignments are made to ATRIB(I), II, XX(I), and ARRAY(I,J). In addition, a special assignment may be made to the variable STOPA to cause the completion of an activity or set of activities. Multiple assignments can be made at an ASSIGN node. The symbol and statement for the ASSIGN node are shown in Figure 3-17.

Basically, each line in the ASSIGN node is a replacement statement with the right-hand side expression involving constants and the variables defined in Tables 3-1 and 3-2. In evaluating the right-hand side expression, multiplication and division are evaluated first, then addition and subtraction. The expression is evaluated from left to right. Parentheses are allowed only to denote subscripts. Note that the statement

ASSIGN, XX(3) = 5.0/10.0*2.0;

sets XX(3) to 1.0. Complex assignments are made by writing function USERF.

ASSIGN,VAR=value, VAR=value, ... ,M;

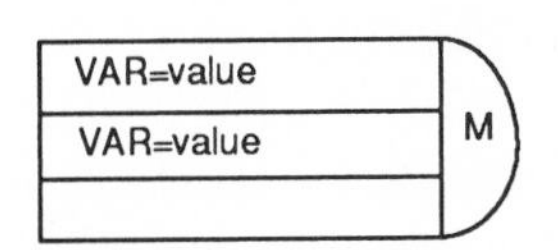

Figure 3-17 ASSIGN node.

3.5.5 GOON Node

The GOON node separates serial activities and acts as a continue type node. The symbol and statement for the GOON node are shown in Figure 3-18. The GOON node is required for sequential activities.

GOON, M;

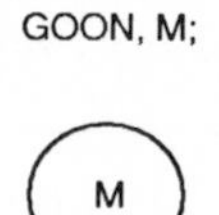

Figure 3-18 GOON node.

3.5.6 COLCT Node

Observations of a SLAM II variable or a time are made at a COLCT node. The observation of time is based on an entity arrival at the COLCT node and may be the time-of-first entity arrival (FIRST), time-of-all entity arrivals (ALL), time between entity arrivals (BETWEEN), or a time interval (INT(NATR)) defined by TNOW-ATRIB(NATR) where TNOW is the time of the entity arrival and ATRIB(NATR) is the value of its NATR*th* attribute.

For each of these variables, estimates for the mean and standard deviation of the observations are obtained. In addition, a histogram of the values collected may be obtained. This is accomplished by specifying a number of cells, NCEL; an upper limit of the first cell, HLOW; and a cell width, HWID. Since histograms are not always requested, a single field identified as H is used on the symbol where it is implied that H represents NCEL/HLOW/HWID. An identifier, denoted ID, is printed on the SLAM II summary report to identify the output associated with each COLCT node. The symbol and statement for the COLCT node are shown in Figure 3-19.

The specification for TYPE may be FIRST, ALL, BETWEEN, INTERVAL(I), or any SLAM II variable.

COLCT, TYPE, ID, NCEL/HLOW/HWID, M;

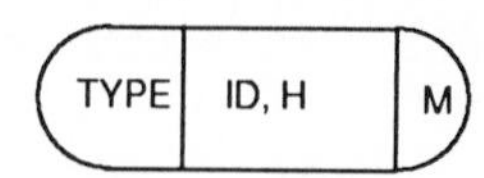

Figure 3-19 COLCT node.

3.6 RESOURCES

Situations arise where an entity requires a resource for a series of activities. SLAM II allows for definitions of resource types using a resource block. The resource capacity is also defined in the resource block and is the number of units available for allocation to entities. An entity waits for a resource at an AWAIT node where both a resource type and a number of units of the resource required by the entity are specified. When an entity arrives at an AWAIT node, it passes through the node if sufficient units of the required resource are available. Otherwise, its flow is halted and it is placed in a file associated with the AWAIT node. The entity is removed from the file when resource units are allocated to it.

To illustrate the use of resources, the manufacturing example presented in Section 3.2 involving a part entity and a machining operation is discussed. When using resources, the machine is explicitly defined as a resource and given a name, say MACHINE. Whenever MACHINE is required by an entity, it must be allocated to the entity at an AWAIT node. This is in contrast to the queue-service activity construct where the machine was implicitly a part of the service activity description. After a MACHINE is allocated to an entity, it must be explicitly taken away from the entity by having the entity pass through a FREE node. In its simplest form, the network segment illustrating the allocation of MACHINE to an entity, the performance of the machining activity, and the returning of MACHINE by the entity is shown in Figure 3-20. This network segment consists

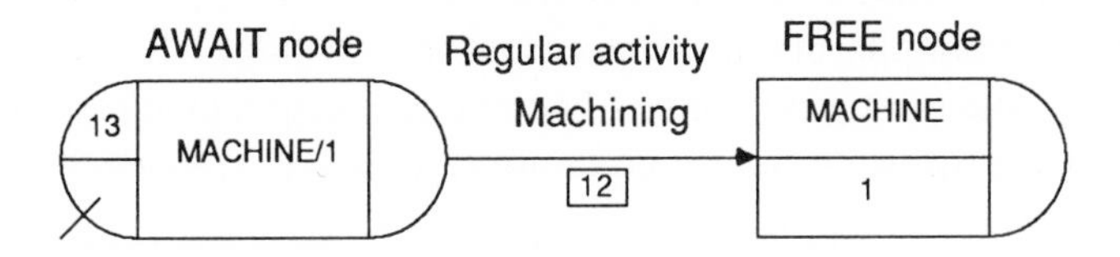

Figure 3-20 Resource-based network segment of a machining operation.

of an AWAIT node, a regular activity, and a FREE node. An entity that arrives to the AWAIT node when one unit of the resource MACHINE is not available, waits in file 13 . If MACHINE is available or when it becomes available, it is allocated to the entity. The entity starts regular activity 12, which models the machining operation. Activity 12 has no restriction on the number of entities that can flow through it concurrently. It is the number of resources of type MACHINE that limits entity flow. Following activity 12, the entity is routed to a FREE node where one unit of MACHINE is made available for reallocation to entities waiting for it. If entities only wait in file 13 for the machine resource then the freed machine is reallocated to the first part entity waiting in file 13. Note that this reallocation of the resource is analogous to a service activity checking its preceding queue to determine if there is an entity waiting and, if there is one, rescheduling the service activity.

A complete network model for the machining operation is shown in Figure 3-21. A resource block is used to define the resource MACHINE. The capacity of the resource, 1 in this illustration, is defined before the double set of lines. Following the double set of lines, file 13 is specified as the file where entities

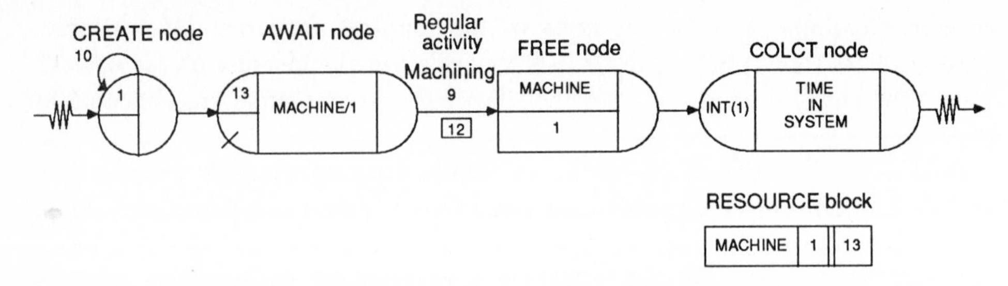

Figure 3-21 Resource-based model of a machining operation.

wait for one unit of MACHINE to become available. If entities waited in other files for MACHINE then they would also be listed in the resource block. In Figure 3-21, it is seen that the AWAIT node and FREE node are used as elements in a total network with entities arriving to them in the same manner as with the basic SLAM II node types.

An entity may require that a resource be allocated to it during several operations. For example, the machine resource is required during both a setup operation and a machining operation. The network segment shown in Figure 3-22 models this situation. A machine resource is allocated to a part entity at the

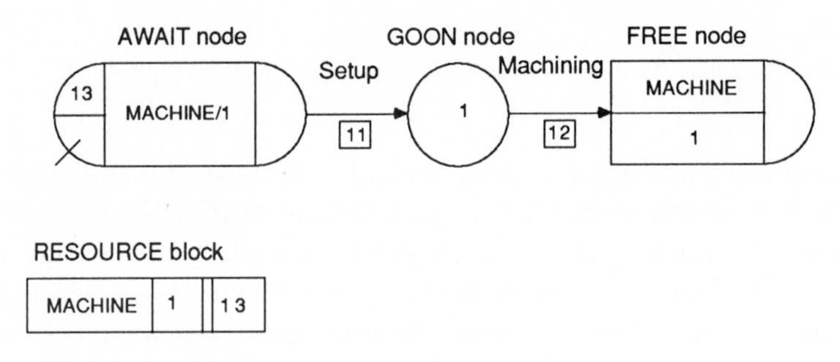

Figure 3-22 Resource usage in sequential activities.

AWAIT node, and a setup operation, activity 11, is performed. The machining operation is then performed as modeled by activity 12. Following machining, the machine resource is freed at the FREE node. The freed machine is reallocated if an entity is waiting in file 13. The entity is removed from file 13, and a setup operation would be started on it. The part entity that arrived to the FREE node continues on its route through the network based on activities emanating from the FREE node.

An embellishment to this model is to make the setup time a function of part type. The network segment in Figure 3-23 accomplishes this. Attribute 2 is

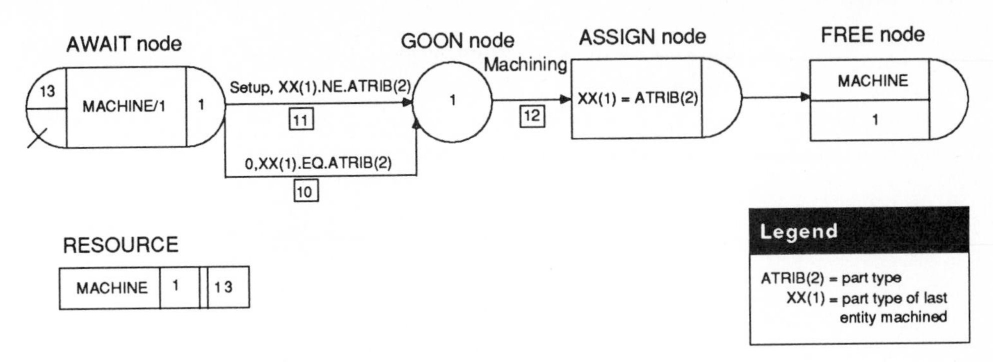

Figure 3-23 Resource usage with dependent setup times.

defined as a part type number and would be set at an ASSIGN node prior to the network segment shown in Figure 3-23. The global variable XX(1) is used to maintain the part type of the last entity machined by the machine resource by setting the value of XX(1) to ATRIB(2) at an ASSIGN node following activity 12. After a machine is allocated to a part entity at the AWAIT node either activity 10 requiring no setup time is performed or activity 11 requiring a setup is performed. No setup time is required if the last part type machined is the same as the part type of the next entity to be machined, that is, if XX(1) is equal to ATRIB(2). This is the condition for activity 10. Otherwise, a setup time is required as shown for activity 11.

In some systems, a resource is used in different modes. For example, the specification given in Table 2-3 indicated that machines in an FMS need to be calibrated using a test cube. The model for including the machining operation and the machine calibration is accomplished by allocating the machine resource in two separate network segments, as shown in Figure 3-24. A test cube entity arrives to an AWAIT node where entities wait in file 23. The resource block indicates that there is only one machine in the model and when that machine is freed, it is allocated to entities waiting in file 23 prior to being allocated to entities waiting in file 13. When the machine is allocated to a test cube entity waiting in file 23 then the calibrate operation modeled by activity 20 is performed. Following calibration, the machine is freed and, assuming there is only one test cube entity, the machine would be allocated to a part type entity in file 13 if one is waiting. By properly ordering the file numbers in the resource block, a priority is established for allocating a resource to waiting entities. In Figure 3-24, two disjoint networks are used to model the processing of a part entity and a test cube entity. This allows the flow of the two types of entities to be kept separate.

In some situations, it is desirable to have the different entities waiting in the same file. This is allowed in SLAM II as the same file number can be used at

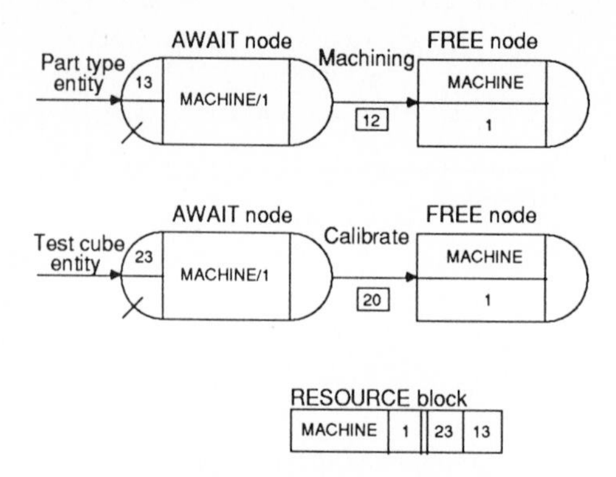

Figure 3-24 Resource usage for two different function.

different AWAIT nodes. For the above example, this would entail identifying file 13 with the AWAIT node in which the test cube entity waits for the machine resource. It is then necessary to set the priority for file 13 so that the test cube entity is at the head of the file and has priority over any waiting part entity. A priority statement would be used to establish the priority or ranking of entities in a file. An important use of this concept occurs when an entity requires more than one unit of a resource. For example, if there were two machines in the system and the calibration of the two machines must be performed sequentially, the test cube entity would then require both machines be allocated to it. Assuming system operation is that a machine is not allocated to a part entity while the test cube entity is waiting, then a model of this operation is given in Figure 3-25. No change is made in the model for the processing of the part entity. In the network segment for the test cube entity, the file of the AWAIT node is changed to 13 and the number of units of the machine resource required is changed to 2. It is assumed that file 13, the test cube entity is ranked higher than the part entities. When a machine becomes available at a FREE node, the first entity in file 13 is checked to determine if a machine can be allocated to it. If the test cube entity is waiting then the freed machine is only allocated if the other machine is available. When both are available, the test cube entity is removed from file 13 and proceeds over activity 20 to calibrate one of the machines. The test cube entity then frees the one machine calibrated, and it is allocated to a part entity if one is waiting. Activity 21 is then started to calibrate the second machine. After the second machine is calibrated, it is freed for allocation to a waiting part entity. In Figure 3-25, the capacity of the resource machine has been changed to 2, and the only file containing entities waiting for machine resources is file 13. An entity's priority in file 13 is established by its attribute 2 value with low values having a higher rank. This is coded in SLAM II as LVF(2) in the PRIORITY statement.

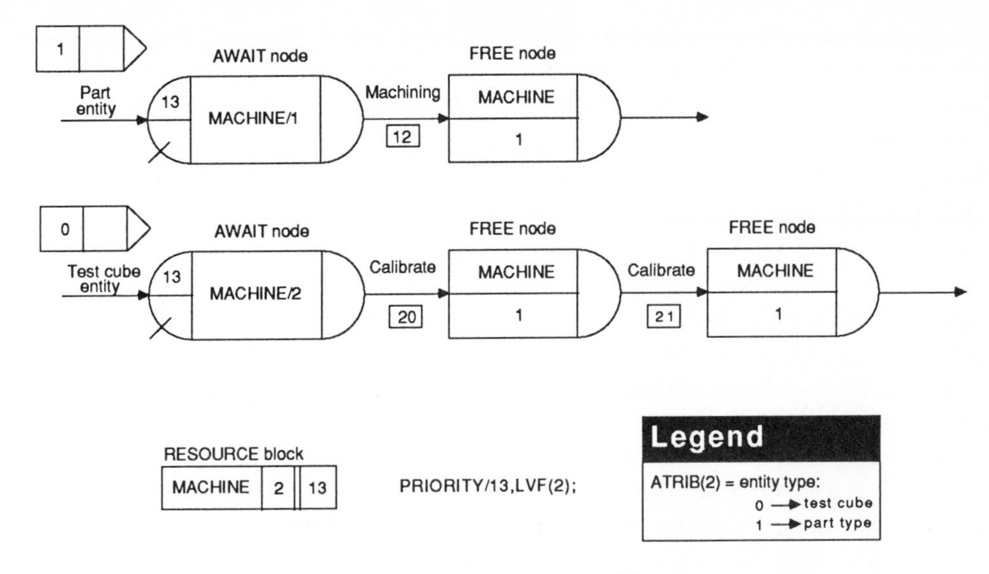

Figure 3-25 Resource usage for different functions with a common file.

In some situations, preventive maintenance is performed on a resource which makes it unavailable to entities. This is conceptually different than the situation with the test cube entity. In such a case, a separate physical entity like the test cube does not flow through the system to use the resource. This situation is modeled in SLAM II by introducing an information entity that passes through an ALTER node to request a change in the capacity of the resource. In Figure 3-26, a network segment shows the creation of an information entity that provides a timing mechanism for requesting preventative maintenance actions on the machine resource. The entity is routed from the CREATE node to an ALTER node where a unit of the machine resource capacity is requested to be made not available. If a machine is free then the capacity of the machine resource is reduced by one. If no machine is free, the request for reduction in machine availability is posted so that the next machine freed is not reallocated but is used to satisfy the capacity reduction request.

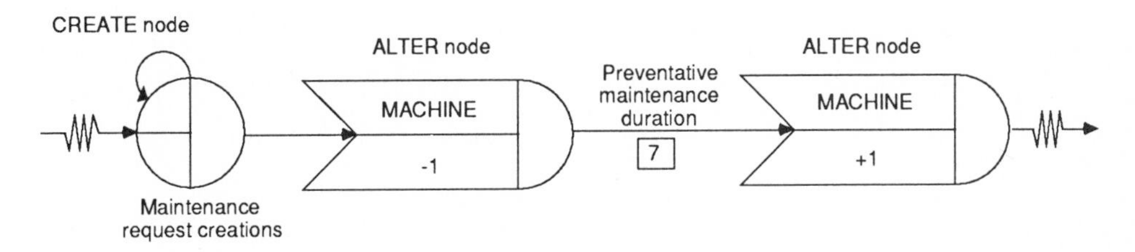

Figure 3-26 Disjoint network for preventative maintenance.

The information entity that arrived to the ALTER node is routed from it at its time of arrival, that is, the information entity does not wait until the request for resource reduction is satisfied. In Figure 3-26, the information entity is delayed by activity 7 which models the preventative maintenance operation. Following the preventative maintenance operation, the information entity is routed to the ALTER node where the capacity of the machine resource is increased by 1. This model assumes that preventative maintenance is to be completed in a specified time period following the request for preventative maintenance. In this way, preventative maintenance is performed during a prescribed period of time or not performed at all. It will not be performed if no machine is available during the scheduled preventative maintenance time. Following the second ALTER node, the information entity is terminated.

The utilization of resources is the time they are allocated to entities divided by the total time the resources are available. By altering the time that a resource is available the denominator of the utilization ratio is decreased. In the maintenance example, this is the way that utilization statistics are computed to omit the time the resource is being maintained. In the situation modeled with the test cube entity, calibration time is considered as part of the utilization time.

In some situations, it is desired to acquire a resource when one is not available. This is accomplished in SLAM II through the use of a PREEMPT node. An entity arriving to a PREEMPT node takes a resource away from an entity and allocates it to the entity arriving to the PREEMPT node. Preemption may be needed for an important part type or to model machine failures. The model segment in Figure 3-27 shows an information entity arriving to a PREEMPT node to model the failure of the machine resource. The function of the PREEMPT node is to reallocate the machine to the arriving entity and to take it away from the entity it is processing. The PREEMPT node can specify the disposition of the entity preempted. In Figure 3-27, since no rerouting instructions are provided on the PREEMPT node, SLAM II assumes that the entity preempted should be returned to the file where it acquired the resource, and that it should be placed first in that file. In addition, when the preempted entity is reallocated the resource, it

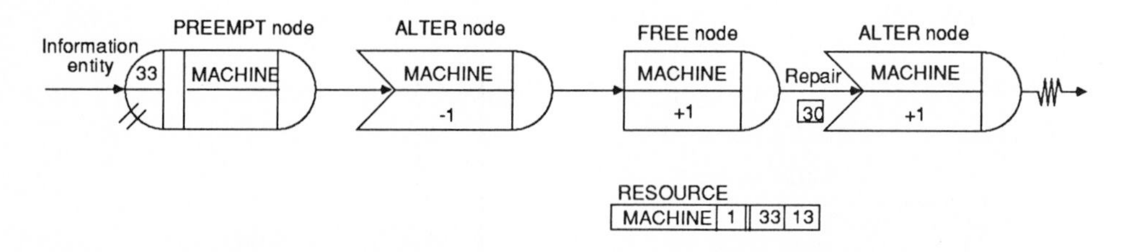

Figure 3-27 PREEMPT node illustration.

should be placed back in the operation from which it was preempted.

In Figure 3-27, the information entity after preempting the MACHINE is routed to an ALTER node to decrease the capacity of the resource preempted, since it is not in use. After the ALTER node, the machine resource is freed to satisfy the capacity decrease requested at the ALTER node. The repair activity for the machine is then modeled as activity 30. After the repair activity, the machine resource is placed back in operation by increasing the capacity of the resource by 1 at an ALTER node.

3.6.1 Resource Block

The RESOURCE block identifies a resource name or label, RLBL; the initial resource capacity, CAP; and the order in which files associated with AWAIT and PREEMPT nodes are to be polled to allocate freed units of the resource. The word block is employed because entities do not flow through it. The RESOURCE block symbol and statement are shown in Figure 3-28.

RESOURCE/ RLBL(CAP), IFLs;

RLBL	CAP	IFL1	IFL2

Figure 3-28 Resource block.

The SLAM II variable NRUSE(RES) maintains the number of units of resource RES that are in use. NNRSC(RES) is the number of units of RES currently available. Statistics are collected on resource utilization and availability and are printed as part of the SLAM II summary report for each resource.

3.6.2 AWAIT Node

AWAIT nodes are used to store entities waiting for UR units of resource RES. When an entity arrives at an AWAIT node and the resource units required are allocated to it, the entity passes through the node and is routed according to the M-number of the AWAIT node. Regular activities emanate from an AWAIT node. If the entity has to wait at the node, it is placed in file IFL. The symbolism and statement for the AWAIT node are shown in Figure 3-29. Normally, RES is specified by a resource label RLBL. The file number, IFL, queue capacity, QC, and blocking and

AWAIT(IFL/QC), RES/UR,
BLOCK or BALK(NLBL), M;

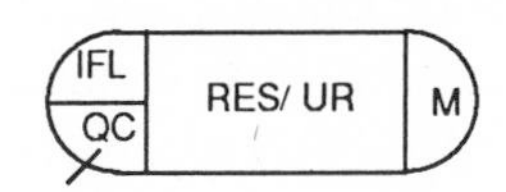

Figure 3-29 AWAIT node.

balking specifications are identical to those used for QUEUE nodes. For AWAIT nodes, the same file number can be associated with more than one AWAIT node. This allows entities to wait in the same file at different AWAIT nodes in the network.

Resource requirements involving multiple resources and conditions may be specified at an AWAIT node by using ALLOC(I) for the RES field. When this is done, the user-written subroutine ALLOC(I,IFLAG) is called, as is discussed in Chapter 5. The GWAIT node described in Chapter 19 can be used for allocating multiple and alternate resources to entities.

3.6.3 FREE Node

A FREE node releases UF units of a resource RES when an entity arrives to it. The freed units are then allocated to entities waiting in PREEMPT and AWAIT nodes in the order prescribed by the files listed with the RESOURCE block. The entity arriving to the FREE node is then routed in accordance with the M-number associated with the FREE node. The symbol and statement for the FREE node are shown in Figure 3-30.

FREE, RES/ UF, M;

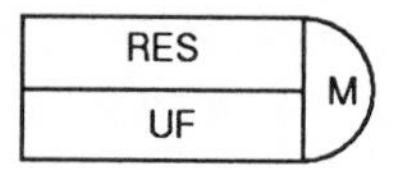

Figure 3-30 FREE node.

3.6.4 ALTER Node

The ALTER node is used to change the capacity of resource RES by CC units. If CC is positive, the number of available units is increased. If CC is negative, the capacity is decreased. The symbol and statement for the ALTER node are shown in Figure 3-31. When the ALTER node is used to decrease capacity, the change is invoked only if a sufficient number of units of the resource are not in use. If this is not the case, the capacity is reduced to the current number in use. Further reductions then occur when resources are freed at FREE nodes. In no case will the capacity of a resource be reduced below zero. Requests for capacity reductions when the capacity is zero are ignored.

ALTER, RES/ CC, M;

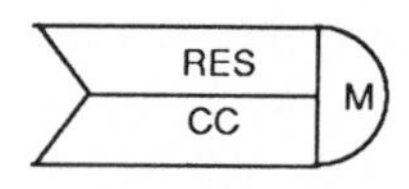

Figure 3-31 ALTER node.

3.6.5 PREEMPT Node

A PREEMPT node usurps one unit of a resource from an entity and allocates it to the entity arriving at the PREEMPT node. If the entity with the resource obtained it at an AWAIT node, preemption will always be attempted. If the resource was obtained at a preempt node, the preemption is attempted only if the priority of the arriving entity is greater than the priority of the entity with the resource. The symbol and statement for the PREEMPT node are shown in Figure 3-32.

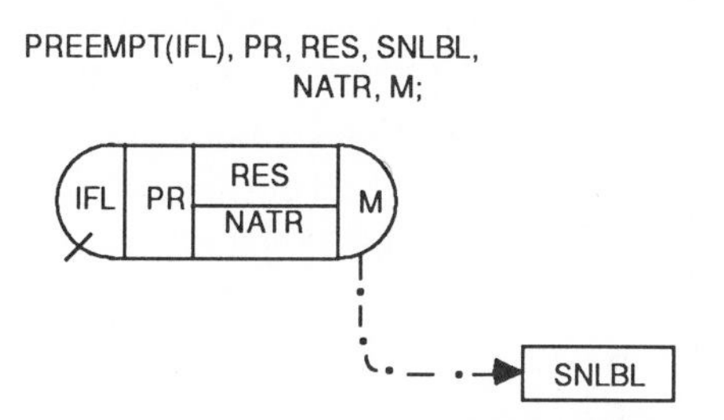

Figure 3-32 PREEMPT node.

The definitions of IFL and RES for the PREEMPT node identify the file number and the requested resource. The priority, PR, is specified as LOW(K) or HIGH(K), where K is an attribute number. The incoming entity will attempt to preempt another entity if its K*th* attribute gives it a higher priority. A preemption attempt is not invoked if the resource is currently in use by an entity that is (1)being processed in a service activity, (2) waiting in a file, or (3) performing an activity with an indefinite duration (REL or STOPA). Entities that do not invoke a preemption wait for the resource in file IFL.

A preempted entity is routed to a send node, which is specified by the label SNLBL. The time remaining to process the preempted entity is stored as its NATR*th* attribute. If no send node label is specified, then the preempted entity is returned to the AWAIT or PREEMPT node where it was allocated the resource. At that node, it is inserted in its file as the first entity waiting for the resource. When the resource is reallocated to the preempted entity, it is placed back in the activity from which it was preempted, with its remaining processing time as the duration for that activity.

As described previously, there are restrictions on when a preemption may be invoked. First, preemptions are only allowed for resources having a capacity of one unit. Second, an entity holding a resource that currently is in a file will not be preempted. Also, if the entity is in a service activity or an activity of indefinite duration, it will not be preempted. The reason for these restrictions is the large combinatorial problem that results from having to determine which of a group of resources to preempt. If it is necessary to model the preemption of a resource whose capacity is greater than one, then the SLAM II interface to user-written inserts should be used.

3.7 GATES

In SLAM II, a GATE is used to stop entity flow. A GATE is either open or closed. Examples of the use of gates are (1) to model shipment entities that arrive during the night for processing the next day, and (2) car entitites that wait for a signal to change. Entities pass through a gate by being routed to an AWAIT node. If the GATE associated with the AWAIT node is closed, the entity waits in a file until the GATE is opened. If a gate is open, the arriving entity passes through the AWAIT node. A GATE is opened by routing an entity through an OPEN node. It is closed when an entity is routed through a CLOSE node. The files in which entities wait for a GATE to open are defined in a GATE block. When a gate is opened, all entities waiting at the gate are routed from the AWAIT nodes associated with the gate.

3.7.1 GATE Block

A GATE block defines a GATE by its label GLBL, the initial status of the GATE, and the file numbers of AWAIT nodes where entities wait for the gate to open. The symbol and statement for the GATE block are shown in Figure 3-33, and the use of a gate with an AWAIT node is shown in Figure 3-34.

GATE/GLBL, OPEN or CLOSE, IFLs;

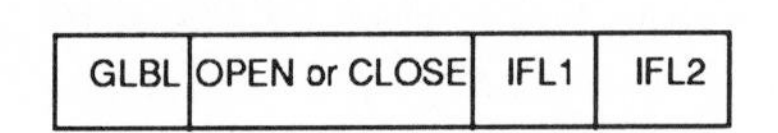

Figure 3-33 GATE block.

AWAIT(IFL/QC), GATE,BLOCK or BALK(NLBL), M;

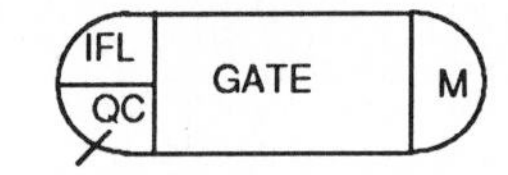

Figure 3-34 AWAIT node for a gate.

3.7.2 OPEN Node

An OPEN node is used to open a GATE. Each entity arriving to an OPEN node causes GATE to be opened. When a gate is opened, all entities waiting for GATE are removed from the files associated with the AWAIT nodes for GATE. The entity that caused GATE to be opened is routed from the OPEN node. The symbol and statement for the OPEN node are shown in Figure 3-35.

OPEN, GATE, M;

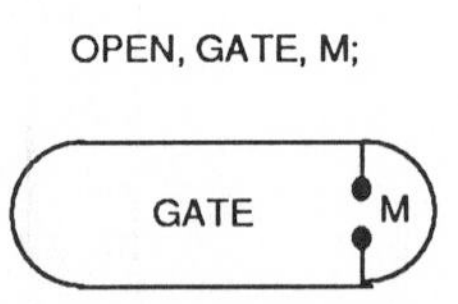

Figure 3-35 OPEN node.

3.7.3 CLOSE Node.

A CLOSE node is used to close a GATE. An entity arriving to a CLOSE node causes the GATE referenced to be closed. An entity arriving at the CLOSE node is routed in accordance its M-number. The symbol and statement for the CLOSE node are shown in Figure 3-36.

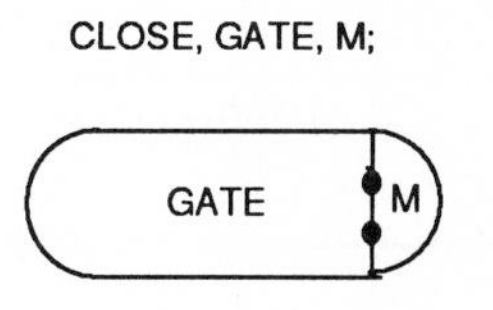

Figure 3-36 CLOSE node.

3.8 LOGIC AND DECISION NODES

There are six nodes in SLAM II that perform logic and decision operations. The ACCUMULATE node accumulates a specified number of entities into a single entity. The BATCH node generalizes the concept of the ACCUMULATE node and allows the identity of the individual entities of a batch to be retained. The UNBATCH node reenters the individual entities of a batch into the network. The MATCH node causes entities to wait until a group of entities with a common characteristic are at QUEUE nodes preceding the MATCH node. The DETECT node creates an entity when a SLAM II variable crosses a threshold value. The SELECT node routes entities to QUEUE nodes, from QUEUE nodes, and to service activities. One of the queue selection rules for the SELECT node is ASSEMBLY, which assembles entities from different QUEUE nodes.

Logic and decision nodes perform operations on an entity or a set of entities. This differs from the probabilistic and conditional routing by an activity, which prescribes how to route an entity from a node.

3.8.1 ACCUMULATE Node

The ACCUMULATE node causes one entity of a group of entities to be routed from it. The ACCUMULATE node can be used to put raw material entities in a tote entity or part entities on a pallet entity. For example, as a unit of raw material arrives, it is routed to an ACCUMULATE node. When the number of raw material entities fills the tote, a tote entity is routed from the ACCUMULATE node. These functions are shown in Figure 3-37. On the left side of the ACUMULATE node are two 3's, which specify the number of incoming entities required to release the node for the first time and the number required for all subsequent releases. In the middle of the node, the criterion FIRST indicates that the attributes of the first arriving entity will be given to the entity that is routed from the ACCUMULATE node.

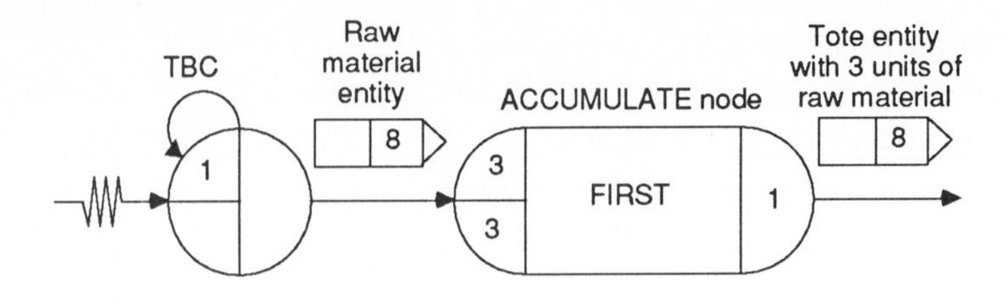

Figure 3-37 Accumulating three raw material entities.

The CREATE node in Figure 3-37 specifies attribute 1 as the creation time of the raw material entity. The FIRST criterion indicates that the tote entity will have the time of creation of the first raw material entity that goes into the tote. If the time that the tote entity is routed from the ACCUMULATE node is desired as the mark time for the tote entity, then LAST should be specified as the criterion for saving the attributes of the raw material entity.

The ACCUMULATE node has many uses in project planning and control situations where entities represent signals and timing pulses. For example, in a construction project, activities are either being performed or not. When they are being performed, the amount of time to finish the activity is decreasing. In SLAM II networks, this corresponds to an entity flowing over an activity and completing its traversal of the activity when time to perform the activity has expired. The entity arrival at a node is a signal that the activity has been completed. ACCUMULATE nodes can be used to count the number of activities completed and provide a logic mechanism to require all prerequisite activities to be performed before starting an activity. In Figure 3-38, an ACCUMULATE node is shown to require the activities of ***buy machine*** and ***remove old machine*** before starting the activity *install machine*. In this illustration, two entities representing two different activities are required to release the ACCUMULATE node for the first time. The infinity on the left hand side of the node indicates that the ACCUMULATE node will not be released again. The requirement for two different activities is not part of the ACCUMULATE node but is a characteristic of project planning networks in which activities are only performed once. An entity flowing over a project activity is a signal that the activity is ongoing. When the activity is completed, the signal entity arrives at the node. If a test is made after the installation of

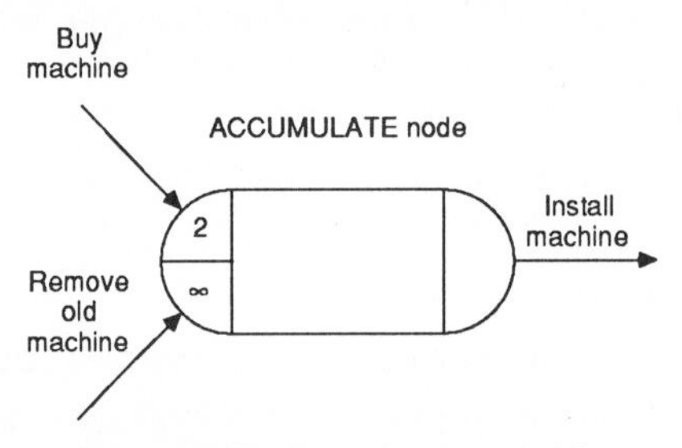

Figure 3-38 Precedence activities.

the machine and if there is a possibility to reinstall the machine, then the subsequent requirement for releasing the ACCUMULATE node would be changed to 1. The network shown in Figure 3-39 models this reinstallation requirement.

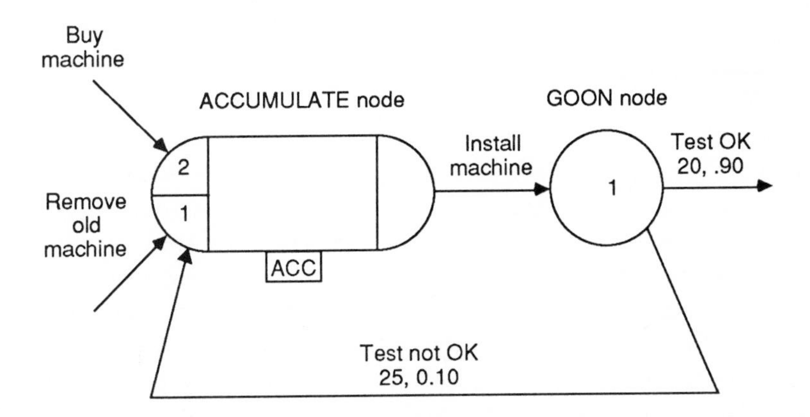

Figure 3-39 ACCUMULATE node with feedback.

In Figures 3-38 and 3-39, no specification is made as to the attributes of the incoming entities to assign to the entity leaving the ACCUMULATE node. This frequently occurs when processing entities representing signals. The default criterion for saving attributes is to save the attributes of the last arriving entity. It should also be noted that the ACCUMULATE node of Figure 3-37 processes entities arriving on a single activity whereas the ACCUMULATE nodes in Figures 3-38 and 3-39 process entities arriving on different activities. The ACCUMULATE node is not designed to differentiate between these two situations. The symbol and statement for the ACCUMULATE node are shown in Figure 3-40. The release specification is given by FR, SR, and SAVE, where FR is a first release requirement, SR is a subsequent release requirement , and SAVE is a criterion for saving attributes. The six possible SAVE criteria are:

1. Save the attributes of the first entity arriving to the node (FIRST).
2. Save the attributes of the entity that releases the node (LAST).
3. Save the attributes of the entity with the highest value of attribute I (HIGH(I).
4. Save the attributes of the entity with the lowest value of attribute I (LOW(I)).

ACCUMULATE, FR, SR, SAVE, M;

Figure 3-40 ACCUMULATE node.

5. Create a new entity whose attributes are the sum of the attributes of all entities that contributed to the new entity (SUM).
6. Create a new entity whose attributes are the product of the attributes of all entities that contributed to the new entity (MULT).

3.8.2 BATCH Node

The BATCH node combines entities until a specified threshold level is reached and then releases a single entity referred to as a batched entity. As an illustration of the BATCH node, consider that a pallet is to be loaded with entities such that the weight of the pallet does not exceed one thousand pounds. The weight of a part is defined as attribute 2 of an entity. In the processing of the pallet, it is necessary to know the weight of all the part entities on the pallet.

The BATCH node shown in Figure 3-41 produces a pallet entity from incoming part entities. In the middle of the node, on the top line, the value 1000 specifies a threshold condition for a batched entity based on an attribute of the individual part entities. The 2 on the same line indicates that attribute 2 is the variable which contains the value to be summed toward the threshold value. As part entities arrive to the BATCH node, their attribute 2 values are summed until the threshold value of 1000 is exceeded at which time a batched entity is created to model a pallet entity. The pallet entity does not include the last entity that arrived as it would cause the weight of the pallet entity to exceed the threshold. The specification in the middle of the BATCH node indicates that attribute 2 of the batched entity is to be set to the sum of all attribute 2 values of the part entities included in the pallet entity. All other attributes of the batched entity are to be taken from the last part entity to arrive at the node. The bottom line of the BATCH node specifies that the individual part entities are to be retained and that attribute 3 of the pallet entity is to be set by SLAM II as a pointer attribute. An UNBATCH node will be used to reestablish the part entities and to

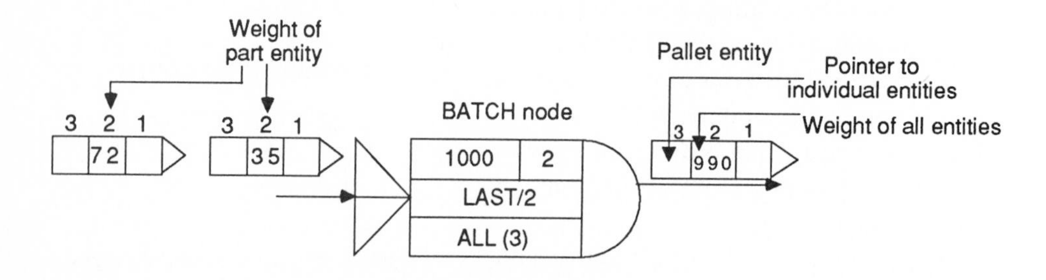

Figure 3-41 BATCH node to build a pallet entity.

insert them into the network at a future time. The triangle on the left side of the node is used to specify the number of different entity types from which multiple batches can be made. An attribute of a part entity specifies the batch in which it is to be included. The symbol and statement for the BATCH node are shown in Figure 3-42. NBATCH is the total number of batches that may be accumulated concurrently at the BATCH node. NATRB is the number of the attribute that specifies the batch for an arriving entity; that is, the value of ATRIB(NATRB) is to be the same for entities in a batch. A secondary use for ATRIB(NATRB) is to cause the batch to be released when the value of this attribute is the negative of the batch number. For this case, the arriving entity is included in the batch. THRESH is the threshold value. NATRS is the number of the attribute that contains the value to be summed and then tested against the threshold. When this sum is greater than or equal to THRESH, a batched entity is formed and released from the BATCH node. If NATRS is not specified, then THRESH is the number of entities that are necessary to constitute a batch.

BATCH, NBATCH/ NATRB, THRESH, NATRS, SAVE, RETAIN, M;

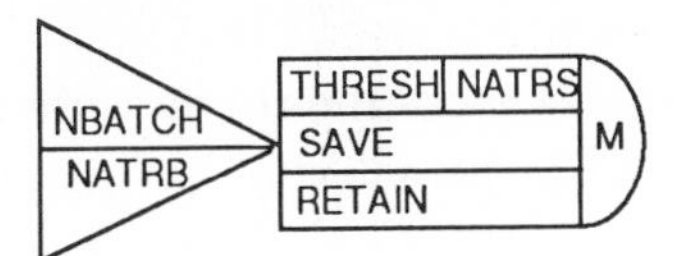

Figure 3-42 BATCH node.

SAVE is used to specify a criterion for defining the attributes of the batched entity. The criterion specifies which entity in the batch is to be used as a basis for the attributes of the batched entity. The options for the criterion are (1) the entity that arrived first for the batch, FIRST; (2) the entity that arrived last for the batch, LAST; (3) the entity with the lowest value of attribute I, LOW(I); and (4) the entity with the highest value of attribute I, High(I). In addition to specifying a criterion, a list of attribute numbers can be given for which the sum is used as the value of the corresponding attribute of the batched entity. For example, FIRST/3,5 specifies that attribute 3 and attribute 5 of the batched entity are to be the sum of the attribute 3 values and attribute 5 values of the entities included in the batch. All other attribute values for the batch entity are taken from the first entity that is placed in the batch.

RETAIN indicates that individual entities included in the batch should be retained for future insertion in the network. The specification ALL(NATRR) saves all the individual entities, and causes an internal reference pointer to be placed in ATRIB(NATRR) of the batched entity. Reference to NATRR enables the individual entities to be retrieved at an UNBATCH node. If it is not necessary to retain the individual entities, then the RETAIN field should be specified as NONE.

3.8.3 UNBATCH Node.

To reinsert the individual part entities into the network, a batched entity is routed to an UNBATCH node which references attribute NATRR. For example, when the pallet entity created in Figure 3-41 arrives to the UNBATCH node shown in Figure 3-43, each part entity that was placed on the pallet is inserted into the network on the activity that follows the UNBATCH node.

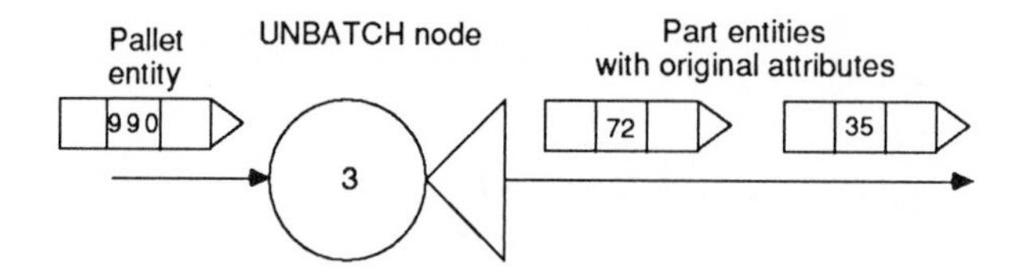

Figure 3-43 UNBATCH node to unload entities from a pallet entity.

The symbol and statement for the UNBATCH node are shown in Figure 3-44. The value of attribute NATRR specifies whether each individual entity of the batch is to be routed from the UNBATCH node. If ATRIB(NATRR) is set by SLAM II at a BATCH node, each of the individual entities of the batch is released from the UNBATCH node and the arriving batched entity is terminated. If the arriving entity is not a batched entity, then ATRIB(NATRR) defines the number of identical entities to be routed from the UNBATCH node. This use of the UNBATCH node models a entity splitting or disaggregation operation. In this case, the attributes of an entity routed from an UNBATCH node are the values of the attributes of the arriving entity.

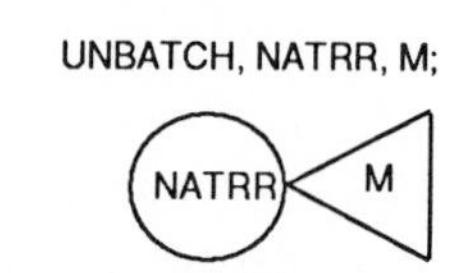

Figure 3-44 UNBATCH node.

3.8.4 MATCH Node

MATCH nodes require entities residing in a set of QUEUE nodes preceding the MATCH node to have the same value for a specified attribute. When this occurs, the MATCH node removes each appropriate entity and routes it to a node specified as the route node associated with that QUEUE node. Thus, each entity is routed individually. The node pairs associated with a MATCH node are specified by QUE/ NOD. The symbol and statement for the MATCH node are shown in Figure 3-45.

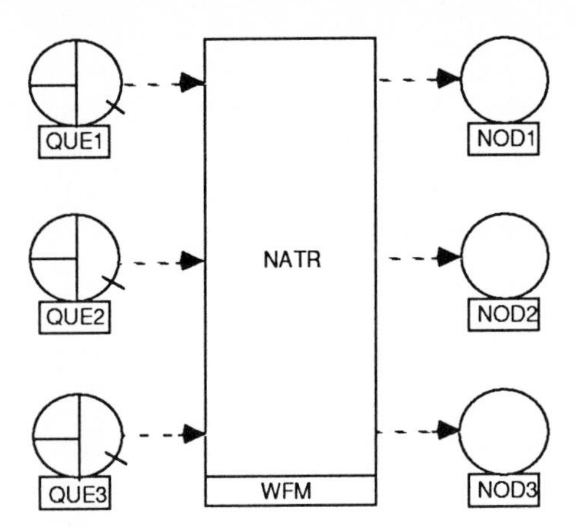

Figure 3-45 MATCH node.

Only QUEUE nodes may precede a MATCH node. A QUEUE node to route node transfer is made when a match occurs. If no route node is specified for a QUEUE node, the entity in that QUEUE node is terminated when a match is made. The attribute number, on which the match is based, is specified as NATR within the MATCH node symbol.

Consider the network segment shown in Figure 3-46 where a radio entity, taken from an airplane, is required to be installed back in its original airplane. Both the radio entity and airplane entity travel through network segments (not shown in Figure 3-46) until they arrive at QUEUE nodes QRAD and QAIR. Two attributes are used: attribute 1 is the entity type, and attribute 2 is the tail number of the airplane. The MATCH node MTN models the process of matching a radio

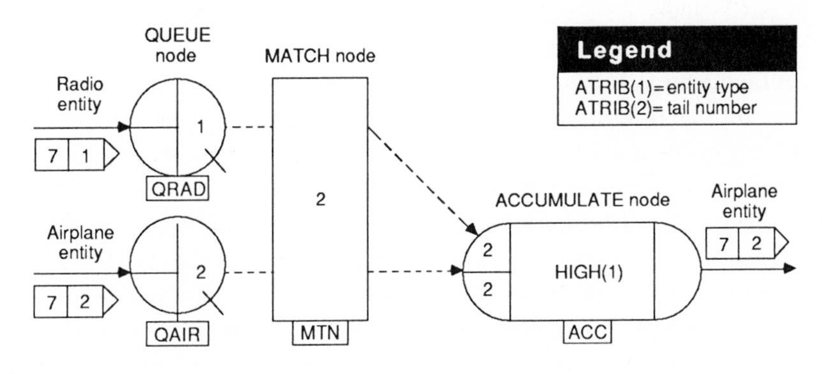

Figure 3-46 Placing a radio back into its original aircraft.

entity and an airplane entity that have the same value for the tail number attribute. When a match occurs, the radio entity is removed from node QRAD and routed to node ACC. The airplane entity is removed from node QAIR and routed to node ACC. At ACCUMULATE node ACC, the entities are combined and the attributes of the entity with the higher value of attribute 1 is selected as the entity to be routed from node ACC. In this instance, the airplane entity has a higher value for attribute 1.

3.8.5 DETECT Node

A DETECT node is used to create an entity when a SLAM II variable crosses a threshold value. The entity created has all its attribute values equal to 0. It is routed from the DETECT node in accordance with the node's M-number. The symbol and statement for the DETECT node are shown in Figure 3-47. The fields for a DETECT node define the crossing variable, XVAR, the direction of the desired crossing detection, XDIR, a threshold, VALUE, and the value of an interval, TOL, beyond the threshold value for which a detection of a crossing is considered within tolerance. The creation of an entity is considered a release of the DETECT node. The duration of an activity may be keyed to such a release. In this way, an entity is held in an activity until a prescribed state of the model is detected. DETECT nodes are used to monitor the number of entities in files and activities, and the number of resources available or in use.

DETECT, XVAR, XDIR, VALUE, TOL, M;

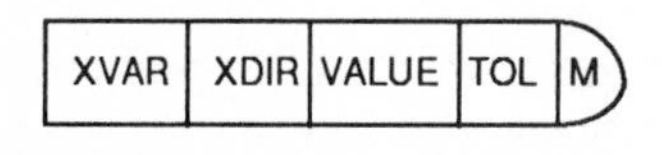

Figure 3-47 DETECT node.

3.8.6 SELECT Node

In manufacturing systems, there are many instances when parts wait in separate buffers for one or more machines to become available. Since different parts may be stored in the queues and different machines may be used to process the parts, selection rules are necessary to decide the next part to be processed when a machine becomes available. A selection is also necessary when a part arrives to a queue and more than one machine is available to process the part. The SELECT node of SLAM II provides the capability for making these decisions. A diagram of a 2-queue, 3-machine work center is shown in Figure 3-48. The SLAM II network segment that models this 3-machine work center is shown in Figure 3-49. The two queue nodes and three service activities are connected

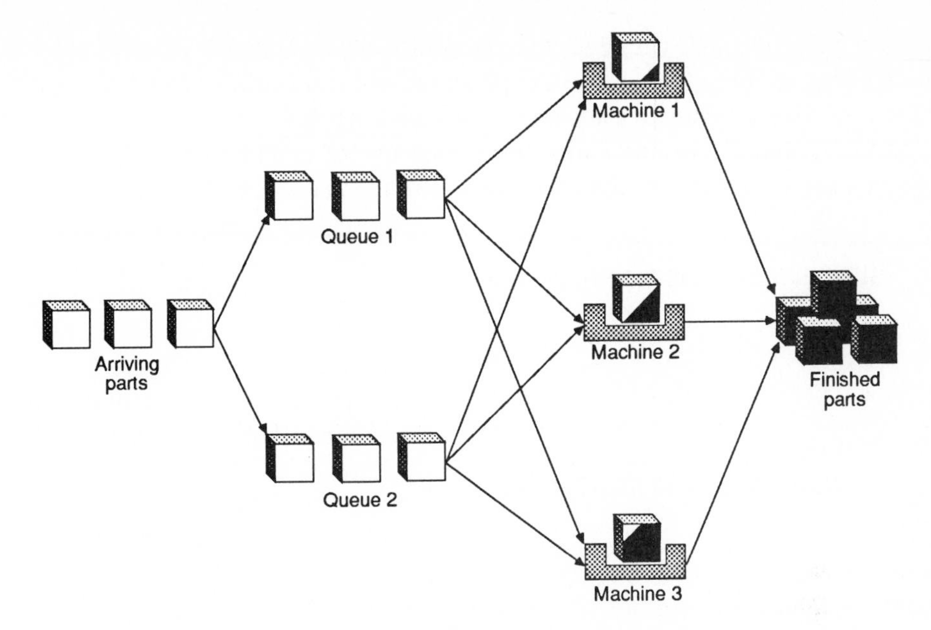

Figure 3-48 Diagram of a multiple-machine work center.

through SELECT node SEL. When a part entity arrives to either one of the queues and a service activity is idle, the SELECT node transfers the part entity to the idle machine for processing. If more than one machine is idle, the selection rule LIT is used to make a choice between the available machines. LIT is an abbreviation for selecting the machine with the longest idle time. When a machine completes processing a part entity, the SELECT node looks backward

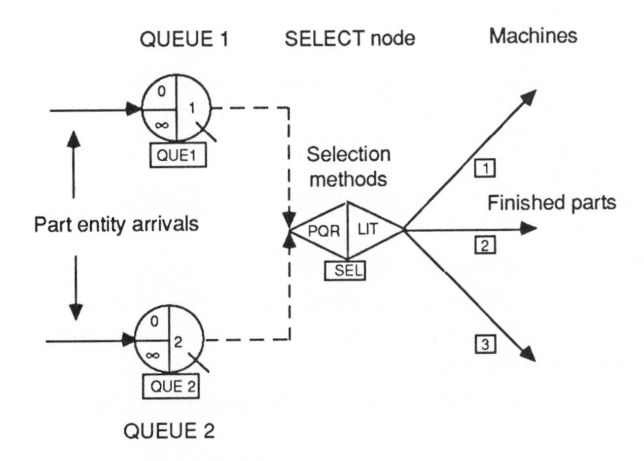

Figure 3-49 Model of a multiple-machine work center.

and examines the QUEUE nodes to determine if a part entity is waiting. If parts are waiting in both queues, then the preferred order selection rule POR is used. This rule tries to select a part entity from node QUE1 first.

SELECT nodes route an entity to or from one of a set of QUEUE nodes to one of a set of servers or both. To accomplish the routing at the SELECT node, the modeler chooses a queue selection rule (QSR) and/or a server selection rule (SSR). The symbol and statement for the SELECT node are shown in Figure 3-50. In the SELECT statement, the QLBLs are the QUEUE node labels associated with the queue selection rule. The QUEUE nodes could be either before or after the SELECT node. The queue selection and server selection rules available in SLAM II are listed in Tables 3-4 and 3-5.

SLBL SELECT, QSR, SSR, QLBLs;

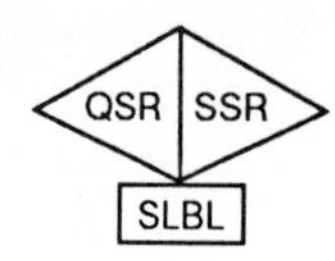

Figure 3-50 SELECT node.

In Table 3-4, the ASM or assembly rule differs from the other rules in that it combines two or more entities into an assembled entity. For assembly, at least one entity must be in each QUEUE node before the SELECT node. For entities

Table 3-4 Priority Rules of SELECT Nodes for Selecting a Queue

Code	Definition
POR	Select in a preferred order.
CYC	Cyclic selection.
RAN	Random selection.
LAV	Select QUEUE node with the largest average number of entities in it to date.
SAV	Select QUEUE node with the smallest average number of entities in it to date.
LWF	Select QUEUE node for which the waiting time of its first entity is the longest.
SWF	Select QUEUE node for which the waiting time of its first entity is the shortest.
LNQ	Select QUEUE node with current largest number of entities in it.
SNQ	Select QUEUE node with current smallest number of entities in it.
LRC	Select QUQUE node with the largest remaining unused capacity.
SRC	Select QUEUE node with the smallest remaining unused capacity.
ASM	Assembly mode option: all incoming queues must contribute one entity.
NQS(N)	User-written function to select a QUEUE node.

Table 3-5 Priority Rules of SELECT Nodes for Selecting a Server

Code	Definition
POR	Select in a preferred order.
CYC	Select servers in a cyclic manner.
LBT	Select the server with the largest amount of busy time to date.
SBT	Select the server with the smallest amount of busy time to date.
LIT	Select the server that has been idle for the longest period of time.
SIT	Select the server that has been idle for the shortest period of time.
RAN	Select randomly according to preassigned probabilities.
NSS(N)	User-written function to select a server.

assembled by a SELECT node, a save attribute criterion is used to specify which entity's attributes are assigned to the assembled entity. The save concept is similar to that presented earlier for ACCUMLATE nodes. The save criterion may be specified as HIGH(I), LOW(I), SUM, or MULT.

3.9 INTERFACE NODES

Two nodes in SLAM II provide direct interfaces to user-written subroutines. The EVENT node causes subroutine EVENT to be called each time an entity arrives at it. Subroutine EVENT is written by the user to perform model-specific logic using the SLAM II functions and subroutines described in Chapter 5. To insert entities into a network, an ENTER node is provided. Entities created by a user are routed to the ENTER node by a call to subroutine ENTER.

3.9.1 EVENT Node

The EVENT node causes subroutine EVENT(JEVNT) to be called every time an entity arrives at the EVENT node. The symbol and the input statement for the EVENT node are shown in Figure 3-51. The value of JEVNT is an event code to be passed as an argument to subroutine EVENT. Subroutine EVENT maps the event code JEVNT onto the appropriate event logic coding. In coding the event logic, the modeler has access

EVENT, JEVNT, M;

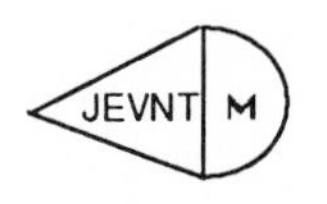

Figure 3-51 EVENT node.

to the SLAM II provided subprograms for performing commonly encountered functions, such as access to SLAM II variables, random sampling, file manipulation, and data collection.

When an entity arrives at an EVENT node, the SLAM II processor loads the attributes of the arriving entity into the ATRIB vector prior to the call to subroutine EVENT(JEVNT). Following the return from subroutine EVENT(JEVNT), the SLAM II processor assigns the values in ATRIB as the attributes of the entity exiting from the EVENT node.

3.9.2 ENTER Node

The ENTER node allows the insertion of an entity into the network from a user-written subroutine. The symbol and statement for the ENTER node are shown in Figure 3-52. Each ENTER node has a user-assigned integer code NUM and an M-number. The ENTER node is released following a return from a user-written routine in which a call was made to subroutine ENTER(NUM, A) where NUM is the numeric code of the ENTER node being released and A is a vector containing the attributes of the entity to be inserted at the ENTER node.

ENTER, NUM, M;

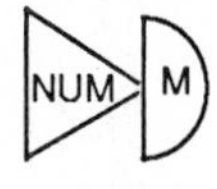

Figure 3-52 ENTER node.

3.10 SUMMARY

An alphabetic listing of SLAM II network symbols is given in Table 3-6. A SLAM II network model is developed by connecting the elements of Table 3-6 into a representation of a system's operation.

3.11 EXERCISES

3-1. Develop a SLAM II model for the following situation. Assembled television sets move through a series of testing stations in the final stage of their production. At the last of these stations, a control setting on the TV sets is tested. If the setting is found to be functioning improperly, the offending set is routed to an adjustment station, where the setting is adjusted. After adjustment, the television set is sent back to the last inspection station, where the setting is again inspected. Television sets passing the final inspection phase, whether for the first time or after one or more routings through the adjustment station, are routed to a packing area. The

Table 3-6 SLAM II Network Symbols

Symbol	Description	Symbol	Description
ACCUMULATE FR SR SAVE M	Accumulates a set of entities into a single entity	FREE RES UF M	Makes resources available for reallocation
ACTIVITY DUR, PROB or COND A	Specifies delay (operation) time and entity routing	GATE GATE OPEN or CLOSE IFL1 IFL2	Logical switch definition and initial status
ALTER RES CC M	Changes the capacity of a resource	GOON M	Continuation node
ASSIGN VAR Value : M	Assigns values to attributes or global system variables	MATCH IQ QC IFL NATR OLBL NLBL	Holds entities in QUEUE nodes until a match on an attribute is made
AWAIT IFL QC RES/UR or GATE M	Holds entities until a resource is available or a gate is open	OPEN GATE M	Opens a gate
BATCH NBAT NATR THRESH NATRS SAVE RETAIN M	Accumulates multiple sets of entities	PREEMPT IFL PR RES NATR M SNLBL	Preempts a resource
CLOSE GATE M	Closes a gate	QUEUE IQ QC IFL or IQ QC IFL	Holds entities until a server becomes available
COLCT TYPE ID,H M	Collects statistics and histograms	RESOURCE RES CAP IFL1 IFL2	Resource definition and initial capacity
CREATE TBC TF MA MC M	Creates entities	SELECT QSR SSR	Selects among queues and servers based on prescribed rules
DETECT XVAR XDIR VALUE TOL M	Creates (generates) an entity when a variable value reaches a prescribed threshold	SERVICE ACTIVITY DUR, PROB N A	Specifies delay (operation) time for servers
ENTER NUM M	Entry point for entity insertion from user-written subprogram	TERMINATE TC or TC	Terminates the routing of entities
EVENT JEVNT M	Transfer of control to user-written subprogram	UNBATCH NATRR M	Restores members of a batched set

time between arrivals of television sets to the final inspection station is uniformly distributed between 3.5 and 7.5 minutes. Two inspectors work side by side at the final inspection station. The time required to inspect a set is uniformly distributed between 6 and 12 minutes. On the average, 85% of the sets pass inspection and continue on to the packing department. The other 15% are routed to the adjustment station, which is manned by a single worker. Adjustment of the control setting requires between 20 and 40 minutes, uniformly distributed. Based on the model, list five decisions that are incorporated into the model that can be analyzed using the model. Estimate performance tendencies that you feel would be associated with changes caused by decisions affecting the model.

Embellishments: Modify the SLAM II model of the inspection and adjustment stations to accommodate the following changes:

(a) An arrival of television sets to the inspection station involves two television sets being inspected.

(b) The adjuster routes 40% of the adjusted sets directly to packing and 60% to the inspectors.

(c) By adding a step to the inspection process, it is felt that the probability of sending a set to the adjustor can be decreased to 0.10; the added step takes 5 minutes. Redraw the network to indicate these changes.

3-2. A paint shop employs six workers who prepare jobs to be spray painted. The preparation operation is lengthy compared to the spraying operation; hence, only two spraying machines are available. After a worker completes the preparation of a job, he or she proceeds to the spraying machine, where the worker must wait, if necessary, for a free spraying machine. The preparation time is normally distributed with a mean of 20 minutes and a standard deviation of 3 minutes. Spraying time is uniformly distributed between 5 and 10 minutes. A SLAM II model of this situation is to be developed to obtain estimates of the utilization of the workers and the spraying machines for five 8-hour days. Also to be determined is the length of time required to prepare and paint a job. It is assumed that jobs to be prepared and painted are always available to the workers.

Embellishment: Include in the model a drying operation that is to be performed after painting. Drying requires 15 minutes and does not require a worker. Include this operation in your SLAM II network. With this addition, what changes would you expect in the statistical quantities of interest?

3-3. A machine shop operates 8 hours a day, 5 days a week. There are 54 machines in the machine shop, and management maintains a work force of 50 operators. The machines are subject to failure, and when a machine fails, another machine, if available, is put into service. A failed machine is serviced by a repairperson, if one is available. Travel times by the operator to a different machine, and the repairperson to a failed machine are considered negligible. An aggregate model of this situation is to be built. Data have been collected on all machines as a group. The time between failures for any machine has been determined to be exponentially distributed, with a mean of 157 hours. Repair times are also based on aggregated data. The time required to repair any failed machine for any repairperson has been determined to be uniformly distributed between 4 and 10 hours. Develop a SLAM II model to estimate the average number of operators busy, the average number of repairpeople busy, the number of machines in a backup status, and the number of machines waiting for repair.

Embellishments

(a) Develop the SLAM II network for the situation described, assuming that four machines are initially in a failed state.

(b) Develop a cost structure that will allow management to decide the number of repairpeople required to service failed machines. Redevelop the network to obtain the data to support the management decision process for determining the number of repairpeople to keep on the payroll.

3-4. Explain the differences among the terms accumulation, assembly, and matching. Specify why different SLAM II concepts are required to model these operations. Develop one example of each operation in a practical situation.

3-5. Build a series of hierarchical models using SLAM II concepts and symbols for a manufacturing facility. At the most aggregate level, consider the facility to be a single server that processes orders. As a first level of disaggregation, consider that three operations are involved in processing orders: (1) order receipt, (2) producing a product, (3) and packing and shipping. At the next level of disaggregation, build SLAM II models for portraying order receipt, production, and packing and shipping. Discuss how each of the models may be used to solve different managerial problems.

3-6. Compare the similarities and differences associated with each of the following sets of SLAM II constructs: (a) service activity and resource; (b) conditional branching and SELECT node routing; (c) entity representing a resource and a resource; (d) a SLAM II network model and a simulation program.

3-7. Describe each of the rules employed at a SELECT node in your own terms. Identify situations in which you think these rules would be appropriate. Develop two rules that are not available to be specified at a SELECT node.

3-8. In SLAM II, there are 20 node types. Discuss the symbolism used within SLAM II from a human factors standpoint. Discuss how you would explain the SLAM II symbols to the following individuals: (a) your supervisor; (b) the dean of the business school; (c) your spouse or potential date; (d) a 13 year-old sibling.

3.12 REFERENCES

1. Archer, N., "Choosing a Language for an MBA System Modeling and Simulation Course", *Simuletter*, July, 1985, pp. 3-16.
2. Banks, J., and J. Carson, "Process-Interaction Simulation Languages", *Simulation*, May, 1985, pp. 225-235.
3. Fishman, G. S., *Principles of Discrete Event Simulation*, Wiley, New York, 1978.
4. O'Reilly, J. J., and W. R. Lilegdon, *SLAM II Quick Reference Manual*, Pritsker & Associates, West Lafayette, IN, 1986.
5. Pritsker, A. A. B., *Modeling and Analysis Using Q-GERT Networks, 2nd* ed., Wiley, New York, 1979.
6. Pritsker, A. A. B., *Introduction to Simulation and SLAM II*, *3rd* ed., Wiley and Systems Publishing, New York and West Lafayette, IN, 1986.

4 SLAM II INPUTS AND OUTPUTS

4.1 INTRODUCTION

An analysis of a SLAM II network is performed by the SLAM II processor using simulation techniques. The SLAM II processor is integrated with SLAMSYSTEM [4] and TESS [5] which are computer systems for supporting modeling and decision making. SLAM II networks define the model of the system. Control statements are used to define the experimental design and analysis to be performed on the network. In this chapter, the simulation procedures employed to analyze a SLAM II model are discussed. SLAM II input procedures and output reports are described, and the possible inputs and outputs from SLAMSYSTEM and TESS are discussed. Extensive documentation is available on these subjects [2, 3, 4, 5]. In other chapters, only inputs and outputs from SLAM II are presented.

4.2 THE SLAM II PROCESSOR

The SLAM II processor simulates the flow of entities through a network by changing, as required, the status of the variables in a model whenever an entity arrives at a node. These changes are made in accordance with the functions performed at the node. The SLAM II processor begins an analysis by identifying the CREATE nodes in the network. At each CREATE node, an entity is generated, marked, and then routed over activities emanating from the node. The time spent in an activity by an entity is simulated in accordance with the duration specified for the activity. An event, corresponding to the arrival of the entity at the end node of the activity, is scheduled and placed on an event calendar.

When all CREATE nodes have been considered, the first event on the event calendar is removed, and an arrival to the node of the event is processed. If the node is not released, that is, it requires more incoming entities, no further action is taken, and the simulation is advanced to the time of the next event on the calendar. If the node is released, the functions of the node are performed using the node's decision logic. Examples of such functions are collection of observations at a COLCT node, value assignments at an ASSIGN node, and resource allocation at an AWAIT node. The entity is then routed along the activities emanating from the node just released.

If more than one activity is performed following the node, identical entities are routed along each activity. If the activities following the node have probabilities specified, a selection of one activity is made using a pseudorandom number. For conditional routing, activities are taken if the condition prescribed for the activity is satisfied. For each activity taken, an activity duration is computed, and the entity or a clone is scheduled to arrive at the end node at the current simulation time plus the activity duration. This arrival-of-entity event is placed on the event calendar. After all activities have been evaluated, the next event is removed from the event calendar and the process is repeated.

Each time a next event is removed from the event calendar, the time of the event is compared to a total time allocated for the simulation. If the time of the next event exceeds the total time, the simulation run is completed. Also, when an entity arrives at a TERMINATE node, the simulation is ended if that node's terminate count has been reached. If the run is not completed, simulation processing continues. When a run is completed, summary statistics are computed and printed.

When an entity arrives at a QUEUE or AWAIT node, the dispositions of the arriving entity and the server or resource are determined. First, a check is made to see if the file associated with the node is full, that is, at its queue capacity. If it is, the entity either balks from the node or blocks its current service activity. If it balks and there is a balk node prescribed, the entity is scheduled to arrive at the balk node immediately. If blocking occurs, the service activity that just completed processing the entity is not made available for processing another entity.

If the file is not full but the servers or resources are all in use, the entity is placed in the file according to the queue-ranking rule specified for the file. If a server or resource is available, the entity is scheduled to complete the activity emanating from the QUEUE or AWAIT node. Throughout the simulation, statistics are maintained on the number of entities in files and the utilization of servers and resources.

When an entity completes a service activity, not only must the entity be routed from the end node of the service activity, but the disposition of the service activity must be considered. If no entities are waiting in the QUEUE node or nodes preceding the service activity, the service activity is made idle. If entities are waiting, a QUEUE node is selected and the entity ranked first in its file is removed and scheduled to arrive at the end node of the service activity. If the file was at its maximum capacity, a check is also made to unblock any service activities preceding the QUEUE node.

When an entity arrives to a FREE or ALTER node, the node is always released. The disposition of the resource units associated with the node is made by allocating them to entities waiting in files as specified by the appropriate resource block. The processing of entity arrivals to other SLAM II nodes follows this pattern, always taking into account the function of the node.

4.3 NETWORK INPUTS

A SLAM II network model may be developed in a graphical form or as a set of network statements. The network or the statement sequence specifies how entities flow through the model. A statement for a symbol provides the information necessary to make status changes based on entity flow. The preparation and format conventions for network statements is presented in Section 4.3.1. The building of a graphical network model for input to the SLAM II processor is accomplished using SLAMSYSTEM or TESS. A discussion of graphical input procedures is given in Section 4.3.2. Both SLAMSYSTEM and TESS employ nonprocedural input procedures and only require the placement of graphical symbols on a screen and the filling out of forms to provide the parameter values for each symbol. SLAMSYSTEM uses a windows environment to provide a friendly interface for SLAM II.

4.3.1 Network Statement Formats

On input to the SLAM II processor, a set of network statements is preceded by a NETWORK statement and followed by an ENDNETWORK statement. The NETWORK statement is

NETWORK;

and denotes the start of network statements to the SLAM II processor. An ENDNETWORK statement denotes an end to all network statements. All the network statements of SLAM II are presented in Table 4-1, including the default values for each field.

The network input statements use four delimiters to separate values:

Table 4-1 SLAM II network Statement Types

Statement Form	Statement Defaults (ND = no default)
Nodes	
ACCUM,FR,SR,SAVE,M;	ACCUM,1,1,LAST,∞;
ALTER,RES/CC,M;	ALTER,ND/ND,∞
ASSIGN,VAR = value,VAR = value,...,M;	ASSIGN,ND = ND,ND = ND,...,∞;
AWAIT(IFL/QC),RES/UR or GATE,BLOCK or BALK(NLBL),M;	AWAIT(first IFL in RLBL's or GLBL's list/∞),ND/1,none,∞;
BATCH,NBATCH/NATRB,THRESH, NATRS,SAVE,RETAIN,M;	BATCH,1/none,ND,entity count,LAST,none,∞;
CLOSE,GATE,M;	CLOSE,ND,∞;
COLCT(N),TYPE or VARIABLE, ID,NCEL/HLOW/HWID,M;	COLCT(ordered),ND,blanks,no histogram/0./1
CREATE,TBC,TF,MA,MC,M;	CREATE,∞,0,no marking,∞,∞;
DETECT,XVAR,XDIR,VALUE,TOL,M;	DETECT,ND,ND,ND,0,∞;
ENTER,NUM,M;	ENTER,ND,∞;
EVENT,JEVNT,M;	EVENT,ND,∞;
FREE,RES/UF,M;	FREE,ND/1,∞;
GOON,M;	GOON,∞;
MATCH,NATR,QLBL/NLBL,...,M;	MATCH,ND,ND/no routing, ND/no routing,...,∞;
OPEN,GATE,M;	OPEN,ND,∞;
PREEMPT(IFL)/PR,RES,SNLBL, NATR,M;	PREEMPT(first IFL in RLBL's list)/no priority,ND,AWAIT node where transaction seized resource,none,∞;
QUEUE(IFL),IQ,QC,BLOCK or BALK(NLBL),SLBLs;	QUEUE(ND),0,∞,none,none,none;
SELECT,QSR/SAVE,SSR,BLOCK or BALK(NLBL),QLBLs;	SELECT,POR/none,POR,none,ND;
TERMINATE,TC;	TERMINATE,∞;
UNBATCH,NATRR,M;	UNBATCH,ND,∞;
Blocks	
GATE/GLBL,OPEN or CLOSE,IFLs/repeats;	GATE/ND,OPEN,ND/repeats;
RESOURCE/RLBL(IRC),IFLs/repeats;	RESOURCE/ND(1),ND/repeats;
Regular Activity	
ACTIVITY/A,duration,PROB or COND,NLBL;ID	ACTIVITY/no ACT number,0.0,take ACT,ND;blank
Service Activity	
ACTIVITY(N)/A, duration,PROB, NLBL; ID	ACTIVITY(1)/no ACT number,0.0,1.0,ND; blank

1. Commas are used to separate fields.
2. Slashes (virgules) are used to allow an optional or normally defaulted value to be contained as a second value in a field. Slashes may also be used to indicate that a set of fields is to be repeated.
3. Parentheses are used to indicate a capacity or associated file.
4. A semicolon is used to end a statement.

All node types have been given verbs as names since the SLAM II modeling approach requires decisions and logical functions to be performed at the nodes of the network.

4.3.2 Graphical Network Building

Graphical methods for building SLAM II networks interactively eliminate the requirement to code the network statements [4, 5]. The graphical network is translated into a statement model which may be transmitted for analysis by any computer that has a SLAM II processor. A SLAM II model can be copied and used as a starting point for other models. Each model is given a name which allows for easy reference and reuse. Editing procedures are available to add, insert, and delete network elements and for changing and moving symbols and parameters. Zoom and scroll features are included.

In SLAMSYSTEM, a network builder with a window interface is used to build a SLAM II network interactively. Figure 4-1 shows a symbol window superimposed on a network window. This screen is obtained by selecting ADD from the pop-down menu under the edit function of the network builder window. In the symbol window, the AWAIT node is shown in reverse video to indicate that it is the current option being selected. The AWAIT node symbol is displayed on the right. After selecting the AWAIT node, a form is presented that contains the fields for the AWAIT node, as shown in Figure 4-2. Initially, default values are given for each field on the form. The form is completed by selecting a field and typing the desired input for the field. The form can be filled in at the time the AWAIT node is selected or it can be edited at a later time. After clicking on the OK option, SLAMSYSTEM requests a location for the AWAIT node. Given a location, it connects the AWAIT node to the other SLAM II symbols in the network with its fields filled in with current values.

A network created with SLAMSYSTEM is stored for integration with other SLAM II statements and models, that is, control statements and program inserts A list of network model names is maintained to facilitate their retrieval. Networks can be copied and given different names to provide a starting point for new models.

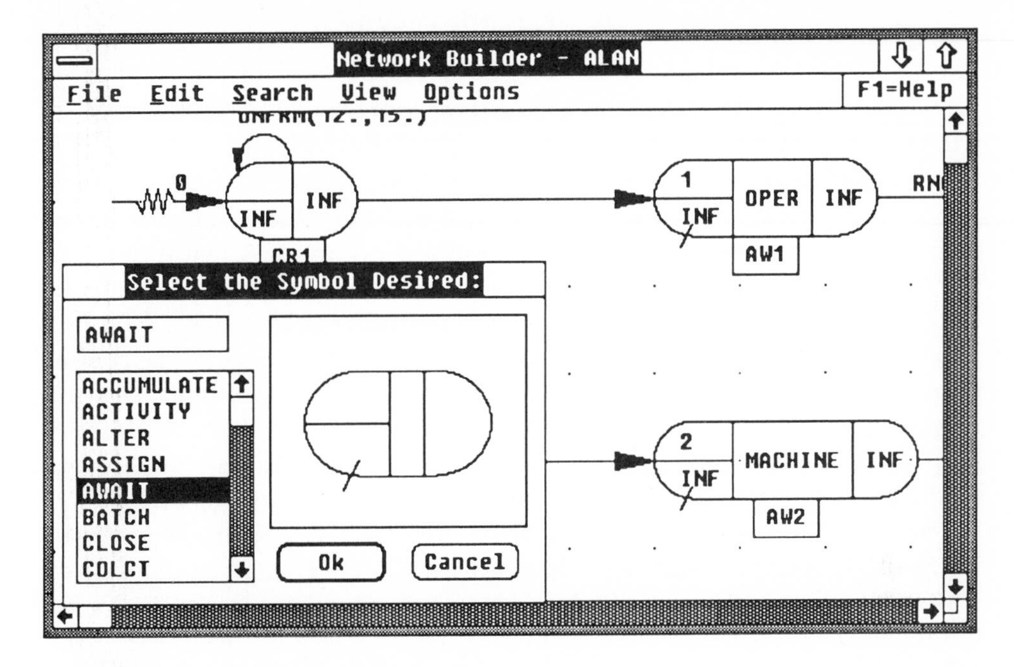

Figure 4-1 Network builder windows.

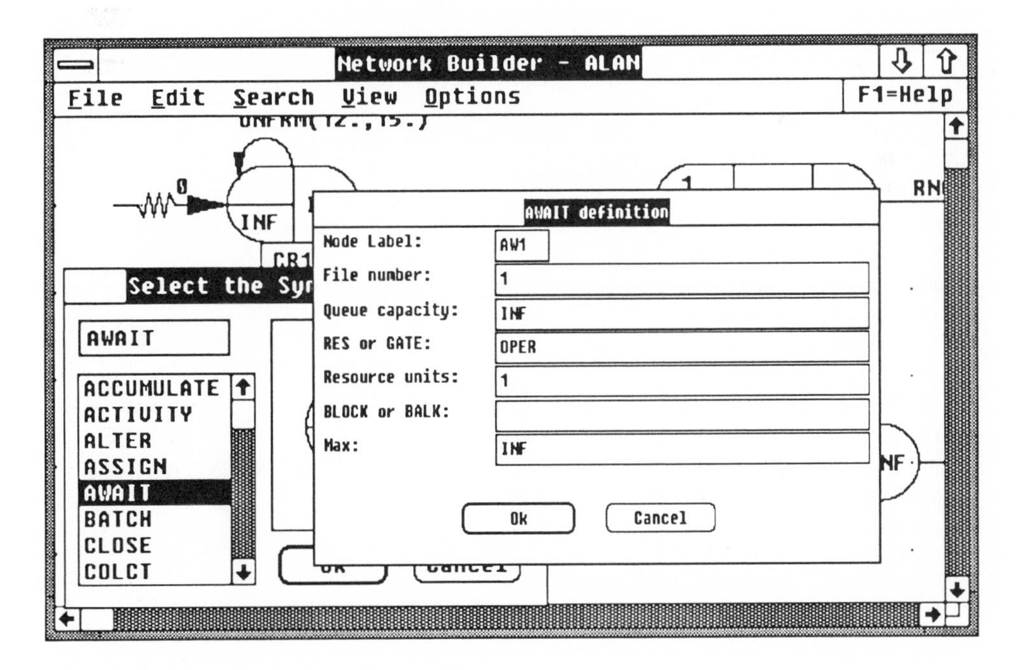

Figure 4-2 Window for AWAIT node fields.

4.4 SIMULATION CONTROL STATEMENTS

Control statements are used in conjunction with network models to execute and document SLAM II simulation programs. A list of the control statement types is presented in Table 4-2 in alphabetical order. In this discussion, only the most frequently used fields are discussed in detail. These statements may be prepared using forms [4, 5]. When a particular control statement is selected, a form is presented, with each of the fields of the control statement listed and the default values for that field displayed as the current value. Each field that requires a change is then edited to complete the control statement. The control statements are not, in general, sequence dependent but must adhere to the following: (1) the GEN statement must be the first statement, the LIMITS statement must be the second statement, and the FIN statement must be the last, (2) the network statements must be immediately preceded by the NETWORK statement and immediately followed by the ENDNETWORK statement, (3) an INITIALIZE statement must precede ENTRY and MONTR statements, and (4) a MONTR statement with the TRACE option, which includes a nodelist, must follow the ENDNETWORK statement.

Table 4-2 SLAM II Control Statements

Statement form
ARRAY(IROW,NELEMENTS)/initial values;
ENTRY/IFL,ATRIB(1),ATRIB(2),...,ATRIB(MATR)/repeats;
EQUIVALENCE/SLAM II variable,name/repeats;
FIN;
GEN,NAME,PROJECT,MO/DAY/YEAR/,NNRNS,ILIST,IECHO,IXQT/IWARN, IPIRH,ISMRY/FSN,IO;
INITIALIZE,TTBEG,TTFIN,JJCLR/NCCLR,JJVAR,JJFIL;
INTLC,VAR=value,repeats;
LIMITS,MFIL,MATR,MNTRY;
MONTR,option,TFRST,TSEC,variables;
NETWORK,SAVE or LOAD,device;
PRIORITY/IFL,ranking/repeats;
RECORD(IPLOT),INDVAR,ID,ITAPE,P or T or B,DTPLT,TTSRT,TTEND,KKEVT;
SEEDS,ISEED(IS)/R,repeats;
SEVNT,JEVNT,XVAR,XDIR,VALUE,TOL;
SIMULATE;
STAT,ICLCT,ID,NCEL/HLOW/HWID;
TIMST,VAR,ID,NCEL/HLOW/HWID;
VAR,DEPVAR,SYMBL,ID,LOORD,HIORD;

4.4.1 GEN Statement

The GEN statement

GEN,NAME,PROJECT,MONTH/DAY/YEAR,NNRNS,ILIST,IECHO,
IXQT/IWARN,IPIRH,ISMRY/FSN,IO;

provides general information about a simulation run or runs. Included on the GEN statement are the analyst's name, a project identifier, date, number of simulation runs, and report options. NAME and PROJECT are both alphanumeric fields and are used for output reports to identify the analyst and the project. Blanks are significant within alphanumeric fields. The MONTH, DAY, and YEAR are entered as integers separated by slashes. The SLAM II variable NNRNS is entered as an integer, has a default value of 1, and denotes the number of simulation runs to be made. The next six fields are specified as YES or NO. The output obtained from a YES input follows:

ILIST	A numbered listing of all input statements is printed including error messages if any.
IECHO	A summary of inputs is printed.
IXQT	Execution is attempted if no input errors are detected.
IWARN	A warning message is printed when an entity is destroyed before reaching a TERMINATE node.
IPIRH	The heading INTERMEDIATE RESULTS is printed prior to execution of each simulation run.
ISMRY	The SLAM II Summary Report is printed.

If a summary report is to be printed, a report frequency field, FSN, is used. The options are F, S or N for the report frequency where F, after the first run only; S, after the first and last runs; or N, an integer, specifying after every N*th* run. The last field of the GEN statement is IO, which specifies the number of columns for output reports. The options are 72 or 132 columns, which are typically used for terminal and line printer outputs respectively. As an example of the control forms input to SLAMSYSTEM, the form to complete the GEN statement is presented in Figure 4-3. Behind the GENERAL statement window is a window displaying the current list of control statements. HELP is available to obtain a definition of any field. The statement which is currently being edited, the GEN statement, is shown in reverse video. This window is shown in the foreground in Figure 4-4. In this section, the definitions of the SLAM II control statements are presented with an emphasis on the important aspects and uses of the controls.

```
Control Builder - SIMPLE
File  Edit  Search                                    F1=Help
GEN,NORDLUND,COURSE FMS ANIMATION,9/13/1988,1,Y,Y,Y/Y,Y,Y/1,132;
LIM,1,3,50;
INIT,,25.;
REC,TNOW,TIME,18;
VAR,NNQ(1),N,NUMBER IN QUEUE 1;

GENERAL definition
Name:          NORDLUND
Project:       COURSE FMS ANIMATION
Date:          9 / 13 / 1988     Number of r 1
Input:         Y        Echo:     Y       Execution:  Y
Warning:       Y        Heading:  Y       Summary:    Y
F or S or N:   1
Width:         132
               Ok            Cancel
```

Figure 4-3 Control statement windows.

```
Control Builder - SIMPLE
File  Edit  Search                                    F1=Help
GEN,NORDLUND,COURSE FMS ANIMATION,9/13/1988,1,Y,Y,Y/Y,Y,Y
LIM,1,3,50;

GENERAL definition
Name:          NORDLUND

Help
GEN STATEMENT

The GEN statement is the first input statement
in a SLAM II model.  This statement provides
general information identifying the model,
the number of executions, and output format.

Previous   Next   Index   Print   Ok
```

Figure 4-4 Help for GEN control statement.

4.4.2 LIMITS Statement

The format of the LIMITS statement is

LIMITS, MFIL, MATR, MNTRY;

The LIMITS statement is used to specify integer limits on the largest file number used (MFIL), the largest number of attributes per entity (MATR), and an estimate of the largest number of concurrent entries in all files (MNTRY).

4.4.3 PRIORITY Statement

The format of the PRIORITY statement is

PRIORITY/ IFILE, ranking/ repeats;

The PRIORITY statement is used to specify the criterion for ranking entities within a file. There are four possible specifications for ranking:

FIFO Entities are ranked based on their order of insertion with early insertion given priority.

LIFO Entities are ranked based on their order of insertion, with late insertions given priority.

HVF(K) Entities are ranked high value first based on the value of the K*th* attribute.

LVF(K) Entities are ranked low value first based on the value of the K*th* attribute.

4.4.4 STAT Statement

Each variable for which observation statistics are collected external to the network by a call to subroutine COLCT must be defined with a STAT statement. In calls to subroutine COLCT (described in Chapter 5), the variable is identified by the code ICLCT. The fields of the STAT statement are

STAT, ICLCT, ID, NCEL/ HLOW/ HWID;

where ICLCT is an integer code, ID is an alphanumeric identifier, and NCEL/ HLOW/ HWID are histogram parameters specifying the number of interior cells, the upper limit for the first cell, and the width of each cell, respectively.

4.4.5 TIMST Statement

The format for the TIMST statement is

TIMST, VAR, ID, NCEL/ HLOW/ HWID;

where VAR is a SLAM II time-persistent variable such as XX or NNQ, ID is an alphanumeric identifier, and NCEL/ HLOW/ HWID provide the SLAM II specification for a histogram.

4.4.6 EQUIVALENCE Statement

An EQUIVALENCE statement allows textual names to be used for SLAM II variables. The format for the EQUIVALENCE statement is

EQUIVALENCE/ SLAM II variable or value, name/ repeats;

The SLAM II variables ATRIB, II, XX,and ARRAY, or a SLAM II random variable or a constant value may be used on the EQUIVALENCE statement. Making of a name equivalent to a SLAM II variable or value provides for increased model readability.

4.4.7 ARRAY Statement

The ARRAY statement is used to initialize one row of the two-dimensional SLAM II variable ARRAY. The number of elements in a row of ARRAY can vary. The format for the ARRAY statement is

ARRAY(IROW, NELEMENTS)/ initial values/ repeats;

where IROW is an integer constant defining the row for which initial values are being provided, NELEMENTS is the number of elements in this row, and initial values are constants to be inserted in the order of the columns for the row.

Elements of ARRAY may be referenced on a SLAM II network wherever a SLAM II variable is allowed. ARRAY subscripts may be constants, II, XX and ATRIB.

4.4.8 INTLC Statement

The INTLC statement is used to assign initial values to SLAM II variables. The format for the INTLC statement is

INTLC, VAR = value, repeats;

where VAR can be XX or ARRAY.

4.4.9 INITIALIZE Statement

The INITIALIZE statement is used to specify the beginning time (TTBEG) and ending time (TTFIN) for a simulation and initialization options for clearing statistics, initializing variables, and initializing files. The format for the INITIALIZE statement is

INITIALIZE, TTBEG, TTFIN, JJCLR/ NCCLR, JJVAR, JJFIL;

where the last three fields are specified as YES or NO and are normally defaulted to YES.

If JJCLR is specified as YES, NCCLR specifies the number of the collect variable up to which clearing is to be performed. If JJCLR is specified as NO, NCCLR specifies the collect variable number up to which clearing is not to be performed. If JJVAR is specified as YES, TNOW is initialized to time TTBEG, and XX is reinitialized before each simulation run. If the field is specified as NO, the initializations are not performed. IF JJFIL is specified as YES, the filing system is initialized before the beginning of each simulation run, otherwise it is not.

4.4.10 SEEDS Statement

The purpose of the SEEDS statement is to permit the user to specify the starting unnormalized random number seed for the random number streams available within SLAM II and to control the reinitialization of streams for multiple simulation runs. The format for the SEEDS statement is

SEEDS, ISEED(IS)/ R, repeats;

The seeds are entered as integers (ISEED) with the stream number (IS) of the seed given in parentheses. If the stream number is not specified, then stream numbers are assigned based on the position of the seed. The first seed is for stream 1, the second for stream 2, and so on. The reinitialization (R) of each stream is controlled by specifying YES or NO following a slash. If R is not included, the default case is assumed and the seed values are not reinitialized. If the SEEDS input statement is not included, the SLAM II processor uses default seed values.

4.4.11 MONTR Statement

The MONTR statement is used to obtain displays of intermediate simulation results or to clear statistical arrays. The format for the MONTR statement is

MONTR, Option, TFRST, TSEC, Variables;

where TFRST is the time for the first execution of the option, and TSEC is a time for successive executions or the completion of the option. The times TFRST and TSEC default to TTBEG and infinity, respectively. If TSEC is defaulted, the MONTR option is executed only at time TFRST. However, if TSEC is specified, the MONTR option is executed at time TFRST and, except for the TRACE and INTERACTIVE options, every TSEC time units. The following five MONTR options are available:

INTERACTIVE Provides for interactive variable examination, modification, and storage of current model status.

SUMRY Causes a SLAM II Summary Report to be printed.

FILES Causes the file section of the SLAM II Summary Report to be printed, and a listing of all entities in the files to be printed.

CLEAR Causes all statistical arrays to be reset.

TRACE Causes the starting and stopping of detailed tracing of each entity as it moves through the network. The trace starts at time TFRST and ends at TSEC. A list of nodes can be provided as follows: TRACE (nodelist). This causes an output only when one of the nodes in the list is processed. Also, for the TRACE option, a list of variables to print on the trace may be defined. Traces can be used to provide the inputs for animations (see Section 4.5.1).

MONTR statements apply only to a single run and must be restated for each run.

4.4.12 SEVNT Statement

The SEVNT statement is used to detect the crossing of a SLAM II variable against a threshold. A tolerance is specified for detecting the crossing. The crossing can be in the positive, negative, or both directions. The SEVNT statement is a control statement that is analogous to the DETECT node in network models. The format for the SEVNT input statement is

SEVNT, JEVNT, XVAR, XDIR, VALUE, TOL;

where JEVNT is a user supplied event code, XVAR is a SLAM II variable, XDIR specifies the direction of crossing with X→ either direction; XP→ positive direction; XN→ negative direction; VALUE is the crossing threshold, and TOL is a numeric value that specifies the tolerance within which the crossing is to be detected. The SEVNT statement can be used, for example, to detect when a queue length reaches a prescribed value and when all servers are idle. When a crossing is determined, subroutine EVENT(JEVNT), described in Chapter 5, is called.

4.4.13 RECORD Statement

The RECORD statement provides general information concerning the values to be recorded at periodic times during a run. The RECORD statement specifies the independent variable, the storage medium, and detailed specifications concerning the type and time interval for the output reports. The basic format for the RECORD statement is

RECORD, INDVAR, ID, ITAPE, P or T or B, DTPLT, TTSRT, TTEND;

The definitions and default values for the fields of the RECORD statement are gaven in Table 4-2.

4.4.14 VAR Statement

A set of VAR statements is used in conjunction with a RECORD statement to define the dependent variables to be recorded. The format for the VAR statement is

VAR, DEPVAR, SYMBL, ID, LOORD, HIORD;

Table 4-2 Fields of the RECORD Statement

Variable	Definition	Default
INDVAR	Name of the independent variable	Required
ID	Alphanumeric identifier for the independent variable	Blanks
ITAPE	Device number to store recorded variables	0
P or T or B	An output format: **P**lot, **T**able or **B**oth	P
DTPLT	Time between successive plot lines	5.0
TTSRT	Start time for recording	TTBEG
TTEND	End time for recording	TTFIN

The definitions and default values for the fields for the VAR statement are given in Table 4-3.

4.4.15 SIMULATE and FINISH Statements

The SIMULATE statement consists of the single field, SIMULATE; or its abbreviation SIM; The SIMULATE statement is used when making multiple simulation runs. One simulation run is executed for the statements preceding the SIMULATE statement. Following each SIMULATE statement, the user may insert an updated set of control statements.

The FINISH statement consists of the single field, FINISH; or its abbreviation FIN; It denotes the end to all SLAM II input statements.

Table 4-3 Fields of the VAR Statement

Variable	Definition	Default
DEPVAR	Name of the dependent variable	Required
SYMBL	Symbol to identify the dependent variable	Blank
ID	Alphanumeric identifier for the dependent variable	Blanks
LOORD	Low ordinate specification: a value; the minimum observed, MIN; or MIN rounded down to a nearest multiple of IVAL, MIN(IVAL)	MIN
HIORD	High ordinate specification: a value; the maximum observed, MAX; or MAX rounded up to a nearest multiple of IVAL, MAX(IVAL)	MAX

4.5 OUTPUT VARIABLE TYPES

The output variables of SLAM II fall into three categories: (1) time persistent, (2) observations, and (3) counts. Examples of time-persistent variables are the number of entities in a file and the number of resources in use. These variables have a value until a change occurs, hence the name, time-persistent variables. Examples of observations are the time a node is released and the transit time of an entity as it journeys from one node to another. Observations are samples of variables used to estimate statistical measures. The third type of variable represents a count of the number of times a specific event occurs, for example, the number of entities balking from a QUEUE node.

A single run of the SLAM II processor produces values of these types of variables. At the most fundamental level, a trace provides the start time and the end time for each activity performed during a run. The trace contains the information required to obtain estimates on all the variables described, since it contains every status change that occurs in a run. Saving a trace is equivalent to the current industrial practice of maintaining records on the time of every event occurrence during the operation of a system. Trace information can be displayed pictorially to show the flow and storage of entities through the network. SLAMSYSTEM and TESS provide for such animations. Because trace data can be voluminous, it is common to summarize the data during a run by computing statistical estimates relating to the operation of the system.

4.5.1 Animation Display

To show the changes in state over time, animations are used. An ***animation*** is a visual display of trace data on a diagram depicting the facilities of a system. To illustrate, three snapshots at times 171, 184, and 190 from a TESS animation of a model of a system that makes castings is given in Figure 4-5 [5]. At time 171, the snapshot states that 50 castings have been introduced and 36 have been produced. There are 12 castings in the model of which 7 are being worked on by 4 mills, the deburring machine, and two inspectors. Two are in the post-process inventory at the mill station and 2 are in the preprocess inventory of the assembly station. The 12th casting is being transported by a crane from the deburr station to the inspect station. The assembly station is down which is indicated by single-cross hatching.† A crane is moving to the postprocess inventory at the mill station to pick up a casting. This is indicated by a crane icon in the RESP CRANES box where RESP is an abbreviation for responding.

† The snapshots were originally in color and used color changes to show status changes.

At time 184, the snapshot shows that one of the four mills is down, and the assembly station is still down. Only one of the inspectors is busy. There are 2 castings waiting to move from the postprocess inventory at the inspect station to the assembly station. For this run, the capacity of the preprocess inventory at the assembly station was set at 4, and, at time 184, it is full. In the model, castings waiting in the postprocess inventory are not moved until space is available in the buffer of the next station. At time 184, there are 15 castings in the system with 5 being worked on, 8 in inventory, and 2 being transported.

At time 190, all the stations except the rework station are busy and 3 cranes are moving. One crane is moving to pick up a casting from the postprocess inventory at the mill station and the other 2 cranes are transporting castings to the inspect and assembly stations. From the above discussion, the operation of the model can be understood in manufacturing terms. By watching an animation, an understanding of model and system procedures is obtained.

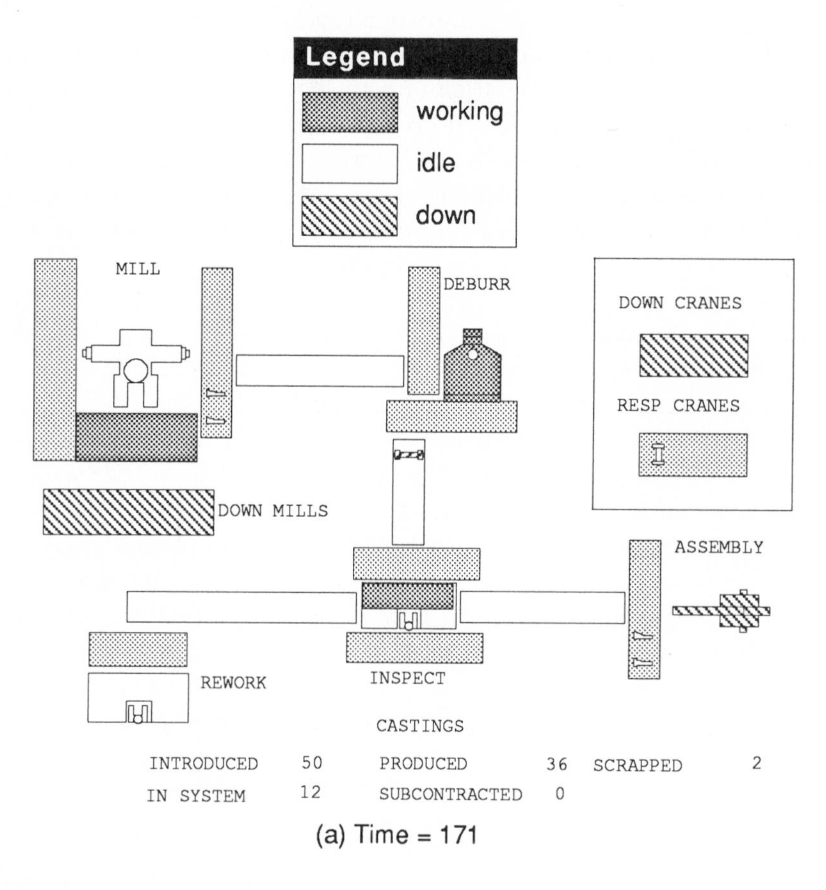

(a) Time = 171

Figure 4-5 Animation snapshots.

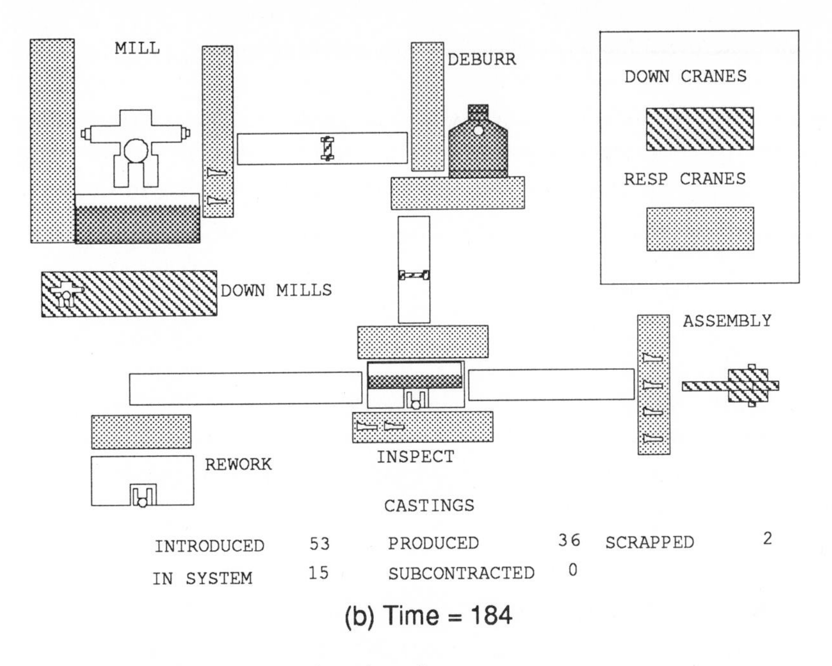

(b) Time = 184

Figure 4-5 Animation snapshots (continued).

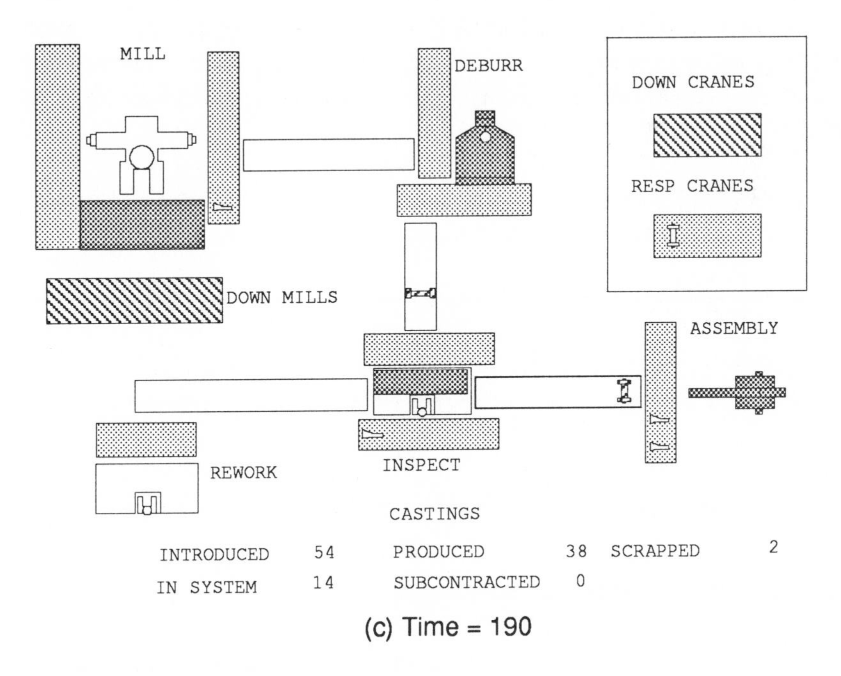

(c) Time = 190

Figure 4-5 Animation snapshots (concluded).

4.5.2 Statistical Estimates for Time-persistent Variables

For a given run, a time-persistent variable appears as in Figure 4-6. To obtain an estimate of the average number, each value is multiplied by the fraction of time it has that value during the run. The equation for making this computation is

$$\bar{n}_T = \sum_{n=0}^{\infty} n f_n$$

where $\bar{n}_T$ is the average over T time units and f_n is the fraction of time that the variable had the value n. f_n is equal to the sum of the time intervals for which the variable had the value n divided by the total length of the simulation run. In the example of Figure 4-6,

$$f_0 = t_1/T;$$
$$f_1 = [(t_2-t_1) + (t_4-t_3) + (t_{10}-t_9)]/T$$
$$f_2 = [(t_3-t_2) + (t_5-t_4) + (t_7-t_6) + (t_9-t_8)]/T$$
and $$f_3 = [(t_6-t_5) + (t_8-t_7)]/T$$

In addition to the average, other values printed for a run are the minimum and maximum values observed during the run and the value of the variable at the end of the run (referred to as the current number on the summary report). For files, an estimate of the average time spent in the file by an entity is also printed.

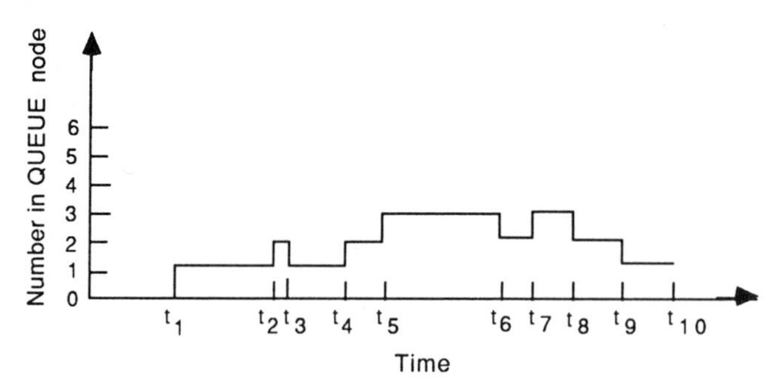

Figure 4-6 Example time history of number in a file.

4.5.3 Statistical Estimates for Resources

Resource utilization is the time-integrated number of resources in use divided by the time the resource could have been used. Statistical estimates of resource utilization and resource availability are computed and printed. Both values are necessary as the capacity of a resource may change through the use of ALTER nodes. For a given run, the number of resources in use at the end of the run, the maximum number of resources simultaneously in use during the run, the number of resources available at the end of the run, and the maximum number of resources simultaneously available are recorded and printed.

4.5.4 Statistical Estimates for Nodes

Statistics based on observations are collected at nodes or through calls to subroutine COLCT. If statistics are for the transit time of an entity to a COLCT node, the identifier of the COLCT node is its label on the summary report. For these variables, an average, which is the sum of the observations divided by the number of observations, is computed. Also computed and printed for these variables are the standard deviation of the observations and the number of observations. The observed values can be highly correlated, since sequential node release times (or sequential entity transit times) can be highly correlated. For example, the value of the twentieth entity's transit time, is probably related to the value of the nineteenth entity's transit time since there is a high likelihood that they resided in files concurrently. Thus, although the standard deviation provides information concerning the spread of the observations, the value observed and computed for a single run should not be used for setting confidence intervals on the theoretical mean. Because of the high correlation anticipated between observed values, it is suggested that statistical estimates of the variance of the average be obtained. The most common approach is to make multiple runs, since each run is an independent replication of the experiment. Alternative methods for estimating the variance of the sample mean are available [1, 3].

4.5.5 Statistical Estimates for Count Variables

Statistical estimates are computed for the number of occurrences or counts of events. One such variable is the number of entities balking from a QUEUE node. A normalized value for this variable is calculated by dividing the total number of entities balking by the time period for the simulation run. In this way, an estimate of the average balking rate is obtained. The average balking rate is important, as it specifies the arrival process to the node to which the entities are routed when they balk. In many instances, entities leave the system and the average balking rate can be used to model the arrival process to other network models that portray the system to which the entity is balking.

Another count variable is the number of times an activity has been completed. The value of this variable is printed at the end of a run for each numbered activity. These counts are useful in the analysis of traffic flow patterns. Activity counts also indicate whether or not an activity was ever completed during a run.

4.5.6 Statistical Estimates over Multiple Runs

If only one value for a variable is observed for each of multiple runs, then a COLCT node can be used to average the observations. In this case, COLCT variables should not be cleared between runs (see specification on the INITIALIZE statement). An alternative procedure is to call subroutine COLCT with an average value at the end of each run. Subroutine COLCT is a SLAM II support routine that is described in Chapter 5.

4.6 HISTOGRAMS

To transform observed values into a more manageable form, a common practice is to group them into classes or cells. The data are then summarized by tabulating the number of observations that fall within each cell. This type of table is called a frequency distribution table or, if graphically presented, a histogram. A histogram provides a clear picture of the range and frequency of data. Typically, when preparing histograms, the cumulative frequency is also obtained by successively adding the frequencies in each cell.

In SLAM II, histograms are produced on request for COLCT, STAT, or TIMST statements. An example of a histogram output is presented in Figure 4-7. From Figure 4-7, it is seen that four columns of data are presented and a graph is generated for the histogram. The four columns of data are:

OBSV FREQ — The number of times the variable of interest is within a specified range.

RELA FREQ — The relative frequency with which a variable has a value in a specified range.

CUM FREQ — The cumulative frequency, with which the variable is less than or equal to the upper value of the range.

UPPER CELL LIMIT — The value that defines the upper value of the cell. All cells include the upper value in their range.

```
                              **HISTOGRAM NUMBER  5**
                                  PROJ. COMPLETION

OBSV      RELA      CUML         UPPER
FREQ      FREQ      FREQ       CELL LIMIT     0        20        40        60        80       100
                                              +    +    +    +    +    +    +    +    +    +    +
0         0.000     0.000      0.1500E+02     +                                                 +
0         0.000     0.000      0.1550E+02     +                                                 +
1         0.002     0.002      0.1600E+02     +                                                 +
5         0.013     0.015      0.1650E+02     +*                                                +
6         0.015     0.030      0.1700E+02     +*C                                               +
6         0.015     0.045      0.1750E+02     +*C                                               +
14        0.035     0.080      0.1800E+02     +** C                                             +
29        0.072     0.153      0.1850E+02     +****   C                                         +
30        0.075     0.228      0.1900E+02     +****      C                                      +
37        0.093     0.320      0.1950E+02     +*****          C                                 +
37        0.093     0.413      0.2000E+02     +*****               C                            +
31        0.078     0.490      0.2050E+02     +****                    C                        +
40        0.100     0.590      0.2100E+02     +*****                        C                   +
33        0.083     0.673      0.2150E+02     +****                             C               +
23        0.058     0.730      0.2200E+02     +***                                 C            +
19        0.047     0.778      0.2250E+02     +**                                    C          +
32        0.080     0.858      0.2300E+02     +****                                      C      +
21        0.052     0.910      0.2350E+02     +***                                          C   +
9         0.023     0.933      0.2400E+02     +*                                             C  +
9         0.023     0.955      0.2450E+02     +*                                              C +
4         0.010     0.965      0.2500E+02     +*                                              C +
14        0.035     1.000         INF         +**                                               C
                                              +    +    +    +    +    +    +    +    +    +    +
400                                           0        20        40        60        80       100

                        **STATISTICS FOR VARIABLES BASED ON OBSERVATION**

                          MEAN          STANDARD      MINIMUM       MAXIMUM       NUMBER OF
                          VALUE         DEVIATION     VALUE         VALUE         OBSERVATIONS

ROJ. COMPLETION   0.2068E+02    0.2103E+01    0.1566E+02    0.2705E+02        400
```

Figure 4-7 Histogram illustration.

4.7 SLAM II OUTPUT REPORTS

This section describes the output reports which are generated by the SLAM II processor. They include an input listing, Echo Report, Trace Report, and Summary Report.

4.7.1 Statement Listing, Input Error Messages, and Echo Report

The SLAM II processor interprets each input statement and performs extensive checks for possible input errors. If the variable ILIST on the GEN statement is specified as YES or defaulted, the processor prints out a listing of the input statements. Each statement is assigned a line number and, if an input error is detected, an error message is printed immediately following the statement where the error occurred. All input errors are treated as fatal errors in SLAM II; that is, no execution is attempted if one or more input errors are detected.

The SLAM II Echo Report provides a summary of the simulation model as interpreted by the SLAM II processor. This report is useful during the debugging and verification phases of model development. SLAMSYSTEM and TESS provide error messages interactively during input preparation.

4.7.2 Trace Report

A Trace Report is requested by a MONTR statement using the TRACE option and causes a report summarizing each entity arrival event at a node to be printed. The Trace Report provides the progress of a simulation by printing, for each entity arrival event, the event time, the label and type of node at which the entity is arriving, and the attributes of the arriving entity. In addition, a summary of activities released from the node is printed, including the duration of the activity started and the end node of the activity. A Trace Report can be obtained either during or after a simulation run.

4.7.3 SLAM II Summary Report

The SLAM II Summary Report displays the statistical results for the simulation and is printed in accordance with the FSN and IO field specifications of the GENERAL statement. The report consists of a general section followed by the statistical results for the simulation categorized by type. The first category of statistics is for variables based on observations and includes the statistics collected within network models by COLCT nodes. The second category of statistics is for time-persistent variables as defined by TIMST statements. This is followed by statistics on files and the event calendar. The next two categories correspond to statistics collected on regular and service activities, respectively. The last category is for resource and gate statistics. The printout of histograms and plots follows the statistical summaries. An example of a SLAM II Summary Report is shown in Figure 4-8. A SLAM II Summary Report includes only those categories of statistics that are applicable to the particular simulation and therefore may include none, some, or all of the categories shown.

4.7.4 Graphic Output Illustrations

This section illustrates the types of graphical outputs that can be obtained from simulations runs [5]. Graphical presentation capabilities have expanded tremendously during the past five years and this trend is expected to continue. Since the intent of this book is not to present methods of presenting graphic displays, the case studies do not dwell on output graphics. This section is

S L A M I I S U M M A R Y R E P O R T

SIMULATION PROJECT CLOTHES MANUFACTURER BY PRITSKER

DATE 2/28/1986 RUN NUMBER 1 OF 1

CURRENT TIME 0.2400E+04
STATISTICAL ARRAYS CLEARED AT TIME 0.0000E+00

STATISTICS FOR VARIABLES BASED ON OBSERVATION

	MEAN VALUE	STANDARD DEVIATION	MINIMUM VALUE	MAXIMUM VALUE	NUMBER OF OBSERVATIONS
TIM BTW SHIPMTS	0.1079E+03	0.2815E+02	0.5505E+02	0.1714E+03	20
VOL IN CRATE	0.1173E+03	0.1520E+02	0.1000E+03	0.1440E+03	21
$VAL IN CRATE	0.9531E+03	0.1239E+03	0.8100E+03	0.1176E+04	21

FILE STATISTICS

FILE NUMBER	LABEL/TYPE	AVERAGE LENGTH	STANDARD DEVIATION	MAXIMUM LENGTH	CURRENT LENGTH	AVERAGE WAITING TIME
1	AWAIT	0.8059	1.5475	7	0	23.0259
2	AWAIT	0.5965	0.8617	3	0	17.0421
3	CALENDAR	4.1071	0.7667	7	5	8.1328

RESOURCE STATISTICS

RESOURCE NUMBER	RESOURCE LABEL	CURRENT CAPACITY	AVERAGE UTILIZATION	STANDARD DEVIATION	MAXIMUM UTILIZATION	CURRENT UTILIZATION
1	SHIRTS	148	63.6587	34.0118	124	124
2	PANTS	136	65.5304	37.8929	135	135

RESOURCE NUMBER	RESOURCE LABEL	CURRENT AVAILABLE	AVERAGE AVAILABLE	MINIMUM AVAILABLE	MAXIMUM AVAILABLE
1	SHIRTS	24	9.8501	0	27
2	PANTS	1	1.9255	0	9

Figure 4-8 Example of the SLAM II summary report.

included to illustrate the types of graphical displays of data that should be considered when modeling and simulating.

A range chart displays the average, minimum, and maximum of a set of observations. In Figure 4-9, a range chart for 10 runs is shown. In Figure 4-10, a range chart for three separate variables is displayed. In Figure 4-11, a bar chart of utilizations of three stations are presented with each station having five possible states. This chart provides information on the utilization and type of utilization of each station as well as a comparison among stations.

Another way of presenting the status of a station is a pie chart. In Figure 4-12, a pie chart of mill utilization is shown which indicates that the mills are

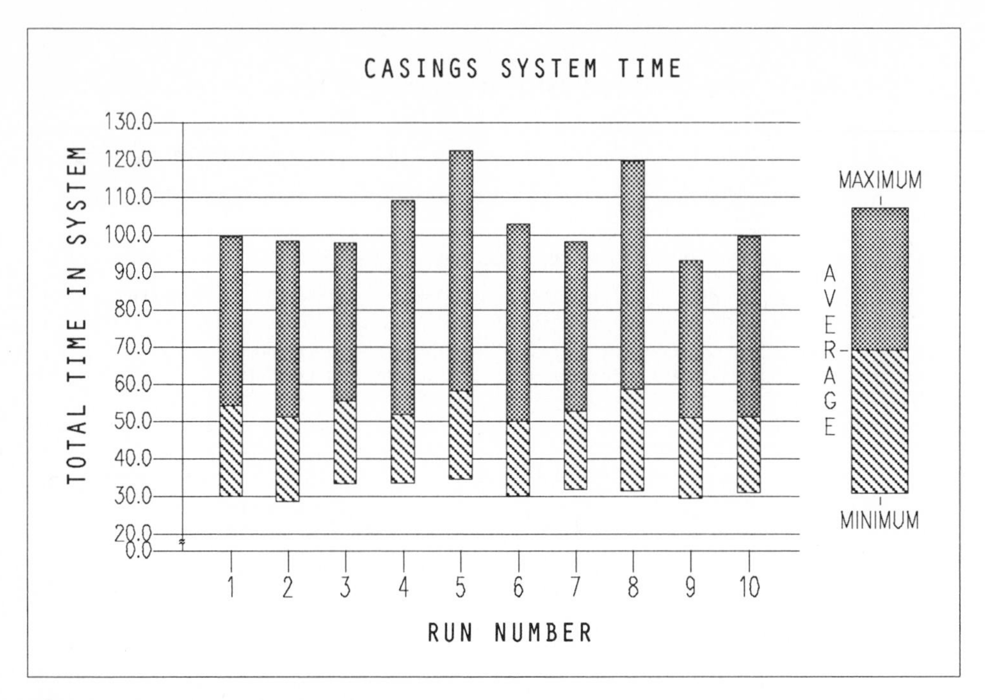

Figure 4-9 Range chart for 10 runs.

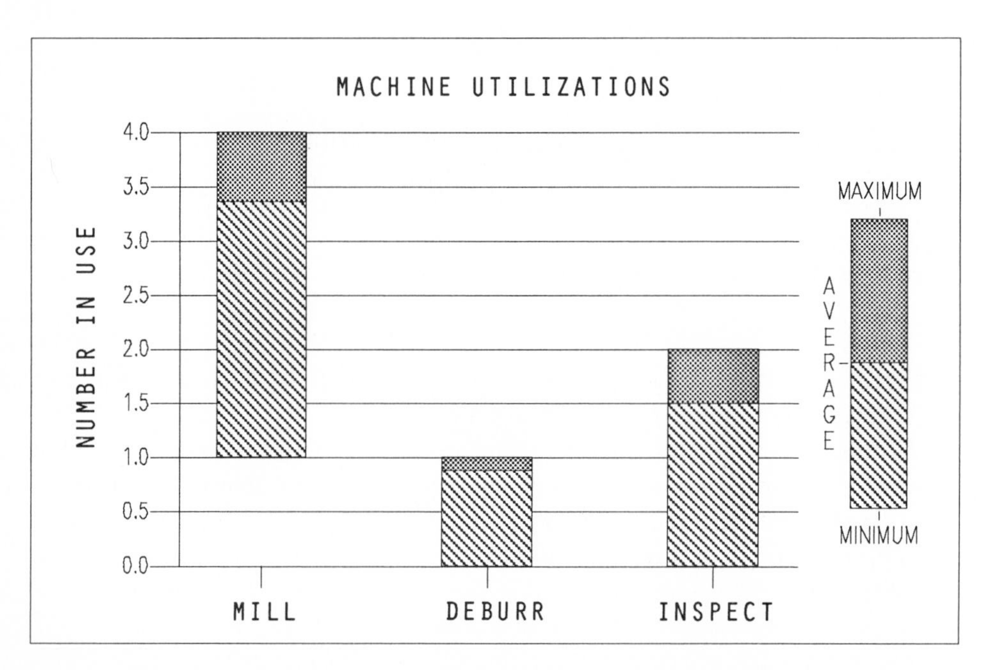

Figure 4-10 Range chart for mill, deburr, and inspect stations.

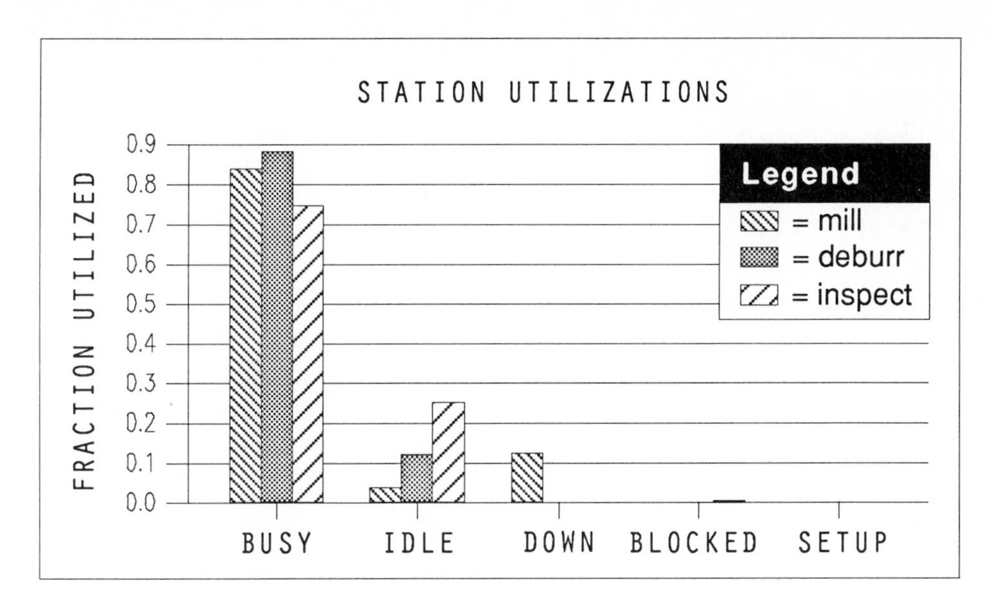

Figure 4-11 Bar chart of mill, deburr, and inspect station utilizations.

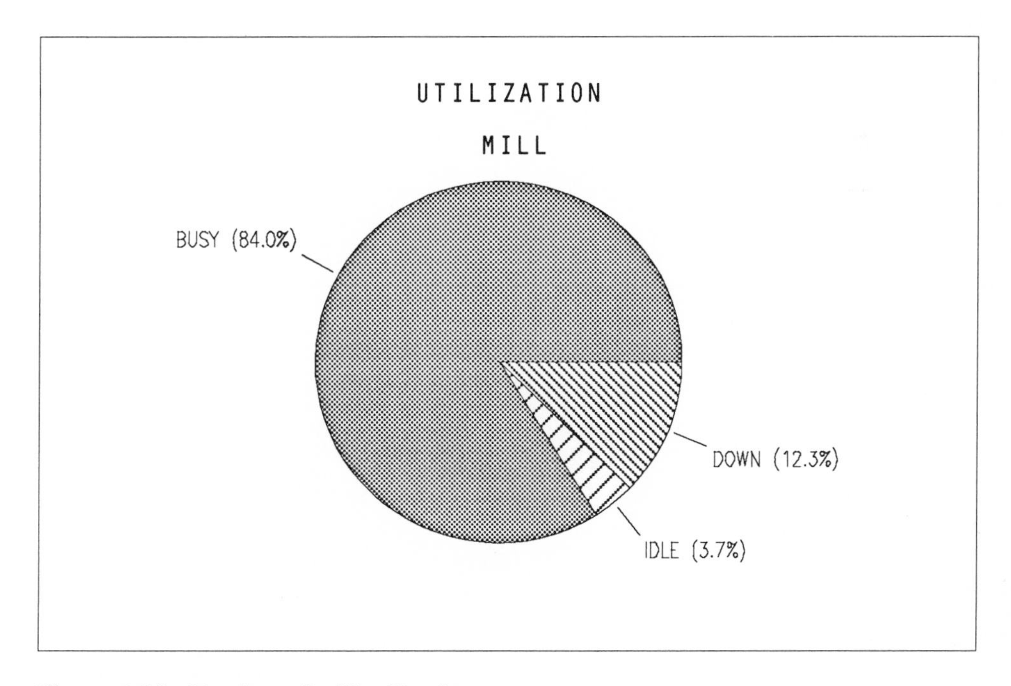

Figure 4-12 Pie chart of mill utilization.

busy 84 per cent of the time, down 12.3 per cent of the time, and idle 3.7 per cent of the time. The pie chart of Figure 4-12 presents a portion of the data displayed in the bar chart of Figure 4-11.

For presentation purposes, the graphs of data can be combined and displayed together. This is shown in Figure 4-13 where information on mill preprocess inventory is shown in three different ways. In the upper left hand corner, a range chart shows statistics on mill preprocess inventory along with statistics on preprocess inventory at the deburr and inspect stations. In the upper right hand corner, a histogram provides the amount of time that the mill preprocess inventory is equal to a specific value. On the bottom of the graph, a plot of the preprocess inventory overtime is presented. The graphs in Figure 4-13 summarize the observed data in three different ways. The plot provides all the data in graphical form. The histogram categorizes the data into cells for presentation. The range chart uses the computations described in Section 4.5.2 and presents the results of theses computations.

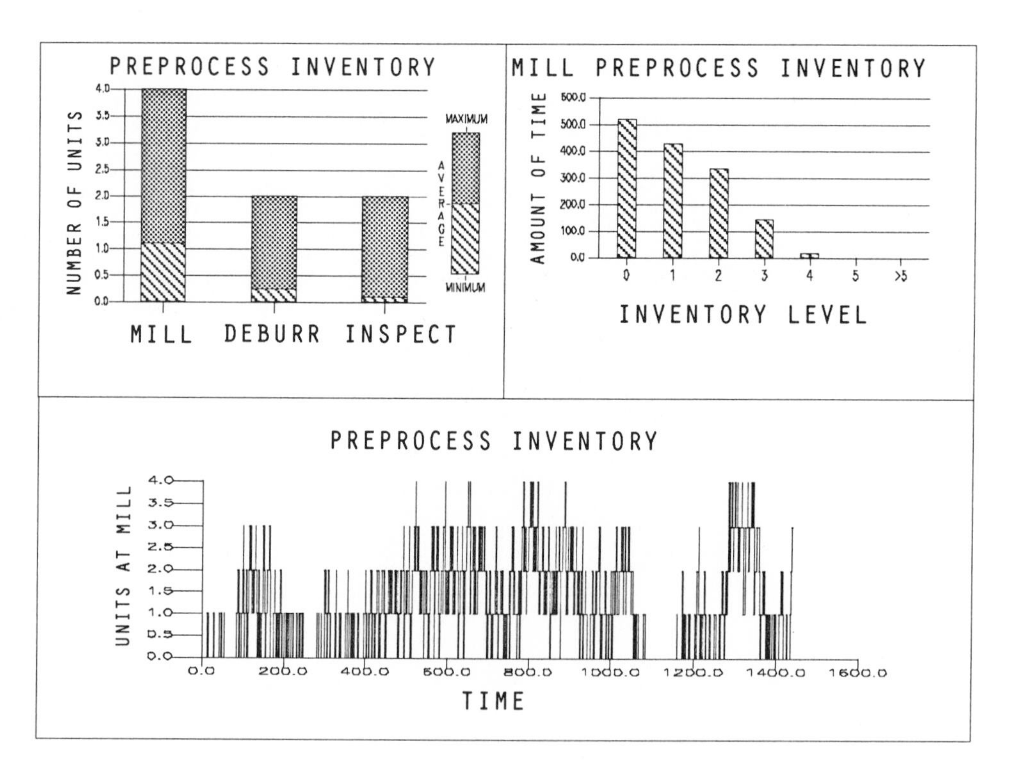

Figure 4-13 Three displays of mill preprocess inventory.

4.8 STORING SLAM II OUTPUTS

During the last two decades, significant progress has been made in the management and control of data through the use of database systems. The use of database techniques for storing data obtained from the SLAM II processor can be employed to great advantage. The use of database retrieval allows a SLAM II user to make comparisons among the outputs from different scenarios, to perform analysis of experiments that have been designed in accordance with statistical design procedures, and to employ the outputs of various SLAM II models as inputs to other simulation, statistics, and optimization programs. The procedures for saving SLAM II outputs are described elsewhere [4, 5].

4.9 SUMMARY

This chapter describes the SLAM II processor, inputs, and outputs. SLAM II statements are universal in that they can be read by the SLAM II processor on any machine. In this way, many front end programs can be developed for SLAM II which are computer-dependent or application dependent.

Each SLAM II statement is uniquely defined by the first three letters of the statement name. Each field in the statement is separated by a comma [,], slash [/], or offset in parentheses [( )]. The end of a statement is indicated by a semicolon (;). Comments may follow the semicolon.

Output statistics are automatically generated for service activities, resources, gates, and files. Statistics for variables based on observation are obtained from the use of COLCT nodes or calls to subroutine COLCT. Time-persistent statistics are obtained through the use of TIMST statements. Histograms may be specified for both observations and time-persistent variables. Plots are obtained if RECORD and VAR input statements are used. The output reports may include the summary report, a listing of the statement model, an echo report, and a trace of the activity and node occurrences may be obtained. The input and output values for a SLAM II model may be stored for future processing within SLAMSYSTEM and TESS, and presented in various graphic displays.

4.10 EXERCISES

4-1. Draw a flowchart that depicts the logic and processing involved when the SLAM II processor is invoked.

4-2. Describe the roles of the following input delimiters within the SLAM II processor: blank, comma, asterisk, slash, parenthesis, and semicolon.

4-3. Prepare a record to define the initial random number seeds to accomplish the following: for stream 1, the default seed is to be reinitialized at the beginning of each run; for stream 2, the seed value for the first run is to be 7816923, and the number is not to be reset at the beginning of each run; the seed value for stream 7 is to be defaulted, and the default value is to be used at the beginning of each run; for stream 10, the seed value is to be 2943712 on the first run and 9672133 on the second run, and no initialization is to be performed after the second run.

4-4. Prepare the input required to perform an analysis using SLAM II for a single-queue, single-server system in which the interarrival and service times are exponentially distributed with mean values of 5 and 4, respectively. There is a queue capacity of 10, and balking entities are lost to the system. One run is to be made for 2000 time units to estimate the time in the system for an entity, the server utilization, and the number of entities waiting for service.

Embellishment: Prepare the inputs to (a) plot the number in the queue over time, (b) draw a histogram of the frequency with which 0, 1, 2, ... entities are in the queue, and (c) obtain a summary report every 200 time units.

4-5. Draw a graph that portrays the number in a system over a 20-time unit period in which five customers arrive. The following pairs of numbers represent the arrival and departure times for each of the five customers: (1,5); (3,6); (7,12); (9,19); and (10,20). Demonstrate for this situation that the average number in the system is equal to the average time in the system times the arrival rate. Repeat the demonstration for the average number in the queue and the average time in the queue.

4-6. Explain why the central limit theorem applies when statistical estimates are obtained over multiple runs.

4-7. Give the rationale for the correctness (or incorrectness) of the following statements:

(a) As the elapsed time for a run increases, larger expected values should be obtained for the estimates of the maximum idle time and the maximum busy time for a server.

(b) An estimated value obtained from multiple runs facilitates the use of standard statistical hypothesis-testing procedures.

(c) Histograms for COLCT nodes are always a part of the SLAM II summary report.

(d) A negative value is never specified for the cell width of a SLAM II histogram.

(e) The method of computing server utilization depends on the number of parallel servers represented by an activity.

4-8. Develop a set of commands that you would like to see in a query language to obtain data from a database in which the SLAM II outputs are stored. Design a set of formats for presenting the data to managers, engineers, and statistical analysts. Indicate which commands should be available to all users.

4.11 REFERENCES

1. Bratley, P., B. L. Fox, and L. E. Schrage, *A Guide to Simulation*, Springer-Verlag, New York, 1983.
2. O'Reilly, J. J., and W. R. Lilegdon, *SLAM II Quick Reference Manual,* Pritsker & Associates, West Lafayette, IN, 1986.
3. Pritsker, A. A. B., *Introduction to Simulation and SLAM II, 3rd ed.*,, Wiley and Systems Publishing, New York and West Lafayette, IN, 1986.
4. Pritsker & Associates, Inc., SLAMSYSTEM Us*er's Guide*, West Lafayette, IN, 1988.
5. Standridge, C. R., and A. A. B. Pritsker, *TESS:The Extended Simulation Support System*, Wiley, New York, 1986.

5 SLAM II SUPPORT ROUTINES

5.1 INTRODUCTION†

SLAM II networks can model a wide class of systems to resolve many problem situations without requiring computer programming. Flexibility, however, is required for a general purpose simulation language. SLAM II provides for the modeling of complex or unique system characteristics through user-written coded inserts. This chapter describes the procedures for a user of SLAM II to integrate FORTRAN subprograms into a SLAM II network model.†† Standard subprograms are included in SLAM II for accessing values of SLAM II variables and for performing SLAM II functions. SLAM II also contains direct support for discrete event and continuous model building [1, 2]. The emphasis in this book is network modeling, and only those features of SLAM II used in conjunction with network models are presented in detail.

5.2 MODELING STRATEGY

As a modeling procedure, it is recommended that a model be built without programming inserts. This may require that system details be modeled in the aggregate. However, by approaching a problem in this manner, a segregation of network constructs and computer programming in model building can be made. This simple procedure can significantly decrease the design and development time inherent in any computer modeling effort. By making the programming effort subservient to the modeling task, a clear definition of the require-

† This chapter describes advanced capabilities of SLAM II and may be bypassed on first reading.

†† The C language may also be used with SLAM II.

ments and functions that need to be programmed usually results. In fact, the structure imposed by SLAM II networks, which specifies where programming inserts may be made, can be of tremendous help in building models of complex situations. As in the case with most methodologies that impose structure, it is at the cost of flexibility. SLAM II networks, however, have the structure to allow for the orderly development of a model, yet provide sufficient flexibility to enable the building of models that have the fidelity to solve the problem being studied. An added benefit of SLAM II is the easy transition from a network model to a combined SLAM II model incorporating discrete event and continuous worldview constructs [2, 3].

5.3 SLAM II NETWORK INTERFACES

Many network interfaces are designed into SLAM II. The variables ARRAY, XX, II, ATRIB, and TNOW provide for the transfer of values of SLAM II variables between a network and subprograms. For the latter four variables, the SCOM1 labeled COMMON block is used to transfer these values:

```
COMMON/SCOM1/ATRIB(100),DD(100),DDL(100),DTNOW,II,MFA,MSTOP,NCLNR
1,NCRDR,NPRNT,NNRUN,NNSET,NTAPE,SS(100),SSL(100),TNEXT,TNOW,XX(100)
```

Definitions of the other variables in SCOM1 are:

MFA	Pointer to the first available location in the SLAM II filing system.
MSTOP	MSTOP is set to -1 to immediately stop a simulation run.
NCLNR	The file number of the event calendar.
NCRDR	The unit number from which SLAM II input statements are read.
NNRUN	The number of the current simulation run.
NPRNT	The unit number to which SLAM II output is written.
NTAPE	The unit number of a scratch tape or device.
TNEXT	The time of the next scheduled discrete event.

The two-dimensional global variable ARRAY generalizes the concept of the XX vector. Function GETARY and subroutines PUTARY and SETARY are provided for accessing and changing the values in ARRAY. Each of the functions NNACT, NNCNT, NNQ, NNRSC, NRUSE, and NNGAT provides access to the value of a network variable.

Function USERF provides a means for writing code to compute a value to be used in a network model. USERF can be used in any location in a network model to replace a SLAM II variable. Subroutines FREE, ALTER, OPEN, and CLOSX are used to change the status of network elements from user-written FORTRAN subprograms. Subprograms ALLOC, SEIZE, NQS, and NSS provide a method for incorporating advanced decision logic when modeling resource allocations and server and queue selection procedures.

The EVENT node is the direct interface between a network model and subroutine EVENT. Within subroutine EVENT, entities in files may be manipulated through the use of SLAMII file-processing routines. This also applies to events on the event calendar. An entity in an activity, whose duration is specified by STOPA, may be made to complete the activity from subroutine EVENT by invoking subroutine STOPA. Inserting entities into a network is accomplished by calling subroutine ENTER, which places an entity in a network at an ENTER node.

At the beginning of each simulation run, the SLAM II processor calls subroutine INTLC to enable the modeler to set initial conditions and to insert entities into files. All the subroutines and functions included within SLAM II may be used at this time. Subroutine OTPUT is called at the end of each simulation run and is used for end-of-run processing, such as clearing out files and printing special results for the simulation run.

5.4 ILLUSTRATIONS OF SLAM II INTERFACES

This section provides illustrations of subroutine INTLC, subroutine OTPUT, function USERF, and subroutine EVENT. These illustrations introduce some of the SLAM II support routines that are described in the latter sections of this chapter. The illustrations provide examples of how to increase the flexibility of network models through programming interfaces.

5.4.1 Subroutine INTLC Illustration

Figure 5-1 presents an example of subroutine INTLC, which is called at the beginning of each run. The global SLAM II variable XX(1) is set equal to the run number NNRUN. XX(1) can then be used on the network to activate different processing capabilities based on the simulation run number. The second part of subroutine INTLC inserts an entity into file 1. Atribute 1 of the entity is set to 1, and attribute 2 is set equal to a sample from a uniform distribution between 10 and 20. The statement CALL FILEM(1,ATRIB) puts these attributes as an entity in file 1. In this way, attribute values can be given to entities that initially reside in the files of a network model.

```
      SUBROUTINE INTLC
      COMMON/SCOM1/ATRIB(100),DD(100),DDL(100),DTNOW,II,MFA,MSTOP,NCLNR
     1,NCRDR,NPRNT,NNRUN,NNSET,NTAPE,SS(100),SSL(100),TNEXT,TNOW,XX(100)
C**** SET GLOBAL VARIABLE 1 TO RUN NUMBER
      XX(1) = NNRUN
C**** INSERT ENTITY INTO FILE 1 WITH ATTRIBUTES OF 1
C**** AND A SAMPLE FROM A UNIFORM DISTRIBUTION
      ATRIB(1) = 1
      ATRIB(2) = UNFRM(10.,20.,1)
      CALL FILEM(1,ATRIB)
      RETURN
      END
```

Figure 5-1 Subroutine INTLC illustration.

5.4.2 Subroutine OTPUT Illustration

Subroutine OTPUT is called at the end of every simulation run. Typically, it is used for end-of-run processing including the printout of specialized statistical reports. A sample version of subroutine OTPUT is given in Figure 5-2. For this illustration, statistics for file 1 and the current entities in file 1 are printed out by a call to subroutine PRNTF(1). In the second part of subroutine OTPUT, statistics on resource IR are printed out if its average resource utilization is greater than 0.8. The average resource utilization for resource IR is obtained from the SLAM II function RRAVG(IR). The printing of resource statistics is accomplished by calling subroutine PRNTR(IR). The printing of the resource statistics is done for the first ten resources defined by resource blocks, that is, for values of IR from 1 to 10.

```
      SUBROUTINE OTPUT
C**** PRINT STATISTICS FOR FILE 1 AND CURRENT ENTITY LIST
      CALL PRNTF(1)
C**** PRINT STATISTICS FOR ANY OF 10 RESOURCES WHOSE
C**** AVERAGE USAGE IS GREATER THAN 0.8
      DO  10   IR = 1,10
C**** RRAVG(IR) IS AVERAGE UTILIZATION OF RESOURCE IR
      IF (RRAVG(IR).GT.0.8) THEN
C**** SUBROUTINE PRNTR(IR) PRINTS STATISTICS FOR RESOURCE IR
            CALL PRNTR(IR)
      ENDIF
   10 CONTINUE
      RETURN
      END
```

Figure 5-2 Subroutine OTPUT illustration.

5.4.3 Function USERF(IFN) Illustration

USERF(IFN) is a SLAM II interface that provides a means for replacing a SLAM II variable on a network by a user function that calculates a value for the variable. Figure 5-3 shows a network segment with four different uses of USERF and the FORTRAN code to obtain values of variables replaced by USERF.

The first use defines the duration of activity 1 as USERF(1). In function USERF(IFN), a correspondence is made between IFN and the statement number by using a computed GOTO statement. For IFN = 1, the duration of activity 1 is set to 15 if gate 1 is open and to 25 if it is not. The value of NNGAT(1) is 0 if gate 1 is open.

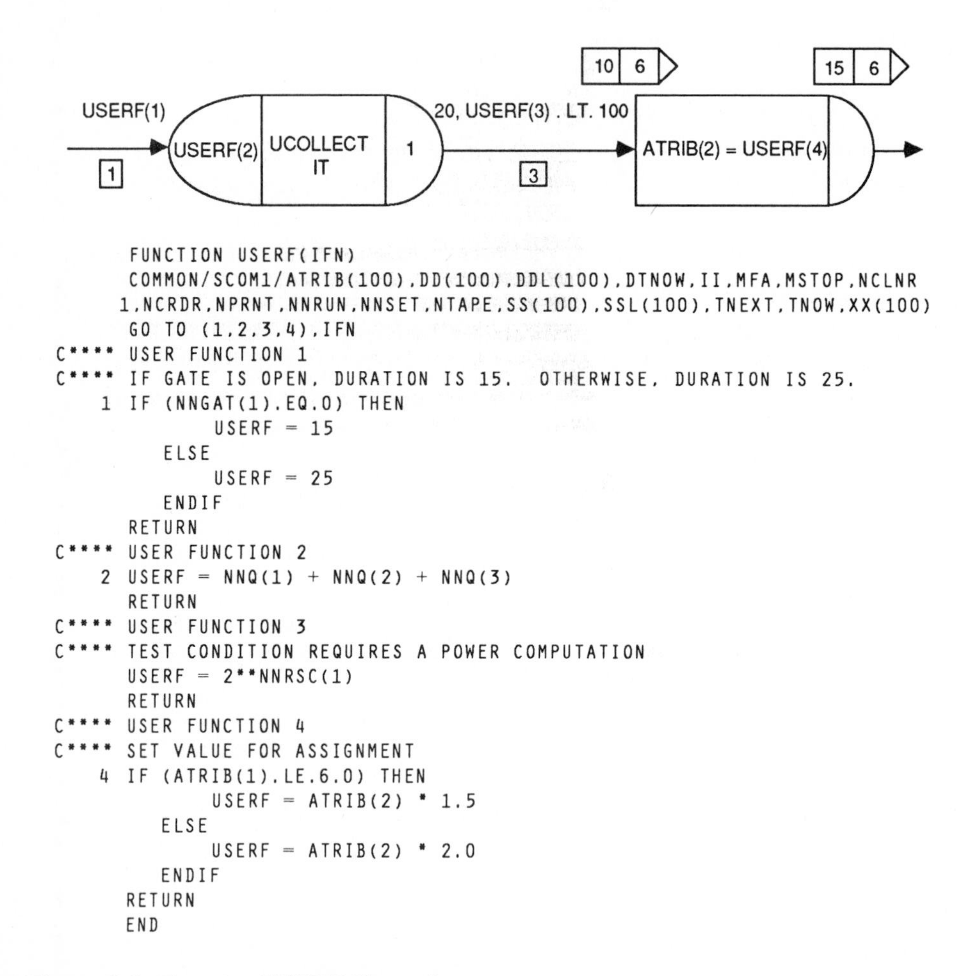


```
      FUNCTION USERF(IFN)
      COMMON/SCOM1/ATRIB(100),DD(100),DDL(100),DTNOW,II,MFA,MSTOP,NCLNR
     1,NCRDR,NPRNT,NNRUN,NNSET,NTAPE,SS(100),SSL(100),TNEXT,TNOW,XX(100)
      GO TO (1,2,3,4),IFN
C**** USER FUNCTION 1
C**** IF GATE IS OPEN, DURATION IS 15.  OTHERWISE, DURATION IS 25.
    1 IF (NNGAT(1).EQ.0) THEN
             USERF = 15
         ELSE
             USERF = 25
         ENDIF
      RETURN
C**** USER FUNCTION 2
    2 USERF = NNQ(1) + NNQ(2) + NNQ(3)
      RETURN
C**** USER FUNCTION 3
C**** TEST CONDITION REQUIRES A POWER COMPUTATION
      USERF = 2**NNRSC(1)
      RETURN
C**** USER FUNCTION 4
C**** SET VALUE FOR ASSIGNMENT
    4 IF (ATRIB(1).LE.6.0) THEN
             USERF = ATRIB(2) * 1.5
         ELSE
             USERF = ATRIB(2) * 2.0
         ENDIF
      RETURN
      END
```

Figure 5-3 Function USERF illustration.

The second use of USERF is at a COLCT node to define the value of an observation. In function USERF at statement 2, USERF is set equal to the sum of the number of entities in files 1, 2, and 3. Every time an entity arrives at the COLCT node, an observation of the number of entities in files 1, 2, and 3 is made.

The third use of USERF is to calculate a value to be used as part of a condition for an activity specification. When IFN is 3, the value of USERF is set equal to 2 raised to the power NNRSC(1). NNRSC(1) is the number of resources being used of resource type 1. Resource type numbers are defined by the order in which the resource blocks are listed in the SLAM II statements. On SLAM II networks, only the arithmetic operations of addition, subtraction, multiplication, and division are allowed. For more complex calculations, function USERF is used.

The fourth use of USERF is to obtain a value of a variable to be used in an assignment. At statement 4, a value is obtained that is equal to 1.5 times the value of attribute 2 if attribute 1 is less than or equal to 6. Otherwise, the value of USERF is set to 2 times the value of attribute 2. The values of attribute 1 and attribute 2 are passed to function USERF through the SLAM II COMMON block SCOM1. At the ASSIGN node shown in the network segment, the value of USERF is given to the second attribute of the entity passing through the ASSIGN node. A representation of an entity passing through the ASSIGN node is also given in Figure 5-3, which shows that attribute 1 is equal to 6. Therefore, the value of attribute 2 of this entity after the ASSIGN node is 1.5 times its incoming value, that is, 1.5*10=15.

5.4.4 EVENT and ENTER Nodes and Subroutines Illustration

An EVENT node is used to perform complex operations that are not available through the use of network node types. When an entity arrives at an EVENT node, a call is made to subroutine EVENT with an event number as an argument. An ENTER node is used to provide a means to insert entities into the network from programs written by a modeler. In Figure 5-4, an EVENT node and an ENTER node are illustrated, and the code for subroutine EVENT is shown which pertains to the two nodes. The event code is specified as 3 and the ENTER node number as 6.

In subroutine EVENT, the event number is referenced as JEVNT. Only the code for event number 3 is shown in this illustration. When the event number is 3 then the variable NINQ is set equal to the number of entities in file 5, that is, the SLAM II variable NNQ(5). If there are more than 10 entities in file 5, subroutine EVENT is used to remove those entities after the 10*th* one and to reinsert them into the network at ENTER node 6. For example, if there are 13

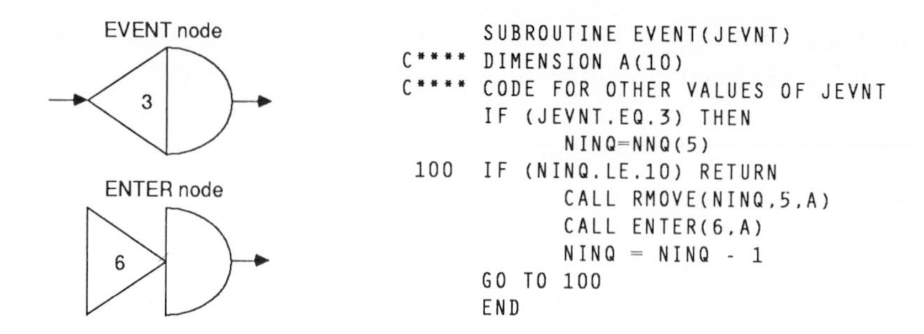

Figure 5-4 EVENT and ENTER illustration.

entities in file 5 then NINQ is initially set to 13. The statement CALL RMOVE(NINQ, 5, A) removes the 13*th* entity in file 5 and places the attributes of that entity in the vector A. The statement CALL ENTER(6, A) causes the entity with attributes defined by the vector A to be routed to the ENTER node whose number is 6. The number in queue variable, NINQ, is then decreased by 1. The code starting at statement 100 is then processed again. The above process is repeated until the number of entities in file 5 is reduced to 10. If the number in the queue is not greater than 10, a return is made from subroutine EVENT, and the entity arriving to the EVENT node is routed from it.

5.5 ACCESSING NETWORK VALUES

Values of network-related variables are obtained using the SLAM II functions listed in Table 5-1.

Table 5-1 Functions for Accessing the Status of an Activity, Gate, or Resource.

Function	Definition
NNACT(I)	Number of active entities in activity I at current time
NNCNT(I)	Number of entities that have completed activity 1
NNGAT(I)	Status of gate number I at current time: 0 → open, 1 → closed
NRUSE(I)	Current number of resource type I in use
NNRSC(I)	Current number of resource type I available
NNBLK(I, IFILE)	Number of entities in activity I blocked by file IFILE

On a SLAM II network, ARRAY is a two-dimensional table for global values. The values in ARRAY are accessed by using function GETARY and are changed by using subroutines PUTARY and SETARY. Definitions of these subprograms for storing and retrieving values from ARRAY are given in Table 5-2.

Table 5-2 Subprograms for Storing and Retrieving Values from ARRAY

FUNCTION GETARY(IR, IC)	Returns the value of row IR, column IC of ARRAY
SUBROUTINE PUTARY(IR, IC, VAL)	Sets the value of row IR, column IC of ARRAY to VAL
SUBROUTINE SETARY(IR, VALUE)	Sets the values of row IR of ARRAY to the values in the vector VALUE

5.6 OBTAINING VALUES ASSOCIATED WITH STATISTICAL ESTIMATES

Table 5-3 lists 29 functions that provide access to values associated with statistical quantities computed during a simulation run. The argument to the SLAM II statistical calculation functions is always a numeric value. For resources, SLAM II or the modeler assigns a resource number to each resource statement in a network statement model. For collect variables, if not specified by the user, SLAM II assigns a sequential numeric code for each COLCT statement, starting with the first number above the highest user-assigned number prescribed on a STAT statement.

For time-persistent variables, both the order in which the TIMST input statements appear in the input and the variable type determine the number to be used in the functions beginning with the letters TT. SLAM II assigns the statistics index by numbering all TIMST statements referencing discrete variables sequentially and then continues the numbering for TIMST statements referencing continuous variables.

5.7 SAMPLING FROM PROBABILITY DISTRIBUTIONS

SLAM II provides FORTRAN function statements to sample from commonly encountered distributions such as the exponential, normal, and beta, as well as from user-defined distributions. The list of functions is summarized in Table 5-4. Note that the functional names are the same as those employed in network models except for DPROB (where the letter N has been dropped), and the argument specifications are identical to those specified in Table 3-1. The reader should note that these are FORTRAN subprograms, and all arguments must be

Table 5-3 Statistical Calculation Functions

Function[a]	Description
Statistics for variables based on observations (COLCT)	
CCAVG(ICLCT)	Average value of variable ICLCT
CCSTD(ICLCT)	Standard deviation of variable ICLCT
CCMAX(ICLCT)	Maximum value of variable ICLCT
CCMIN(ICLCT)	Minimum value of variable ICLCT
CCNUM(ICLCT)	Number of observations of variable ICLCT
Statistics for time persistent variables (TIMST)	
TTAVG(ISTAT)	Time integrated average of variable ISTAT
TTSTD(ISTAT)	Standard deviation of variable ISTAT
TTMAX(ISTAT)	Maximum value of variable ISTAT
TTMIN(ISTAT)	Minimum value of variable ISTAT
TTPRD(ISTAT	Time period for statistics on variable ISTAT
TTTLC(ISTAT)	Time at which variable ISTAT was last changed
Queue statistics	
FFAVG(IFILE)	Average number of entities in file IFILE
FFAWT(IFILE)	Average waiting time in file IFILE
FFSTD(IFILE)	Standard deviation for file IFILE
FFMAX(IFILE)	Maximum number of entities in file IFILE
FFPRD(IFILE)	Time period for statistics on file IFILE
FFTLC(IFILE)	Time at which number in file IFILE last changed
Resource statistics	
RRAVG(IRSC)	Average utilization of resource IRSC
RRAVA(IRSC)	Average availability of resource IRSC
RRSTD(IRSC)	Standard deviation of utilization of resource IRSC
RRMAX(IRSC)	Maximum utilization of resource IRSC
RRPRD(IRSC)	Time period for statistics on resource IRSC
RRTLC(IRSC)	Time at which resource IRSC utilization was last changed
Activity statistics	
AAAVG(IACT)	Average utilization of activity IACT
AAMAX(IACT)	Maximum utilization of activity IACT, or maximum busy time if activity IACT is a single-server service activity
AASTD(IACT)	Standard deviation of the utilization of activity IACT
AATLC(IACT)	Time at which the status of activity IACT last changed
Gate statistics	
GGOPN(IG)	Percent of time that gate IG was open
GGTLC(IG)	Time at which the status of gate IG last changed

[a] The names of the functions are all five letters in length with the first two letters the same and representing one of the six types of variables for which statistics are desired. The last three letters of the function name prescribe the statistic of interest.

Table 5-4 Function Statements to Obtain Random Samples.

Function [a]	Descriptions
DRAND(IS)	Pseudo-random number from random IS
XRN(IS)	Last random number from stream IS
EXPON(XMN, IS)	Sample from an exponential distribution with mean XMN using stream IS
UNFRM(ULO, UHI, IS)	Sample from a uniform distribution in the interval ULO to UHI using stream IS
WEIBL(BETA, ALPHA, IS)	Sample from a Weibull distribution with shape parameter ALPHA and scale parameter BETA using stream IS
TRIAG(XLO, XMODE, XHI, IS)	Sample from a triangular distribution in the interval XLO to XHI with mode XMODE using tream IS
RNORM(XMN, STD, IS)	Sample from a normal distribution with mean XMN and standard deviation STD using stream IS
RLOGN(XMN, STD, IS)	Sample from a lognormal distribution with mean XMN and standard deviation STD using stream IS
ERLNG(EMN, XK, IS)	Sample from an Erlang distribution which is the sum of XK exponential samples each with mean EMN using stream IS
GAMA(BETA, ALPHA, IS)	Sample from a gamma distribution with parameters BETA and ALPHA using stream IS
BETA(THETA, PHI, IS)	Sample from a beta distribution with parameters THETA and PHI using stream IS
NPSSN(XMN, IS)	Sample from a Poisson distribution with mean XMN using stream IS
DPROB(CPROB, VALUE, NVAL, IS)	Sample from a user-defined discrete probability function with cumulative probabilities and associated values specified in arrays CPROB and VALUE, with NVAL values using stream IS

[a] Definitions of the arguments are given in Table 3-1.

specified with the correct type of FORTRAN variable (integer or real). In particular, the stream number cannot be defaulted as is the case for network statements. For DPROB, the first two arguments are vectors containing cumulative probabilities and their associated variable values. Function DPROB differs from the network discrete probability function, DPROBN, in that for DPROBN, two row numbers of the SLAM II variable ARRAY are specified as the first two arguments.

5.8 CHANGING THE STATUS OF RESOURCES, GATES, AND ACTIVITIES

SLAM II subroutines are provided for freeing resources, altering resource capacities, opening and closing gates, and stopping an activity. Subroutine FREE(IR, N) releases N units of resource number IR. The freed units of the resource are made available to waiting entities according to the order of the file numbers specified in the RESOURCE block of the network model. Subroutine ALTER(IR, N) changes the capacity of resource number IR by N units. In the case where the capacity of the resource is decreased below current utilization, freed resources are not reallocated when they become available. When the capacity is reduced to zero, additional capacity reduction requests have no effect.

SLAM II provides the capability to open and close gates using FORTRAN statements. Subroutine OPEN(IG) opens gate number IG. Subroutine CLOSX(IG) closes gate number IG.

In complex systems, the length of a specific activity may not be known a priori but may depend on the dynamics of system operation. One way of modeling such an indefinite activity duration is to specify the activity duration as STOPA(NTC) where NTC is a code that may be entity related. NTC allows the modeler to stop selectively a specific entity undergoing an activity without stopping other entities undergoing the same activity. The mechanism for stopping the activity for a network entity from FORTRAN code is to call subroutine STOPA(NTC). Such a call causes an end of activity event to occur for every entity that is being processed by an activity whose duration was specified as STOPA(NTC).

5.9 FILE AND EVENT CALENDAR ACCESS

A file provides the mechanism for storing entities and their attribute values in a prescribed ranking with respect to other entities in the file. The file NCLNR is used for maintaining events on the event calendar. Associated with a file is a ranking criterion that specifies the procedure for ordering entities within the file. Thus, each entity in a file has a rank that denotes its position in the file relative to the other members in the file. A rank of 1 indicates that the entity is first in the file. Possible ranking criteria for files are first in, first out (FIFO); last in, first out (LIFO); high value first (HVF) based on an attribute value; and low value first (LVF) based on an attribute value. The ranking criterion for a file is assumed to be FIFO unless otherwise specified using the PRIORITY statement described in Chapter 4.

In SLAM II, files are distinguished by integer numbers assigned to the files by the user. SLAM II automatically collects statistics on each file and provides the function NNQ(IFILE), which returns the number of entities in file IFILE. In the following subsections, a description is given of how to insert and remove entities from files and how to schedule an event onto the event calendar.

5.9.1 SUBROUTINE FILEM(IFILE, A)

Subroutine FILEM(IFILE, A) inserts an entity with attributes specified in the array A into file IFILE. The entity's rank in the file is determined by SLAM II based on the priority specified for the file. If IFILE is associated with a network node, the SLAM II processor does not directly insert the entity into the file, but processes the entity as an arrival at the node immediately following the return from the subroutine in which subroutine FILEM was invoked. Thus, the value of the number of entities in a file obtained from function NNQ is not change until after a return is made to the SLAM II processor.

5.9.2 Subroutine RMOVE(NRANK, IFILE, A)

Subroutine RMOVE(NRANK, IFILE, A) removes the NRANK*th* entity from file IFILE and places its attributes into the vector A. The dimension of A must be set greater than or equal to MATR + 3, where MATR is the number of attributes specified on the LIMITS statement. For example, the statement CALL RMOVE(3,7,ATRIB) removes the third entity from file 7 and redefines ATRIB with the attributes of the removed entity.

5.9.3 Scheduling Events

The SLAM II processor completely relieves the user of the responsiblity for chronologically ordering events. The user can schedule events to occur, and SLAM II causes subroutine EVENT to be called at the appropriate time in the simulation. In network models, events are scheduled automatically. By calling subroutine SCHDL(JEVNT, DTIME, A), event JEVNT is placed on the event calendar to occur at TNOW + DTIME. Attributes associated with the event are specified by the vector A. The values of these attributes are put in ATRIB at the time the event occurs, and subroutine EVENT is called. Thus, ATRIB at the time of event processing contains the values that were in the buffer array A at the time that the event was scheduled.

5.9.4 Additional File Manipulation Routines

A list of additional file-manipulation subprograms is given in Table 5-5.

Table 5-5 List of Additional File Manipulation Subprograms

Function	Description[a]
LOCAT(IRANK, IFILE)	Returns pointer to location of entry with rank IRANK in file IFILE
MMFE(IFILE)	Returns pointer to first entry (rank 1) in file IFILE
MMLE(IFILE)	Returns pointer to last entry in file IFILE
NNQ(IFILE)	Returns number of entries in file IFILE
NPRED(NTRY)	Returns pointer to the predecessor entry
NSUCR(NTRY)	Returns pointer to the successor entry
PRODQ(NATR, IFILE)	Returns the product of the values of attribute NATR for each entry in file IFILE
SUMQ(NATR, IFILE)	Returns the sum of the values of attribute NATR for each entry in file IFILE
Subroutine	Description
LINK(IFILE)	Inserts entry whose attributes are stored in MFA in file IFILE
ULINK(NRANK, IFILE)	Removes entry with rank NRANK from file IFILE without copying its attribute values

[a] A pointer value of zero indicates that no entry exists that satisfies the desired function.

5.10 ALLOCATING RESOURCES

If ALLOC(I) is specified in the field normally used for a resource label at an AWAIT node, then SLAM II calls subroutine ALLOC(I, IFLAG) whenever an entity arrives at the AWAIT node or a resource is freed that can be allocated to an entity waiting in the file of the AWAIT node.

Subroutine ALLOC is coded to determine which, if any, entity in a file may proceed. SLAM II loads ATRIB with the attributes of the first entity in the AWAIT node file prior to calling ALLOC.

The two arguments for subroutine ALLOC are:

1. A user code, I, to differentiate calls to ALLOC from different AWAIT nodes.

2. A flag, IFLAG, set by the user to inform SLAM II of the entity to which an allocation has been made. If IFLAG is set to zero, no allocation has been made.

In subroutine ALLOC, if it is determined that an entity should proceed from the AWAIT node, the user seizes the appropriate resources (discussed later) and, through the setting of IFLAG, communicates to SLAM II whether it should remove the entity from the file of the AWAIT node. IFLAG should be set to plus or minus the rank of the entity to be removed from the AWAIT node file for processing in the network. A positive IFLAG value indicates that the ATRIB buffer has been set in ALLOC to the attribute values to be associated with the entity removed from the file. If IFLAG is set to the negative of the rank, the entity is removed and no change is to be made to its attributes. If no allocation is possible, IFLAG should be set to 0.

When the resources are to be allocated at an AWAIT node, the number of units NU of resource IR is seized through a call to subroutine SEIZE(IR,NU). Subroutine SEIZE then sets NU units of resource type IR busy and updates the statistics for that resource. Through the use of subroutines ALLOC and SEIZE, complex resource allocation rules can be included in simulation models. In coding subroutine ALLOC, all FORTRAN coding procedures and SLAM II subprograms can be used except that no operations on the event file are allowed. In addition, subroutines FREE and ALTER may not be called to change the number of resources available from subroutine ALLOC. A resource allocated in subroutine ALLOC may not be preempted.

5.11 SELECTING QUEUES AND SERVICE ACTIVITIES

In a network model, SELECT nodes are used to route an entity to one of a set of parallel queues or to select an entity from a set of parallel queues for processing by a server. By specifying NQS(N), the user-written function NQS with argument N is used as the queue selection rule for a SELECT node. The SLAM II user writes function NQS to execute the desired queue selection logic and returns the file number of the selected queue as the value assigned to NQS. When queue selection is done by the user, it is the user's responsibility to ensure that a feasible choice is made. Whenever there are no feasible choices, the user sets NQS to zero to indicate that no queue was selected.

A user-coded server selection rule is associated with a SELECT node by specifying NSS(N), where N is a numeric index that allows the user to differentiate between different user-written selection rules. Function NSS is called when an entity arrives at an empty QUEUE node associated with a SELECT node whose server selection rule is NSS. In function NSS, an activity number is assigned to NSS to indicate the service activity selected. The service activity selected must have an available server to process the newly arriving entity. If all servers following the SELECT node are busy, NSS should be returned as zero in which case the arriving entity is placed in the QUEUE node to which it arrived. The service activity numbers following the SELECT node are known to the user as they are associated with the activity statements following the SELECT node in the network input statements.

When a value of NSS is returned, SLAM II takes the appropriate action of either putting the arriving entity into a QUEUE node or making the selected service activity busy by scheduling an end-of-activity event.

5.12 REPORT WRITING SUBROUTINES

A set of subroutines is available to obtain summary reports or sections of a summary report. The output from each of these subroutines corresponds to a specific section of the SLAM II Summary Report. To differentiate between histograms for observed variables at COLCT nodes and histograms for time-persistent variables, two printing subroutines are provided. These are PRNTH(ICLCT) for observed variables and PRNTB(ISTAT) for time-persistent variables. A description of the subroutines is presented in Table 5-6.

Table 5-6 Report Writing Subroutines.

Subroutine	Description[a]
SUMRY	Prints the SLAM II Summary Report
PRNTF(IFILE)	Prints statistics and the contents of file IFILE
PRNTC(ICLCT)	Prints statistics for ICLCT*th* COLCT variable
PRNTH(ICLCT)	Prints a histogram for ICLCT*th* COLCT variable
PRNTP(IPLOT)	Prints a plot and/or table for plot/table number IPLOT
PRNTT(ISTAT)	Prints statistics for ISTAT*th* time-persistent variable
PRNTB(ISTAT)	Prints a histogram (bar chart) for time-persistent variable ISTAT*th*
PRNTR(IRSC)	Prints statistics for IRSC*th* resource
PRNTA	Prints statistics for activities
PRNTG(IG)	Prints statistics for IG*th* gate

[a] If an argument of 0 is used, statistics for all possible values are printed.

5.13 GENERAL SUPPORT SUBPROGRAMS

There are many concepts associated with simulation that are supported by SLAM II indirectly through input statement types or by support subprograms. A list of these support subprograms and a description of their use is given in Table 5-7. In addition, the SLAM II Interactive Execution Environment (IEE) provides commands for debugging and verification. By selecting the INTERACTIVE option on the MONTR control statement, the commands in Table 5-8 are made available.

Table 5-7 Support Subprograms

Subroutine	Purpose
CLEAR	Clear statistics
ERROR(KODE)	Error interface
NNLBL(IDUM)	Return node label for current event
NNUMB(IDUM)	Return activity number for current event
NNVNT(IDUM)	Return event number for current event
TRACE	Start trace
UERR(KODE)	User-written error interface
UMONT(IT)	User-written monitor interface
UNTRA	Stop trace

Table 5-8 Interactive Execution Environment (IEE) Commands

Command	Purpose
ADVANCE	Advance simulation time
BREAKPOINT	Set an interrupt point for the simulation
CALL	Call a SLAM II support routine
CANCEL	Cancel a breakpoint
CONTINUE	Continue the simulation
DIARY	Create a log of IEE inputs and outputs
EXAMINE	Examine SLAM II variables
HELP	Provide on-line help messages
LOAD	Load previously saved model status
SAVE	Save current model status
SET	Change SLAM II variables
STATUS	View simulation and IEE status
STEP	Advance time by a number of events
STOP	End the simulation
TYPE	View the model source code

5.14 SUMMARY

This chapter describes the procedures for including programming inserts into SLAM II network models. With such procedures, there is no limit to the flexibility of modeling available with SLAM II. Table 5-9 is a listing of the tables included in this chapter to facilitate locating information about SLAM II subprograms.

Table 5-9 List of Tables in Chapter 5

Table	Title
5-1	Functions for accessing the status of an activity, gate ,or resource
5-2	Subprograms for storing and retrieving values from ARRAY
5-3	Statistical calculation functions
5-4	Function statements to obtain random samples
5-5	List of additional file manipulation subprograms
5-6	Report writing subroutines
5-7	Support subprograms
5-8	Interactive Execution Environment (IEE) commands

5.15 EXERCISES

5-1. Discuss the orientation and viewpoint required of a modeler when defining a system in terms of the SLAM II symbols. Discuss the differences between structural information requirements of modeling and the detailed data required to characterize activities. Indicate the constructs necessary to portray procedures and functions within a simulation model.

5-2. Explain why it is not necessary to have a precise problem formulation including all data elements when building a SLAM II model. Compare this approach to the one used when developing an analytic model in which the problem formulation and all data descriptions must be precisely stated before the modeling effort can begin.

5-3. Describe three locations where management decision making can be included within a SLAM II model. Discuss the limitations with regard to the types of management decision making that can be included within SLAM II models.

5-4. Discuss differences in the use of FUNCTION USERF(IFN) and SUBROUTINE EVENT(JEVNT).

5-5. Write SUBROUTINE INTLC to create 3 entities and place them at ENTER node 2 at the beginning of a simulation run. Set the first 3 attributes of each entity to 1, 2, and 3, respectively.

5-6. Write statements to perform the following functions. (Only the statements, not their location in any subprogram, are required.)

(a) Access all attribute values of the entity currently being processed and store them in the vector ATT.
(b) Set the *3rd* attribute of the current entity to a value of 10.
(c) Access the value of ARRAY(2, ATRIB(4)) where ATRIB(4) is the *4th* attribute of the entity currently being processed when the statement is invoked. Store this value in a local variable AT3.
(d) If the number of entities in activity 7 is greater than 5, cause an entity to arrive at ENTER node 10 in 5 time units with its first two attribute values being 17 and 4, respectively.
(e) If the remaining capacity of resource 3 is less than 2, stop activity 7 whose duration is specified as STOPA(6).
(f) If the average number in file 8 is currently greater than 5, increase the capacity of resource 3 by 2 units.
(g) If the average utilization of service activity 1, whose corresponding QUEUE node is node 3, is greater than 10, cause all entities waiting to be processed by server 1 to be routed to QUEUE node Q2 with a time delay of 10 units.

5-7. Build a model of the one-server, single-queue situation where the service time remaining on an entity changes based on the number of entities waiting. If a new arrival causes more than five entities to be in the queue, the service time remaining is decreased by 0.10 minute for each waiting customer greater than five, including the new arrival. In no case, however, can the remaining service time be decreased by more than 50% and the remaining service time can be decreased only once; that is, if two entities arrive during the servicing of an entity, only the first arrival causes a decrease in the remaining service time. Assume that interarrival times are exponentially distributed with a mean of 2 time units, and service time is exponentially distributed with a mean of 1.7 time units.

5-8. Jobs arrive at a machine tool on the average of one per hour. The distribution of these interarrival times is exponential. During normal operation, the jobs are processed on a first-in, first-out basis. The time to process a job is normally distributed with a mean of 0.5 hour and a standard deviation of 0.10 hour. Processing times are never less than 0.25 hour or greater than 2 hours. In addition to the processing time, there is a setup time that is uniformly distributed between 0.2 and 0.5 hour. Jobs that have been processed are routed to a different section of the shop and are considered to have left the machine tool area.

The machine tool experiences breakdowns during which time it can no longer process jobs. The time between breakdowns is exponentially distributed with a mean of 20 hours. When a breakdown occurs, the job being processed is removed from the machine tool and, after a 0.1-hour delay, is placed at the head of the queue of jobs waiting to be processed. The service time for the job preempted by the machine breakdown is the remaining service time plus an additional setup time that is again uniformly distributed.

When the machine tool breaks down, a repair process is initiated, which is accomplished in three phases. Each phase is exponentially distributed with a mean of 0.75 hour. Since the repair time is the sum of independent and identically distributed exponential random variables, the repair time is Erlang distributed.

Build a SLAM II model to analyze the machine tool, to obtain information on the utilization of the machine tool, and to calculate the time required to process a job. Make five 6000-time unit runs. Obtain outputs from each run.

5.16 REFERENCES

1. O'Reilly, J. J. and W. R. Lilegdon, *SLAM II Quick Reference Manual*, Pritsker & Associates, West Lafayette, IN, 1986.
2. Pritsker, A. A. B., *Introduction to Simulation and SLAM II*, *3rd* ed., Wiley and Systems Publishing, New York and West Lafayette, IN, 1986.
3. Pritsker, A. A. B., *The GASP IV Simulation Language*, Wiley, New York, 1974.

6 DECISION AND RISK ANALYSIS

6.1 INTRODUCTION

A decision situation consists of a choice from a set of alternatives, A_i, in the face of uncertain future states S_j. If the manager makes decision A_i and future state S_j happens, the outcome O_{ij} occurs. In making a choice of one of the alternatives, a decision maker either explicitly or implicitly assigns a value or values to the possible outcomes associated with the alternative. The values are defined by $V(O_{ij})$. The manager may also make a judgment about the possibility of future state S_j occurring. p_j is defined as this probability. With this terminology, decision making can be viewed within the framework of the table presented in Figure 6-1.

		Future state			
		S_1	S_2	. . .	S_F
		p_1	p_2	. . .	p_F
Alternative	A_1	O_{11}	O_{12}	. . .	O_{1F}
	A_2	O_{21}	O_{22}	. . .	O_{2F}
	⋮	⋮	⋮		⋮
	A_C	O_{C1}	O_{C2}	. . .	O_{CF}

Figure 6-1 General decision framework: $V(O_{ij})$ is the value associated with outcome O_{ij}

The value associated with an outcome can be one or more performance measures. The performance measures may be known with certainty or only known probabilistically. Common performance measures for outcomes are profit, cost, probability of bankruptcy, reliability, and probability of success.

Faced with a decision situation as presented in Figure 6-1, the manager needs a procedure for choosing an alternative. The selection of an alternative is referred to as a choice decision. Many principles of choice for selecting alternatives in light of diverse projected future states have been proposed. Examples of principles of choice are to select the alternative that has the highest expected value associated with the outcomes over all future states, to select the alternative that maximizes the minimum value associated with any future state, and to select the alternative that has the highest expected value for the future state with the highest chance of occurring [7]. Each principle of choice provides a procedure for transforming the probabilities of future states into a metric that is used as the basis for selecting an alternative. The actual occurrence of a future state is based on chance and is referred to as a chance decision.

The preceding discussion considers only a single choice whose outcome is determined through a probabilistic mechanism. Sequential decision making involves multiple-choice decisions determined sequentially. Each selection of an alternative results in both the value associated with the outcome and a new decision situation. This can be thought of within the context of the decision framework presented in Figure 6-1 as each outcome having a value or values and a pointer to another decision table. Typically, decision making is not conceived as continuing ad infinitum, but rather a final decision is one in which all the values of the outcomes can be specified in terms of the performance measures of interest; that is, no further pointers to decision tables are contained at the last decision stage.

Networks, and SLAM II in particular, play an important descriptive and analysis role for sequential decision making. In this chapter, network models of sequential decision situations are presented. Because decisions in many instances involve costs and profits directly, rather than costs that are a function of activity duration, Section 6.2 describes replacing the time variable on the network with a cost variable. The remainder of the chapter describes decision trees, decision networks, and risk assessment.

6.2 THE USE OF COST IN PLACE OF TIME IN SLAM II NETWORKS

In some network problems, it is convenient to interpret a branch delay as a cost of performing an activity. When employing cost in place of time in a SLAM II network, the same logical operations apply. However, care is required when formulating a model to make sure that the SLAM II concepts are feasible. A few illustrations should be sufficient to indicate the benefits and potential pitfalls of a cost, rather than a time, orientation.

Activities in series represent the summing of activity costs. For example, when an entity traverses the three activities in series shown in Figure 6-2, a sample of the sum of the three costs ($c_1 + c_2 + c_3$) is collected at the COLCT node. A probability distribution may be used to specify each activity cost. As with time specifications in SLAM II, a negative value cannot be associated with the cost of an activity.

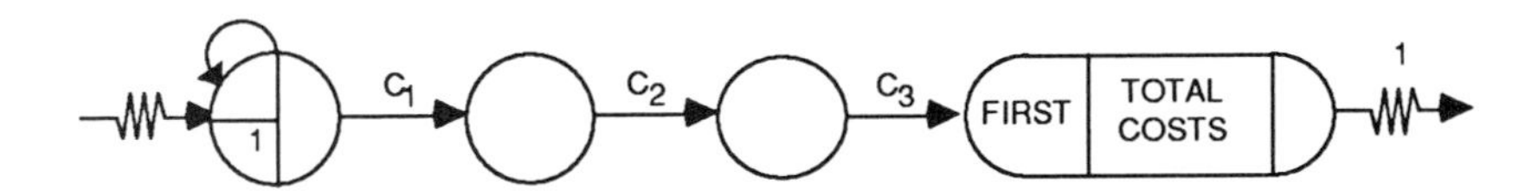

Figure 6-2 Three activities in series.

The minimum of a set of activity costs is modeled by activities in parallel, with only a single observation at a COLCT node required to terminate a run. For two activities, this is modeled as shown in Figure 6-3.

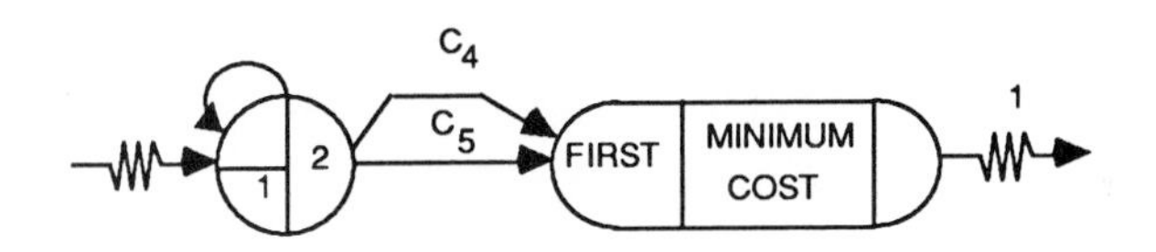

Figure 6-3 Observing the minimum of two activities.

The maximum of a set of activity costs is also modeled by activities in parallel, but with an ACCUMULATE node that requires all activities incident to the node to be completed. The network in Figure 6-4 illustrates this for two cost activities in parallel.

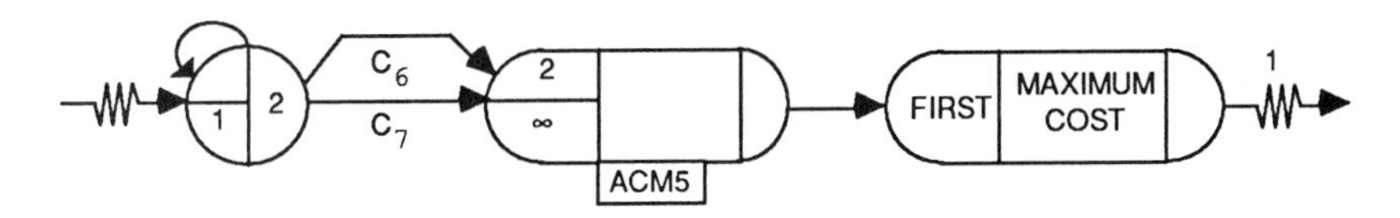

Figure 6-4 Observing the maximum of two activities.

A weighted sum of costs can be modeled by a probabilistic routing of entities. The network in Figure 6-5 models the selection of cost c_8 with a probability of p_8 and the selection of cost c_9 with a probability of p_9. On any run, the cost is either c_8 or c_9. In N runs, if c_8 occurs NNCNT(8) times and c_9 occurs NNCNT(9) times, then NNCNT(8)/N and NNCNT(9)/N are the observed values of the fraction of runs on which activities 8 and 9 were performed.

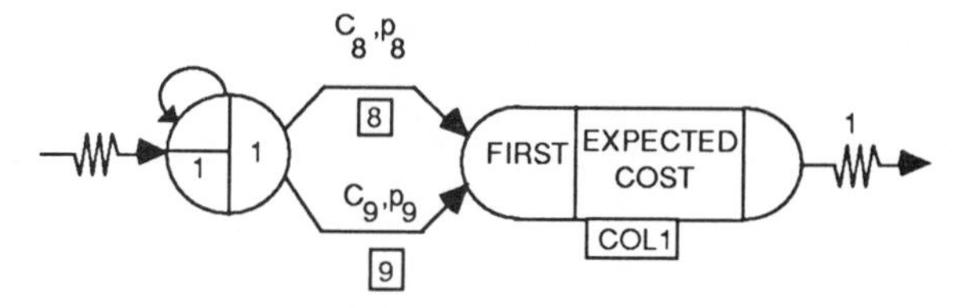

Figure 6-5 Observing the weighted sum of two activities.

An average of costs can be modeled by having each activity release its ending node and then collecting ALL statistics at the COLCT node. The network in Figure 6-6 represents the observing of all costs in order to obtain an averaging of costs for two activities. In this situation, the average on a run is taken over the number of entities arriving to the COLCT node.

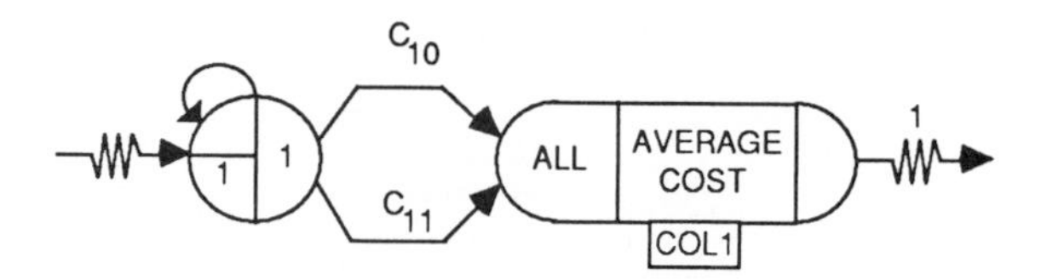

Figure 6-6 Observing the total cost to obtain an average cost.

A queueing situation may be based on cost. The network in Figure 6-7 which can be interpreted as c_1 dollars being expended before a next arrival to the network occurs. A cost is associated with service, and a next service cannot be started until a cost of c_2 is expended on the entity being served. In this case, INTERVAL (I) statistics at the COLCT node represent the total service cost while an entity is in the system.

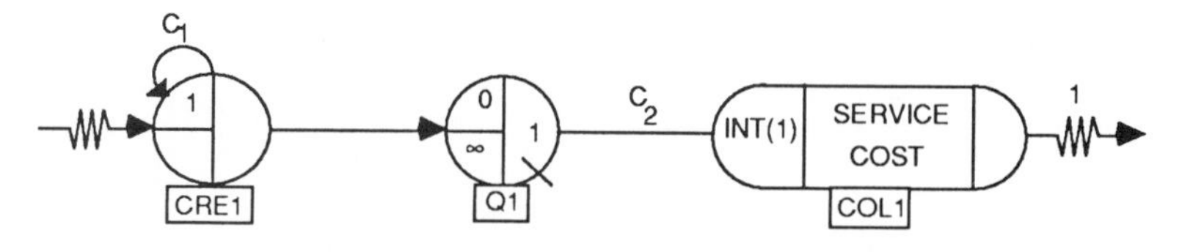

Figure 6-7 Cost-based queueing network.

Clearly, care is required when employing cost concepts directly on SLAM II networks. A standard procedure for including costs is to relate them to activity durations. By developing user functions or by using ASSIGN nodes, specified cost equations that are dependent on the activity durations can be used to accumulate costs [6, 8, 12].

6.3 DECISION TREES

A decision tree has a special network form in which each node represents a decision point at which a single activity emanating from the node is selected. Both chance and choice decisions can be modeled in a decision tree. A *chance decision* corresponds to probabilistic branching. A *choice decision* corresponds to conditional branching.

An important characteristic of a tree is that there is a unique path from the CREATE node to any other node in the network. An illustration of a decision tree modeled in the SLAM II syntax is shown in Figure 6-8. From the figure

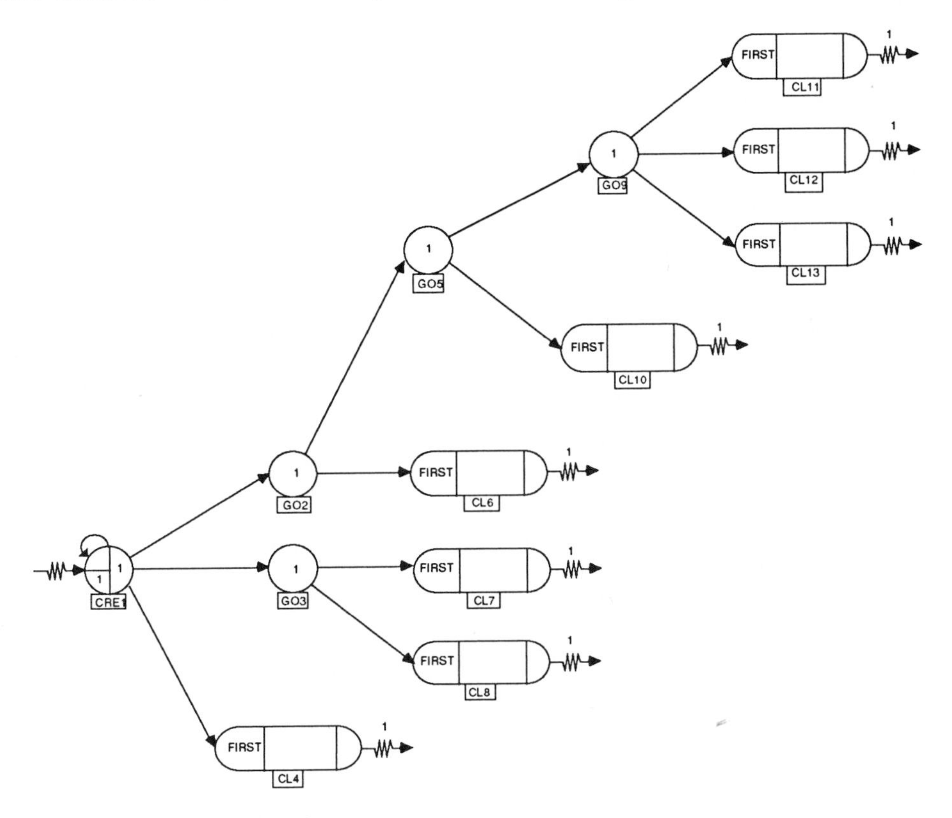

Figure 6-8 SLAM II decision tree.

it is seen that a SLAM II decision tree has the following properties: a single branch is incident to each node; each node can be released only once and requires a single arrival for its release; a single branch is taken from each node; and no restriction on activity time is made. Multiple COLCT nodes are associated with the tree, and each COLCT node represents a different outcome since it can only be reached by one path.

A SLAM II model of a decision tree represents the time and/or cost to reach a terminating condition. For each branch emanating from a chance decision node, a probability of the branch being selected (the future state occurring) and the value for the outcome associated with the branch are specified. The ending node for a branch represents the next decision point.

A special, but often studied, case of decision trees involves only chance decisions. Such trees are referred to as probabilistic decision trees. They are modeled in SLAM II by having all branching done probabilistically. The form of a probabilistic decision tree modeled in SLAM II is presented in Figure 6–9.

Activities are never performed concurrently in decision trees. That is, only a single activity is ever ongoing at a given time. Because of this simple structure,

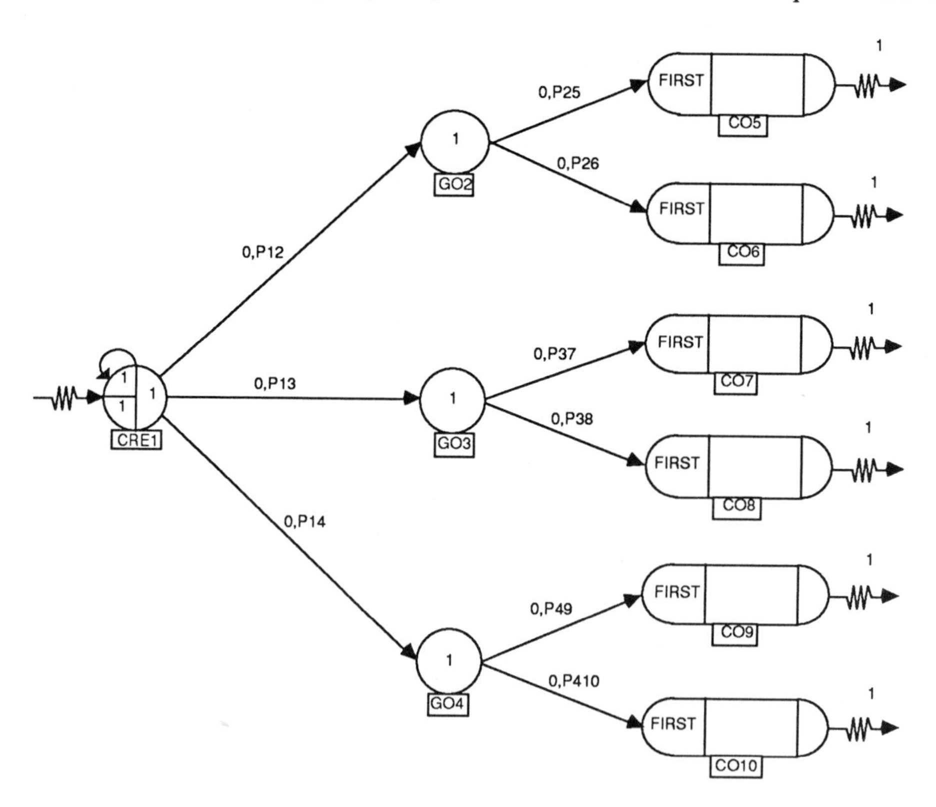

Figure 6-9 Probabilistic decision tree.

it is often possible to compute analytically the probabilities, release times, and costs associated with each COLCT node. For probabilistic decision trees involving no attribute assignments, the following statements can be made:

1. The probability of releasing a COLCT node is the product of the probabilities of the branches on the path leading to the node.
2. The time (cost) to reach a COLCT node is the sum of the time (cost) on the branches on the path leading to the node.

SLAM II becomes an attractive tool for modeling decision tree processes when conditions, as well as probabilities, govern the branches to be taken. This will occur when the decision to be made at a node is not independent of the path to reach a node. That is, the current decision is not independent of previous decisions. Attribute assignments at ASSIGN nodes and conditional branching features of SLAM II facilitate the representation of the decision tree processes for which analytic solutions are not readily available.

6.4 DECISION NETWORKS

Decision networks generalize decision trees by permitting more than one branch to be incident to a node and by allowing multiple CREATE nodes. A decision structure is still present because only a single branch is selected at each node. However, a decision network allows the possibility of different decision selections leading to the same outcome.

6.5 DECISION TREE MODEL FOR NEW PRODUCT INTRODUCTION

To illustrate decision tree analysis procedures, a network representation of the introduction of a new product is presented. This example is an adaptation of a decision tree analysis presented in [4], which is also discussed in [12]. The decision process involves sequential decisions regarding whether a product should be introduced regionally or nationally. First, a decision as to whether to market the product directly to a national market or to start with a regional marketing strategy is required. If initial marketing is done regionally, a second decision is needed involving whether to remain regional or to market nationally based on the observed regional response to the product. The SLAM II network to model this situation is shown in Figure 6-10. The network model portrays the sequence of decisions to be made and not how to make the decisions, which would require a more complex network than the one illustrated here.

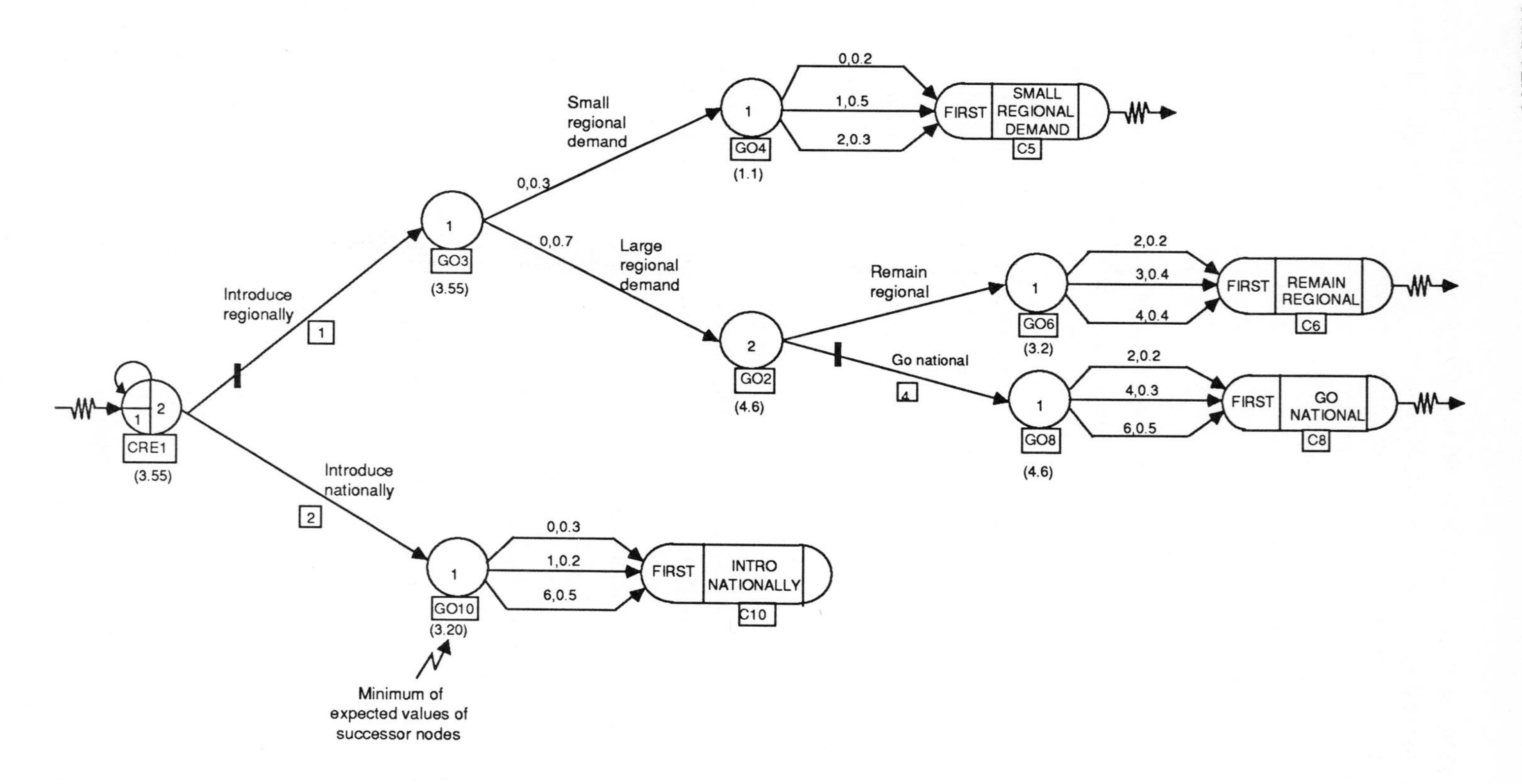

Figure 6-10 SLAM II model of decision tree related to introducing a new product.

The choice decision to be made following CREATE node CRE1 requires the initial marketing strategy to be either to introduce the product on a regional or national basis. In the network, both alternatives are selected so that a comparison of outcomes may be made. In this example, the returns from the consequences of all decisions are represented by activity duration values, and the chance events are represented by probabilities on the branches. If the product is introduced regionally, the branch to GOON node GO3 is taken. The output side of node GO3 represents the chance decision that there will be a small regional demand (probability of 0.3) or a large regional demand (probability of 0.7). If a small regional demand occurs, node GO4 is released. The return for regional marketing with a small demand is zero, which can occur with a probability of 0.2. There is a probability of 0.5 that the return will be 1, and a 0.3 probability that the return will be 2, if the demand is small regionally. The parallel branches for node GO4 to COLCT node C5 sample from this probability distribution.† No further decision making is involved when a small regional demand occurs.

When a large regional demand occurs, the branch from node GO3 to node GO2 is taken. At node GO2, a decision to remain regional or to introduce the product nationally is made. If the decision is to remain regional, node GO6 is released, followed by node C6. The probabilistic branches from node GO6 to node C6 represent the distribution of expected returns, given that the decision at node GO2 is to keep the product regional and there is a large regional demand for the product. If the decision at node GO2 is to introduce the product nationally after it is first introduced regionally, node GO8 is released and then node C8 is released. The expected return obtained from this marketing strategy is then associated with the statistics collected at node C8.

Note that the returns based on the decision processes are represented by activity values (duration times) in a manner similar to the costs described in Section 6.2. In SLAM II, all such returns must be positive. If any negative returns were associated with the decision, ASSIGN nodes and SLAM II global variables would be used to translate these values such that only positive values were associated with the activities. When a translation of values is made, the final result must be adjusted for each translation. Alternatively, function USERF and/or subroutine OTPUT could be used to collect the cost or to obtain statistics.

Before discussing the SLAM II analysis for this model, consider the standard method for evaluating the expected return from the decision network. The procedure starts at each terminating COLCT node where the expected returns

† The three parallel branches all describe the same activity and, hence, are allowed to be incident to a single node without violating the decision tree characteristics. Clearly, a single activity could be used to represent the three parallel branches.

associated with nodes whose emanating activities lead to this COLCT node are computed. This process is repeated through backward recursion until a choice decision node is encountered. At a choice decision node, the largest expected return is determined of all nodes immediately following the choice decision node. The largest of these expected values is then associated with the choice decision node, which assumes that the decision maker is maximizing expected returns. In this manner, expected returns for all of the nodes in the network are obtained, and eventually an expected value for the entire network is associated with the CREATE node. The expected return for the CREATE node is then designated as the expected return for the decision tree. Since a selection for each choice decision node has been made in order to compute the expected return, it is a simple matter to trace from the CREATE node through the network to identify each choice decision that should be made in order to maximize the expected return for the network. Below each node in Figure 6-10, the expected return is given. A heavy vertical bar is shown on an activity following each choice decision node to indicate the selection that should be made using the maximum expected return criterion at each of the these nodes.

The procedure for using SLAM II to analyze a decision tree network requires that the choice decisions be replaced with nodes that employ deterministic branching. This modification allows all paths in the tree to be evaluated so that all possible choices can be assessed. The SLAM II statement model is given in Figure 6-11 and the outputs for the requested 1000 runs with deterministic branching at CREATE node CRE1 and GOON node GO2 are shown in Figure 6-12. The termination condition for a run is not specified, so each run is ended after all activities are completed.

In Figure 6-12, each path in a decision tree is evaluated and estimates are provided for its probability of being selected and the average and standard deviation of the return if it is selected. Each path represents a sequence of decisions or outcomes for a policy at the choice nodes. Thus, the values associated with node C4 are for the choice decision to introduce regionally followed by the chance outcome that a small regional demand occurs. The values associated with node C4 in the SLAM II output correspond to the values associated with node GO4 in a standard decision tree analysis. Similarly, the maximum of the values for nodes C6 and C8 is associated with node GO2. In this manner, the standard decision tree analysis could be performed using SLAM II outputs.

SLAM II may be used more advantageously in decision analysis when the decision logic to be modeled is more complex. The SLAM II syntax provides capabilities not available in decision trees through the use of conditional branching based on time, attribute values, and node realizations that can capture the significant aspects of the real decision process.

```
GEN,HAMMESFAHR,DECISION TREE,1/5/88,1000,N,N,,N,Y/1000;
LIMITS,,,50;
NETWORK;
CRE1   CREATE,,,,,2;        INITIATE PRODUCT DECISION TREE
       ACT/1,,,G03;;        INTRODUCE REGIONALLY
       ACT/2,,,G010;;       INTRODUCE NATIONALLY
G03    GOON,1;              DETERMINE REGIONAL DEMAND
       ACT,,.3,G04;         SMALL REGIONAL DEMAND
       ACT,,.7,G02;         LARGE REGIONAL DEMAND
G04    GOON,1;              DETERMINE RETURN FOR SMALL REGIONAL DEMAND
       ACT,0,,.2,C4;        RETURN IS 0
       ACT,1,,.5,C4;        RETURN IS 1
       ACT,2,,.3,C4;        RETURN IS 2
C4     COLCT(1),FIRST,C4 SMALL REG DEMAND;   SMALL REGIONAL DEMAND
       TERM;
;
G02    GOON,2;              LARGE REGIONAL DEMAND
       ACT/3,,,G06;;        REMAIN REGIONAL
       ACT/4,,,G08;;        GO NATIONAL
G06    GOON,1;              DETERMINE RETURN FOR LARGE REGIONAL DEMAND
       ACT,2,,.2,C6;        RETURN IS 2
       ACT,3,,.4,C6;        RETURN IS 3
       ACT,4,,.4,C6;        RETURN IS 4
C6     COLCT(2),FIRST,C6 REMAIN REGIONAL;    LARGE REGIONAL DEMAND
       TERM;
;
G08    GOON,1;              DETERMINE RETURN IF GO NATIONAL
       ACT,2,,.2,C8;        RETURN IS 2
       ACT,4,,.3,C8;        RETURN IS 4
       ACT,6,,.5,C8;        RETURN IS 6
C8     COLCT(3),FIRST,C8 GO NATIONAL;        LARGE REGIONAL DEMAND
       TERM;
;
G010   GOON,1;              DETERMINE RETURN FOR INTRO NATIONAL DECISION
       ACT,0,,.3,C10;       RETURN IS 0
       ACT,1,,.2,C10;       RETURN IS 1
       ACT,6,,.5,C10;       RETURN IS 6
C10    COLCT(4),FIRST,C10 INTRO NATIONALLY;  INTRODUCE NATIONALLY ONLY
       TERM;
       END;
INIT,0,,N;
FIN;
```

Figure 6-11 SLAM II input statements for new product introduction model.

6.6 SLAM II ANALYSIS OF NEW PRODUCT INTRODUCTION

The decision tree approach employs the criterion of largest expected value. This criterion does not take into account the distribution of returns associated with different decision policies. When a decision process is to be exercised once

```
S L A M   I I   S U M M A R Y   R E P O R T

SIMULATION PROJECT DECISION TREE                BY HAMMESFAHR

DATE  1/ 5/1988                                 RUN NUMBER 1000 OF 1000

CURRENT TIME   0.4000E+01
STATISTICAL ARRAYS CLEARED AT TIME  0.0000E+00
```

STATISTICS FOR VARIABLES BASED ON OBSERVATION

	MEAN VALUE	STANDARD DEVIATION	MINIMUM VALUE	MAXIMUM VALUE	NUMBER OF OBSERVATIONS
C4 SMALL REG DEM	0.1086E+01	0.6825E+00	0.0000E+00	0.2000E+01	301
C6 REMAIN REGION	0.3187E+01	0.7761E+00	0.2000E+01	0.4000E+01	699
C8 GO NATIONAL	0.4492E+01	0.1609E+01	0.2000E+01	0.6000E+01	699
C10 INTRO NATION	0.3245E+01	0.2827E+01	0.0000E+00	0.6000E+01	1000

Figure 6-12 SLAM II output for new product introduction model.

or only a few times, it is important to know the probability that the return may be small even if an optimal decision policy is employed. SLAM II provides such information.

Decision alternatives can be viewed as competitors, where the largest return from a decision policy is the winner of the competition. In a network form, this implies parallel paths leading to a node. The path with the largest return at the terminating node constitutes the best policy. Viewed in this light, a sequence of decisions is represented by an inverted tree with many CREATE nodes and a single terminating node. The alternative decisions are represented by the activities that lead into a node, and the activity that causes the node to be released is the choice to be made.

A SLAM II network model of the decision alternatives for the new product introduction example as described in Section 6-5 is shown in Figure 6-13, with the corresponding input statement given in Figure 6-14. The node labels assigned in this network have been selected to agree, in the main, with those presented in Figure 6-10. ACCUMULATE nodes ACM1 and ACM2 represent the choice decisions where the last entity to arrive at each of these nodes represents the desirable decision policy for a particular simulation run.† The selection of an alternative at a choice decision node is reflected in the number of times incoming activities release its following COLCT node. When returns are equal for a set of alternative activities preceding an ACCUMULATE node,

† If a minimization problem is being considered, the first release requirement for choice decision nodes 1 and 2 would be 1.

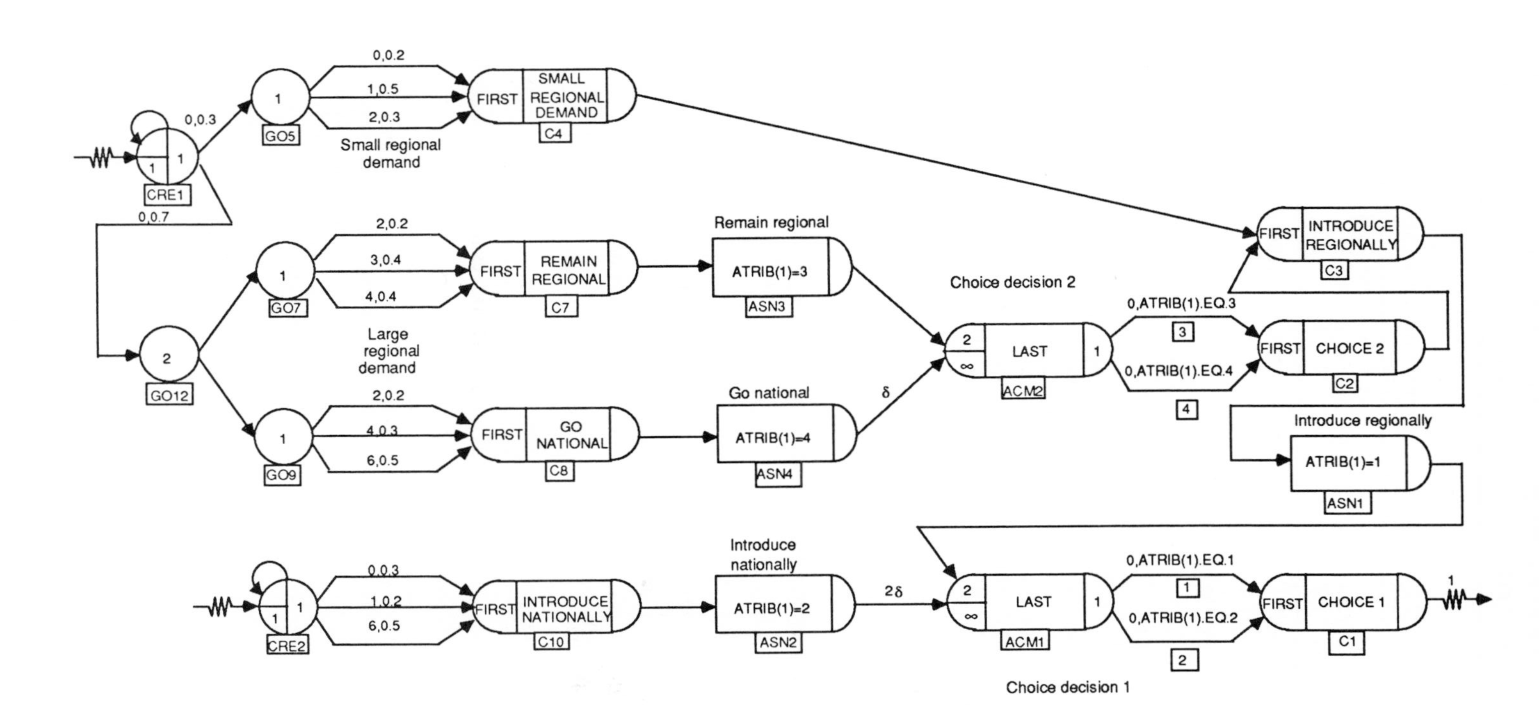

Figure 6-13 SLAM II model of new product introduction decision (δ = .00001).

```
GEN,HAMMESFAHR,PRODUCT DECISION,1/5/88,1000,N,N,,N,Y/1000;
LIMITS,,1,50;
NETWORK;
CRE1  CREATE,,,,1,1;                INITIATE REGIONAL INTRODUCTION DECISION
      ACT,,.3,G05;                  SMALL REGIONAL DEMAND
      ACT,,.7,G012;                 LARGE REGIONAL DEMAND
G05   GOON,1;                       DETERMINE SMALL REGIONAL RETURNS
      ACT,0.,.2,C4;                 RETURN 0
      ACT,1.,.5,C4;                 RETURN 1
      ACT,2.,.3,C4;                 RETURN 2
C4    COLCT,FIRST, SMALL REG DEMAND; SMALL REGIONAL DEMAND
      ACT,,,C3;
;
;     LARGE REGIONAL DEMAND
;
G012  GOON,2;                       REGIONAL OR NATIONAL FOR REGIONAL INTRO
      ACT,,,G07;                    STAY REGIONAL
      ACT,,,G09;                    GO NATIONAL
G07   GOON,1;                       DETERMINE RETURN FOR STAY REGIONAL DECISION
      ACT,2.,.2,C6;                 RETURN 2
      ACT,3.,.4,C6;                 RETURN 3
      ACT,4.,.4,C6;                 RETURN 4
C6    COLCT(2),FIRST,C6 REMAIN REGIONAL; LARGE REGIONAL DEMAND
ASN3  ASSIGN,ATRIB(1)=3.;           COUNT FOR ACTIVITY 3 (STAY REGIONAL DECISION)
      ACT,,,ACM2;                   PASS LARGEST RETURN
;
;     NATIONAL DECISION
;
G09   GOON,1;                       DETERMINE RETURN FOR GO NATIONAL DECISION
      ACT,2.,.2,C8;                 RETURN IS 2
      ACT,4.,.3,C8;                 RETURN IS 4
      ACT,6.,.5,C8;                 RETURN IS 6
C8    COLCT(3),FIRST,C8 GO NATIONAL;LARGE REGIONAL DEMAND
ASN4  ASSIGN,ATRIB(1)=4.;           COUNT FOR ACTIVITY 4 (GO NATIONAL DECISION)
      ACT,.00001,,ACM2;             GIVE PRIORITY TO NATIONAL DECISION IF TIED
ACM2  ACCUM,2,10,LAST,1;            PASS ATRIB(1) VALUE FOR LAST ENTITY ARRIVAL
      ACT/3,,ATRIB(1).EQ.3.,C2;;    COUNT STAY REGIONAL DECISIONS
      ACT/4,,ATRIB(1).EQ.4.,C2;;    COUNT GO NATIONAL DECISIONS
C2    COLCT(5),FIRST,C2 CHOICE 2; GO NATIONAL/STAY REGIONAL
C3    COLCT(6),FIRST,C3 INTRO REGIONALLY; REGIONAL RETURN
ASN1  ASSIGN,ATRIB(1)=1.;           COUNT FOR ACTIVITY 1 (REGIONAL INTRO)
      ACT,,,ACM1;                   PASS RETURN FOR REGIONAL INTRODUCTION
;
;     CREATE NATIONAL ENTITY
;
CRE2  CREATE,,,,1,1;                INITIATE NATIONAL INTRODUCTION DECISION
      ACT,0.,.3,C10;                RETURN 0
      ACT,1.,.2,C10;                RETURN 1
      ACT,6.,.5,C10;                RETURN 6
C10   COLCT(4),FIRST,C10 INTRO NATIONALLY; RETURN FOR NATIONAL INTRO
ASN2  ASSIGN,ATRIB(1)=2.;           PASS RETURN FOR NATIONAL INTRODUCTION
      ACT,.00002,,ACM1;             GIVE PRIORITY TO NATIONAL INTRODUCTION;
ACM1  ACCUM,2,10,LAST,1;            PASS ATRIB(1) VALUE FOR LAST ARRIVAL
      ACT/1,,ATRIB(1).EQ.1.,C1;;    COUNT FOR REGIONAL INTRODUCTION
      ACT/2,,ATRIB(1).EQ.2.,C1;;    COUNT FOR NATIONAL INTRODUCTION
C1    COLCT(7),FIRST,C1 CHOICE 1; INTRODUCE NATIONALLY/REGIONALLY
      TERM,1;
      END;
INIT,0,,N;
FIN;
```

Figure 6-14 SLAM II input statements of new product introduction decision.

a tie-breaking criterion is necessary to make the choice. For this example, it is assumed that it is preferred to go "national" if, at any time, the returns are equal. This is modeled in the SLAM II network by assigning a small return (0.00001 indicated by δ) to those branches representing the national choice. At each subsequent decision stage, this value is doubled to again break ties.

Conditional branching is used following each ACCUMULATE node to count the number of times each choice decision is made over multiple runs of the network. For example, following node ACM2, activity 3 is taken if the choice is to remain regional, given that the product was introduced regionally and a large regional demand occurred. Likewise, the branch for activity 4 represents a decision to go national, given that the product was first introduced regionally. Activity 1, following node ACM1, represents the choice decision to introduce the product regionally. The decision to introduce the product nationally is represented by activity 2. The number of times that activities 1, 2, 3, and 4 are completed is computed in subroutine OTPUT at the end of each simulation run. At the end of the last run, the counts for the activities representing the optimal decisions are printed, as shown in Figure 6-15.

ACTIVITY NUMBER	BEST DECISION COUNT	FRACTION BEST
1	436.	0.4360
2	564.	0.5640
3	127.	0.1270
4	572.	0.5720

Figure 6-15 Counts of optimal decisions.

In 1000 runs of the new product introduction, a higher return was achieved 436 times if the product was introduced regionally, whereas in 564 runs the optimal (or equivalent) decision was to introduce the product nationally. Thus, 56.4% of the time, the better choice is to introduce the product nationally. There were 699 runs in which choice decision 2 was made, and in 572 of the runs it was best to go national. This represents an 81.8% chance of obtaining a greater return when making the "go national" decision. These percentages are important because they specify the fraction of time that the highest possible return is obtained.†

Subroutine OTPUT is shown in Figure 6-16, where function NNCNT(NA) is used to determine the number of entities completing activity NA. The number of times that activity NA is the desirable choice is stored in COUNT(NA), and

† A theoretical (complete enumeration) analysis of this model indicates that 16% of the time the returns are equal for choice decision 2 and 22.3% of the time the returns are equal for choice decision 1.

```
      SUBROUTINE OTPUT
C
      COMMON/SCOM1/ATRIB(100),DD(100),DDL(100),DTNOW,II,MFA,MSTOP,NCLNR
     1,NCRDR,NPRNT,NNRUN,NNSET,NTAPE,SS(100),SSL(100),TNEXT,TNOW,XX(100)
C
      DIMENSION COUNT(4)
      DATA XRUNS/1000/
C
      DO 10 NA = 1,4
         COUNT(NA) = COUNT(NA) + NNCNT(NA)
   10 CONTINUE
C
      IF (NNRUN .GE. XRUNS) THEN
         WRITE(6,15)
         DO 20 NA = 1,4
            XCOUNT = COUNT(NA)/XRUNS
            WRITE(6,25) NA,COUNT(NA),XCOUNT
   20    CONTINUE
      ENDIF
      RETURN
C
   15 FORMAT(///22X,4HBEST/5X,8HACTIVITY,7X,8HDECISION,7X,8HFRACTION/
     16X,6HNUMBER,10X,5HCOUNT,10X,4HBEST//)
   25 FORMAT(I10,7X,F10.0,5X,F10.4)
C
      END
```

Figure 6-16 Subroutine OTPUT for SLAM II model of new product introduction.

the fraction of the times the decision is made over all runs is computed as XCOUNT. The SLAM II output reports for this example are shown in Figure 6-17. The statistics for the nodes of the network provide additional information. The values for nodes C4, C6, C8, and C10 provide estimates of the expected return, standard deviation, and the minimum and maximum values associated with each of the choice decisions. Using expected values, the choice decision 2, to remain regional or go national, is made by comparing the values of node C6 and node C8. Since the expected value for node C8 is higher (4.49 versus 3.19), the decision to go national would be made. From the analysis made in the preceding paragraph, the estimated probability of this being the best possible decision is 0.818.

The statistics associated with node C2 indicate that, if the correct decision is always made, the expected return from choice decision 2 is 4.78 with a standard deviation of the return of 1.25. The statistics for node C3 provide the expected return if choice decision 1 is to introduce regionally and the correct choice for decision 2 is always made. This value of 3.67 is higher than 3.55 as given in Figure 6-4 since the correct decision at node C2 is presumed. At node C1, the average return is estimated for introducing the new product. This value could be used to compare the expected return for this product to the

```
S L A M   I I   S U M M A R Y   R E P O R T

SIMULATION PROJECT PRODUCT DECISION            BY HAMMESFAHR

DATE  1/ 5/1988                                RUN NUMBER 1000 OF 1000

              CURRENT TIME   0.4000E+01
              STATISTICAL ARRAYS CLEARED AT TIME  0.0000E+00

              **STATISTICS FOR VARIABLES BASED ON OBSERVATION**
```

	MEAN VALUE	STANDARD DEVIATION	MINIMUM VALUE	MAXIMUM VALUE	NUMBER OF OBSERVATIONS
C4 SMALL REG DEMAND	0.1086E+01	0.6825E+00	0.0000E+00	0.2000E+01	301
C6 REMAIN REGIONAL	0.3187E+01	0.7761E+00	0.2000E+01	0.4000E+01	699
C8 GO NATIONAL	0.4492E+01	0.1609E+01	0.2000E+01	0.6000E+01	699
C2 CHOICE 2	0.4780E+01	0.1253E+01	0.2000E+01	0.6000E+01	699
C3 INTRO REGIONALLY	0.3668E+01	0.2027E+01	0.0000E+00	0.6000E+01	1000
C10 INTRO NATIONALLY	0.3245E+01	0.2827E+01	0.0000E+00	0.6000E+01	1000
C1 CHOICE 1	0.4849E+01	0.1827E+01	0.2000E-04	0.6000E+01	1000

Figure 6-17 Outputs from SLAM II model of new product introduction decisions.

expected returns for other products that might be under consideration. Thus, if the decision maker had complete information about each possible return associated with this product, the average expected return for the product would be 4.85 With this information, a budget can be established for research on better forecasting methods.

This example demonstrates the following key concepts associated with decision networks:

1. The decision process can be broken down into understandable elements, which leads to a comprehensive treatment of the choices available.
2. The expected return at a choice decision and the probability of making a correct choice are important variables in the decision process.
3. With discrete returns, equivalent choices (ties) can affect the percentage of times an alternative is selected.
4. The incremental return from increased information can be quantified.

6.7 RISK ASSESSMENT

Risk is a performance measure that is used to evaluate alternatives. ***Risk*** is usually defined as the expected value of an undesirable outcome, that is, the product of the magnitude of an undesirable outcome multiplied by the frequency of its occurrence. A special case that is often reported in the literature defines the magnitude of an undesirable outcome as unity. This results in risk being defined as the probability of occurrence of an undesirable outcome.

With decision trees and decision networks, terminating nodes define outcomes and the probability of releasing a terminating node provides an estimate of risk. When the magnitude of an undesirable outcome is included in the definition of risk, a SLAM II user function is employed to obtain the product of the magnitude and the frequency of occurrence of the undesirable outcome.† User-collected statistics provide for statistical estimates over multiple runs. Alternatively, average value computations can be obtained using networks similar to those presented in Section 6.2, where activity cost replaces activity duration on the network. If the risk analysis involves time delays as well as risk computations, a user function for computing average risk is required.

In some risk analyses, the undesirable outcome is defined in terms of time. For example, an undesirable outcome might be the nondelivery of an item by April 15. As another example, the undesirable outcome could be the completion of a product development by a competitor before a like product is developed. In these situations, risk is computed using SLAM II by defining two disjoint networks, each having a terminating node. In the network describing the product development, one terminating node represents the desirable outcome. In the other network, the TERMINATE node represents the undesirable outcome. In multiple runs, the fraction of runs in which the undesirable terminating node is reached represents the risk. If a value can be associated with this node, the product of the probability and the value can be formed and a generalized risk value computed.

There is a close association between risk analysis and project planning, since both are concerned with the probabilities and time associated with success and failure. The main difference is that a risk analysis is oriented to the development of plans and their assessment. The use of SLAM II for project planning is described in Chapters 7 and 8.

6.8 SUMMARY

This chapter presents concepts with respect to decision modeling and risk analysis. The use of cost as the additive parameter, instead of time, is described. The methodology associated with decision trees, decision networks, and risk analysis is presented, and examples are given where appropriate. Procedures are described to estimate the value of complete information in a decision situation. Such procedures permit the assessment of the worth of information gathering functions. Throughout the chapter, examples are given that demonstrate the communication and analysis capabilities of a network approach to decision and risk analysis.

† Standard terminology in risk analysis would refer to an undesirable event.

6.9 EXERCISES

6-1. Formulate the various principles of choice described in Section 6-1 in mathematical form.

Embellishment: If the outcome from an alternative future state presented in Figure 6-1 is a pointer to another decision framework, show how the mathematical forms derived in this exercise would be altered.

6-2. Express the decision to buy a new piece of machinery in the decision framework presented in Figure 6-1; that is, provide specific definitions for each component of the decision framework.

6-3. Consulting firms are constantly bidding on contracts by preparing proposals. Define the alternative and possible future states for a consulting firm with regard to the bidding to obtain a contract. Specify and discuss the principle of choice that you think is used by most consulting firms.

6-4. Build a SLAM II network to estimate the distribution of the cost of performing three consecutive projects, where the estimated costs for the first project are normally distributed with a mean of $10,000 and a standard deviation of $1000; the cost of the second project is exponentially distributed with a mean of $9000; and the cost of the third project is triangularly distributed with a modal value of $16,000, a minimum of $12,000, and a maximum of $25,000.

6-5. Convert a decision situation with which you are familiar into a SLAM II network model and run the model to obtain the values of the outcomes.

6-6. Explain in your own words the difference between the decision-making model presented in Figure 6-4 and the model representation shown in Figure 6-6.

6-7. An oil company needs to decide whether or not to drill at a given location before its option expires [12]. There are many uncertainties, such as cost of drilling and the value of the deposit. Records are available for other drillings in the area. More information can be obtained about the geophysical structure at the site by conducting a sounding. The following information is available. Drilling costs are estimated at $1,000,000 and might lead to payoffs (deposit values) that can be classified as follows:

Big	$4,000,000
Moderate	2,000,000
Small	1,250,000

If a sounding is not taken, it is estimated that the probability of a big payoff is 20%, a moderate payoff is 50%, and a small payoff is 30%. A sounding costs $150,000 and the probabilities of the various sounding results are: big, 0.25; moderate, 0.50; and small, 0.25. From past experiences, the conditional probabilities of an actual result given a sounding result are as follows:

Sounding	Actual Result		
Result	Big	Moderate	Small
Big	0.60	0.30	0.10
Moderate	0.20	0.70	0.10
Small	0.10	0.30	0.60

(a) Perform a standard decision tree analysis of the decision problem.
(b) Perform a SLAM II analysis of the decision problem to obtain the distribution of payoffs.
(c) Determine the worth of increasing the diagonal probabilities by 0.10.

6.10 REFERENCES

1. Austin, L. M., and J. R. Burns, *Management Science: An Aid for Managerial Decision-Making*, Macmillan, New York,1985.
2. Eisner, H., *Computer-Aided Systems Engineering*, Prentice-Hall, Englewood Cliffs, NJ, 1987.
3. Emshoff, J. R., and R. L. Sisson, *Design and Use of Computer Simulation Models*, Macmillan, New York, 1970.
4. Hespos, R. F., and P. A. Strassman, "Stochastic Decision Trees for the Analysis of Investment Decisions," *Management Science*, Vol. 11, No. 10, August 1965, pp. B244-259.
5. Keeney, R. L., and H. Raifa, *Decision With Multiple Objectives: Preferences and Value Trade-offs*, Wiley, New York, 1976.
6. Moore, L. J., D. F. Scott, and E. R. Clayton, "GERT Analysis of Stochastic Systems," *Akron Business and Economic Review*, 1974, pp. 14-19.
7. Morris, W. T., *Analysis of Management Decision Making*, Richard D. Irwin, Chicago, IL, 1964.
8. Pritsker, A. A. B., and C. E. Sigal, *The GERT IIIZ User's Manual,* Pritsker & Associates, West Lafayette, IN, 1974.
9. Radford, K. J., *Complex Decision Problems: An Integrated Strategy for Resolution*, Reston Publishing, Reston, VA, 1977.
10. Sigal, C. E., "Stochastic Shortest Route Problems," Ph.D. dissertation, Purdue University, West Lafayette, IN, December 1977.
11. Watters, L. J., and M. J. Vasilik, "A Stochastic Network Approach to Test and Checkout," *Proceedings, Fourth Conference of Applicaton of Simulation*, 1970, pp. 113-123.
12. Whitehouse, G. E., *Systems Analysis and Design Using Network Techniques*, Prentice-Hall, Englewood Cliffs, NJ, 1973.

7 PROJECT PLANNING: BASIC CONCEPTS

7.1 INTRODUCTION

Larger and larger projects are being conceived, designed, and developed. As projects become larger, organizational and managerial problems increase. Individual efforts are no longer feasible, and teams of workers, machines, and materials must be integrated to plan, organize, and manage today's systems. This tendency toward larger projects results in the need for systems engineers and managers to plan, direct, and control the activities and resources allocated to the projects. A great deal of forethought is required to plan a large-scale endeavor. A major deterrent to good planning is the communication of ideas and concepts that are initially nebulous and possibly ill formed. Out of these concerns, graphical representations of projects for planning purposes have developed.

Many textbooks and monographs describe the fundamentals of project planning and management [2, 4, 9, 16]. Wiest and Levy [16] give an excellent historical perspective of both the organizational and technical aspects involved in project planning. The Project Management Institute (PMI) is the professional society for individuals whose main field of interest is project planning. The project planning field has matured over the years, and the use of project management techniques for the planning of construction projects, space vehicle developments, manufacturing systems, transportation conveyances, and the like is well documented.

The first techniques developed for project management were PERT (Program Evaluation and Review Technique) and CPM (Critical Path Method). These network techniques provide a communication vehicle for describing large projects in a network form. However, the modeling versatility of these techniques is limited. In addition, their analysis procedures are inaccurate because of the many assumptions required when attempting to compute, analytically, measures of performance associated with project planning[7]. GERT and Q-GERT were developed to extend the modeling capability of PERT and CPM [13, 15], and these techniques provide the basis for the network concepts of SLAM II.

In this chapter, emphasis is placed on analyzing networks for project planning purposes and on demonstrating the SLAM II modeling and analysis capabilities that are beyond those provided in the PERT and CPM techniques.

7.1.1 Project Planning Terminology and Performance Measures

Although network models are discussed throughout this book, the network terminology specific to project planning is defined in this section. The fundamental concept in project planning is an activity. An activity is any portion of a project that requires time, or resource elements, such as labor, paper work, and machines. Commonly used terms synonymous with an activity are a task and a function. Typically, activities are graphically represented by branches. When activities are represented by branches, project planning models are referred to as activity-on-branch networks.†

In project planning, an event is an instantaneous point in time that marks the beginning or ending of an activity. In networks, nodes are used to represent events. Precedence between activities means that the ending event for one activity must occur before the starting event for another activity. A precedence relation is shown in network form by two activities in series separated by a node. The precedence activity is said to be incident to the node, and the activity that requires the precedence activity is said to emanate from the node. Every activity emanating from a node requires each activity incident to that node to be completed before the activity may be started.

The type of precedence described in the preceding paragraph is referred to as a direct or immediate precedence. Indirect precedence is modeled by having a sequence of activities, that is, a path segment. Thus, if activity 4 has activity 3 as a precedence activity, and activity 3 has activity 2 as a precedence activity, activity 4 has activity 2 indirectly as a precedence activity. Such indirect precedence relations are inherent in a network model. Fundamentally, a network is a graphical representation of a project plan that shows the immediate and indirect precedence relations among all the activities defined for the project.

A list of all activities on a project is referred to as a work breakdown structure. The work breakdown structure organizes the activities in tabular form and defines the activities in terms of the time required to perform the activity, the

† SLAM II employs an activity-on-branch representation. When activities are represented by nodes, or blocks, the models are referred to as activity-on-node networks or as precedence diagrams. In activity-on-node networks, branches are used only to show the precedence between activities.

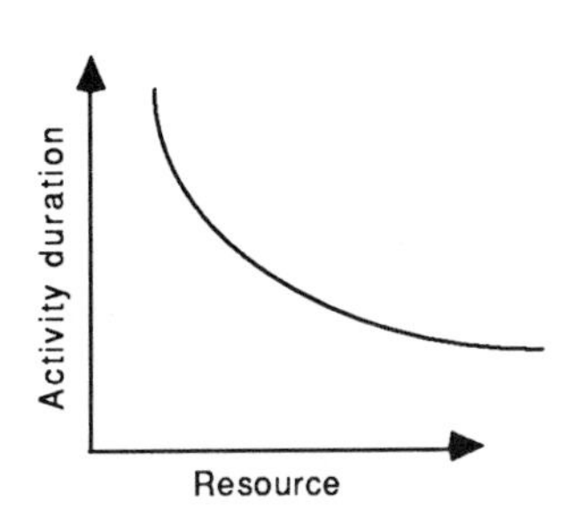

Figure 7-1 Tradeoff curve.

cost of performing the activity, and any associated resources required to perform the activity. For advanced network modeling, a probability that the activity will need to be performed is included. Also, for advanced project planning models, the change in duration due to increased resources allocated to the activity is described. A representative activity duration-resource allocation curve is shown in Figure 7-1 which illustrates that shorter activity durations are obtained through increased resource allocations. Typical resources associated with an activity are workers, materials, machines, and money.

To illustrate project planning terminology, refer to the simplified network in Figure 7-2. For this network, there are five nodes and seven activities. Directly from the network, it is seen that activity 1 must precede activities 3 and 4. Also, the precedence activities to activity 6 are activities 2, 3, and 5. Clearly, activities 1 and 4 must also precede activity 6, even though they are not incident to node 3. The event time associated with node 3 is the largest of the completion times of the activities incident to node 3. For PERT and CPM networks, it is required that the number of release requirements for a node be equal to the number of incident activities.

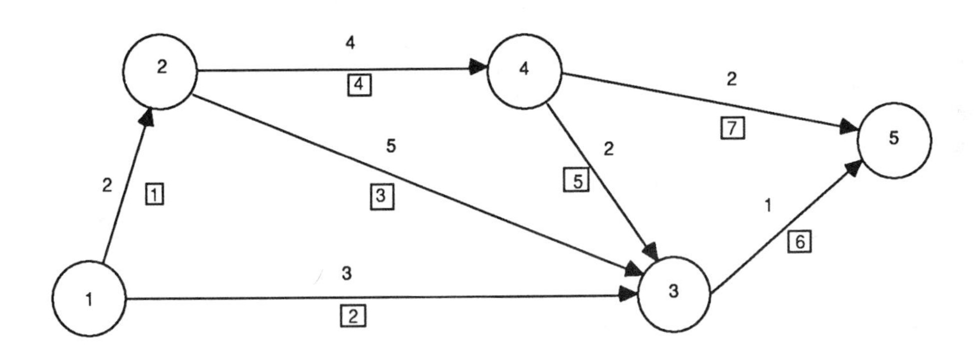

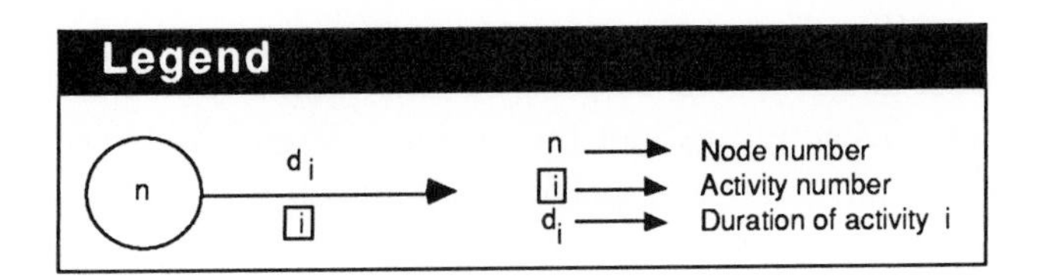

Figure 7-2 Illustrative PERT project planning network.

Another way to display the project is to use a Gantt chart. In a Gantt chart, each activity is displayed on a time line, with the start time of an activity restricted to being after the completion of all precedence activities. The Gantt chart for the network of Figure 7-2 is shown in Figure 7-3. For convenience, the start and end node numbers have been placed on the Gantt chart in circles, and the activity numbers have been shown in squares. The Gantt chart shows the time phasing of the activities, but does not provide an adequate display of precedence. From the Gantt chart, it can be determined that the project duration is nine time units. Using this example, performance measures associated with project planning are defined next.

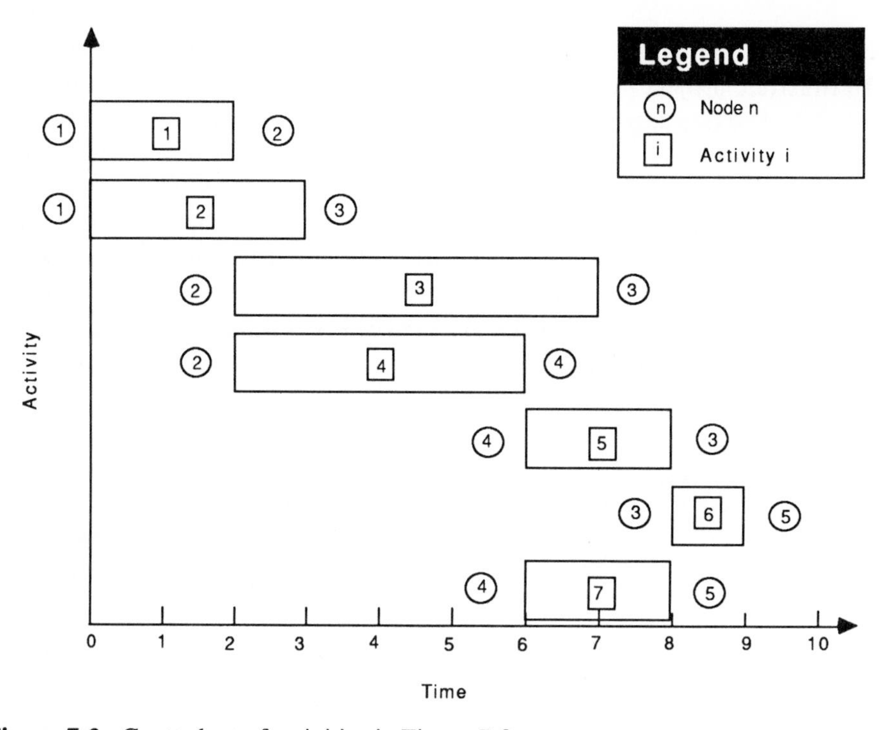

Figure 7-3 Gantt chart of activities in Figure 7-2.

The project completion time is typically the most important measure of project success. If activity cost and resource use are assumed as constant, the total project cost and resource use can be determined from information on the completion times of activities and the project. If cost and resource use are a function of when an activity is performed, or how an activity is performed, additional performance measures associated with cost and resource use need to be computed.

A delay in the start of an activity may not delay the project. From the Gantt chart, the start time of activity 7 can be delayed one time unit without increasing the project duration. This leeway is called slack, or float, in project planning. Different definitions of slack and the use of slack in project scheduling are discussed in Chapter 9.

A sequence of activities from the starting node to the ending node is called a ***path***. When activity durations are constants, there exists at least one path such that each activity on the path has no slack. Such a path is referred to as a ***critical path***. The critical path method is a procedure for identifying a critical path and for computing slack times for activities not on the critical path. The PERT technique may be used to perform such an analysis.

In some project planning studies, significant events are defined, which are referred to as ***milestones***. A milestone is included on the network by a node, and the time that the milestone occurs is also a measure of project performance.

7.2 GRAPHICAL REPRESENTATION OF PROJECTS

The network of a repair project shown in Figure 7-4 is a typical graphical project representation. At the beginning of the project, three parallel activities are performed that involve the disassembly of power units and instrumentation [1], the installation of a new assembly [2], and the preparation for a repair check [3]. Cleaning, inspection, and repairing the power units [4] and calibrating the instrumentation [5] may be done only after the power units and instrumentation have been disassembled. Thus, activities 4 and 5 must follow activity 1 in the network. Following the installation of the new assembly [2] and after the instruments have been calibrated [5], a check of interfaces [6] and a check of the new assembly [7] can be made. The repair check [9] can be made after both the assembly is checked [7] and the preparation for the repair check [3] have been completed. The assembly and test of power units [8] can be performed following the cleaning and maintenance of the power units [4]. The project is considered completed when all nine activities are completed. Since activities 6, 8, and 9 have no successor activities, the completion of all three of these activities signifies the end of the project. This is indicated on the network by having activities 6, 8, and 9 incident to node 6, the terminating node for the project.

One of the first steps in project planning is to construct a network, as has been done for the example in Figure 7-4. Construction of the network requires the activities of the project to be defined and, typically, organized into work elements, that is, a work breakdown structure. The network is a graphical portrayal of the activities, including precedence relations, and provides a vehicle for communication during planning and analysis. The next step is to direct attention to the time durations of the activities. Data are collected and each

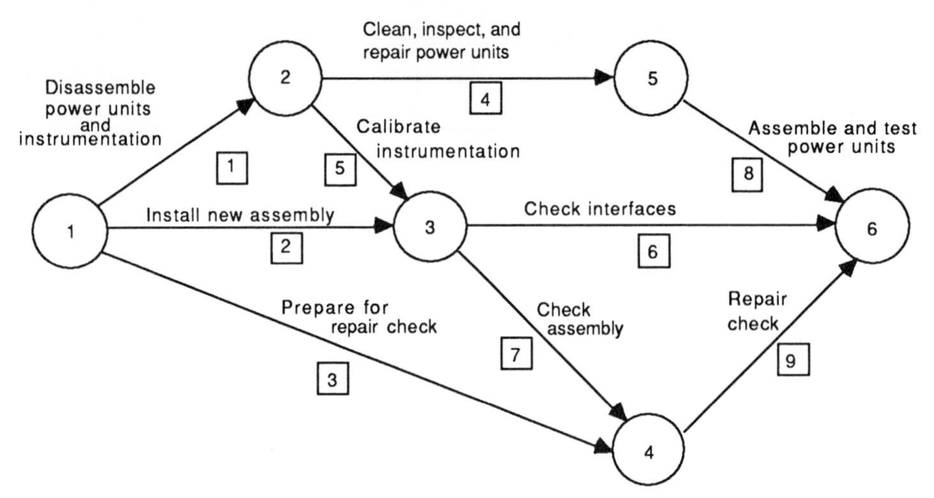

Figure 7-4 Network model of a repair project.

activity duration is quantitatively described. Frequently, a constant duration or a set of three values (most likely, optimistic, and pessimistic duration) is used for such a description. Guidelines for the selection of an appropriate distribution function have been developed [13], and programs for fitting distributions to data sets are available [5,11].†

To illustrate a simulation analysis for the repair project described in Figure 7-4, a corresponding SLAM II network is shown in Figure 7-5. A single entity is generated at the CREATE node to initiate the project. This insures that each activity on the network is only traversed once during any simulation run. Furthermore, each node is released when the number of incoming entities equals the number of incoming activities (branches). These conditions allow each node to be is released just once, and when this occurs, all its preceding activities have been completed.

Activities 1, 2, and 3 are scheduled at time 0 since they emanate from the CREATE node. The time duration for each activity is assumed to be characterized by a triangular distribution (TRIAG). The three parameters for the TRIAG function are given on the branch. Hence, the optimistic, expected, and pessimistic values for activity 1 are 1, 3, and 5, respectively. At the completion of activity 1, the time of the first release (FIRST statistics) is collected at COLCT node N2 and activities 4 and 5 are initiated. ACCUMULATE node N3 requires two incoming entities before it is released, and it waits to be released until both activity 2 and activity 5 are completed. When this occurs, the following COLCT

† Early in the design process, the characterization of time durations may be subjectively determined by an analyst rather than by an analysis of data. This facilitates the modeling process and helps identify total data collection needs. However, characterizations established in this manner must be verified.

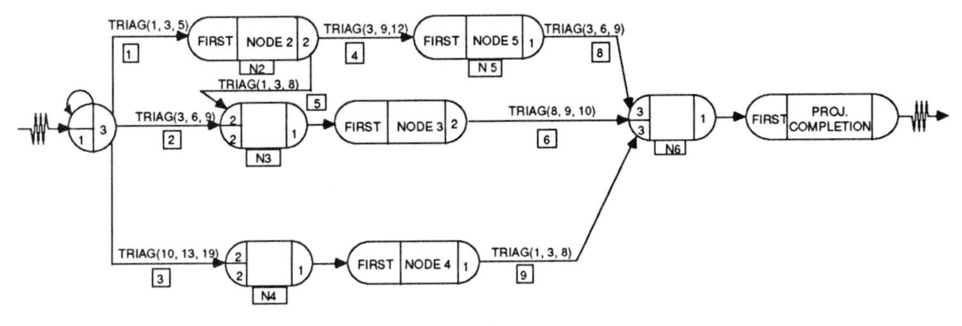

Figure 7-5 SLAM II network of repair project.

node records the time of the release of node N3 and then starts activities 6 and 7. Likewise, node N4 is not released until activities 3 and 7 are completed. The time of release for node N4 is recorded at its following COLCT node and activity 9 is started. The time of completion of activity 4 is recorded at COLCT node N5, after which activity 8 is initiated. When activities 6, 8, and 9 are completed, ACCUMULATE node N6 is released, and the time required to complete the project is recorded by a COLCT node at the end of the network. In this way, statistics on the release of node 6 estimate the project duration.

7.3 PROJECT COMPLETION TIME DISTRIBUTION

In project planning, a project's duration time is the most important network measure. Referring to the example network in Figure 7-5, the completion time of the project corresponds to the time that ACCUMULATE node N6 is released. Since each network activity is a random variable, the time to release each event node is also a random variable. To compute statistics on EVENT node release times, the SLAM II modeler includes COLCT nodes observing FIRST statistics immediately following each ACCUMULATE node. This is shown in the statement model given in Figure 7-6.

Figure 7-7 illustrates the SLAM II output for 400 simulated observations for project completion time. On the average, the project is completed in 20.68 days. The shortest completion time in 400 observations is 15.66 days, and the longest is 27.05 days. The standard deviation is 2.103 days. The standard error of the mean project completion times is estimated by dividing the standard deviation by the square root of the number of observations, that is, 2.103/20 or 0.11 day. Therefore, a 95% confidence interval for the mean project completion time (approximately two standard errors from the mean, assuming a normal distribution) is 20.46 to 20.90 days. This confidence interval has a probability of 0.95 of containing the true mean of the network model's completion time.

```
GEN,OREILLY,PERT NETWORK,3/15/1982,400,,NO,,NO,YES/400;
;
;     PERFORM 400 ITERATIONS
;
LIM,,1,700;
NETWORK;
        CREATE,,,,1,3;
        ACT/1,TRIAG(1.,3.,5.,1),,N2;      DISASSEMBLE
        ACT/2,TRIAG(3.,6.,9.,2),,N3;      INSTALL
        ACT/3,TRIAG(10.,13.,19.,3),,N4;   PREPARE
N2      COLCT,FIRST,NODE 2,10,0.0,0.5;
        ACT/4,TRIAG(3.,9.,12.,4),,N5;     INSPECT & REPAIR
        ACT/5,TRIAG(1.,3.,8.,5),,N3;      CALIBRATE
N3      ACCUM,2,2;
        COLCT,FIRST,NODE 3,20,3,0.5;
        ACT/6,TRIAG(8.,9.,16.,6),,N6;     INTERFACES
        ACT/7,TRIAG(4.,7.,13.,7),,N4;     ASSEMBLY
N4      ACCUM,2,2;
        COLCT,FIRST,NODE 4,20,10.,0.5;
        ACT/9,TRIAG(1.,3.,8.,9),,N6;      REPAIR
N5      COLCT,FIRST,NODE 5,20,12.,0.5;
        ACT/8,TRIAG(3.,6.,9.,8),,N6;      ASSEMBLY
N6      ACCUM,3,3;
        COLCT,FIRST,PROJ. COMPLETION,20,15.,0.5;
        TERM;
        ENDNETWORK;
INIT,,,NO;
FIN;
```

Figure 7-6 SLAM II statement model for repair project.

Referring to the histogram in Figure 7-8, several interpretive statements can be made about the project completion time based on the 400 network runs. The far right column gives the upper bound of each cell in the histogram. Observed relative and cumulative frequencies for each cell are given in tabular form. Relative frequency is plotted with an asterisk. The letter C is used to plot cumulative frequency. Referring to the tenth cell of the histogram, it is noted

```
                    S L A M   I I   S U M M A R Y   R E P O R T

              SIMULATION PROJECT PERT NETWORK              BY OREILLY

              DATE  3/15/1982                              RUN NUMBER  400 OF  400

              CURRENT TIME    0.1906E+02
              STATISTICAL ARRAYS CLEARED AT TIME   0.0000E+00

                        **STATISTICS FOR VARIABLES BASED ON OBSERVATION**
                       MEAN         STANDARD        MINIMUM        MAXIMUM        NUMBER OF
                       VALUE        DEVIATION       VALUE          VALUE         OBSERVATIONS

NODE 2               0.3029E+01    0.7893E+00     0.1160E+01     0.4821E+01         400
NODE 3               0.7377E+01    0.1408E+01     0.4092E+01     0.1151E+02         400
NODE 4               0.1601E+02    0.1971E+01     0.1101E+02     0.2116E+02         400
NODE 5               0.1123E+02    0.1942E+01     0.5404E+01     0.1582E+02         400
PROJ. COMPLETION     0.2068E+02    0.2103E+01     0.1566E+02     0.2705E+02         400
```

Figure 7-7 Final summary statistics for 400 simulation runs of repair project.

```
                              **HISTOGRAM NUMBER  5**
                                 PROJ. COMPLETION

OBSV    RELA    CUML         UPPER
FREQ    FREQ    FREQ      CELL LIMIT      0        20        40        60        80       100
                                       +    +    +    +    +    +    +    +    +    +    +
0       0.000   0.000     0.1500E+02    +                                                  +
0       0.000   0.000     0.1550E+02    +                                                  +
1       0.002   0.002     0.1600E+02    +                                                  +
5       0.013   0.015     0.1650E+02    +*                                                 +
6       0.015   0.030     0.1700E+02    +*C                                                +
6       0.015   0.045     0.1750E+02    +*C                                                +
14      0.035   0.080     0.1800E+02    +** C                                              +
29      0.072   0.153     0.1850E+02    +****   C                                          +
30      0.075   0.228     0.1900E+02    +****     C                                        +
37      0.093   0.320     0.1950E+02    +*****        C                                    +
37      0.093   0.413     0.2000E+02    +*****            C                                +
31      0.078   0.490     0.2050E+02    +****                C                             +
40      0.100   0.590     0.2100E+02    +*****                   C                         +
33      0.083   0.673     0.2150E+02    +****                       C                      +
23      0.058   0.730     0.2200E+02    +***                          C                    +
19      0.047   0.778     0.2250E+02    +**                             C                  +
32      0.080   0.858     0.2300E+02    +****                              C               +
21      0.052   0.910     0.2350E+02    +***                                   C           +
9       0.023   0.933     0.2400E+02    +*                                      C          +
9       0.023   0.955     0.2450E+02    +*                                         C       +
4       0.010   0.965     0.2500E+02    +*                                         C       +
14      0.035   1.000        INF        +**                                                C
                                       +    +    +    +    +    +    +    +    +    +    +
400                                     0        20        40        60        80       100


                    **STATISTICS FOR VARIABLES BASED ON OBSERVATION**

                        MEAN          STANDARD        MINIMUM        MAXIMUM       NUMBER OF
                        VALUE         DEVIATION       VALUE          VALUE         OBSERVATIONS

PROJ. COMPLETION     0.2068E+02     0.2103E+01     0.1566E+02     0.2705E+02         400
```

Figure 7-8 Statistics and histogram for project completion time distribution.

that of the 400 network realizations there are 37 project completion times greater than 19 days, but less than or equal to 19.5 days. The relative frequency of this observation is 0.093, or 9.3%. The cumulative frequency associated with 19.5 days is 0.320. Based on this cumulative frequency, the estimated probability of the project being completed within 19.5 days is 0.320. The estimated probability of the project taking more then 19.5 days is 1 - 0.320 = 0.680. Other probability estimates can be obtained for any of the values listed in the histogram.

7.4 STATISTICS ON ACTIVITY START TIMES

The release of any of the network nodes corresponds to the start of the activities emanating from its following COLCT node. The output for all the COLCT nodes is shown in Figure 7-7. The statistics collected and the interpretive statements that can be made for each node are similar to those described for the terminating COLCT node labeled PROJ. COMPLETION. These statistics provide valuable scheduling information on the probable start dates of the project's activities.

From the project network in Figure 7-5, the release of node N2 corresponds to the initiation of power unit cleaning [4] and instrumentation calibration [5]. The average start date is 3.029. The earliest start date is 1.160, and the latest start date is 4.821. Scheduling in the light of this uncertainty is difficult for a project manager. Therefore, the manager must be prepared for change and have contingency plans ready. The degree of uncertainty in project planning is quantified by the start-time statistics presented in Figure 7-7. Further analysis of start times and finish times is presented in Chapter 9.

7.5 STATISTICS ON PROJECT INTERVALS

It is often of interest to estimate the time lapse between two project milestones. For example, in the repair project, suppose that information is desired on the elapsed time between the completion of the power unit disassembly 1 and the start of the repair check 9. This time may be estimated through the use of interval statistics. To collect the desired statistics for this situation, the network in Figure 7-5 requires the following modifications:

1. Attribute 1 is initially marked at the CREATE node for each run. Since each run starts at time zero, ATRIB(1) = 0.
2. An ASSIGN node is inserted between node N2 and activity 5 to set attribute 1 equal to the current simulation time, TNOW, when an entity passes through the ASSIGN node.
3. The SAVE specification for ACCUMULATE nodes N3 and N4 is set to HIGH(1) to save the value of ATRIB(1) set in the ASSIGN node.
4. An additional COLCT node is added after node N4 to collect the interval statistics based on ATRIB(1) using INT(1).

Each time the COLCT node collecting interval statistics is released, the elapsed time is recorded corresponding to the interval of time between the completion of activity 1 and the start of activity 9. A partial network that includes the network modifications described is shown in Figure 7-9.

The SAVE specification for an ACCUMULATE node can affect the observations when collecting interval statistics in project networks. For example, in Figure 7-9 the values collected at the COLCT node C4 depend on the path of the entity whose attributes are saved at node N4. With a HIGH(1) specification, the attributes of the entity that took the longest time to reach node N4 are saved. If the SAVE specification at node N4 is LOW(1), then the attributes of the entity

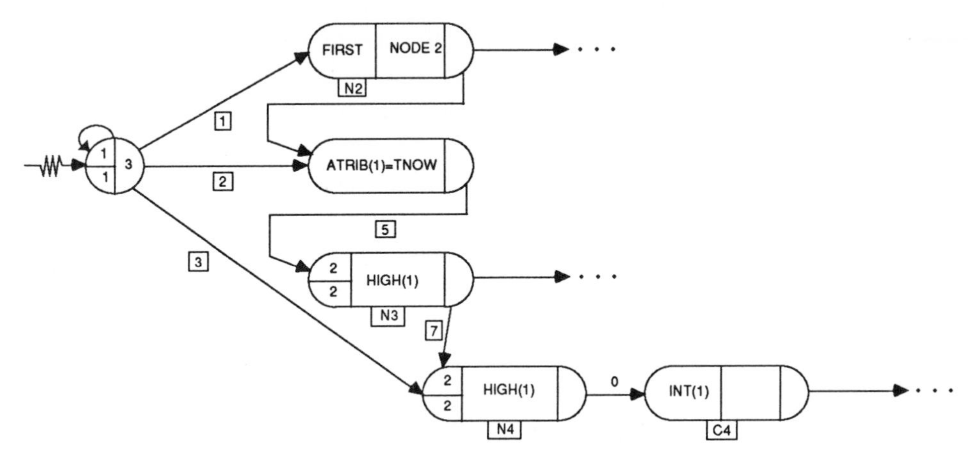

Figure 7-9 Illustration of marking and interval statistics.

with the smallest path transition time from the CREATE node to node N4 are saved. This situation clearly indicates that care must be exercised in the specification of the SAVE criterion for the ACCUMULATE nodes and the placement of the corresponding COLCT nodes when modeling projects and collecting interval statistics.

Multiple-interval statistics can be obtained by using additional attributes. For example, if an additional ASSIGN node is placed immediately following node N3 assigning TNOW to ATRIB(2), the SAVE specification for ACCUMULATE node N6 set to HIGH(2), and a COLCT node placed at the end of the network in Figure 7-5 to collect INT(2) observations, then interval statistics on the elapsed time from the release of node N3 until the end of the project would be reported on the SLAM II summary report.

7.6 CRITICALITY INDEXES

As defined in Section 7.1, a critical path is the set of activities that have no slack. In a project network, activities on the ***critical path*** are denoted as ***critical activities***. From a project management standpoint, the progress of critical activities should be monitored carefully if delays in meeting project due dates are to be avoided. Alternatively, identification of these activities provides a means for evaluating which activities to expedite in order to expedite the total project.

The concept of a single critical path, as assumed in PERT analysis, is inadequate for networks with random activity durations. Current practice is to allocate additional resources to critical activities to make the critical path duration similar to other path durations. Repeated allocations of this type cause a network's duration to be based on multiple paths and less related to a single critical path. Therefore, the critical path and the set of critical activities may vary from network realization to network realization. Thus, the relative frequency, or probability, that an activity is on the critical path is important. This probability measure is called a ***criticality index*** [14].

In simulating a project network, an entity is routed over each parallel activity. When milestones occur, the last entity arriving at a node represents the activity that is critical for the milestone. To accomplish the storing of potential critical activities, an attribute of the entity traversing an activity is assigned the activity number of the activity. ATRIB(I) is employed for this purpose where the value of I depends on the number of previous activities that the entity has passed over. ATRIB(1) is used to maintain the number of previous activities. In an EVENT node before each activity, the value of ATRIB(1) is increased by 1, I is set to ATRIB(1), and ATRIB(I) is given the activity number of the next activity started. The initial value of ATRIB(1) is zero. By using the SAVE criterion LAST at each ACCUMULATE node in the network, the attributes of the entity that released the ACCUMULATE node are maintained.

7.7 EXAMPLE OF CRITICALITY INDEX ESTIMATION USING SLAM II

In this section, the repair project example is used to illustrate the computation of criticality indexes using SLAM II. For convenience, the network in Figure 7-4 is presented again as Figure 7-10.

The SLAM II network model to obtain criticality indexes for the repair project is shown in Figure 7-11. An EVENT node precedes each activity. As was done previously, ACCUMULATE nodes are used where multiple activities converge, and the number of releases required for these nodes is set equal to the number of incoming branches. The SAVE criterion for each ACCUMULATE node is specified as LAST so that the attributes of the last entity arriving at the node are given to entities routed from the node. One run of the project is completed after COLCT node N6 has been released where FIRST statistics are collected.

At the start of a run, attribute 1 is set to 1 at an ASSIGN node that follows the CREATE node. The created entity is transformed into three entities following the ASSIGN node by having an M-number equal to 3. Each entity is routed to an EVENT node which causes subroutine EVENT(NXTACT) to be called. In EVENT (NXTACT), ATRIB(2) is assigned the event code NXTACT, which, by design, is also the number of the next activity that the entity is beginning.

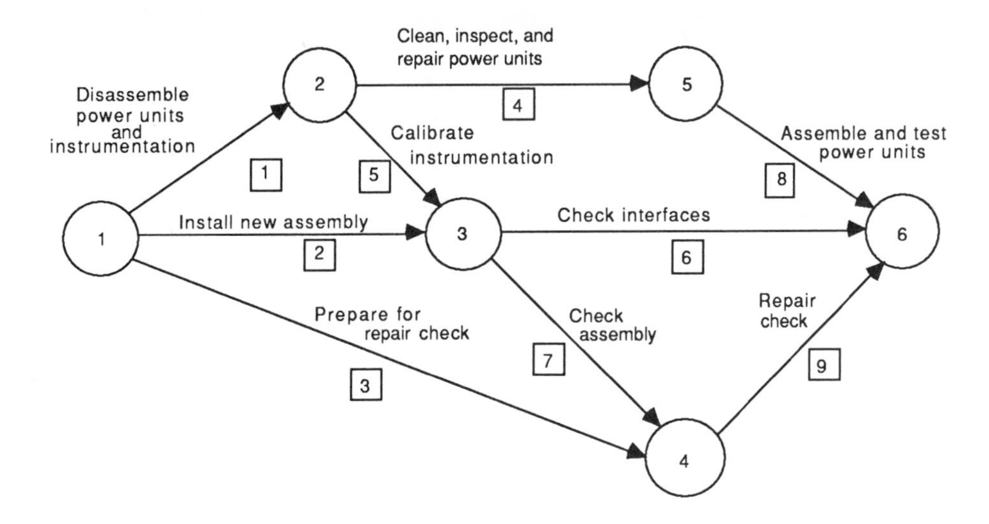

Figure 7-10 PERT network representation of a repair project.

Subroutine EVENT is shown in Figure 7–12. If NXTACT is less than 10, then a next activity is initiated and the updating of the ATRIB vector is performed. If NXTACT is 10, then the project is completed, and the attributes of the entity that released the last milestone, node A689, are used to identify the critical path.† The number of times an activity is critical is stored in the vector XCRIT.

Ten thousand runs of the network were requested on the GEN statement. After all runs are completed, the value of the variable XCRIT(I) is the number of times activity I is observed to be on a critical path. After the final run, the criticality index for each activity is computed as the number of runs the activity is on a critical path divided by the number of runs made (10,000). The activity number and criticality index are then printed out.

The output of the criticality indexes for each activity of the network is shown in Figure 7-13. The criticality index for activity 1 is 0.6016, indicating that on 6016 of the 10,000 runs it was on a critical path. Based on the values in Figure 7-13, every activity is critical at least once. To decrease the project duration, the activities with the highest criticality indexes should be examined for possible improvement. Improvement in these activities will not necessarily decrease the project duration, however, since there is a (1 - criticality index) probability that the activity will not be on a critical path for any one run of the project.

† Since the model presented here is for illustrative purposes, ties are not considered.

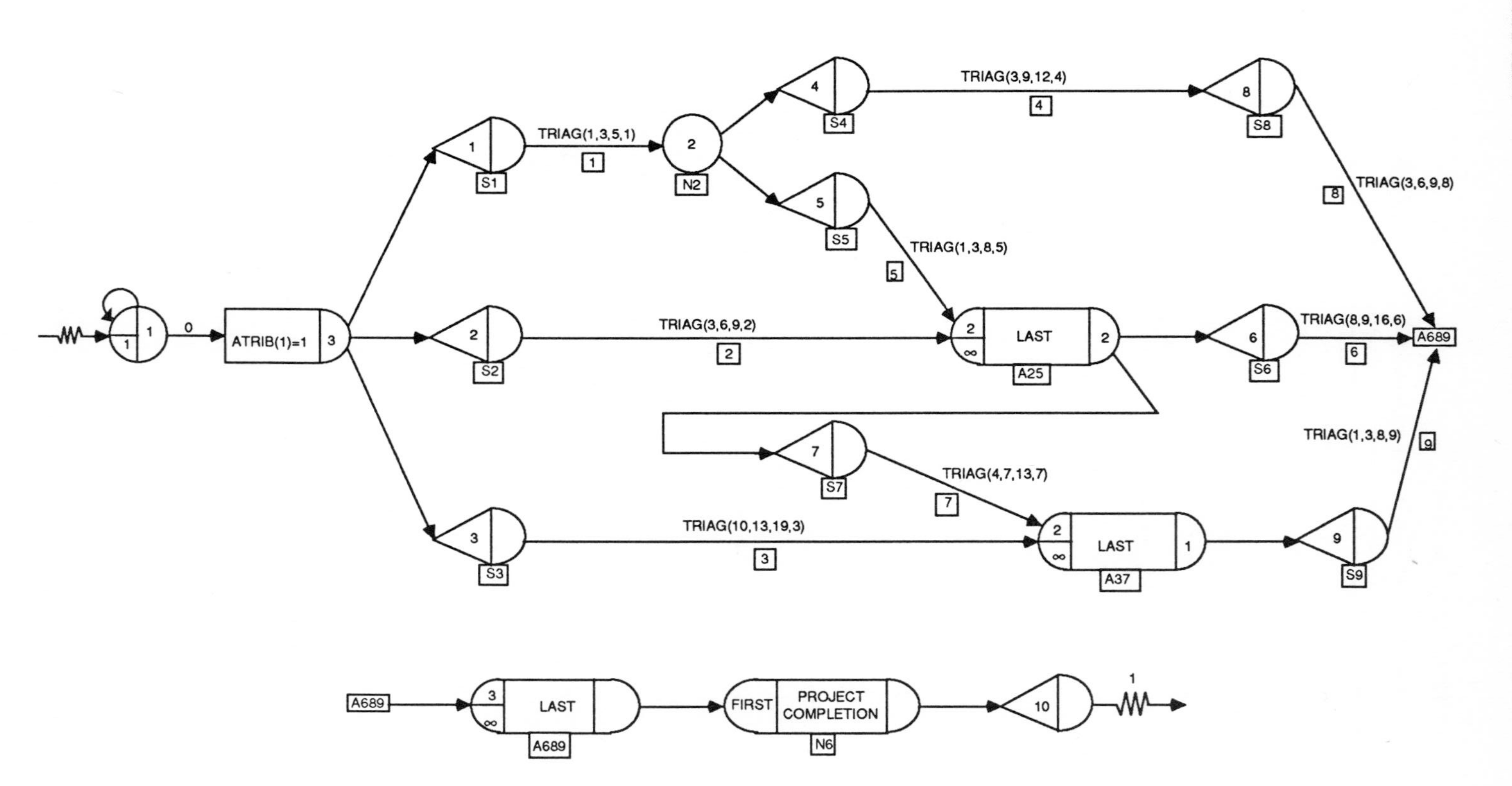

Figure 7-11 SLAM II network to obtain criticality indexes.

```
      SUBROUTINE EVENT (NXTACT)
C
      INTEGER NXTACT
C
      COMMON/SCOM1/ATRIB(100),DD(100),DDL(100),DTNOW,II,MFA,MSTOP,NCLNR
     1,NCRDR,NPRNT,NNRUN,NNSET,NTAPE,SS(100),SSL(100),TNEXT,TNOW,XX(100)
C
      DIMENSION XCRIT(9)
      INTEGER I,INUM
      REAL YCRIT
      INTEGER NEXT
      DATA XRUNS /10000/
      DATA (XCRIT(I),I=1,9) /9*0.0/
C
      IF (NXTACT .LT. 10) THEN
         ATRIB(1) = ATRIB(1)+1
         NEXT = ATRIB(1)
         ATRIB(NEXT) = NXTACT
      ELSE
C****    ADD TO NUMBER OF TIMES AN ACTIVITY IS CRITICAL
         IMAX = ATRIB(1)
         DO 10 I = 2,IMAX
            INUM = ATRIB(I)
            XCRIT(INUM) = XCRIT(INUM)+1.0
   10    CONTINUE
C
         IF (NNRUN .GE. XRUNS) THEN
            DO 30 I = 1,9
               YCRIT = XCRIT(I)/XRUNS
               WRITE(6,100) I,YCRIT
   30       CONTINUE
         ENDIF
      ENDIF
C
      RETURN
C
  100 FORMAT(15X,35H THE CRITICALITY INDEX FOR ACTIVITY,I4
     1,3H IS,F10.6)
      END
```

Figure 7-12 Subroutine EVENT to estimate criticality indexes.

```
               THE CRITICALITY INDEX FOR ACTIVITY   1 IS  0.601600
               THE CRITICALITY INDEX FOR ACTIVITY   2 IS  0.205100
               THE CRITICALITY INDEX FOR ACTIVITY   3 IS  0.193300
               THE CRITICALITY INDEX FOR ACTIVITY   4 IS  0.130900
               THE CRITICALITY INDEX FOR ACTIVITY   5 IS  0.470700
               THE CRITICALITY INDEX FOR ACTIVITY   6 IS  0.241900
               THE CRITICALITY INDEX FOR ACTIVITY   7 IS  0.433900
               THE CRITICALITY INDEX FOR ACTIVITY   8 IS  0.130900
               THE CRITICALITY INDEX FOR ACTIVITY   9 IS  0.627200
```

Figure 7-13 Criticality indexes for activities in the repair project.

7.8 PROJECT COST ESTIMATION

Suppose that each activity in the repair project example has an associated cost, consisting of a setup cost and variable cost, that is related to the activity time. In this situation, it is of interest to compute the total project cost. This total cost is a random variable, as it depends on activity times that are random variables. In this situation, it might be desirable to obtain statistics and a histogram on project cost as well as project completion time. With cost statistics, answers could be formulated to the following management questions:

1. What is the expected project cost?
2. What is a cost range such that the expected project cost is within the range 95% of the time?
3. What is the probability that the cost will exceed a specified number of dollars?

Applications are given in Chapter 10 that provide answers to these questions. In this chapter, only techniques to model cost considerations using the maintenance repair example are illustrated. The network shown in Figure 7-11 may be used for collecting the desired cost statistics. Changes are only necessary in specifying activity durations and in subroutine EVENT.

To compute the cost associated with an activity, the setup cost and variable cost for the activity are established in DATA statements in function USERF, where USERF is set to the time to perform the activity. Let the vectors SETC(N) and VARC(N) represent the setup cost and variable cost for activity N. The following statement is added in function USERF:

```
TOTALC=TOTALC+SETC(N)+VARC(N)*USERF
```

This statement obtains a running total cost for all activities on the network.

When subroutine EVENT is called at the end of a run, event code 10, a sample is collected of the total cost for performing the project. This is accomplished by making the following call to subroutine COLCT:

```
CALL COLCT(TOTALC,1)
```

where the index 1 is used so that the cost output is printed first on the summary report. The STAT statement required including a histogram specification is

```
STAT,1,TOTAL COST,20/100/10;
```

From this example, it can be seen that SLAM II provides an enhanced technique for the analysis and management of projects that greatly expands those available through PERT and CPM project evaluation methods. Yet, only basic concepts relating to project evaluation through network simulation have been presented. Additional concepts and advanced applications for project evaluation and management analysis through network simulation are presented in the next three chapters.

7.9 SUMMARY

The terminology and performance measures used in project planning are presented in this chapter. The procedures for the modeling of a project in a network form are illustrated. Methods for estimating the project completion time distribution, activity criticality indexes, and project costs are described.

7.10 EXERCISES

7-1. Employ SLAM II to estimate the project completion time distribution and the criticality index for the networks in Figure 7-14. Where feasible, use analytic techniques to develop the requested values. Compare the results with the PERT estimate for the project completion time distribution.

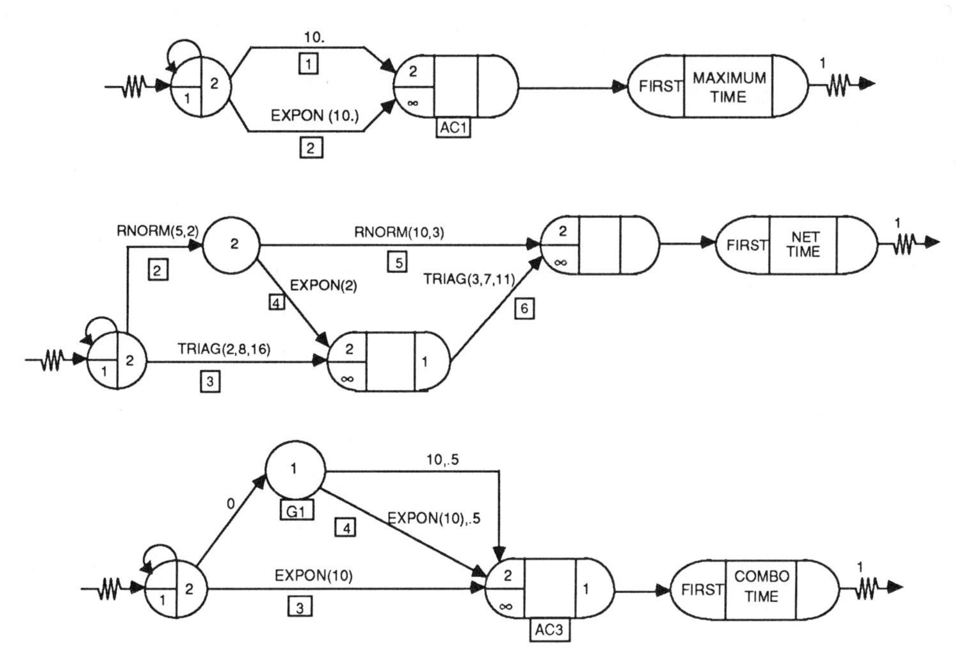

Figure 7-14 Networks for Exercise 7-1.

7-2. For the repair project network shown in Figure 7-5, develop Gantt charts assuming that each activity time is the earliest time, the modal time, and the latest time. From these Gantt charts, specify the range of possible start times for each activity in the network.

7-3. Specify the effects on project completion time that you would anticipate if an exponential distribution were used in place of the triangular distribution for the example in Section 7.3. The mean of the exponential distribution should be set equal to the mean of the triangular distribution, that is, the average of the three parameters of the triangular distribution.

7.11 REFERENCES

1. Archibald, R. D., and R. L. Villoria, *Network-Based Management Systems (PERT/CPM)*, Wiley, New York, 1968.
2. Archibald, R. D., *Managing High-Technology Programs and Projects*, Wiley, New York, 1976.
3. Halpin, D. W., and W. W. Happ, "Digital Simulation of Equipment Allocation for Corps of Engineering Construction Planning," U.S. Army, CERL, Champaign, IL, 1971.
4. Horowitz, J., *Critical Path Scheduling*, Ronald Press, New York, 1976.
5. Law, A. M., and W. D. Kelton, *Simulation Modeling and Analysis*, McGraw-Hill, New York, 1982.
6. Lee, S. M., G. L. Moeller, and L. A. Digman, *Network Analysis for Management Decisions: A Stochastic Approach*, Kluwer-Nijhoff Publishing, Boston, MA, 1982.
7. MacCrimmon, K. R., and C. A. Ryavec, "An Analytical Study of the PERT Assumptions," *Operations Research*, Vol. 12, 1964, pp. 16-38.
8. Malcolm, D. G., and others, "Application of a Technique for Research and Development Program Evaluation," *Operations Research*, Vol. 7, 1959, pp. 616-669.
9. Moder, J., C. R. Phillips, and E. W. Davis, *Project Management with CPM, PERT, and Precedence Diagramming*, 3rd ed., Van Nostrand Reinhold, New York, 1970.
10. Moore, L. J., D. F. Scott, and E. R. Clayton, "GERT Analysis of Stochastic Systems," *Akron Business and Economic Review*, 1974, pp. 14-19.
11. Musselman, K. J., W. R. Penick, and M. E. Grant, *AID: Fitting Distributions to Observations: A Graphical Approach*, Pritsker & Associates, West Lafayette, IN, 1981.
12. Pritsker, A. A. B., *The GASP IV Simulation Language*, Wiley, New York, 1974.
13. Pritsker, A. A. B., *Modeling and Analysis Using Q-GERT Networks*, 2nd ed., Wiley and Pritsker & Associates, New York and West Lafayette, IN, 1979.
14. van Slyke, R. M., "Monte Carlo Methods and the PERT Problem," *Operations Research*, Vol. 11, September-October 1963, pp. 839-860.
15. Whitehouse, G. W., "Extensions, New Developments, and Applications of GERT: Graphical Evaluation and Review Technique," Ph.D. dissertation, Arizona State University, 1966.
16. Wiest, J., and F. Levy, *Management Guide to PERT-CPM*, Prentice-Hall, Englewood Cliffs, NJ, 1969.

8 PROJECT PLANNING: CHANCE AND CHOICE CONCEPTS

8.1 INTRODUCTION

Project managers are often faced with considerable project uncertainties and risks. These uncertainties arise due to external factors such as regulatory or government agency actions, financing difficulties, and weather. Internal factors that make project planning a nondeterministic activity are site location problems, design alternatives, unusual construction constraints, and construction change orders. In addition, there are technological considerations whose outcomes are not known with certainty, such as technology improvement, contractor and subcontractor performance variability, test and checkout procedures, and rejection possibilities. Network models provide the means to model these types of uncertainty. In particular, SLAM II provides the concepts of probabilistic and conditional branching, random activity times, and different nodal release requirements for this pur-pose. This represents a significant modeling improvement beyond the capabilities of PERT- and CPM-type networks. Detailed descriptions of new modeling and analysis concepts needed for project planning in the face of uncertainty are given in [6] and [7].

8.2 PERT SHORTCOMINGS AND SLAM II CAPABILITIES

In Chapter 7, only a small subset of SLAM II capabilities was needed to develop a project network model because PERT-type networks are constrained by the following modeling assumptions:

1. The number of activity completions required to release a node is equal to the number of branches ending at the node.

2. All branching is done on a deterministic basis.
3. No cycles (feedback) are allowed in the network.
4. Projects are always completed successfully, as the concept of failure is nonexistent. Thus, a PERT/CPM analysis deals entirely with the time (or other additive variable) at which nodes of the network are realized.
5. No explicit storage or queueing concepts are available.

The following sections demonstrate the modeling flexibility available in SLAM II to eliminate these modeling assumptions and the SLAM II constructs that can be employed for planning purposes when uncertainty exists. A description of applied projects that have involved one or more of these constructs is provided in Section 8.6.

8.3 ACTIVITY FAILURE AND ITS RAMIFICATIONS

Often in project planning, one must account for the possibility that an activity may end unsuccessfully. Sometimes this outcome requires repetition of the failed activity; at other times it leads to project failure. Each of these cases, including the repeated performance of an activity, are important in planning projects.

8.3.1 Modeling Activity Failure

Suppose that, in the repair project example presented in Chapter 7, the activity "assemble and test power units," activity 8, has only an 80% chance of successful completion. Thus, 20% of the time it is necessary to repeat the activity. The SLAM II network shown in Figure 8-1 models this possibility. GOON node N7 has been inserted into the network. Activity 8 ends at node N7, which selects the next activity probabilistically, that is, upon completion of activity 8, node N7 is released and only one activity is scheduled to start from node N7. Twenty percent of the time, activity 10 is taken, requiring a one-time unit delay before releasing node N5, again which in turn restarts activity 8. Thus, activity 10 represents a decision to repeat activity 8. Note that activity 10 can cause subsequent releases of node N5. In this way, activity 8 is repeated until activity 11 is taken, which is one of the conditions required before node N6 can be released. A project's completion time can increase significantly when failure and the reperformance of activities are possible.

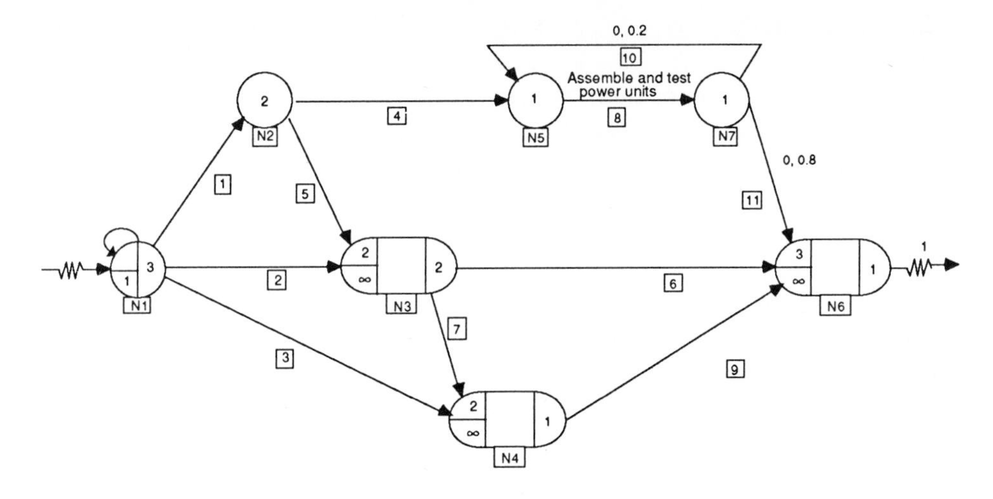

Figure 8-1 Repair project with possible repeated activity failure.

8.3.2 Project Failure

Another aspect of activity failure is that it may represent an inability to complete a project. In Figure 8-2, a SLAM II network is presented that models project failure when node N8 is released. Node N8 is released when activity 10 is selected; that is, the branching at node N7 probabilistically selects activity 10 and not activity 11. The selection of activity 10 occurs in one out of five runs, on the average, corresponding to the 0.20 probability assigned to activity 10. Project failure could also be modeled for other activities in the network.

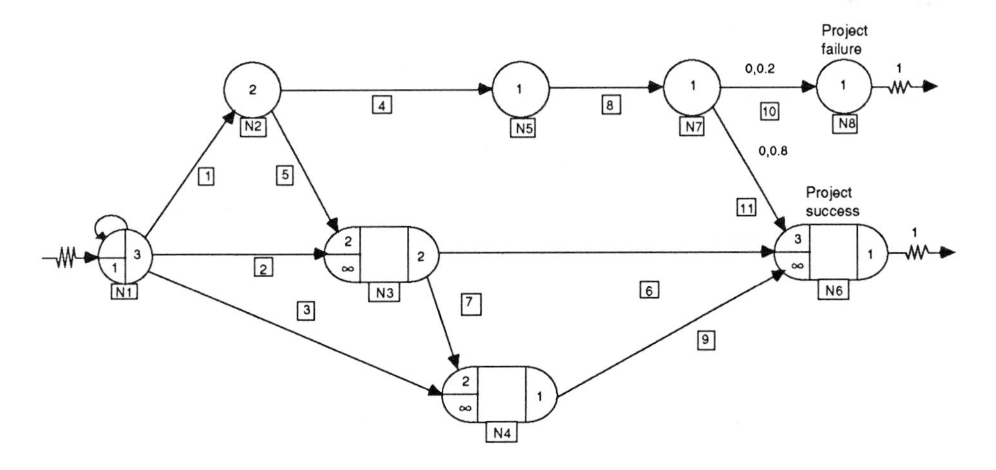

Figure 8-2 Repair project with possible project failure .

In Figure 8-3, nodes have been inserted into the network following activities 4 and 6, which determine if the project fails based on the completion of these activities. The addition of node N9 and activities 12 and 13 is similar to the addition of node N7 and activities 10 and 11 described previously. This addition demonstrates how to model multiple locations where failure may occur probabilistically. The addition of node N10 with conditional branching models project failure, which is defined as activity 6 not being completed by time 100. Thus, at node N10, activity 14 is taken if activity 6 is not completed by time 100. Activity 14 causes node N8 to be released which results in project failure.

The example presented in Figure 8-3 also illustrates the concept of multiple activities incident to a node (node N8) for which any activity completion causes the node to be released. This models the logical OR input specification for a node. If a 2 were specified for the first release requirement for node N8, a decision process would be modeled where two out of three of the failure conditions are required in order to declare the project a failure. When doing this, it is necessary to ensure that the project could be completed. For this example, this can be done by changing the first release requirement for node N6 from 3 to 2. Thus, if any two of the three branches leading to node N6 are completed, a project success occurs; otherwise, the project fails.

8.3.3 Repeated Performance of an Activity

In Section 8.3.1, methods to model the repetition of an activity are described. This concept can be modified to have the possibilities of repeating the activity

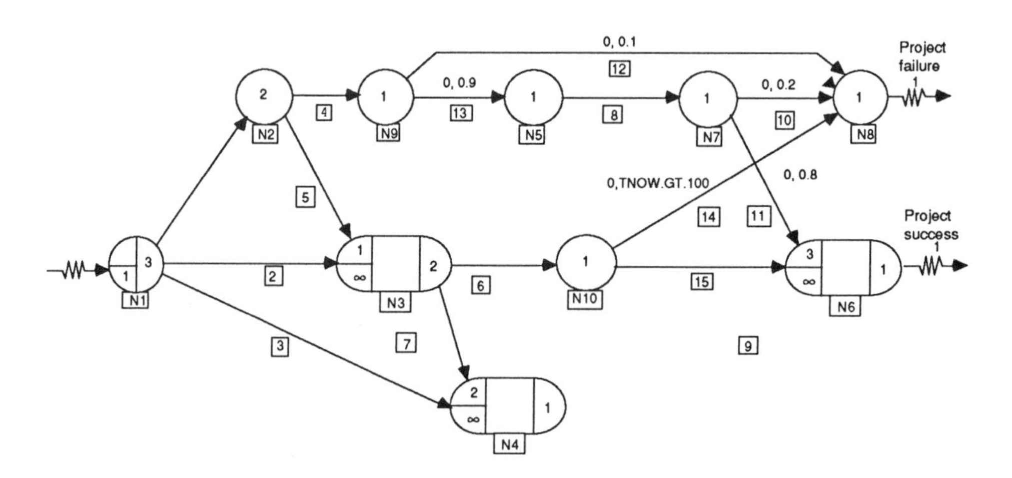

Figure 8-3 SLAM II project network with multiple failure specifications.

lessen or increase as more repetitions occur. This is accomplished by modifying the probabilities associated with a repetition and maintaining these probabilities as attributes on which branching is based. Alternatively, an attribute value could count activity repetitions and the repeating of the activity allowed only if the attribute value is less than a prescribed number of repetitions. This attribute value would be set using an ASSIGN node to increase the value of attribute by 1 [ATRIB(1) = ATRIB(1) + 1]. An illustration of this branching concept is shown in Figure 8-4. In this network segment, activity 16 can only be selected twice; that is, when attribute 1 is 1 and when attribute 1 is 2. When node N20 is released the third time, attribute 1 is set equal to 3 and the condition for activity 16 is not satisfied.

Also shown in Figure 8-4 is the changing of an activity time based on the number of repeats. At node N20, attribute 2 is decreased by 0.5 time unit each time an entity passes through it. Attribute 2 defines the time delay for activity 15. Assuming that attribute 2 is set prior to the release of node N19, the time to perform activity 15 is equal to the original value of attribute 2 for the first time, attribute 2 minus 0.5 the second time, and the original value of attribute 2 minus 1 the third time. More complex changes in activity times can be accomplished through the use of user functions, which allow the modeling of learning effects, stress factors, and other time-dependent activity functions.

The concept of modifying activity times based on the number of activity repetitions can be generalized to model interdependent activity times for different activities. In the preceding example, a dependence has been inserted on the network with regard to the time to perform activity 15 based on previous performance times for activity 15. In an analogous way, other activities could be affected by the time to perform activity 15 through the incrementing or decrementing of the value of the attribute describing the time to perform activity 15 (in this case attribute 2) or through the assignment of a new attribute value, which could be a function of attribute 2. The use of the number of times an activity, I, is completed, NNCNT(I), is useful for this purpose.

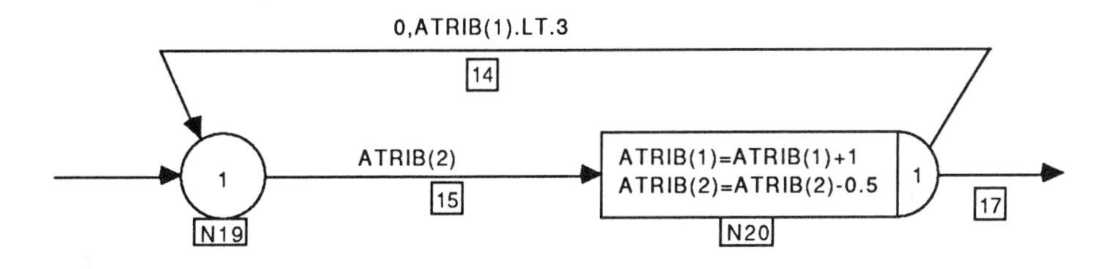

Figure 8-4 Limited activity repetitions with activity time modification.

8.4 REPRESENTING DECISION LOGIC IN PROJECTS

Frequently, the occurrence of an event requires a decision as to which activity or set of activities to initiate from a group of candidate activities. In the planning process, normally only the group of candidate activities can be listed and the criteria specified upon which the decision will be made. Since these criteria are normally related to time or the status of other network activities, the outcome of the scheduling decision cannot be predicted with certainty. SLAM II allows this decision logic to be incorporated into a network model through the use of conditional branching. In this way, a model directly incorporates the impact of decision logic on planning alternatives.

Consider an example where experiment I causes a project to fail if phase 1 fails twice. In Figure 8-5, a decision process is represented at node N18 as the event identifying the occurrence of the second failure. The branching conditions are shown on the activities emanating from node N18. If the time into the project is greater than 100 when node N18 is released, that is, TNOW.GT.100, the branch to node N17 is selected and a second failure of Phase I leads to project failure.

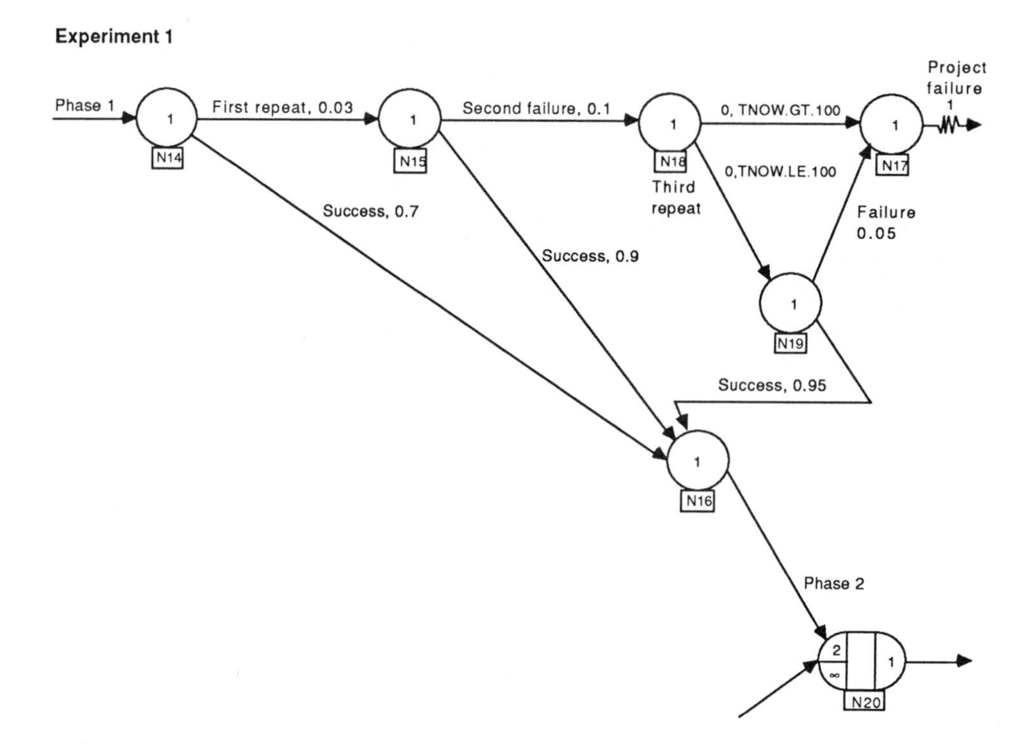

Figure 8-5 Network illustration for time-dependent logic at event node.

However, if the time is less than or equal to 100, TNOW.LE.100, the branch to node N19 is selected and phase 1 is repeated for a third time. A third failure of phase 1 leads to project failure; a successful third attempt leads to phase 2. Hence, a decision has been represented that is a function of time. It is not known how long it will take to reach node N18, but it is known that the successful performance of phase 1 must be completed by time 100 if the project is to be considered successful. Conditional branching models this decision logic and allows the investigation of its impact on the planning process. An alternative representation of this decision logic using feedback concepts is given in Figure 8-6.

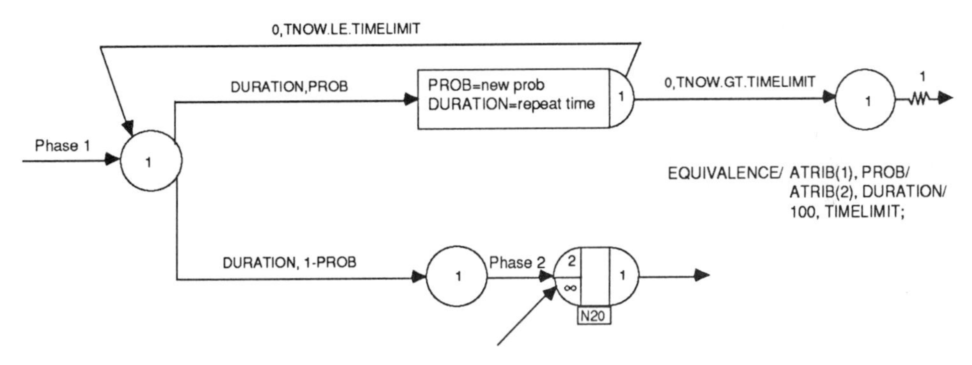

Figure 8-6 Alternative representation of Figure 8-5.

Decision logic at a node may also be dependent of the status on other events. Consider the network in Figure 8-7, where the decision on the second failure of phase 1 is modified to test the status of experiment II. If node N13 has been released, experiment II has been completed. In Figure 8-7, the branch to node N17 is selected if XX(1).EQ.1, that is, the ASSIGN node N13 has been released before node N18 is released. The branch to node N19 is selected if node N13 has not been released. Hence, a third attempt of phase 1 is performed only if experiment II has not yet been completed when the second attempt for phase 1 failed. If experiment II is only performed once, the decision variable could be the completion of activity 13 using the condition NNCNT(13).EQ.1.

Decision logic based on the status of other network activities is very important when planning for the potential impact of supporting technology under development. For example, assume that ASSIGN node N50 in Figure 8-8 represents the successful development of a scientific procedure that ensures the success of phase 1. Consequently, the branch from node N18 to node N17 is selected if node N50 has not been released, XX(1).NE.1, by the time node N18 is released. If, however, node N50 has been released, the branch from node N18

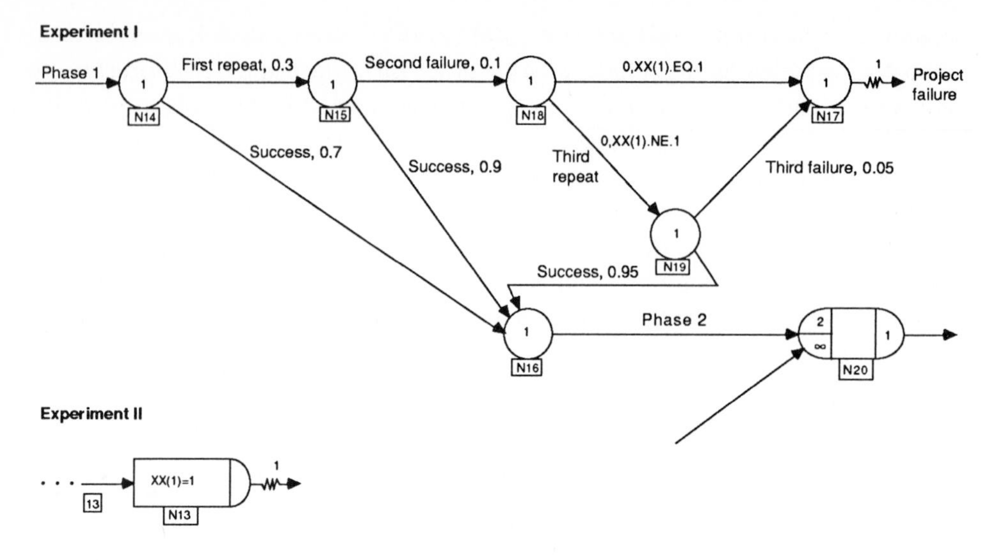

Figure 8-7 Network illustrating an event decision as a function of the occurrence of other events.

XX(1)=1
Scientific procedure developed
N49
N50

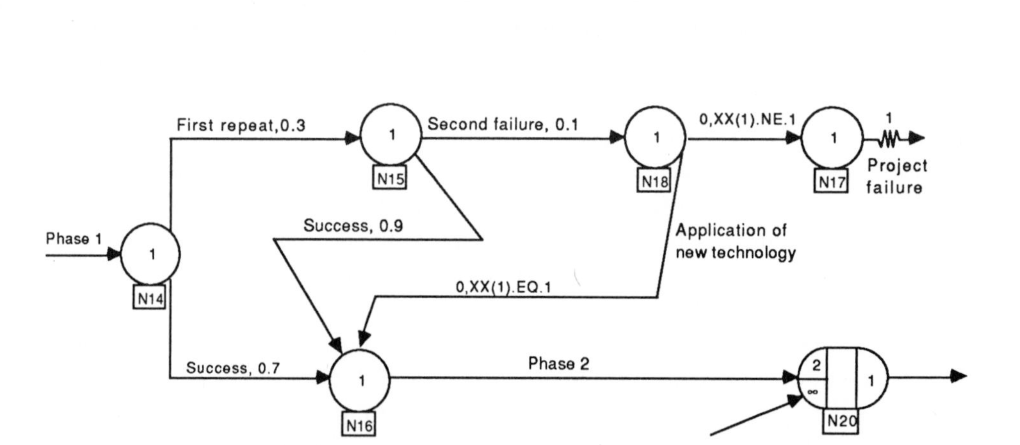

Figure 8-8 Network illustrating an event decision as a function of the completion of a separate network segment.

to node N16 is taken. This activity corresponds to the application of the new technology. Since conditional branching can also be attribute based, a modeler can incorporate additional complexity in the decision process by coding program inserts that describe decision-specific rules.

Another illustration of the use of conditional activities occurs when it is decided that an increased rate of effort for phase III of a project may be desirable. In Figure 8-9, the requisition process for obtaining additional funding is represented by activity 40. If activity 40 is completed prior to the completion of phase 2, then phase 3, option A is performed under authorization for the added funds. If the funds are not received prior to the completion of phase 2, then phase 2, option B is performed under normal funding levels. The network model for this situation depicts the branching conditions allowed following node N26 that are dependent on the completion of activity 40.

The features described in this section illustrate how SLAM II can be used to include memory within a network model. This allows past performance to affect future performance and eliminates the need for Markovian assumptions that are associated with many network-based project modeling techniques.

The use of conditional and probabilistic branching, when coupled with attribute assignments and user functions, provides extensive decision logic for modeling project failure and success conditions. With this capability, there is a chance that some nodes in the network will not be released during a given run. For decision-making purposes, it may be important to have information concerning the probability that a node (milestone) is reached. With the SLAM II processor, a project can be simulated many times (over multiple runs), and the proportion of runs in which a node is released is used to estimate the probability of releasing the node. It is computed as the number of the runs in which the node is released at least once, divided by the total number of runs. Note that a node could be released more than once in a run, but that the probability of nodal release is not affected by such multiple occurrences on a single run.

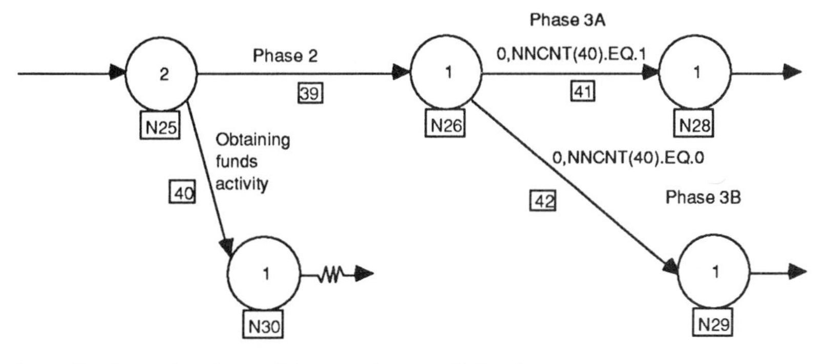

Figure 8-9 Project planning with adaptive modifications.

8.5 IMPLICATIONS OF PROJECT PLANNING IN THE FACE OF RISK AND UNCERTAINTY

Project planning, when confronted with an uncertain future, is a difficult process. It requires a management approach that establishes alternative plans that can be implemented based on the actual historical development of the project. By knowing in advance that specific uncertainties may occur and their possible impact on overall project performance, a project manager would be able to redirect a project when early warning signals indicate that a specified uncertain future state is likely to occur. When uncertainty is considered, better choices of how to redirect a project activity to maintain low risk levels and required margins of safety can be included in project plans. This type of project planning requires a great deal of organization on the part of the project manager. To take uncertainty into account when planning projects, the following steps are suggested:

1. Define the organizational responsibilities of technical/ administrative managers.
2. Conduct training and indoctrination programs to acquaint all levels of project management with the potential alternative plans and the uncertainty associated with the plans.
3. Establish an implementation task team with the responsibility for network planning and status control that reports directly to the project manager.
4. Provide for periodic discussions and critiques of project plans and historical project activity.
5. Develop a definition of the project and alternative plans in terms of a work breakdown structure and work packages. Require an agreement on the definitions used in the work breakdown structure.
6. Assign responsibility for reporting the performance for each work package of the work breakdown structure.
7. Assure that work packages and associated network plans clearly define work to be done before time estimating begins.
8. Assign primary responsibility for time estimation.
9. Require a review of time estimates for any work packages involving any significant degree of uncertainty, and for those activities that have a high probability of being critical, verify the time estimates to a greater degree.
10. Establish updating procedures and due dates for control purposes.
11. Establish decision-making procedures for reallocation of resources and for decision making relative to alternative plans.

8.6 APPLICATIONS OF PROJECT PLANNING UNDER UNCERTAINTY

Brief descriptions are provided in the following sections of project planning under uncertainty. Emphasis is placed on the cause and need for considering the new modeling concepts described throughout this chapter. The general managerial concerns described in Section 8.5 are not included in the presentations, but it should be assumed that such concerns were taken into account.

8.6.1 The Assault Breaker Program

This project planning application description is extracted from a paper by Papageorgiou [4] in which the details of the network model and simulation results are presented. The Assault Breaker Program [2] was a joint project between the Air Force and the Army aimed at developing systems that could detect, locate, and strike enemy armor at ranges well beyond the forward edge of the battle area. Its concept includes a surveillance strike system consisting of an airborne radar to sense second echelon armor, and then guiding aircraft and/or air-to-surface missiles against that armor.

The objective of the Assault Breaker Program is to plan Air Force participation in a series of field technology demonstrations and to accomplish the development planning necessary to support a recommendation for full-scale engineering development.

Complexity and uncertainty are included in the Assault Breaker Program in that it is an accelerated program and the different program segments may not follow the normal routine of planning and development. Another complexity is the interagency character of the program, which requires integration, coordination, and monitoring of the different aspects of the program and all the activities for the participating agencies. This complexity is indicated by the following paragraph [4]:

> The Air Force has developed an airborne moving target indicator and synthetic aperture radar (Pave Mover) capable of locating armor and guiding munitions against the target. This is a segment of the Ground Target Attack Control System which aims at an integrated force management capability to manage and direct friendly forces against second echelon enemy forces. The other segments include the ground target attack control element that aims at real time operational control, weapon and target pairing, and aircraft and weapon guidance, the integration and interface segment that aims at incorporating this control center with existing and planned control and communications elements; and the aircraft/weapons interface segment that aims at developing the hardware

> that interfaces between the radar platform/aircraft and the weapons and direct attack aircraft. The Air Force has also developed air-to-surface munitions while the Army has developed a surface-to-surface missile which is interoperable with the Pave Mover radar.

A network simulation model for the Assault Breaker Program was developed. At the beginning, the detailed structure of the project was nebulous in the minds of the people involved. However, they were forced through the networking process to think about the structure, clarify their perceptions, cross-verify them, and arrive at a consensus. As a result, the original draft of the network went through a number of revisions, with each revision more closely approaching reality. The final network is presented in [4].

A common feature of the network is that decisions based on achieved conditions are used to model the effect of tests on subsequent modifications of a component, and that the chance of further modifications of the component will be different than that before the first modification. Modeling concepts similar to those described in Section 8.4 were used for this project. Another part of the network required conditional branching based on the outcome from a field demonstration. The final network consisted of approximately 500 activities. Data were collected and the network was analyzed using simulation. The following is an extract from the conclusions section of the paper describing this planning subject [4].

> The benefits derived from this analysis were significant. The network development process itself helped the people involved with the project gain a better understanding of its structure and the interfacing of its component parts. The derived network can now serve as a basis for further refinements; briefings on the project for new staff; and further simulation for planning and control.
>
> The network model and simulation was extremely helpful to the program director in planning the future course of the program, estimating the possibility of meeting deadlines, and systematically documenting proposed plans. Given the difficulties involved in estimating the values of the input parameters, sensitivity analysis can be carried out to test the sensitivity of the results to the accuracy of the data. Also, given the uncertainty that surrounds such programs, experimentation can be carried out by modifying the network and observing the effect of the modifications upon the plans.

8.6.2 Project Networks Incorporating Improvement Curve Concepts

A network-based simulation approach for interactive project network analysis was developed by Wolfe, Cochran, and Thompson [5]. The concepts included in this network-based system included activity durations based on improvement curve trends, storage limitations, alternative activities when bottlenecks occurred, probabilistic branching, costs due to material shortages, and personnel turnover. The interactive capability provided a convenient means for inputting, formatting, and editing large data sets associated with planning the construction of large projects such as ships, aircraft, and turbines. Database techniques were used for storing such information and to allow a multiple-run analysis capability.

Wolfe, Cochran, and Thompson [5] present an example of project planning of a ship fabrication and assembly project. They include an analysis of project completion time, resource utilization, and total costs. They conclude that such a network-based interactive computer system fulfills a growing need to incorporate uncertainty and advanced modeling concepts into project planning analysis.

8.6.3 Planning and Scheduling Overhauls of Ships

Overhauls of U.S. Navy ships are performed every five years on the average. With current force levels, this means that over 80 overhauls are ongoing each year. A regularly scheduled overhaul for a Navy ship can cost several hundred thousand dollars for a small tug or hundreds of millions of dollars for a large aircraft carrier. A regular overhaul consists primarily of the following eight types of activities: ripout, repair, shop test, space preparation, installation of repaired equipment, in place testing, systems testing, and sea trial testing. Each type of activity has nondeterministic times, may require that the activities be repeated, may require different resources and space, and, in general, may require the type of modeling for uncertain conditions described in this chapter.

Johnston developed a network model for analyzing regular overhauls of U.S. Navy ships [3]. Included in the model were resources and cost information. A detailed analysis of overhaul costs and time was performed, which included a determination of criticality indexes for the activities associated with the overhaul. The model was developed to answer the following questions:

1. How long does the overhaul take from start to completion?
2. What is the extent of the variability for an overhaul, and what type of distribution does it generally follow?
3. How long does it take to achieve certain key overhaul events such as the end of the dry dock period?
4. When should key events that require external observers/inspectors be scheduled?
5. How does failure of a critical inspection affect overhaul completion?

6. What key repair items are the most critical to timely overhaul completion and are worthy of increased attention?
7. What is the projected overhaul cost, given the variability of the work package?
8. How are the shipyard resources (for example, dry docks and personnel) utilized in the course of an overhaul?

Johnston demonstrated that simulation outputs could be used to answer the foregoing questions and to improve decision making related to the scheduling and performance of regular overhauls. He concluded that uncertainty needed to be taken into account in the planning and analysis of the regular overhaul process.

8.7 SUMMARY

In this chapter, project planning in the face of uncertainty is presented. The SLAM II modeling concepts that support planning under uncertainty are described, and the network modeling procedures for such a purpose are presented. Application areas that employed such techniques are summarized. The need for including uncertainty in project planning is established.

8.8 EXERCISES

8-1. Give a possible rationale as to why the shortcomings listed for PERT and CPM in Section 8.2 were not of importance to its developers.

8-2. Reduce the network shown in Figure 8-1 to one that can be analyzed using PERT techniques. Describe how an analysis would be performed.

8-3. Analyze the network shown in Figure 8-2 by assessing the time to go from node 1 to node 8 and the time to go from node 1 to node 6. Make any assumptions you desire concerning the analysis procedure to be used in assessing the project completion times.

8-4. To illustrate the relationship between conditional branching and nodal modification, perform the following:
   (a) Convert the network in Figure 8-3 to one in which node 10 is replaced by node 11 when time exceeds 100 units.
   (b) Convert the network in Figure 8-8 to one that does not use nodal modification.
   (c) Describe a situation in which conditional branching cannot replace a nodal modification operation.

8-5. Discuss the differences between risk assessment and projects involving decision logic.

8-6. Write a paragraph that elaborates on the steps that were suggested in Section 8.5

for planning projects to take uncertainty into account.

8-7. Write a function USERF that would incorporate improvement curve concepts into a project that was described by a SLAM II network.

8-8. Discuss how a SLAM II model can be used to answer the questions posed in Section 8.6 by Johnston in planning and scheduling the overhaul of ships.

8.9 REFERENCES

1. Federal Power Commission Exhibit EP-237, "Risk Analysis of the Arctic Gas Pipeline Project Construction Schedule," Vol. 167, Federal Power Commission, 1976.
2. Jaglinski, T., *Program Management Plan for Ground Target Attack Control System Assault Breaker*, unpublished document, HQ Electronics Systems Division, Hanscom Air Force Base, Bedford, MA, April 1980.
3. Johnston, R. E., *Discrete Event Simulation as a Management Tool for Planning and Scheduling Overhauls of U.S. Navy Ships*, unpublished masters' project, Purdue University, December 1980.
4. Papageorgiou, J. C., An Application of GERT to Air Force Systems Development Planning, *Project Management: Methods and Studies*, ed. by Burton V. Dean, Elsevier Science Publishers, B.V. (North Holland), The Netherlands, 1985, pp. 237-252.
5. Wolfe, P. M., E. B. Cochran, and W. J. Thompson, "A GERTS-Based Interactive Computer System for Analyzing Project Networks Incorporating Improvement Curve Concepts," *AIIE Transactions*, Vol. 12, No. 1, March 1980, pp. 70-79.
6. Wortman, D. B., and C. E. Sigal, *Project Planning and Control Using GERT*, Pritsker & Associates, West Lafayette, IN, October 1978.
7. Yancey, D. P., and K. J. Musselman, *Critical Statistics in General Project Planning Networks*, Pritsker & Associates, West Lafayette, IN, September 1980.

9 PROJECT PLANNING: SCHEDULING AND RESOURCES

9.1 INTRODUCTION

This chapter presents the issues related to activity scheduling within project planning. Activity scheduling involves the specification of start times for activities and determining the times associated with the achievement of nodes. In project planning, specialized terminology has evolved that facilitates the determination of these quantities. In Sections 9.2 and 9.3, scheduling terminology and scheduling computations for constant activity durations are presented. This is followed by sections that describe the scheduling procedures when activity durations are random variables. Included in these sections are criteria for judging the criticality of an activity with respect to meeting its scheduled start and end times. The last sections of the chapter are devoted to project planning when resources are limited.

9.2 SCHEDULING TERMINOLOGY

In scheduling the activities of a project, it has been found that certain quantities associated with nodes and activities provide helpful information. For example, the earliest time an activity can start provides information that restricts the time horizon associated with the starting of an activity. The ***early start time*** for an activity, abbreviated ES, is equal to the release time of the node that immediately precedes the activity. Another quantity of interest is the ***early finish time*** for an activity, EF, which is the sum of ES and the activity duration.

When all activity durations are constants, all early start times and early finish times can be computed from the precedence relations depicted on the network. The early start times of activities emanating from CREATE nodes are assumed as zero and, hence, their early finish times are equal to their duration. The early start times for other activities are equal to the release time of the node from which they emanate, where the release time is the largest of the early finish times of the activities incident to the node. When a project has a single TERMINATE node, the largest early finish time of all activities incident to the terminating node of the network is the anticipated project duration. This process for computing the project duration is called the ***forward pass***.

Assuming that it is desirable to finish the project as early as possible, the project duration as just computed also represents the latest finish, LF, time for all activities leading into the TERMINATE node. Starting with the latest finish time, a ***late start time***, LS, can be calculated by subtracting the activity duration from the late finish time. Consider now a start node of an activity that leads to the terminating node. The smallest of the late start times of the activities emanating from this node defines the latest release time for the node that does not result in a delay for the project.

This latest time for a node release is then specified as the ***late finish time*** for all activities leading into the node. Given the late finish time, the late start time for each of these activities can be computed. This computation of late finish times and late start times is known as the ***backward pass*** through the network and continues until all CREATE nodes are reached.

After making both the forward pass and the backward pass, the amount of slack, or float, associated with an activity for scheduling purposes can be estimated. There are at least four methods for defining the slack, with the two most commonly used slack values being free slack and total slack.†

Free slack (FS) is the amount of time that an activity started as early as possible can be delayed before it affects the start time of activities emanating from the end node of the activity. This type of slack is local in that a decrease in the amount of slack may not directly affect the total project duration. The computation for free slack is the earliest release time for the end node of an activity minus the activity's early finish time.

A slack value that does directly affect the total project duration is referred to as ***total slack*** (TS). Total slack for an activity is defined as the difference between the late finish time and the early finish time for an activity. If the activity takes additional time units above the total slack, the project will be delayed, because the latest release time for the end node for the activity will be increased.

† Two other slack values are interference slack and safety slack. Interference slack assumes that all predecessor activities are delayed to their latest start times, while succeeding activities are started as early as possible. This type of slack represents the most stringent constraint on the scheduling of activities. Safety slack is defined as the amount of time an activity start can be delayed without delaying the start of successor activities, assuming that all predecessors start as late as possible.

9.3 SCHEDULING COMPUTATIONS

The following nomenclature is used in the formulas that describe the various scheduling computations and in the graphical diagrams.

t_j = mean duration for activity j
ES_j = earliest start time for activity j
EF_j = earliest finish time for activity j
LS_j = latest start time for activity j
LF_j = latest finish time for activity j
TE_n = earliest release time for node n
TL_n = latest release time for node n
TS_j = total slack for activity j
FS_j = free slack for activity j
s(j) = start node number for activity j
e(j) = end node number for activity j

The computations required for the forward pass are:

1. $TE_{source} = 0$
2. $ES_j = TE_{s(j)}$
3. $EF_j = ES_j + t_j$
4. $TE_{e(j)}$ = largest of the EF_j values for all activities with end node e(j)

The backward-pass computations are:

5. $TL_{sink} = TE_{sink}$
6. $LF_j = TL_{e(j)}$
7. $LS_j = LF_j - t_j$
8. $TL_{s(j)}$ = smallest of LS_j values for all activities with start node s(j)

The computations for total slack and free slack are:

9. $FS_j = TE_{e(j)} - EF_j$
10. $TS_j = LF_j - EF_j = LS_j - ES_j$

The computations associated with the scheduling algorithm given are shown on the diagram of Figure 9-1.† Table 9–1 gives the values for each activity. As long as each activity time is assumed to have a constant value, the algorithm for making these computations is straightforward.

† Moder and Phillips [2] refer to the type of network given in Figure 9-1 as a space diagram which contains a large amount of information, and may be confusing to a manager or a decision maker.

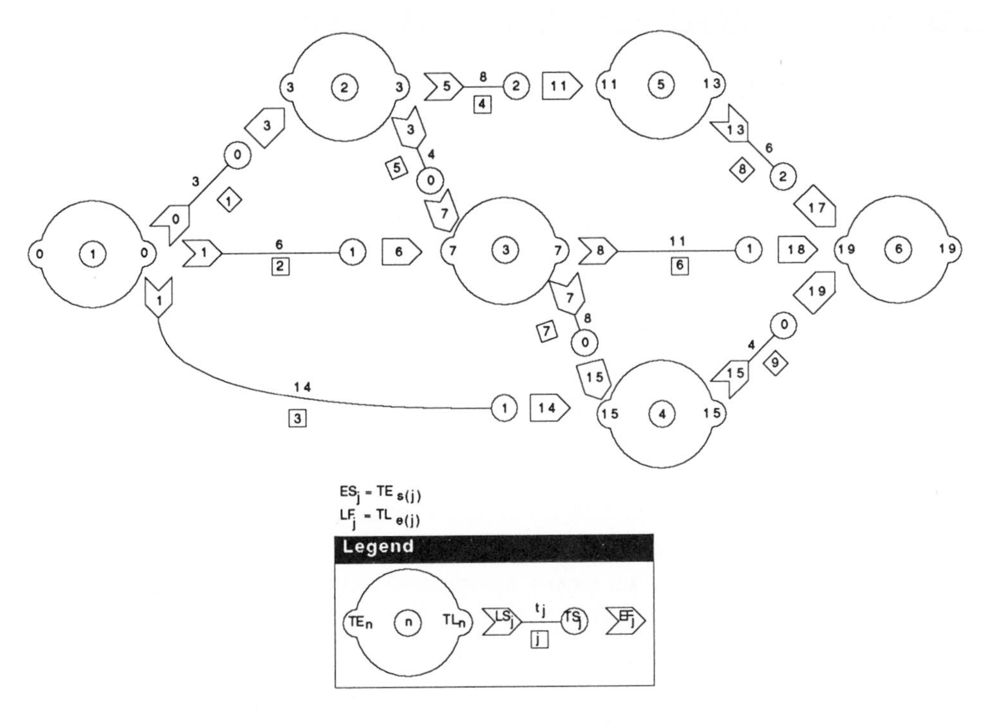

Figure 9-1 Network diagram illustrating scheduling computations for repairman problem.

Table 9-1 Scheduling Computations Based on Expected Values

	Activity					Total	Activity
Number j	Expected Duration t_j	Early Start ES_j	Late Start LS_j	Early Finish EF_j	Late Finish LF_j	Activity Slack TS_j	Free Slack FS_j
1	3	0	0	3	3	0	0
2	6	0	1	6	7	1	1
3	14	0	1	14	15	1	1
4	9	3	5	11	13	2	0
5	4	3	3	7	7	0	0
6	11	7	8	18	19	1	1
7	8	7	7	15	15	0	0
8	6	11	13	17	19	2	2
9	4	15	15	19	19	0	0

9.4 SLAM II APPROACH TO SCHEDULING COMPUTATIONS

When SLAM II is used to simulate a PERT network, the forward-pass computations are made for every run of the network. The node release times are the $TE_{s(j)}$ values for each node, which define the early start times for each activity whose start node is j; that is, $ES_j = TE_{s(j)}$. Since the expected value of the sum of two random variables is the sum of their expected values, the expected activity duration, t_j, can be added to the average ES_j to estimate the average early finish time, EF_j, for an activity. The sample of activity duration values could be used, but by using the expected activity durations directly less variation exists in the early finish time estimates.

SLAM II does not automatically store the information to make the backward pass required to compute the late start, late finish, and slack values. The subroutines necessary to make these computations are presented in the next section. First, a direct approach involving manual computations is presented, which illustrates the underlying procedures involved in performing the scheduling computations. The backward computations involve the analysis of the network starting at the TERMINATE node and moving toward the CREATE node. This can be accomplished with SLAM II by redefining the network such that the TERMINATE node is the CREATE node and reversing the direction of each activity; that is, make the end node of the activity its start node and the start node of the activity its end node. The SLAM II statement model for this reverse project is given in Figure 9-2. A simulation of this reverse network provides the information required to make the backward-pass computations for the original network. The node realization times in this reverse network after subtraction for the project duration represent the TL_n, latest release times, for the network nodes. The latest start and latest finish times can be obtained by subtracting the node realization times from the project duration as computed during the forward pass. To have the same time durations for each run of the network, a separate stream number is used to generate the durations for each activity. This avoids having to store the activity durations for the backward pass. Although this step is not necessary, as statistical variation in the activity durations is to be expected when employing SLAM II, the use of the same random values decreases the variability of the results.

The outputs from running the backward pass using SLAM II are shown in Figure 9-3. By combining these values with those presented in Figure 7-8, Table 9-2 can be prepared.

```
GEN,HAMMESFAHR,REVERSE REPAIR PROJ.,1/15/88,400,N,N,,N,Y/400;
LIMITS,,2,50;
NETWORK;
      CREATE,,,,1,3;                      END NODE 6
      ACT/8,TRIAG(3.,6.,9.,8),,N5;        ACTIVITY 8
      ACT/6,TRIAG(8.,9.,16.,6),,A1;       ACTIVITY 6
      ACT/9,TRIAG(1.,3.,8.,9),,N4;        ACTIVITY 9
N5    COLCT,FIRST,NODE 5,,1;              NODE 5
      ACT/4,TRIAG(3.,9.,12.,4),,A2;       ACTIVITY 4
N4    COLCT,FIRST,NODE 4,,2;              NODE 4
      ACT/7,TRIAG(4.,7.,13.,7),,A1;       ACTIVITY 7
      ACT/3,TRIAG(10.,13.,19.,3),,A3;     ACTIVITY 3
A1    ACCUMULATE,2,100;                   COMBINE ACTIVITIES 6 & 7
N3    COLCT,FIRST,NODE 3,,2;              NODE 3
      ACT/5,TRIAG(1.,3.,8.,5),,A2;        ACTIVITY 5
      ACT/2,TRIAG(3.,6.,9.,2),,A3;        ACTIVITY 2
A2    ACCUMLUATE,2,100;                   COMBINE ACTIVITIES 4 & 5
      COLCT,FIRST,NODE 2,,1;              NODE 2
      ACT/1,TRIAG(1.,3.,5.,1);            ACTIVITY 1
A3    ACCUMULATE,3,100;                   COMBINE ACTIVITIES 1, 2, & 3
      COLCT,FIRST,PROJ. START;            START NODE 1
      TERM,1;
      END;
INIT,0,,N;
FIN;
```

Figure 9-2 SLAM II statement model for reverse repairman network.

```
                    S L A M   I I   S U M M A R Y   R E P O R T

            SIMULATION PROJECT REVERSE REPAIR PROJ.        BY HAMMESFAHR

            DATE  1/15/1988                                RUN NUMBER  400 OF  400

            CURRENT TIME   0.1906E+02
            STATISTICAL ARRAYS CLEARED AT TIME  0.0000E+00


                    **STATISTICS FOR VARIABLES BASED ON OBSERVATION**

                  MEAN         STANDARD      MINIMUM       MAXIMUM      NUMBER OF
                  VALUE        DEVIATION     VALUE         VALUE        OBSERVATIONS

NODE 5            0.5965E+01   0.1240E+01    0.3274E+01    0.8919E+01      400
NODE 4            0.3972E+01   0.1403E+01    0.1011E+01    0.7678E+01      400
NODE 3            0.1273E+02   0.1896E+01    0.8421E+01    0.1789E+02      400
NODE 2            0.1701E+02   0.2127E+01    0.1260E+02    0.2370E+02      400
PROJ. START       0.2068E+02   0.2103E+01    0.1566E+02    0.2705E+02      400
```

Figure 9-3 Simulation results for reverse repairman model.

Table 9-2 Scheduling Computations Based on SLAM II Outputs

Activity			Average					
Number j	Expected Duration t_j	Standard Deviation σ_j	Early Start ES_j	Late Start LS_j	Early Finish EF_j	Late Finish LF_j	Total Slack TS_j	Free Slack FS_j
1	3	0.816	0.00	0.67	3.00	3.67	0.67	0.03
2	6	1.225	0.00	1.95	6.00	7.95	1.95	1.38
3	14	1.871	0.00	2.71	14.00	16.71	2.71	2.01
4	8	1.871	0.00	6.71	11.03	14.71	3.68	0.20
5	4	1.472	3.03	3.95	7.03	7.95	0.92	0.35
6	11	1.764	3.03	9.68	18.38	20.68	2.30	2.30
7	8	1.871	7.38	8.71	15.38	16.71	1.33	0.63
8	6	1.225	11.23	14.68	17.23	20.68	3.45	3.45
9	4	1.472	16.01	16.68	20.01	20.68	0.67	0.67

In Table 9-2, the standard deviation of the activity times is obtained by making the computations required for the triangular distribution associated with each activity.† The early start time for each activity is obtained as the node release time estimated by the simulation of the original network (Figure 7-8). The early finish time is computed as the early start time plus the expected duration. The late finish time for an activity is obtained from the simulation of the reverse network (Figure 9-3). Late start times are computed by subtracting the expected activity duration from the late finish times. The computations of total activity slack and activity free slack are made using the equations given previously.

Comparing the values of start and finish times presented in Tables 9-1 and 9-2, it is seen that all values are larger when using the SLAM II outputs. This is to be expected since the scheduling computations based on expected values can be proved to be optimistic. Thus, even for the small network associated with the repairman problem, improved scheduling for start and end dates can be made based on refined estimates of the start and finish times.

The SLAM II output provides additional information regarding start and finish times. For example, the standard deviations for the early start times and late finish times are available directly from the forward and reverse network simulations. The variance of the early finish time for an activity is equal to the variance of the early start time plus the variance of the activity duration.

† The variance for a triangularly distributed random variable with parameters a, m, and b is

$$\sigma^2 = \frac{a(a - m) + b(b - a) + m(m - b)}{18}$$

Similarly, the variance of the late start time is the sum of the variance of the late finish time plus the variance of the activity duration. These variance estimates are shown in Table 9-3. In deciding when to schedule an activity to be started, both the estimates of the mean start time and the variance of that time should be used. The start time can then be based on the risk associated with being able to achieve a prescribed starting date. In Table 9-2, estimates of the average total slack and average free slack values are also given. An interpretation and use of average slack are required. Because of the random variation, a project can no longer be delayed in accordance with the amount of slack without a probable change in the project completion time, or the start time of the following activities. Such a delay may or may not cause problems. The degree of confidence on activity and project completion times depends on the amount of variation in the slack estimates.

Table 9-3 Variance and Standard Deviation Estimates for Start and Finish Times

Activity No.	Variation		Early Start ES_j		Late Start LS_j		Early Finish EF_j		Late Finish LF_j	
j	σ_j^2	σ_j	s_j^2	s_j	s_j^2	s_j	s_j^2	s_j	s_j^2	s_j
1	0.667	0.816	0.000	0.000	5.191	2.278	0.667	0.816	4.524	2.127
2	1.500	1.225	0.000	0.000	5.095	2.257	1.500	1.225	3.594	1.896
3	3.500	1.871	0.000	0.000	5.468	2.338	3.500	1.871	1.968	1.403
4	3.500	1.871	0.623	0.789	5.038	2.244	4.123	2.031	1.538	1.240
5	2.167	1.472	0.623	0.789	5.762	2.400	2.790	1.670	3.594	1.896
6	3.111	1.764	1.900	1.408	7.534	2.745	5.093	2.257	4.423	2.103
7	3.500	1.871	1.982	1.408	5.468	2.338	5.482	2.341	1.968	1.403
8	1.500	1.225	3.771	1.942	5.923	2.434	5.271	2.296	4.423	2.103
9	2.167	1.472	3.885	1.971	6.590	2.567	6.052	2.460	4.423	2.103

A ranking of the activities by high value of average slack time is a possible method for ordering the activities that could be started later. A more appropriate ranking is based on the ratio of the average slack time to the standard deviation of the activity duration. This ratio provides an indication of the number of standard deviations that the average slack time represents. Higher values of the ratio indicate that there is less likelihood that the average slack time will be exceeded due to the basic variability inherent in the performance of the activity. A low value of the ratio indicates that there is little leeway in the start time for the activity. The ratio and activity ranking based on the ratio are shown in Table 9–4. Also given in Table 9-4 are the criticality indexes computed in Section 7.6. It can be seen that there is a large positive correlation between the ranking of critical activities based on the ratio of average slack to activity duration standard deviation and the criticality index. It is conjectured that this correlation should exist for all networks.

Table 9-4 Activity Criticality Assessment

Activity		Total Slack TS_j		Ratio Index TS_j/σ_j		Criticality Index CI_j	
j	σ_j	Value	Rank	Value	Rank	Value[a]	Rank
1	0.816	0.67	(1, 2)	0.82	4	0.6016	2
2	1.225	1.95	5	1.59	7	0.2051	6
3	1.871	2/71	7	1.45	7	0.1933	7
4	1.871	3.68	9	1.97	8	0.1309	(8, 9)
5	1.472	0.92	3	0.63	2	0.4707	3
6	1.764	2.30	6	1.30	5	0.2419	5
7	1.871	1.33	4	0.71	3	0.4339	4
8	1.225	3.45	8	2.82	9	0.1309	(8, 9)
9	1.472	0.67	(1, 2)	0.46	1	0.6272	1

[a] Source: From Figure 7-10

9.5 PROGRAMMING THE SCHEDULING COMPUTATIONS

The scheduling computations described in Section 9.4 can be made in subroutine EVENT using the SLAM II model presented in Figure 7-7 for computing criticality indexes. Figure 9-4 presents the statement model for the scheduling computations and differs from the model presented in Chapter 7 primarily in the use of XX(N) as the duration for activity N. The value of XX(N) is computed in subroutine EVENT(N). Nine rows of ARRAY are defined where the N*th* row describes the characteristics for activity N. The five values in a row correspond to the START node number, the END node number, and the three parameters for the triangular distribution. Note that the START and END node numbers correspond to the original milestone network. The definitions of the variables employed in the subroutines are shown in Table 9-5. Initial values for the scheduling variables are established in subroutine INTLC, shown in Figure 9-5. All values are set to zero except for the latest release time, which is set to a large number.

```
GEN,HAMMESFAHR,PROJECT SCHEDULING,3/29/88,400,N,N,,N,Y/400;
LIMITS,,8,50;
ARRAY(1,5)/1,2,1,3,5/ (2,5)/1,3,3,6,9/(3,5)/1,4,10,13,19;
ARRAY(4,5)/2,5,3,9,12/(5,5)/2,3,1,3,8/(6,5)/3,6,8,9,16;
ARRAY(7,5)/3,4,4,7,13/(8,5)/5,6,3,6,9/(9,5)/4,6,1,3,8;
NETWORK;
        CREATE,,,,1,1;                        START REPAIR PROJECT
        ASSIGN,ATRIB(1)=1.,3;                 SET ATTRIBUTE 1 TO 1
        ACT,,,S1;
        ACT,,,S2;
        ACT,,,S3;
;
S1      EVENT,1;
        ACT/1,XX(1);                          DISASSEMBLE
        GOON,2;
        ACT,,,S4;
        ACT,,,S5;

S2      EVENT,2;
        ACT/2,XX(2),,A25;                     INSTALL
S3      EVENT,3;
        ACT/3,XX(3),,A37;                     PREPARE
;
S4      EVENT,4;
        ACT/4,XX(4);                          INS,REP
        EVENT,8;
        ACT/8,XX(8),,A689;                    TEST
S5      EVENT,5;
        ACT/5,XX(5),,A25;                     CALIBRATE
;
A25     ACCUMULATE,2,,LAST,2;
        ACT,,,S6;
        ACT,,,S7;
;
S6      EVENT,6;
        ACT/6,XX(6),,A689;                    INTERFACES
S7      EVENT,7;
        ACT/7,XX(7),,A37;                     ASSEMBLY
;
A37     ACCUMULATE,2,,LAST,1;
        EVENT,9;
        ACT/9,XX(9),,A689;                    REPAIR
A689    ACCUMULATE,3,,LAST,1;
        COLCT,FIRST,PROJ. COMPLETION,20/15./.5,1;
        EVENT,10;
        TERM,1;
        END;
INIT,0,,N;
FIN;
```

Figure 9-4 SLAM II statement model for scheduling computations example.

Table 9-5 Variable Definitions for Slack Computation

Variable Name	Definition	Initial Value
SES(I)	Sum of early start times, activity I	0.0
SEF(I)	Sum of early finish times, activity I	0.0
SLS(I)	Sum of late start times, activity I	0.0
SLF(I)	Sum of late finish times, activity I	0.0
STS(I)	Sum of total slack, activity I	0.0
SFS(I)	Sum of free slack, activity I	0.0
TL(N)	Latest release time, node N	10**20
TE(N)	Earliest release time, node N	0.0
ARTIB(1)	Number of activities on the critical path	0.0
ATRIB(J),J>1	Sequence of activities on the critical path	
ARRAY(N,5)	Vector to input start node, end node, and parameters for activity N	See Figure 9-4
NEND	Ending node of current activity	
NSTART	Starting node of current activity	
DUR(I)	Variable for storing duration for activity I	
XLS(I)	Variable for computing latest start time for activity I	
XLF(I)	Variable for computing latest finish time for activity I	
TS(I)	Variable for computing total slack for activity I	
FS(I)	Variable for computing free slack for activity I	

```
      SUBROUTINE INTLC
C
      COMMON/SCOM1/ATRIB(100),DD(100),DDL(100),DTNOW,II,MFA,MSTOP,NCLNR
     1,NCRDR,NPRNT,NNRUN,NNSET,NTAPE,SS(100),SSL(100),TNEXT,TNOW,XX(100)
      COMMON/UCOM1/TE(6),TL(6),DUR(9),ES(9),EF(9),TS(9),FS(9),XLS(9),
     1XLF(9),SES(9),SEF(9),SLS(9),SLF(9),STS(9),SFS(9)
C**** INITIALIZE SUMMING VARIABLES OF FIRST RUN
      IF (NNRUN.LE.1) THEN
         DO 4 I = 1,9
            SES(I) = 0.0
            SEF(I) = 0.0
            SLS(I) = 0.0
            SLF(I) = 0.0
            STS(I) = 0.0
            SFS(I) = 0.0
    4    CONTINUE
      ENDIF
C**** INITIALIZE TL ON EACH RUN
      DO 10 I = 1,6
         TL(I) = 1.E20
   10 CONTINUE
      RETURN
      END
```

Figure 9-5 Subroutine INTLC for scheduling computations example.

The scheduling variables are computed in subroutine EVENT, which is given in Figure 9-6. If N is less than 10, subroutine EVENT draws a sample from a triangular distribution with the parameters associated with activity N. The early release time for the START node is set equal to TNOW. TNOW is also the early start time for activity N. The early finish time of activity N is set equal to the current time plus the duration of activity N, that is, TNOW + XX(N). The attributes of the entity passing through the EVENT node are then updated with N as the next attribute value. The next attribute to use is obtained from ATRIB(1), that is, ATRIB(NEXT) = N, where NEXT = ATRIB(1) + 1.

If the event code is 10, then the network has been simulated and the remaining scheduling calculations can be made. First, the early release time of the ending node of the network is set equal to the early finish time of the activity just completed. Also, for the ending node, the latest release time is set equal to the early release time.

Starting at statement 15, the scheduling values are calculated for all activities leading into node NEND. If an activity ends at node NEND, then its late start time is equal to the late release time for the node minus the duration of the activity. The late release time for the starting node of the activity is equal to the smallest late start time for all activities emanating from the node.

The preceding computations are made for all nodes and activities in the network. The values obtained in a run are combined with the values from other runs in the DO loop ending at statement 30. After all runs are made, average values for the scheduling variables are computed and printed. The outputs from these computations are given in Figure 9-7.

ACTIVITY NUMBER	EARLY START	LATE START	EARLY FINISH	LATE FINISH	TOTAL SLACK	FREE SLACK
1	0.00	0.64	3.03	3.67	0.64	0.00
2	0.00	1.98	5.98	7.95	1.98	1.40
3	0.00	2.82	13.89	16.71	2.82	2.12
4	3.03	6.52	11.23	14.72	3.49	0.00
5	3.03	4.08	6.91	7.95	1.05	0.47
6	7.38	9.62	18.44	20.68	2.24	2.24
7	7.38	8.70	15.39	16.71	1.32	0.62
8	11.23	14.72	17.19	20.68	3.49	3.49
9	16.01	16.71	19.98	20.68	0.70	0.70

Figure 9-7 Scheduling values computed in subroutine EVENT.

9.6 PROJECT PLANNING WITH LIMITED RESOURCES

Throughout the discussion of project planning, precedence between activities is established based on the order in which the activities are performed. When

```
      SUBROUTINE EVENT(N)
C
      COMMON/SCOM1/ATRIB(100),DD(100),DDL(100),DTNOW,II,MFA,MSTOP,NCLNR
     1,NCRDR,NPRNT,NNRUN,NNSET,NTAPE,SS(100),SSL(100),TNEXT,TNOW,XX(100)
C
      COMMON/UCOM1/TE(6),TL(6),DUR(9),ES(9),EF(9),TS(9),FS(9),XLS(9),
     1XLF(9),SES(9),SEF(9),SLS(9),SLF(9),STS(9),SFS(9)
C
      DATA LASTN /6/
      DATA XRUNS /400/
C
      IF (N .LT. 10) THEN
C******* OBTAIN PARAMETERS FOR ACTIVITY TIMES
          XLOW = GETARY(N,3)
          XMOD = GETARY(N,4)
          XHI  = GETARY(N,5)
C
C******* SAMPLE ACTIVITY TIME FOR THE TRIANGULAR DISTRIBUTION
          XX(N) = TRIAG(XLOW,XMOD,XHI,N)
C
C******* OBTAIN START NODE FOR ACTIVITY
          NN = GETARY(N,1)
C
C******* SET START NODE'S EARLIEST RELEASE TIME
          TE(NN) = TNOW
C
C******* DETERMINE ACTIVITY EARLY START AND FINISH TIMES
          ES(N) = TNOW
          EF(N) = TNOW + XX(N)
C
C******* SET INDEX FOR RECORDING ACTIVITIES ON THE ENTITY'S PATH
          ATRIB(1) = ATRIB(1) + 1.
C
C******* SET NEXT ATTRIBUTE TO CURRENT ACTIVITY NUMBER
          NEXT = ATRIB(1)
          ATRIB(NEXT) = N
      ELSE
C******* SET I EQUAL TO THE NUMBER OF LAST ATTRIBUTE FOR CURRENT
C******* ENTITY AND SET J EQUAL TO THE ACTIVITY NUMBER OF THE
C******* LAST ACTIVITY ON THE CRITICAL PATH FOR THIS RUN
          I = ATRIB(1)
          J = ATRIB(I)
C
C******* SET NEND EQUAL TO THE ENDING NODE FOR THE NETWORK
          NEND = LASTN
C******* SET THE EARLIEST FINISH TIME FOR THE NETWORK
          TE(NEND) = EF(J)
C******* SET THE LATEST FINISH TIME FOR THE NETWORK
          TL(NEND) = TE(NEND)
C
C******* LOCATE ALL ACTIVITIES LEADING TO NODE NEND
   15     CONTINUE
```

Figure 9-6 Subroutine EVENT for scheduling computations example.

```
            DO 20 I = 1,9
               NN = GETARY(I,2)
               IF (NN.EQ.NEND) THEN
C**************** SET XLS FOR THIS ACTIVITY
                  XLS(I) = TL(NEND) - XX(I)
C**************** LOCATE THE START NODE NSTART FOR THIS ACTIVITY
                  NSTART = GETARY(I,1)
C**************** SET TL FOR NODE NSTART
                  IF(XLS(I).LT.TL(NSTART))  TL(NSTART) = XLS(I)
               ENDIF
   20       CONTINUE
C
C********** DECREMENT NEND FOR THE NEXT PRECEDING NODE ON THE NETWORK
            NEND = NEND - 1
         IF (NEND.GT.1) GO TO 15
C
C******* LOCATE PRECEDING ACTIVITIES FOR ALL REMAINING NODES ON NETWORK
         DO 30 I = 1,9
C********** CALCULATE XLF, TS, AND FS FOR ALL ACTIVITIES OF THE NETWORK
            XLF(I) = XLS(I) + XX(I)
            TS(I) = XLF(I) - EF(I)
            NN = GETARY(I,2)
            FS(I) = TE(NN) - EF(I)
C********** ADD OBSERVATIONS FOR THIS RUN TO SUMS
            SES(I) = SES(I) + ES(I)
            SEF(I) = SEF(I) + EF(I)
            SLS(I) = SLS(I) + XLS(I)
            SLF(I) = SLF(I) + XLF(I)
            STS(I) = STS(I) + TS(I)
            SFS(I) = SFS(I) + FS(I)
   30    CONTINUE
C
         IF (NNRUN .GE. XRUNS) THEN
C********** CALCULATE AND PRINT AVERAGES FOR THE SIMULATION
            WRITE(6,100)
            DO 40 I = 1,9
               SES(I) = SES(I)/XRUNS
               SEF(I) = SEF(I)/XRUNS
               SLS(I) = SLS(I)/XRUNS
               SLF(I) = SLF(I)/XRUNS
               STS(I) = STS(I)/XRUNS
               SFS(I) = SFS(I)/XRUNS
               WRITE(6,200)I,SES(I),SLS(I),SEF(I),SLF(I),STS(I),SFS(I)
   40       CONTINUE
         ENDIF
      ENDIF
C
  100 FORMAT(///,1X,'ACTIVITY',6X,'EARLY',7X,'LATE',8X,'EARLY',7X,
     1'LATE',8X,'TOTAL',7X,'FREE',/2X,'NUMBER',7X,'START',7X,'START',6X
     2,'FINISH',6X,'FINISH',7X,'SLACK',7X,'SLACK'//)
  200 FORMAT(I6,4X,F10.2,5(2X,F10.2))
      RETURN
      END
```

Figure 9-6 Subroutine EVENT for scheduling computations example (concluded).

activities compete for limited resources, there exists an implicit set of precedence constraints that dictates that activities must be performed sequentially because of the lack of resources that can be allocated simultaneously. The requirement of resources for performing activities requires the use of resource allocation procedures, rules, or algorithms to establish which activities to start when an activity's precedence requirements are satisfied.

SLAM II can be used to model project planning with limited resources. The complexity of the decision processes associated with resource allocation in project networks may require specialized tools. The work of J. Wiest and his development of the SPAR programs [4] has been instrumental in fostering developments in the field. Commercially available packages for allocating resources in PERT-type networks are available [2]. Research on project planning with limited resources for GERT-type networks has also been explored, and specialized GERT programs were developed for this purpose. The research by Wortman on START [5] and Wortman, Duket, Seifert, and Chubb on SAINT [6, 7] has been significant. Q-GERT has also been successfully applied to the resource-constrained project network problem.

9.7 REPAIRMAN MODEL WITH RESOURCE REQUIREMENTS

The network for the repair project is redrawn in Figure 9-7 with the addition of the number of repairmen required to perform an activity included below the branch in a triangle. From Figure 9-8, activities 1 and 2 require two repairmen

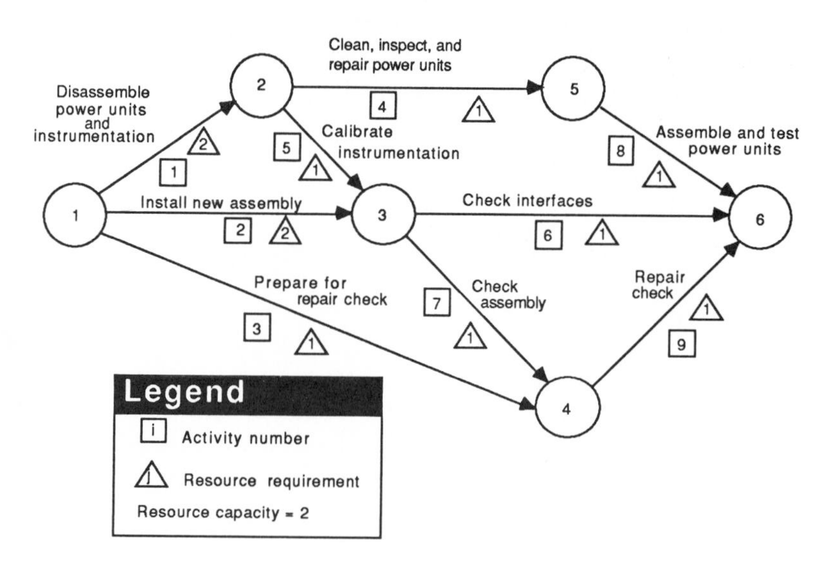

Figure 9-8 Network model of a repair project with resource requirements.

but only one repairman is necessary to perform all other activities. Questions that relate to the analysis of this network are similar to those previously presented, but, in addition, the utilization of the repairmen is of interest. Decision making relates to the total amount of resources to be allocated to the project (in this example the two repairmen) and, once allocated, the manner in which they should be used to perform the activities. Trade-offs clearly exist between resource capacity and project duration.

Figure 9-9 presents a SLAM II model of the repair project with the resource allocation decisions modeled explicitly on the network. Since resources are to be allocated, entities are placed in AWAIT nodes to prevent the start of activities until the resources are available to complete an activity. Activities 1, 2, and 3 follow AWAIT nodes A1, A2, and A3, where the repairman resource is allocated. At the start of a run, an entity is placed in each of these AWAIT nodes. Activities 1 and 2 require two units of the resource REPMEN while activity 3 requires one unit of the resource. For this model, there are only two units of REPMEN available. The first decision process involves the decision to allocate repairmen to one of these three activities. In the resource block shown in Figure 9-9, priority is first given to activity 1, then activity 2, and activity 3 next. For convenience, file numbers of AWAIT nodes and their following activity numbers have been assigned the same values. The two available repairmen are both allocated to the waiting entity in node A1, and activity 1 is started. When activity 1 is completed, two units of the resource are freed at FREE node F1. These two units are then reallocated to the entity waiting in node A2, and activity 2 is initiated; nodal statistics are collected at COLCT node N2, and an entity is placed in AWAIT nodes A4 and A5 to wait for the allocation of resources.

When activity 2 is completed, the repairmen are freed at node F2. One unit of the freed resource is allocated to the entity in node A3 to began activity 3. The other repairman starts working on activity 5 according to the priority established in the RESOURCE block. In addition, the first entity requirement for ACCUMULATE node N3 is satisfied by the completion of activity 2. When either activity 3 or 5 is completed, the freed resource is allocated to the entity waiting in node A4, and activity 4 is initiated. The network model has incorporated a decision to use the same repairman on activity 8 that works on activity 4. Therefore, these two activities are performed sequentially.

When activity 5 is completed, the second requirement for the release of node N3 is satisfied, statistics are collected, and an entity is placed in both nodes A6 and A7. Node A7 is given priority over node A6 for the next available resource. A description of the remainder of the network is similar to the preceding. The project is completed when node N6 is released, which occurs after activities 6, 8, and 9 are completed.

Subroutine OTPUT is used to obtain waiting time and resource statistics over

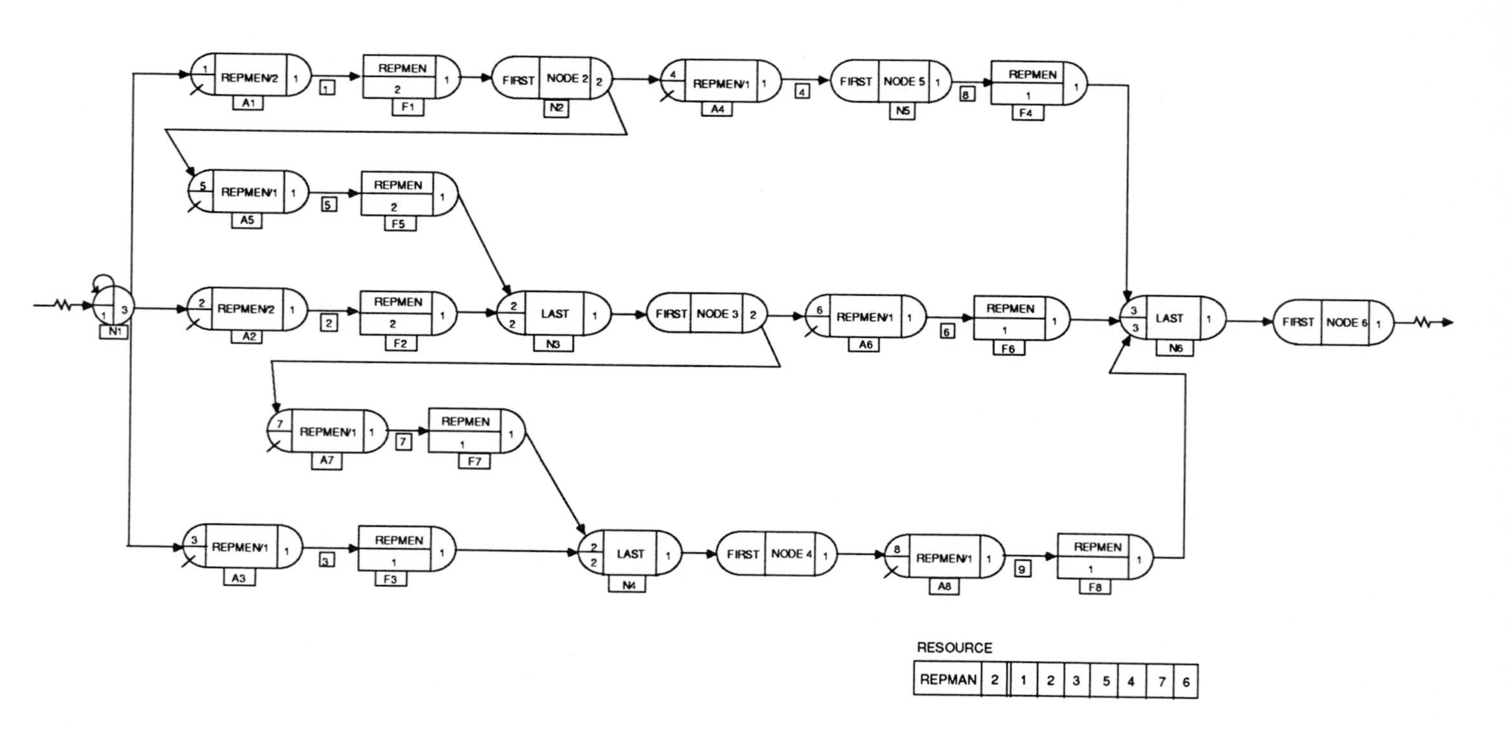

Figure 9-9 SLAM II model of repairman project with resource constraints.

multiple runs and to print out a stylized summary report. The FORTRAN listing for subroutine OTPUT is given in Figure 9-10.

The SLAM II statement model for this illustration is given in Figure 9-11, and demonstrates the large number of decisions involved in project planning under limited resources. The output from the statement model presented in Figure 9-11 is shown in Figure 9-12. The number of repairmen busy is estimated at 1.9; that is, on the average each repairman is busy 95% of the time. With only two repairmen, no free time has been built into the project, and the only idle time occurs at the end of the project. This occurs when one repairman finishes an activity and no other activities are to be performed. Thus, on each project realization one repairman will be busy 100% of the time, while the other has, on the average, 10% idle time at the end of the project. For more complex projects, repairmen would normally wait for other repairmen to become available before activities can be started.

For this project with two repairmen, it is possible to estimate some of the activity starting delay times. For example, the delay time in starting activity 2 is equal to the time to perform activity 1. Since the average time to perform activity 1 is three time units, it is expected that there will be a 3-time unit delay

```
      SUBROUTINE OTPUT
C
      INCLUDE 'PARAM.INC'
C
      COMMON/SCOM1/ATRIB(100),DD(100),DDL(100),DTNOW,II,MFA,MSTOP,NCLNR,
     +              NCRDR,NPRNT,NNRUN,NNSET,NTAPE,SS(100),SSL(100),TNEXT,
     +              TNOW,XX(100)
      DIMENSION XCRIT(9)
C
      CHARACTER*4 NNAME(5),NNPRJ(5)
      COMMON/CCOM5/ NNAME,NNPRJ
C
      COMMON/GCOM5/ IISED(MSTRM),JJBEG,JJCLR,MMNIT,MMON,NNCFI,NNDAY,
     +               NNPT,NNRNS,NNSTR,NNYR,SSEED(MSTRM),LSEED(MSTRM)
C
      INTEGER NUMRSC
C
      DO 10 I = 1,8
         CALL COLCT(FFAWT(I),I)
   10 CONTINUE
C
      CALL COLCT (RRAVG(1),9)
      IF (NNRUN .GE. 400) THEN
         NUMRSC = NNRSC(1)
         WRITE (6,20) NNAME,MMON,NNDAY,NNYR,NUMRSC,NNRUN
         CALL PRNTC(0)
         CALL PRNTH (14)
      ENDIF
C
      RETURN
   20 FORMAT (35X,42HREPAIRMAN MODEL WITH RESOURCE REQUIREMENTS,//,36X,
     +         3HBY ,5A4,2X,5HDATE ,I2,1H/,I2,1H/,I4,//,37X,
     +         11HOUTPUT FOR ,I2,15H REPAIRMEN AND ,I5,5H RUNS)
      END
```

Figure 9-10 Subroutine OTPUT for repairman project with resource requirements.

```
GEN,RUTH WHITIS,REPMEN AND RESOURCES,1/18/88,400,N,N,,N,N;
LIMITS,8,2,100;
STAT,1,AVE DELAY ACT 1;                        FILE 1
STAT,2,AVE DELAY ACT 2;                        FILE 2
STAT,3,AVE DELAY ACT 3;                        FILE 3
STAT,4,AVE DELAY ACT 4;                        FILE 4
STAT,5,AVE DELAY ACT 5;                       FILE 5
STAT,6,AVE DELAY ACT 6;                       FILE 6
STAT,7,AVE DELAY ACT 7;                       FILE 7
STAT,8,AVE DELAY ACT 9;                       FILE 8
STAT,9,AVE UTIL REPMEN;                       RESOURCE 1
NETWORK;
      RESOURCE/REPMEN(2),1,2,3,5,4,7,6,8;  SET REPAIRMEN AT 2
      CREATE,,,,1,3;                       START SIMULATION
      ACT,,,A1;
      ACT,,,A2;
      ACT,,,A3;
A1    AWAIT(1),REPMEN/2,,1;                2 MEN NEEDED FOR JOB 1, PRI 1
      ACT/1,TRIAG(1,3,5,1),,F1;;           ACTIVITY TIME FOR JOB 1
A2    AWAIT(2),REPMEN/2,,1;                2 MEN NEEDED FOR JOB 2, PRI 2
      ACT/2,TRIAG(3,6,9,2),,F2;;           ACTIVITY TIME FOR JOB 2
A3    AWAIT(3),REPMEN/1,,1;                1 MAN NEEDED FOR JOB 3, PRI 3
      ACT/3,TRIAG(10,13,19,3),,F3;;        ACTIVITY TIME FOR JOB 3
F1    FREE,REPMEN/2,1;                     JOB 1 COMPLETE.  FREE 2 MEN
N2    COLCT,FIRST,NODE 2,,2;               COLLECT STATISTICS (JOB 1)
      ACT,,,A4;
      ACT,,,A5;
F2    FREE,REPMEN/2,,1;                    JOB 2 COMPLETE.  FREE 2 MEN
      ACT,,,N3;
F3    FREE,REPMEN/1,,1;                    JOB 3 COMPLETE.  FREE 1 MAN
      ACT,,,N4;
A4    AWAIT(4),REPMEN/1,,1;                1 MAN NEEDED FOR JOB 4, PRI 5
      ACT/4,TRIAG(3,9,12,4),,N5;;          ACTIVITY TIME FOR JOB 4
A5    AWAIT(5),REPMEN/1,,1;                1 MAN NEEDED FOR JOB 5, PRI 4
      ACT/5,TRIAG(1,3,8,5),,F5;;           ACTIVITY TIME FOR JOB 5
F5    FREE,REPMEN/1,,1;                    JOB 5 COMPLETE.  FREE 1 MAN
N3    ACCUMULATE,2,2,LAST,1;               WAIT FOR JOBS 2 & 5 TO FINISH
      COLCT,FIRST,NODE 3,,2;               COLLECT STATISTICS (JOBS 2 & 5)
      ACT,,,A6;
      ACT,,,A7;
A6    AWAIT(6),REPMEN/1,,1;                1 MAN NEEDED FOR JOB 6
      ACT/6,TRIAG(8,9,16,6),,F6;;          ACTIVITY TIME FOR JOB 6
A7    AWAIT(7),REPMEN/1,,1;                1 MAN NEEDED FOR JOB 7
      ACT/7,TRIAG(4,7,13,7),,F7;;          ACTIVITY TIME FOR JOB 7
F7    FREE,REPMEN/1,,1;                    JOB 7 COMPLETE.  FREE 1 MAN
N4    ACCUMULATE,2,2,LAST,1;               WAIT FOR JOBS 3 & 7 TO FINISH
      COLCT,FIRST,NODE 4,,1;               COLLECT STATISTICS (JOBS 3 & 5)
      ACT,,,A8;
N5    COLCT,FIRST,NODE 5,,1;               COLLECT STATISTICS (JOB 5)
      ACT/8,TRIAG(3,6,9,8),,F4;;           ACTIVITY TIME FOR JOB 8
F4    FREE,REPMEN/1,,1;                    JOB 8 COMPLETE.  FREE 1 MAN
      ACT,,,N6;
F6    FREE,REPMEN/1,,1;                    JOB 6 COMPLETE.  FREE 1 MAN
      ACT,,,N6;
A8    AWAIT(8),REPMEN/1,,1;                1 MAN NEEDED FOR JOB 8
      ACT/9,TRIAG(1,3,8,9),,F8;;           ACTIVITY TIME FOR JOB 8
F8    FREE,REPMEN/1,,1;                    JOB 8 COMPLETE.  FREE 1 MAN
N6    ACCUMULATE,3,3,LAST,1;               WAIT FOR JOBS 6, 8, & 9 TO FIN
      COLCT,FIRST,PROJECT COMPLETE,20/30,/1,,1; PROJECT STATISTICS
      TERM,1;                              ALL JOBS FINISHED.  END RUN
      END;
INIT,0,,N;
FIN;
```

Figure 9-11 SLAM II statement model for repairman project with resources.

REPAIRMAN MODEL WITH RESOURCE REQUIREMENTS

BY RUTH WHITIS DATE 1/18/1988

OUTPUT FOR 2 REPAIRMEN AND 400 RUNS

STATISTICS FOR VARIABLES BASED ON OBSERVATION

	MEAN VALUE	STANDARD DEVIATION	COEFF. OF VARIATION	MINIMUM VALUE	MAXIMUM VALUE	NUMBER OF OBSERVATIONS
AVE DELAY ACT 1	0.0000E+00	0.0000E+00	0.9999E+04	0.0000E+00	0.0000E+00	400
AVE DELAY ACT 2	0.3029E+01	0.7893E+00	0.2606E+00	0.1160E+01	0.4821E+01	400
AVE DELAY ACT 3	0.9005E+01	0.1512E+01	0.1679E+00	0.4462E+01	0.1292E+02	400
AVE DELAY ACT 4	0.9855E+01	0.1902E+01	0.1930E+00	0.5375E+01	0.1507E+02	400
AVE DELAY ACT 5	0.5976E+01	0.1284E+01	0.2149E+00	0.3072E+01	0.8955E+01	400
AVE DELAY ACT 6	0.1407E+02	0.1924E+01	0.1368E+00	0.8369E+01	0.2029E+02	400
AVE DELAY ACT 7	0.9795E+01	0.2303E+01	0.2351E+00	0.3591E+01	0.1602E+02	400
AVE DELAY ACT 9	0.3166E+00	0.1016E+01	0.3208E+01	0.0000E+00	0.6236E+01	400
AVE UTIL REPMEN	0.1904E+01	0.5547E-02	0.2913E-02	0.1888E+01	0.1950E+01	400
NODE 2	0.3029E+01	0.7893E+00	0.2606E+00	0.1160E+01	0.4821E+01	400
NODE 3	0.1288E+02	0.2106E+01	0.1635E+00	0.7439E+01	0.1848E+02	400
NODE 4	0.3069E+02	0.2973E+01	0.9685E-01	0.2213E+02	0.3833E+02	400
NODE 5	0.2108E+02	0.2753E+01	0.1306E+00	0.1456E+02	0.2841E+02	400
PROJECT COMPLETE	0.3839E+02	0.3114E+01	0.8112E-01	0.2993E+02	0.4990E+02	400

Figure 9-12 SLAM II output report for repairman model with two repairmen.

in starting activity 2. Similarly, the delay in starting activity 3 is equal to the sum of the times to perform activities 1 and 2, which is 9 time units. The delay in starting activity 5 is the time required to perform activity 2, or approximately 6 time units. It is not possible to estimate directly all the delay times, as they depend on the order in which activities are completed. By reviewing traces or animations of different runs, the possible sequences of activity completions may be observed.

The estimate of the average time to complete the project is 38.39 time units. The estimated time to complete the project without resources is 20.68 time units. Thus, the addition of resource requirements adds approximately 17.71 time units to the project duration. Not only has the average project duration increased, but the range of possible project durations has widened, as indicated by the minimum and maximum values observed. With resource requirements, the range is approximately 20 time units, whereas for the unconstrained resource case it is only 11.4 time units. The histogram presented in Figure 9-13 shows there is a 12% chance that the project will take greater than 42 time units, which is more than double the average project duration if resources are not considered.

To investigate the effect of increasing the number of repairmen, this SLAM II project network was rerun with the capacity of the repairmen resource increased from two to three, four, and five. A summary of the results from these runs is shown in Table 9-6. As expected, the average project completion time decreases and approaches the estimate obtained for the unlimited resource case. Also shown is the decrease in the standard deviation and the minimum and maximum of the project completion times. In analyzing the results of the runs made for the four-repairmen case, it is seen that only activities 3 and 5 are

```
                                   **HISTOGRAM NUMBER 14**

                                       PROJECT COMPLETE

OBSV     RELA      CUML          UPPER
FREQ     FREQ      FREQ       CELL LIMIT       0         20        40        60        80       100
                                               +    +    +    +    +    +    +    +    +    +    +
   1    0.002     0.002      0.3000E+02        +                                                 +
   0    0.000     0.002      0.3100E+02        +                                                 +
   6    0.015     0.018      0.3200E+02        +*                                                +
   6    0.015     0.032      0.3300E+02        +*C                                               +
  15    0.038     0.070      0.3400E+02        +** C                                             +
  26    0.065     0.135      0.3500E+02        +***   C                                          +
  30    0.075     0.210      0.3600E+02        +****      C                                      +
  56    0.140     0.350      0.3700E+02        +*******          C                             +
  55    0.138     0.488      0.3800E+02        +*******                C                       +
  47    0.117     0.605      0.3900E+02        +******                       C                 +
  35    0.087     0.692      0.4000E+02        +****                              C            +
  48    0.120     0.813      0.4100E+02        +******                                  C      +
  27    0.068     0.880      0.4200E+02        +***                                        C   +
  17    0.043     0.923      0.4300E+02        +**                                          C  +
  13    0.032     0.955      0.4400E+02        +**                                           C +
   8    0.020     0.975      0.4500E+02        +*                                             C+
   5    0.013     0.988      0.4600E+02        +*                                             C+
   3    0.007     0.995      0.4700E+02        +                                               C
   1    0.002     0.998      0.4800E+02        +                                               C
   0    0.000     0.998      0.4900E+02        +                                               C
   1    0.002     1.000      0.5000E+02        +                                               C
   0    0.000     1.000         INF            +                                               C
 -                                          +    +    +    +    +    +    +    +    +    +    +
 400                                           0         20        40        60        80       100


                         **STATISTICS FOR VARIABLES BASED ON OBSERVATION**

                          MEAN          STANDARD        MINIMUM         MAXIMUM        NUMBER OF
                          VALUE         DEVIATION       VALUE           VALUE          OBSERVATIONS

      PROJECT COMPLETE  0.3839E+02     0.3114E+01     0.2993E+02      0.4990E+02          400
```

Figure 9-13 Histogram of project completion times for repairman model with two repairmen.

delayed because of limited resources. Thus, when the fifth repairman is employed, no activities are delayed, and the resource-constrained model is equivalent to the unlimited resource situation.

The decrease in average project completion time is obtained at the cost of additional repairmen. Clearly, there are benefits for completing the project

Table 9-6 Effect of Number of Repairmen on Project Duration and Resource Utilization

Number of	Project Completion Time					Resource Utilization
Repairmen	Average	Std. Dev.	Min.	Max.	Average	Ave/Rep
2	38.39	3.11	29.93	49.90	1.90	0.95
3	27.26	2.52	20.50	33.77	2.68	0.89
4	23.24	2.43	17.13	30.45	3.15	0.79
5	20.68	2.10	15.66	27.05	3.53	0.71

earlier, and these benefits must be compared against the increased cost of assigning repairmen to the project. Note that, as the number of repairmen is increased, the utilization of each repairman decreases. This occurs because the repairmen are waiting for precedence activities to be completed before they can start work on their next assigned activity.

A common approach in project planning is to assign resources on an as needed basis. In SLAM II, this would be modeled by initially setting the capacity of the resource to a given value and then altering this capacity. First, the capacity is altered by a positive amount and then by a negative amount to change the resource availability at specific times during the performance of the project.

This analysis has shown that a significant amount of information is available in the SLAM II summary reports to enable project planning when resources are constrained and activity durations are random variables.

9.7.1 Alternative SLAM II Model

An alternative way to model the repair project with limited resources is shown in Figure 9-14. In this model, the decision to allocate resources is accomplished by ranking entities waiting for activities requiring the same number of resource units in a common file. Thus all AWAIT nodes reference either file 1 or file 2. The AWAIT nodes preceding activities requiring one repairman reference file 1, and those that precede activities requiring two repairmen reference file 2. In this model, the activity released first is given priority when allocating resources units.

When the required resource units are allocated to perform an activity, the activity is started. After the activity is performed, other activities are initiated in accordance with the network predecessor relations. Also, following the completion of an activity, the resource units used by that activity are freed at either FREE node F1 or F2. Since it is desired to have the reallocation of freed resources occur after the release of successor activities, a small time delay in freeing the resources at nodes F1 and F2 is included in the network description. This is represented on the activities leading to the FREE nodes by a δ where $\delta = 0.00001$.

The project presented in Figure 9-14 represents a general approach to project planning with limited resources. By using a ranking attribute, different scheduling procedures may be modeled. Possible ranking attributes are: number of following activities, total work content (the product of resources units and time units) of following activities, slack measures, and activity duration. The computation of such priority values for entities could be made at ASSIGN nodes, entered through DATA or ARRAY statements, initialized in subroutine INTLC,

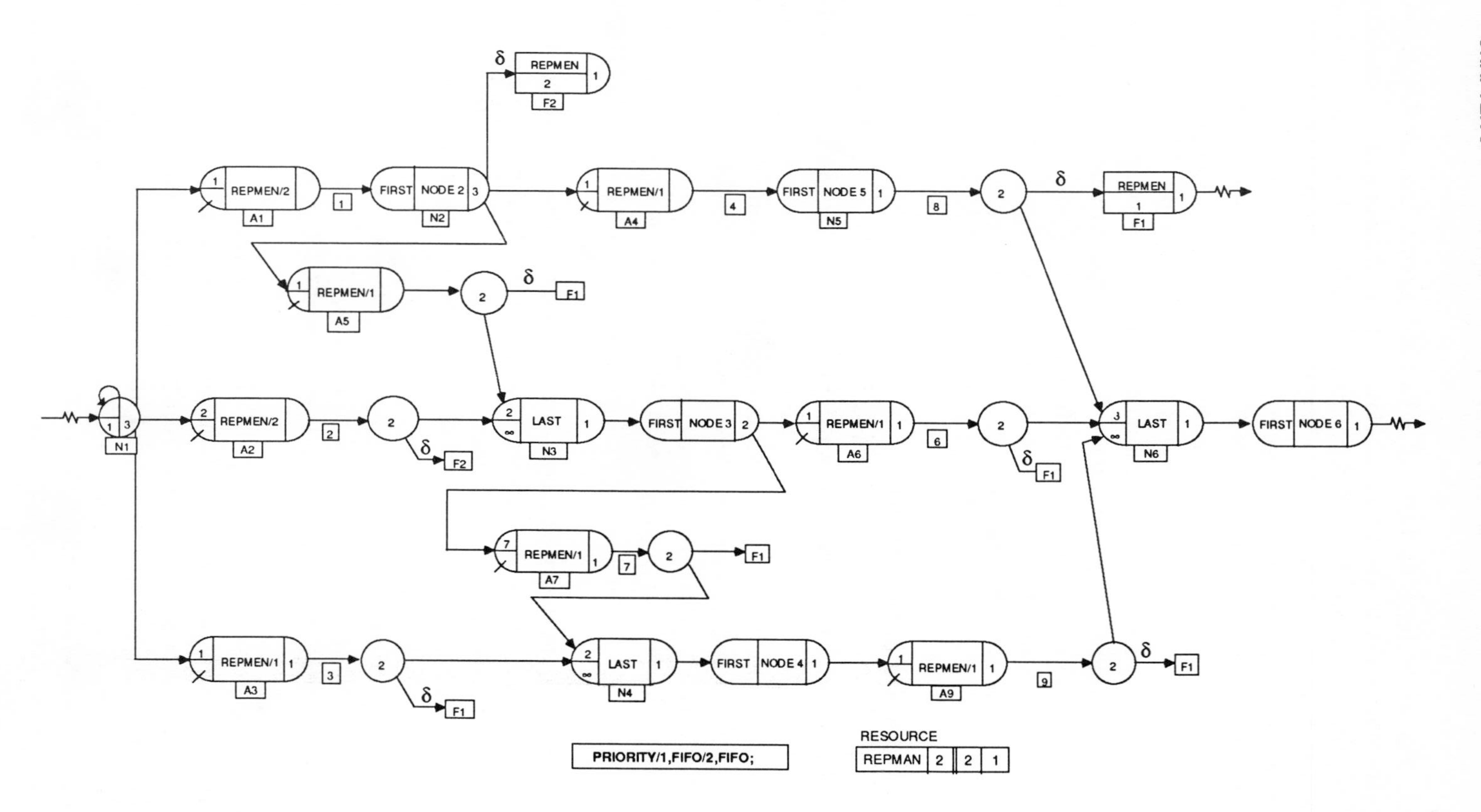

Figure 9-14 Alternative SLAM II model of repairman project with limited resources.

and/or computed at EVENT nodes or in function USERF. These vehicles allow the priority to be dynamically computed.

Another embellishment to the model is to vary the availability of resources and to stop one activity to start another when existing resources of a higher-priority activity are released. This approach to project planning could be used when activities have multiple resource requirements and complex allocation procedures are employed. Through the use of these concepts, complex scheduling rules for modeling project planning with limited resources may be accomplished.

9.8 SUMMARY

In this chapter, scheduling and resource concepts associated with project planning are described. Methods for computing start and finish times and slack values are presented. Procedures and models for performing project planning with limited resources are illustrated. This chapter establishes the foundation for using SLAM II to make project planning decisions.

9.9 EXERCISES

9-1. The following list represents the operations involved in a plant maintenance project to remove and replace a section of a pipe [1]. Draw the PERT network for this project, making appropriate assumptions regarding precedence requirements.

Activity	Description	Expected Duration
1	Prepare materials list	8
2	Deactivate line	8
3	Procure and deliver pipe materials	120
4	Procure and deliver valves	120
5	Procure and deliver paint	16
6	Procure and deliver insulation	16
7	Erect scaffold	12
8	Remove old pipe and valves	36
9	Prefabricate pipe sections (except valves)	40
10	Install prefabricated pipe sections and valves	24
11	Start insulation of pipe and valves	8
12	Pressure test pipe and valves	8
13	Complete insulation	8
14	Paint insulation	16
15	Remove scaffold	4
16	Reactivate line	8

Restrictions:

1. The scaffold is required for removing the old pipe but not for deactivating the line.
2. Operations 11 and 12 need to be done concurrently and must precede operation 13.
3. The pipe can be insulated after it is back in service. Perform a detailed scheduling analysis on the network. Include the computation of slack time, early and late start and finish times, and the variability associated with the completion time.

9-2. Explain the differences in project scheduling when a due date for a project is prescribed versus the situation when one is not prescribed. Interpret the various slack measures for each situation.

9-3. Discuss the assumptions made when computing slack times using PERT/CPM procedures. Give a rationale for using such procedures and assess the impact of the approximations required.

9-4. Discuss the impact of project planning activities, including the allocation of resources to balance path times, on the probability of exceeding the expected completion time (a) when PERT is used and (b) when SLAM II is used.

9-5. For the network of Exercise 9-1, the following resources are required for each activity:

Activity	Plumbers	Helpers
1	0	1
2	1	1
3	0	1
4	0	1
5	0	1
6	0	1
7	0	2
8	1	1
9	1	0
10	1	1
11	1	1
12	1	1
13	1	1
14	0	1
15	0	2
16	1	1
Number available	1	2

In the face of these limited resources, plan the project so as to minimize the total time required to perform the project.

9-6. For the network of Exercise 9-1, perform a detailed scheduling analysis using the SLAM II processor. Compare the results with those obtained in Exercise 9-1.

9-7. For the network of Exercise 9-1, assuming exponentially distributed durations, evaluate the criticality of each activity based on total slack, the ratio index, and the criticality index.

9.10 REFERENCES

1. Horowitz, J., *Critical Path Scheduling*, Ronald Press, New York, 1976.
2. Moder, J. J., C. R. Phillips, and E. W. Davis, *Project Management with CPM, PERT, and Precedence Diagramming*, 3rd ed., Van Nostrand Reinhold, New York, 1970.
3. Petersen, P., "Project Control Systems," *Datamation*, June 1979, pp. 147-162.
4. Wiest, J. D., "A Heuristic Model for Scheduling Large Projects with Limited Resources," *Management Science*, Vol. 13, No. 6, Febuary 1967, pp. B359-B377.
5. Wortman, D. B., "A Simulation Technique for Allocating Resources to Tasks," Master's thesis, Purdue University, West Lafayette, IN, December 1977.
6. Wortman, D. B., S. D. Duket, and D. J. Seifert, "Simulation of a Remotely Piloted Vehicle/Drone Control Facility Using SAINT," *Proceedings, Summer Computer Simulation Conference,* San Francisco, CA, 1975.
7. Wortman, D. B., and others, *New SAINT Concepts and the SAINT Simulation Program*, AMRL-TR-75, Aerospace Medical Research Laboratory, Wright-Patterson Air Force Base, OH, April 1975.

10 PROJECT PLANNING: APPLICATIONS

10.1 INTRODUCTION

This chapter describes three applications of project planning concepts and procedures. The applications selected are from actual case studies reported in the literature. Only the essential elements are presented, and the literature cited should be consulted for complete details. The three applications relate to contract bidding and negotiations, research and development planning, and technological forecasting. The large overlap of the project planning issues in these areas is brought to light in the description of management concerns for each application. Through these applications, the modeling of such concerns and the quantitative assessment of alternative courses of action are demonstrated.

10.2 AN INDUSTRIAL CONTRACT NEGOTIATION PROCESS

This example analyzes the bidding process of a construction firm that specializes in building gasoline plants. The firm's clients are large oil companies, which send out request for bids on new gasoline plants to the construction firm and its competitors. A successful bidding process is essential, but costly and time consuming.

A study by Bird, Clayton and Moore [2] analyzed the bidding process from the viewpoint of the construction firm. The study provided management with a means for assessing the cost and effectiveness of the bidding process. In addition, alternative modes of operation were analyzed for the purpose of improving this key aspect of the business. Although this example involves the construction industry, the managerial concerns presented are common to a large number of industrial negotiation situations. In fact, Moore and Clayton [8] present a network analysis of a similar situation from the vantage point of an oil company's management faced with the task of determining the number of vendors from which to request bids .

10.2.1 System and Management Concerns

To introduce the contract bidding or negotiation process, consider the schematic diagram in Figure 10-1. The process begins with an initial client contract, which, in this case, is an oil company requesting a bid on the construction of a gasoline plant. The firm initiates a preliminary study of the proposed project. This involves the initial reports to management and the preparation of engineering, production, financial, marketing, and purchasing reports. Each of these evaluations may indicate that the proposed project is not in the best interests of the construction firm. It is management policy to decide against bidding if any of the evaluations are negative. When this occurs, the costs of both the preliminary and detailed studies are considered as sunk costs. If, however, all the reports are positive, a management conference is held and more detailed studies ensue. If no negative recommendations are made as a result of the detailed studies, marketing and engineering plans are finalized, and the client is presented with a detailed project proposal. The client may reject the bid, accept it, or request revisions. Acceptance or rejection completes the contract negotiation process. Any stated revisions are incorporated into the proposal, and the client evaluation process is repeated until either acceptance or rejection is obtained.

Each activity described above involves time and money, and each has an impact on the total effectiveness and success of the construction firm. The central questions that relate to the firm's profitability for this phase of its business are as follows:

1. How much time and money does it take to obtain a successful contract?
2. How much time and money is spent in bidding for contracts that are lost?
3. What is the probability of winning or losing a contract?
4. How much time and money are spent in each phase of the bidding process?

A network model assists in answering these questions. The model provides a vehicle for evaluating ways to redesign the current mode of operation. Important questions when considering alternative operational procedures are the following:

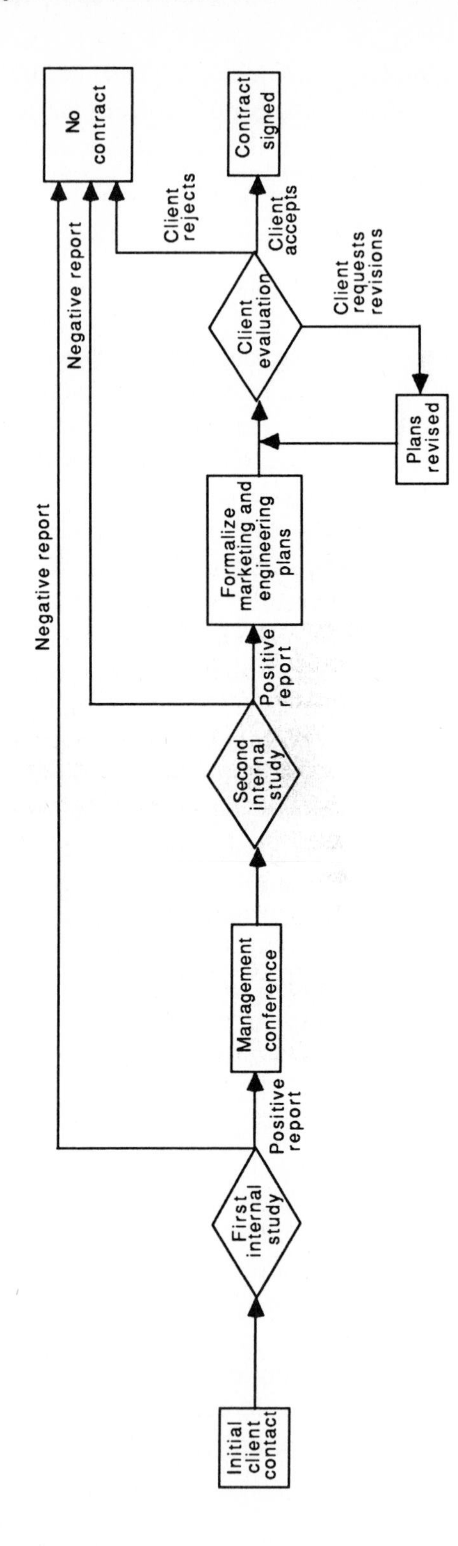

Figure 10-1 Schematic of contract negotiation process.

1. What is the effect on time and costs if initial report activities are resequenced such that activities with a high probability of rejecting a contract precede lower-risk activities?
2. What is the effect on time and costs from an increase in the expenditure of resources in the first report stage? How does this affect the probability of obtaining a contract?

The discussion in this section concentrates on the network logic used to model the random time elements and the probabilistic outcomes of the negotiation process. Emphasis is placed on how network statistics provide information to answer the questions posed.

10.2.2 The SLAM II Model

Figure 10-2 is the SLAM II network model of the negotiation process. Fixed and variable costs are associated with each activity. A fixed cost is charged each time an activity occurs, while a variable cost accumulates over the period of time required to complete the activity. In the following description, activity numbers are given in boxes.

The process begins with the initial contact by a client [1]. Reports to the marketing vice-president [2] and to the president [3] are then made sequentially. Current policy is to perform a preliminary analysis for a project, which consists of five parallel studies: engineering [4], production [5], financial [6], marketing [7], and purchasing [8]. Except for the production report, the evaluation of the reports is performed at the GOON node following the activities, as represented by activities [9], [10], and [13] through [18]. The probabilistic branching permits the modeling of uncertain outcomes in each case. For example, activities 9 and 10 represent the outcomes of the engineering report, which has a 20% chance of recommending against a contract and an 80% chance of initial approval. If any one of the reports is negative, the project is not negotiated further, and the network terminates at COLCT node C22. COLCT node C22, user function USERF(38), and COLCT node C23 represent a no-contract decision. The evaluation of the production schedule is complicated by the possibility of production subcontracting [11]. An unfavorable production subcontract could occur [19], which also leads to a termination of the negotiation process.

When all five reports are favorable, ACCUMULATE node A12 is released and FIRST statistics are collected to obtain estimates on the time required to reach this decision point. Cost data are also collected to estimated the costs of the negotiation process up to the time at which the five favorable reports are obtained. Corporate-level planning conferences are then held based on the five

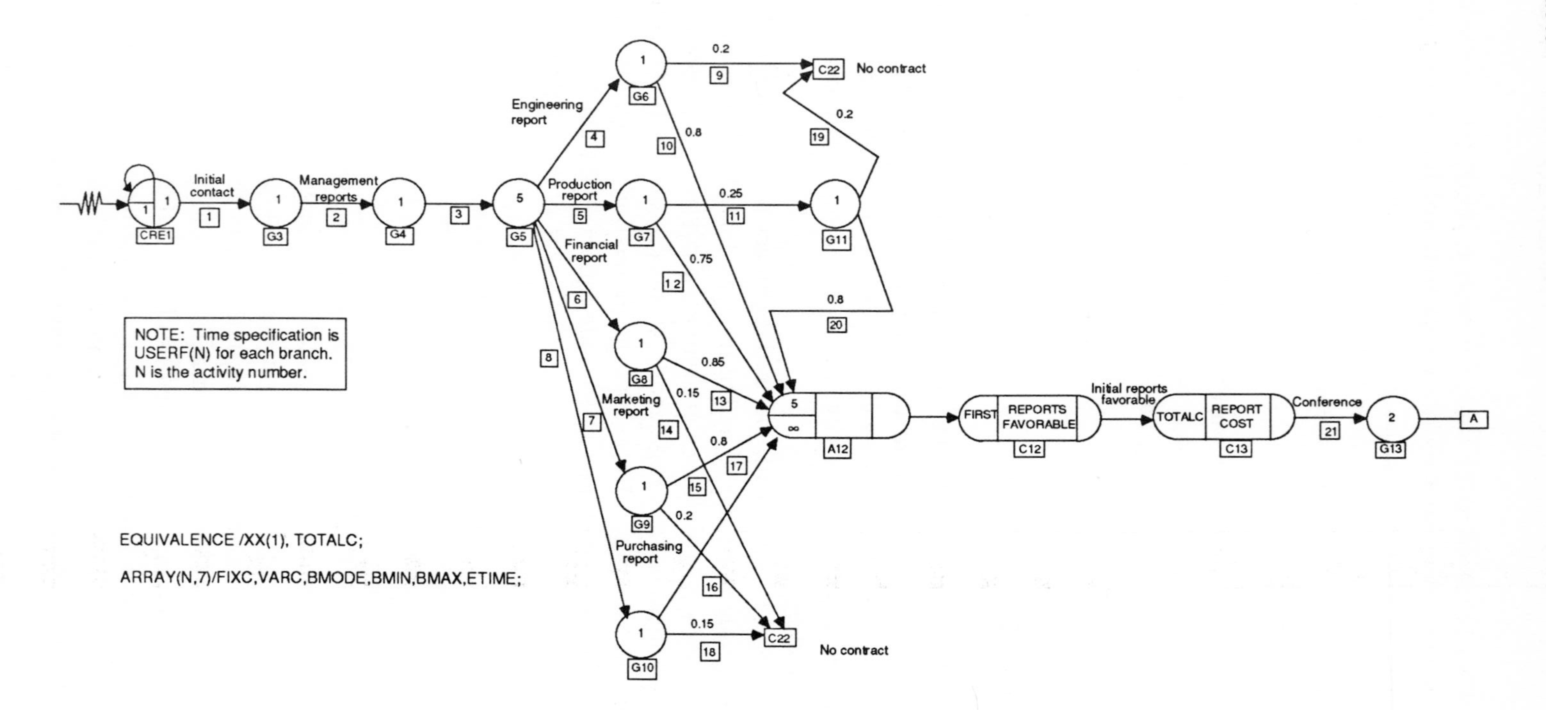

Figure 10-2 SLAM II network for industrial negotiation study.

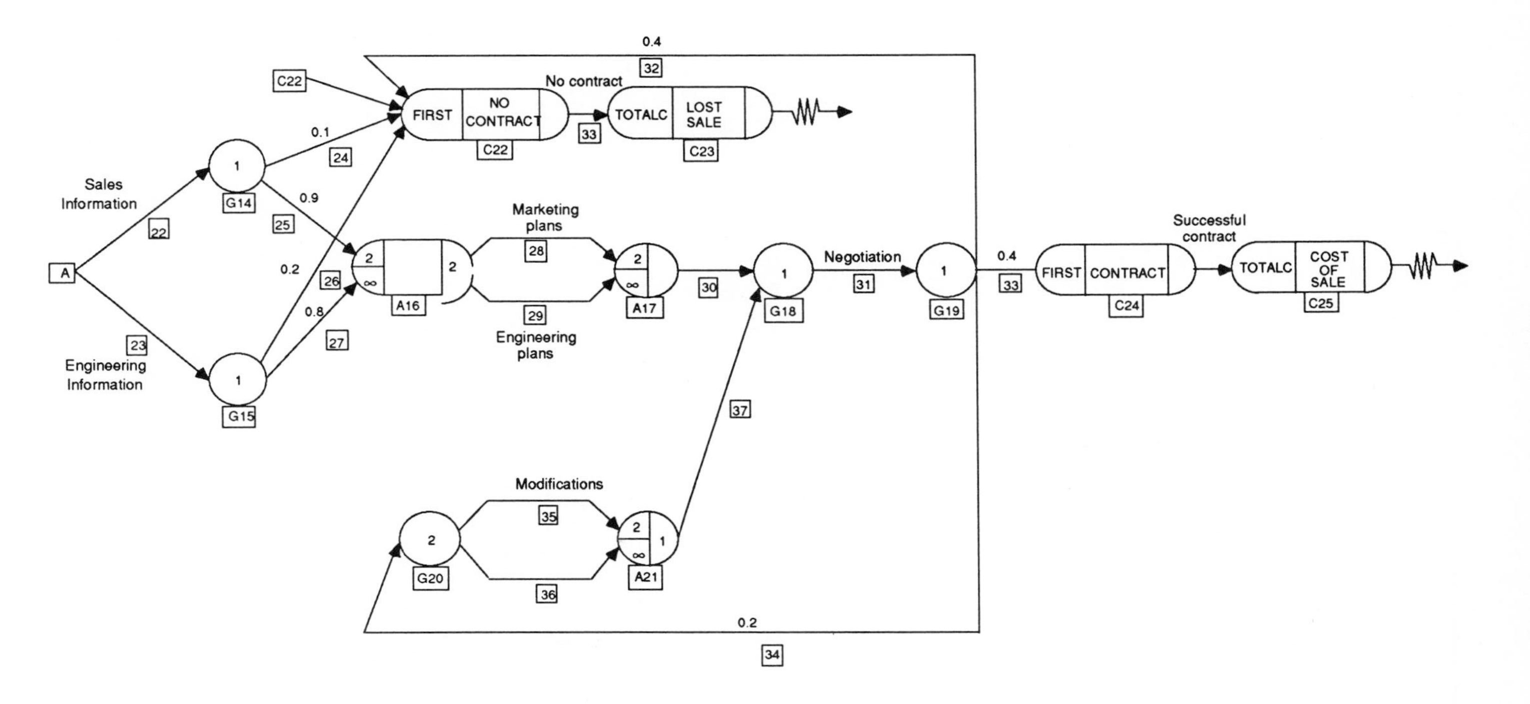

Figure 10-2 SLAM II network for industrial negotiation study (concluded).

reports [21]. Further information is then sought of a sales [22] and engineering nature [23]. If either of these studies results in unfavorable information, the sales negotiation process is terminated following activities [24] and [26]. If positive information is obtained from the studies, marketing negotiation plans [28] and engineering design plans [29] are formulated. Based on these plans, corporate-level strategy is developed in a conference [30]. This strategy is implemented in a negotiation conference with the buying firm [31]. The results of the negotiation conference are no sale [32], contract awarded [33], or further negotiations are.requested by the buyer involving plan modifications [34]. Marketing [35] and engineering [36] modifications are then performed. A corporate-level meeting to evaluate and reconcile the modifications is held [37]. Following this meeting, a return is made to GOON node G18, where the negotiation process with the buyer is resumed [31]. At COLCT node C24, a successful sale has been made, and FIRST statistics are collected of the time required for the entire process. Statistics relating to the total costs of the process are collected at COLCT node C25. At COLCT node C23, statistics are collected for total cost of the negotiation process when a sale does not materialize.

The network model for the negotiation process allows a modeler to include uncertain outcomes as modeled by the probabilistic branching at nodes throughout the network. In addition, recycling of activities is permissible, as illustrated by the sequence of activities [31], [34], [35], [36], and [37], which form a closed loop.

The SLAM II statement model for the sales negotiation process is given in Figure 10-3. The elements of the ARRAY statements are defined as follows:

ARRAY(N,6)/FIXC,VARC,CONSTANT or BMODE,BMIN,BMAX,ETIME;

where N is the activity number, FIXC is the fixed cost, VARC is the variable cost, BMODE is the modal activity time if activity N is from a BETA distribution, BMIN is the minimum activity time, BMAX is the maximum activity time and ETIME is the estimated ending time after a sample has been drawn. If the activity duration is a constant, then CONSTANT is used in place of BMODE, and BMIN and BMAX are zero.

The details involved in obtaining cost estimates from the network model are now described. In general, all activity times and costs estimates are obtained in a user function from data that are stored in an ARRAY statement for each activity. For this study, all activity times are either constant values or samples from a BETA distribution. Cost estimates are obtained by collecting statistics for total cost at COLCT nodes following each milestone in the network. Typically, in a network model, cost estimates for an activity are computed at the start of an activity. Therefore, it is necessary to reduce the cost of an ongoing activity by the variable cost for the time remaining for an activity that is stopped in progress.

```
GEN,HAMMESFAHR,NEGOTIATION MODEL,1/20/88,500,N,N,,N,Y/500;
LIMITS,,2,50;
EQUIVALENCE/XX(1),TOTALC;
ARRAY(1,6)/0.,11.,2.,1.25,4.,0./(2,6)/0.,31.,.5,.25,1.,0.;
ARRAY(3,6)/0.,51.,.25,.25,.5,0./(4,6)/200.,20.,24.,16.,40.,0.;
ARRAY(5,6)/100.,12.,8.,0.,0.,0./(6,6)/50.,10.,16.,0.,0.,0.;
ARRAY(7,6)/50.,10.,8.,0.,0.,0./(8,6)/300.,17.,24.,16.,40.,0.;
ARRAY(9,6)/0.,20.,8.,0.,0.,0./(10,6)/0.,20.,8.,0.,0.,0.;
ARRAY(11,6)/0.,10.,24.,8.,40.,0./(12,6)/0.,20.,8.,0.,0.,0.;
ARRAY(13,6)/60.,20.,10.,0.,0.,0./(14,6)/120.,20.,12.,0.,0.,0.;
ARRAY(15,6)/61.,20.,12.,0.,0.,0./(16,6)/248.,20.,16.,0.,0.,0.;
ARRAY(17,6)/61.,17.,2.,0.,0.,0./(18,6)/61.,17.,2.,0.,0.,0.;
ARRAY(19,6)/0.,51.,2.,0.,0.,0./(20,6)/0.,20.,8.,0.,0.,0.;
ARRAY(21,6)/750.,0.,8.,0.,0.,0./(22,6)/400.,31.,16.,0.,0.,0.;
ARRAY(23,6)/400.,32.,16.,0.,0.,0./(24,6)/0.,73.,3.,1.,18.,0.;
ARRAY(25,6)/0.,0.,0.,0.,0.,0./(26,6)/0.,73.,3.,1.,16.,0.;
ARRAY(27,6)/0.,0.,0.,0.,0.,0./(28,6)/200.,20.,24.,8.,40.,0.;
ARRAY(29,6)/800.,26.,80.,40.,160.,0./(30,6)/0.,73.,2.,1.,8.,0.;
ARRAY(31,6)/400.,73.,6.,2.,16.,0./(32,6)/0.,0.,0.,0.,0.,0.;
ARRAY(33,6)/0.,0.,0.,0.,0.,0./(34,6)/0.,0.,0.,0.,0.,0.;
ARRAY(35,6)/200.,20.,12.,4.,20.,0./(36,6)/800.,26.,40.,20.,80.;
ARRAY(37,6)/0.,73.,1.,0.5,4.,0.;
NETWORK;
CRE1   CREATE,,,,1,1;                    START SIMULATION
       ACT/1,USERF(1);;                  INITIAL CONTACT
G3     GOON,1;                           REPORT TO MARKETING VP
       ACT/2,USERF(2);;                  GENERATE MANAGEMENT REPORTS
G4     GOON,1;                           REPORT TO PRESIDENT
       ACT/3,USERF(3);;                  GENERATE MANAGEMENT REPORTS
G5     GOON,5;                           START DEPARTMENT EVALUATIONS
       ACT/4,USERF(4),,G6;;              ENGINEERING REPORT
       ACT/5,USERF(5),,G7;;              PRODUCTION REPORT
       ACT/6,USERF(6),,G8;;              FINANCIAL REPORT
       ACT/7,USERF(7),,G9;;              MARKETING REPORT
       ACT/8,USERF(8),,G10;;             PURCHASING REPORT
G6     GOON,1;                           ENGINEERING DECISION
       ACT/9,USERF(9),0.2,C22;;          UNFAVORABLE ENGINEERING REPORT
       ACT/10,USERF(10),0.8,A12;;        FAVORABLE ENGINEERING REPORT
G7     GOON,1;                           PRODUCTION DECISION
       ACT/11,USERF(11),0.25,G11;;       UNFAVORABLE PRODUCTION REPORT
       ACT/12,USERF(12),0.75,A12;;       FAVORABLE PRODUCTION REPORT
G8     GOON,1;                           FINANCE DECISION
       ACT/13,USERF(13),0.85,A12;;       FAVORABLE FINANCIAL REPORT
       ACT/14,USERF(14),0.15,C22;;       UNFAVORABLE FINANCIAL REPORT
G9     GOON,1;                           MARKETING DECISION
       ACT/15,USERF(15),0.8,A12;;        FAVORABLE MARKETING REPORT
       ACT/16,USERF(16),0.2,C22;;        UNFAVORABLE MARKETING REPORT
G10    GOON,1;                           PURCHASING DECISION
       ACT/17,USERF(17),0.85,A12;;       FAVORABLE PURCHASING REPORT
       ACT/18,USERF(18),0.15,C22;;       UNFAVORABLE PURCHASING REPORT
G11    GOON,1;                           SUBCONTRACTING DECISION
       ACT/19,USERF(19),0.2,C22;;        UNFAVORABLE SUBCONTRACTING
       ACT/20,USERF(20),0.8,A12;;        FAVORABLE SUBCONTRACTING
A12    ACCUMULATE,5,999,,1;              WAIT FOR ALL REPORTS
C12    COLCT,FIRST,REPORT FAVORABLE,,1;  TIME, ALL REPORTS FAVORABLE
```

Figure 10-3 SLAM II statement model for industrial negotiation study.

```
C13    COLCT,TOTALC,REPORT COST,,1;          COST, ALL REPORTS FAVORABLE
       ACT/21,USERF(21),,G13;;               STAFF CONFERENCE
G13    GOON,2;                               AGREEMENT
       ACT/22,USERF(22),,G14;;               GENERATE SALES DATA
       ACT/23,USERF(23),,G15;;               GENERATE ENGINEERING DATA
G14    GOON,1;                               MARKETING DECISION
       ACT/24,USERF(24),0.1,C22;;            UNFAVORABLE SALES INFORMATION
       ACT/25,USERF(25),0.9,A16;;            FAVORABLE SALES INFORMATION
G15    GOON,1;                               ENGINEERING DECISION
       ACT/26,USERF(26),0.2,C22;;            UNFAVORABLE ENG. INFORMATION
       ACT/27,USERF(27),0.8,A16;;            FAVORABLE ENG. INFORMATION
A16    ACCUMULATE,2,999,,2;                  WAIT FOR SALES AND ENG. REPORT
       ACT/28,USERF(28),,A17;;               DEVELOP MARKETING PLANS
       ACT/29,USERF(29),,A17;;               DEVELOP ENGINEERING PLANS
A17    ACCUMULATE,2,999,,1;                  WAIT FOR MKT. AND ENG. PLANS
       ACT/30,USERF(30),,G18;;               STRATEGY CONFERENCE
G18    GOON,1;                               START NEGOTIATIONS
       ACT/31,USERF(31),,G19;;               NEGOTIATION PROCESS
G19    GOON,1;                               DETERMINE RESULTS OF NEGOTIATION
       ACT/32,USERF(32),0.4,C22;;            UNFAVORABLE NEGOTIATIONS
       ACT/33,USERF(33),0.4,C24;;            FAVORABLE NEGOTIATIONS
       ACT/34,USERF(34),0.2,G20;;            PLAN MODIFICATIONS REQUIRED
G20    GOON,2;                               NEW MKT. AND ENG. DATA NEEDED
       ACT/35,USERF(35),,A21;;               MARKETING MODIFICATIONS
       ACT/36,USERF(36),,A21;;               ENGINEERING MODIFICATIONS
A21    ACCUMULATE,2,2,,1;                    WAIT FOR COMPLETE MODIFICATIONS
       ACT/37,USERF(37),,G18;;               CORPORATE MEETING, RENEGOTIATE
C22    COLCT,FIRST,NO CONTRACT,22,10,,10,,1;  TIME, NO CONTRACT DECISION
       ACT/38,USERF(38),,C23;;               RECOVER PREALLOCATED VAR-COST
C23    COLCT,TOTALC,LOST SALE COST,18,2000,,1000,,1;  COST, LOST SALE
       ACT,,,T1;                             END RUN
C24    COLCT,FIRST,CONTRACT,18,100,,20,,1;  TIME TO SUCCESSFUL CONTRACT
       COLCT,TOTALC,COST OF SALE,12,8000,,1000,,1;  COST, CONTRACT
T1     TERM,1;                               END RUN
       END;
INIT,0,,N;
FIN;
```

Figure 10-3 SLAM II statement model for industrial negotiation study (concluded).

Function USERF(N) is shown in Figure 10-4. The user function number N is the activity number of the activity about to start in the network. Consequently, the data stored in the ARRAY statements can be accessed directly for each activity by using the user function number to reference a row number in ARRAY.

The first statement in function USERF(N) tests if the negotiation process is halted because of an unfavorable report. This occurs when activity 38 is initiated. Ongoing activities are stopped because their reports are no longer required, and the variable cost due to their remaining activity time is not incurred. The variable cost for an activity is added to the total cost when it is initiated (see below): hence, the variable cost for the remaining time must be subtracted from the total cost of the project. This is accomplished by first testing

```
      FUNCTION USERF(N)
      COMMON/SCOM1/ATRIB(100),DD(100),DDL(100),DTNOW,II,MFA,MSTOP,NCLNR
     1,NCRDR,NPRNT,NNRUN,NNSET,NTAPE,SS(100),SSL(100),TNEXT,TNOW,XX(100)
      EQUIVALENCE (XX(1),TOTALC)
      IF (N.EQ.38) THEN
C******* REPORT UNFAVORABLE
C******* RECOVER VARIABLE COST OF ONGOING ACTIVITIES
         DO 20 I = 4,37
C********** DETERMINE IF ACTIVITY IS ONGOING
            IACT = NNACT(I)
            IF (IACT .GT. 0) THEN
C************** AN ONGOING ACTIVITY.
C************** DETERMINE REMAINING ACTIVITY TIME
               ETIME = GETARY(I,6)
               RTIME = ETIME - TNOW
C************** REDUCE TOTAL COST BY UNUSED VARIABLE COST
C************** FOR THIS ACTIVITY
               VARC = GETARY(I,2)
               TOTALC = TOTALC - (VARC * RTIME)
            ENDIF
   20    CONTINUE
         USERF = 0.
      ELSE
C******* DETERMINE ACTIVITY TIME FOR THIS ACTIVITY
         BMAX = GETARY(N,5)
         IF (BMAX .LE. 0) THEN
C********** ACTIVITY TIME IS CONSTANT
            CONST = GETARY(N,3)
            USERF = CONST
         ELSE
C********** SAMPLE ACTIVITY TIME FROM THE BETA DISTRIBUTION
            BMODE = GETARY(N,3)
            BMIN = GETARY(N,4)
C********** DETERMINE MEAN AND STANDARD DEVIATION
            XMEAN = (BMIN + (4.0*BMODE) + BMAX)/6.0
C********** CALCULATE NORMALIZED MEAN
            XMU = (XMEAN - BMIN)/(BMAX - BMIN)
            VAR = 1.0/36.0
C********** CALCULATE THETA AND PHI
            THETA = (((XMU**2)/VAR)*(1.0-XMU)) - XMU
            PHI = THETA*((1.0 - XMU)/XMU)
C********** SAMPLE ACTIVITY TIME
            USERF = (BETA(THETA,PHI,1) * (BMAX - BMIN)) + BMIN
         ENDIF
C******* ESTABLISH ACTIVITY ENDING TIME
         ETIME = TNOW + USERF
C******* STORE ACTIVITY ENDING TIME IN ELEMENT 6 OF ACTIVITY'S ARRAY
         CALL PUTARY(N,6,ETIME)
C******* CALCULATE TOTAL COST FOR THIS ACTIVITY
         FIXC = GETARY(N,1)
         VARC = GETARY(N,2)
         TOTALC = TOTALC + FIXC + (VARC * USERF)
      ENDIF
      RETURN
      END
```

Figure 10-4 Function USERF for negotiation model.

the current status of activities 4 through 37 in a DO loop. (Activities 1, 2, and 3 are not tested because they are always completed during each run.) The SLAM II function NNACT(I) returns the status of activity I. The statement IACT = NNACT(I) sets the variable IACT equal to zero if activity I is not ongoing and to 1 if activity I is currently in progress. When an activity is started, its ending time, ETIME, is placed into ARRAY(I,6). This value is now retrieved, and the current simulation time subtracted from this value to determine the remaining activity time, that is, RTIME = ETIME - TNOW. Total cost is then adjusted for the recovered variable cost of this activity by multiplying the remaining activity time by its variable cost and subtracting the result from total cost: that is, TOTALC = TOTALC - VARC * RTIME. When each activity has been tested, the activity time for activity 38 is set equal to zero [USERF=0.0], control is returned to the SLAM II processor, statistics on total costs are collected at COLCT node 23, and the current run of the network is terminated.

The remainder of function USERF(N) calculates an activity time and uses it to compute costs and to estimate the activity's end time. If ARRAY(N,6) is equal to zero, then the activity time is a constant. In this case, USERF is set equal to ARRAY(N,3) which is accessed using function GETARY(N,3). Otherwise, USERF is sampled from the BETA distribution using the parameters stored in columns 3 through 5 of row N of ARRAY. The ending time, ETIME, for the activity is calculated as TNOW + USERF. This value is stored in column 6 of ARRAY using the statement

```
CALL PUTARY (N, 6, ETIME)
```

As discussed above, ETIME is used to adjust the total cost for the model for any activity in progress when the negotiation process is stopped before a sale is realized.

Total costs, TOTALC, are updated as each activity is started by adding to the current total cost, the fixed cost, FIXC, for the activity, stored in ARRAY(N,1), and the activity's variable cost, VARC, stored in ARRAY(N,2), multiplied by the activity duration (USERF), that is, TOTALC = TOTALC + FIXC + (VARC * USERF).

10.2.3 Model Output and Use

The SLAM II model of the industrial renegotiation process was analyzed for 500 runs of the network. Summaries of the time and cost results are presented in Figure 10-5. The estimate of the probability of losing a contract is 0.832 computed from 416 no-contract outcomes occurring in 500 runs. This indicates that 83.2% of the negotiations end in failure.

SIMULATION PROJECT NEGOTIATION MODEL BY HAMMESFAHR

DATE 1/20/1988 RUN NUMBER 500 OF 500

STATISTICS FOR VARIABLES BASED ON OBSERVATION

	MEAN VALUE	STANDARD DEVIATION	MINIMUM VALUE	MAXIMUM VALUE	NUMBER OF OBSERVATIONS
REPORT FAVORABLE	0.3781E+02	0.4821E+01	0.2889E+02	0.5210E+02	225
REPORT COST	0.3041E+04	0.1482E+03	0.2749E+04	0.3541E+04	225
NO CONTRACT	0.6060E+02	0.5350E+02	0.2217E+02	0.3320E+03	416
LOST SALE COST	0.4879E+04	0.3209E+04	0.2469E+04	0.2128E+05	416
CONTRACT	0.1739E+03	0.3389E+02	0.1166E+03	0.2953E+03	84
COST OF SALE	0.1138E+05	0.1853E+04	0.9057E+04	0.1873E+05	84

Figure 10-5 Summary report for time and costs associated with negotiation model.

Examining the output for favorable reports received at COLCT node C12, it is seen that 45% of the potential projects result in all five reports being favorable. Thus, 55% of the negotiations are turned down for internal reasons. The time estimates indicate that it takes approximately 38 days to decide that negotiations should be carried beyond the internal report phase, and the cost data collected at COLCT node C13 indicate that it costs $3041 for this phase of the process. This information is extremely useful to a decision maker who is attempting to improve the sales negotiation process. It provides trade-off data regarding the possibility of increasing the probability of favorable reports versus the time and costs required to obtain the favorable reports. Such trade-offs can be evaluated by developing alternative network segments leading to node A12. One alternative is the sequence in which the engineering, production, financial, marketing, and purchasing reports should be made. Since engineering and marketing have the highest probability of issuing a negative report, another alternative is to have these two activities performed prior to the other three. Either of these alternatives will increase the time required to reach the decision relative to all reports, but will reduce the total cost.

The summary statistics for NO CONTRACT (lost sale) indicate that it takes 60 days on the average to make this determination. Since COLCT node C12 can be reached from many points in the network, this time should have a wide variability. This is indicated by its standard deviation of 53.5, and its range of 22 to 332 days. The average cost associated with a lost sale is $4879. Comparing this value with project costs when favorable reports are obtained, it is seen that, on the average, $4879 - 3041 = $1838 is expended above the report costs when a lost sale occurs. When the project succeeds, which occurs 16.8% of the time, the negotiation process takes approximately 174 days and costs an estimated $11,380.

The cost histogram associated with COLCT node C23 for lost sales is presented in Figure 10-6. This histogram illustrates that the distribution function has discrete breaks due to the different paths to reach node C23. For example, failed reports occurred on 275 runs (500 - 225). The histogram for lost sale cost shows 275 values in the range of $2000 to $4000; hence, this cluster of values is associated with failed reports. The other values in the histogram are for the costs when a lost sale occurred after favorable reports were received.

Figure 10-7 presents a histogram of the costs associated with a successful negotiation. From the cost histogram, the total costs were greater than $13,000 in 16 of the 84 successful negotiations. Thus, high costs are to be expected in approximately one out of five successful negotiations.

10.3 INDUSTRIAL RESEARCH AND DEVELOPMENT PLANNING

Effective analysis and planning of research and development (R&D) activities is vital to the success of many corporate endeavors. Corporate marketing and production strategies depend heavily on the outcome of R&D projects. Normally, long planning horizons are involved, which, by their very nature, make R&D projects uncertain ventures. An analysis of such projects must account for the probabilistic nature of activity durations, the possibility of alternative strategies, the repeating of activities, and the possibility of project failure. As described in preceding chapters, these features can be modeled in SLAM II. In fact, network simulation has been used as the modeling vehicle for diverse R&D planning activities [1, 9, 13]. An example from [9] is included here to illustrate the use of SLAM II in R&D planning.

10.3.1 The Organization and Its Study Objectives

The organization to be modeled involves two research teams responsible for four R&D projects. The two teams work on different projects, but they begin at the same time and each team is capable of working on each of the four projects. Each of the projects consists of five generic stages:

1. Problem definition
2. Research activity
3. Solution proposal
4. Prototype development
5. Solution implementation

For this example, each stage is represented in an aggregate manner as a single activity. The time and cost values for each stage could be obtained as the output

HISTOGRAM NUMBER 4

LOST SALE COST

OBSV FREQ	RELA FREQ	CUML FREQ	UPPER CELL LIMIT
0	0.000	0.000	0.2000E+04
191	0.459	0.459	0.3000E+04
84	0.202	0.661	0.4000E+04
0	0.000	0.661	0.5000E+04
40	0.096	0.757	0.6000E+04
32	0.077	0.834	0.7000E+04
0	0.000	0.834	0.8000E+04
0	0.000	0.834	0.9000E+04
13	0.031	0.865	0.1000E+05
34	0.082	0.947	0.1100E+05
9	0.022	0.969	0.1200E+05
2	0.005	0.974	0.1300E+05
4	0.010	0.983	0.1400E+05
2	0.005	0.988	0.1500E+05
1	0.002	0.990	0.1600E+05
1	0.002	0.993	0.1700E+05
2	0.005	0.998	0.1800E+05
0	0.000	0.998	0.1900E+05
0	0.000	0.998	0.2000E+05
1	0.002	1.000	INF
416			

STATISTICS FOR VARIABLES BASED ON OBSERVATION

	MEAN VALUE	STANDARD DEVIATION	MINIMUM VALUE	MAXIMUM VALUE	NUMBER OF OBSERVATIONS
LOST SALE COST	0.4879E+04	0.3209E+04	0.2469E+04	0.2128E+05	416

Figure 10-6 Histogram of costs associated with an unsuccessful negotiation.

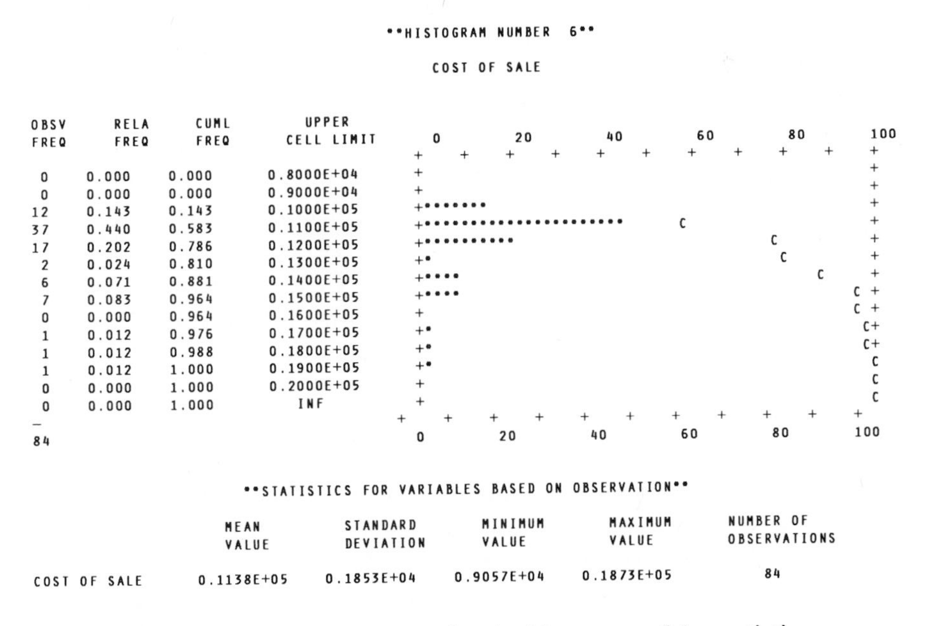

HISTOGRAM NUMBER 6

COST OF SALE

OBSV FREQ	RELA FREQ	CUML FREQ	UPPER CELL LIMIT
0	0.000	0.000	0.8000E+04
0	0.000	0.000	0.9000E+04
12	0.143	0.143	0.1000E+05
37	0.440	0.583	0.1100E+05
17	0.202	0.786	0.1200E+05
2	0.024	0.810	0.1300E+05
6	0.071	0.881	0.1400E+05
7	0.083	0.964	0.1500E+05
0	0.000	0.964	0.1600E+05
1	0.012	0.976	0.1700E+05
1	0.012	0.988	0.1800E+05
1	0.012	1.000	0.1900E+05
0	0.000	1.000	0.2000E+05
0	0.000	1.000	INF
84			

STATISTICS FOR VARIABLES BASED ON OBSERVATION

	MEAN VALUE	STANDARD DEVIATION	MINIMUM VALUE	MAXIMUM VALUE	NUMBER OF OBSERVATIONS
COST OF SALE	0.1138E+05	0.1853E+04	0.9057E+04	0.1873E+05	84

Figure 10-7 Histogram of costs associated with a successful negotiation.

values from a detailed model of each stage. SLAM II facilitates this hierarchical modeling approach.

The study objective is to obtain estimates of the probability of success, cost, and time duration for the project. These estimates can be used to answer the following specific questions relating to research and development projects:

1. What is the chance of the project succeeding or failing?
2. When can a conclusion be reached concerning the impact of the project?
3. What is the range of costs to be expected on a successful or an unsuccessful project?
4. What is the effect of slowing down or speeding up the project through different budget allocations?

10.3.2 The SLAM II Model

A SLAM II network of the five stages is shown in Figure 10-8. The two possible project outcomes are represented by COLCT nodes C6 and C7. The five stages are represented sequentially; however, feedback branches emanating after GOON nodes G2, G4, and G5 represent the possibility of repeated activities. Probabilistic branching from node G2 represents two possible outcomes: problem redefinition required or problem definition acceptance. In the latter case, the activity representing the research stage can begin. After the research activity, an evaluation process ensues. The outcome of this activity is modeled by the branching probabilities following node G4. The possible outcomes include a determination of project failure (branch to node C7), a return to the problem definition stage (branch to node G1), a return to research activity (branch to node G2), or the initiation of prototype development (branch to node G5). Since the prototypc developed may not be adequate, a redevelopment activity is represented by the loop around node G5. If the prototype is acceptable, the next activity is solution implementation. COLCT nodes C6 and C7 provide estimates of the time required and probability for successful project completion and unsuccessful completion, respectively. Costs associated with each activity may also be obtained, as illustrated in the previous example.

For multiple R&D projects, it is necessary to link several networks of the type described in Figure 10-8. As one project ends (either by failure or successful completion), branching from the project ending node to the next project occurs. For the study described here, two research teams are employed, and it is necessary to include logic in the network that specifies the assignment of a team to a project. The SLAM II model of the multiteam, multiproject R&D process is shown in Figure 10-9.

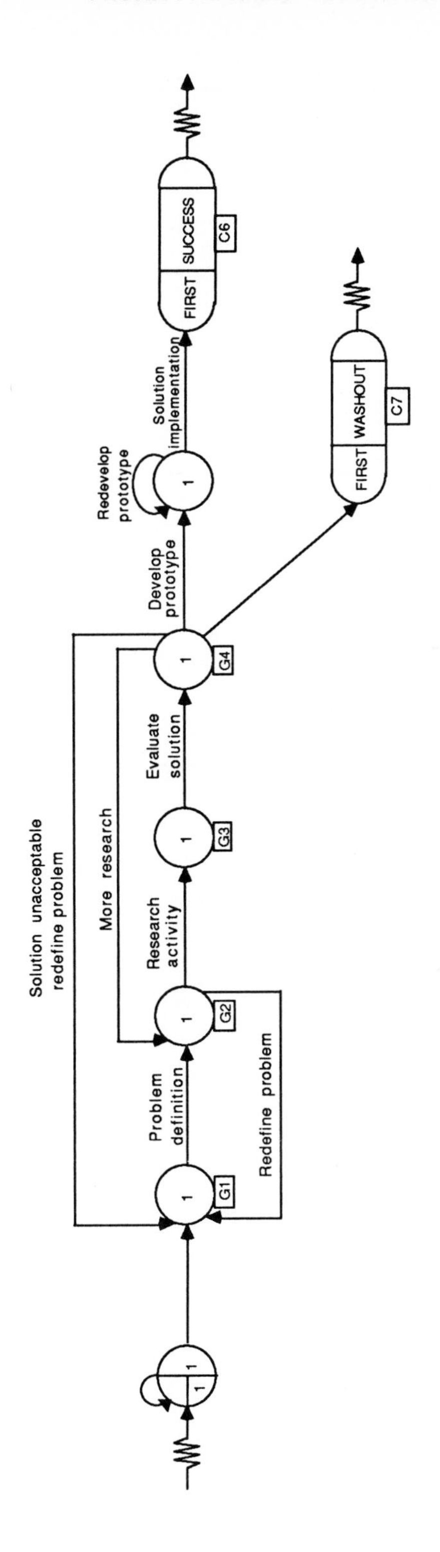

Figure 10-8 SLAM II network of the stages in an R&D project.

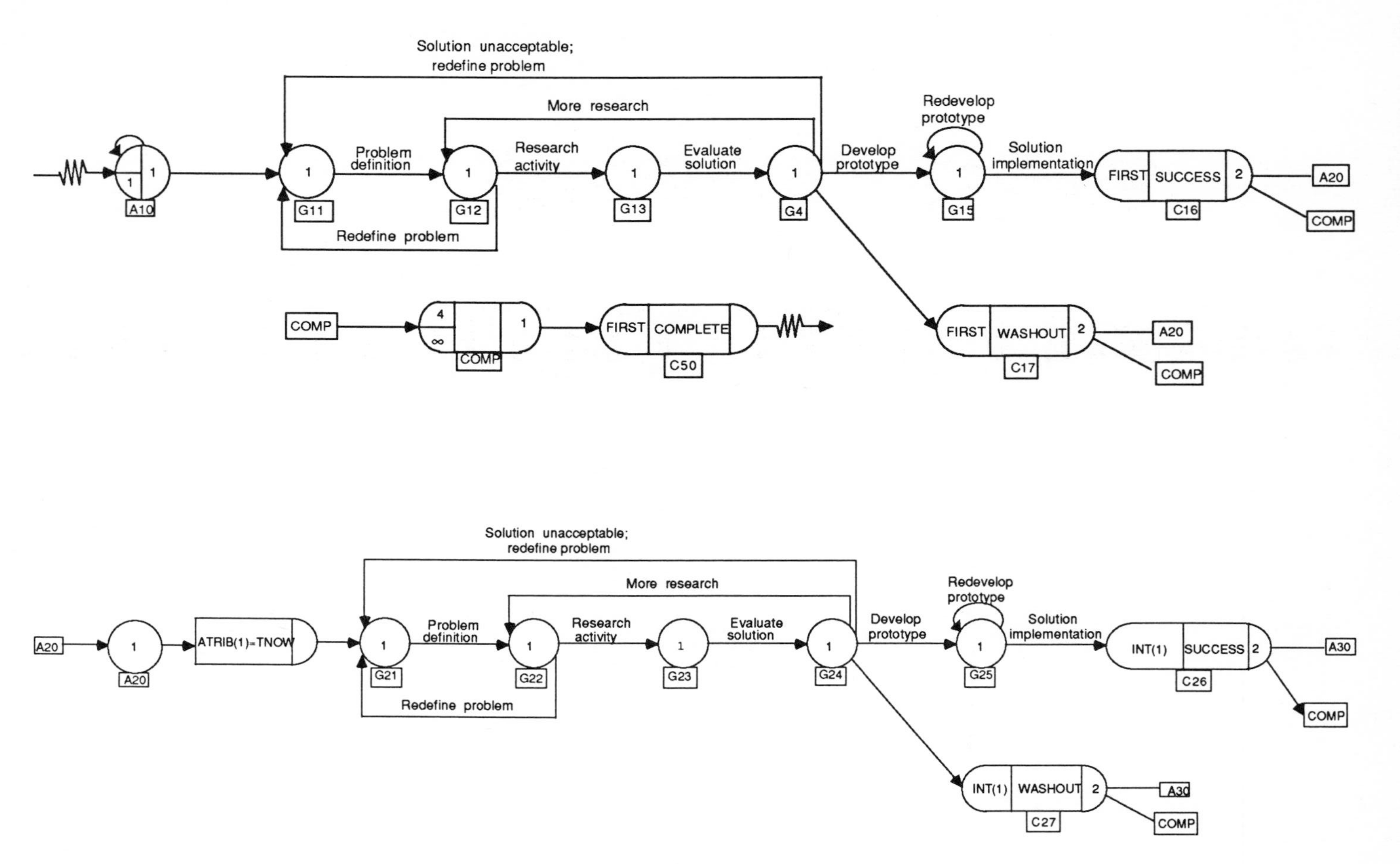

Figure 10-9 SLAM II network of multiteam, multiproject team R&D process.

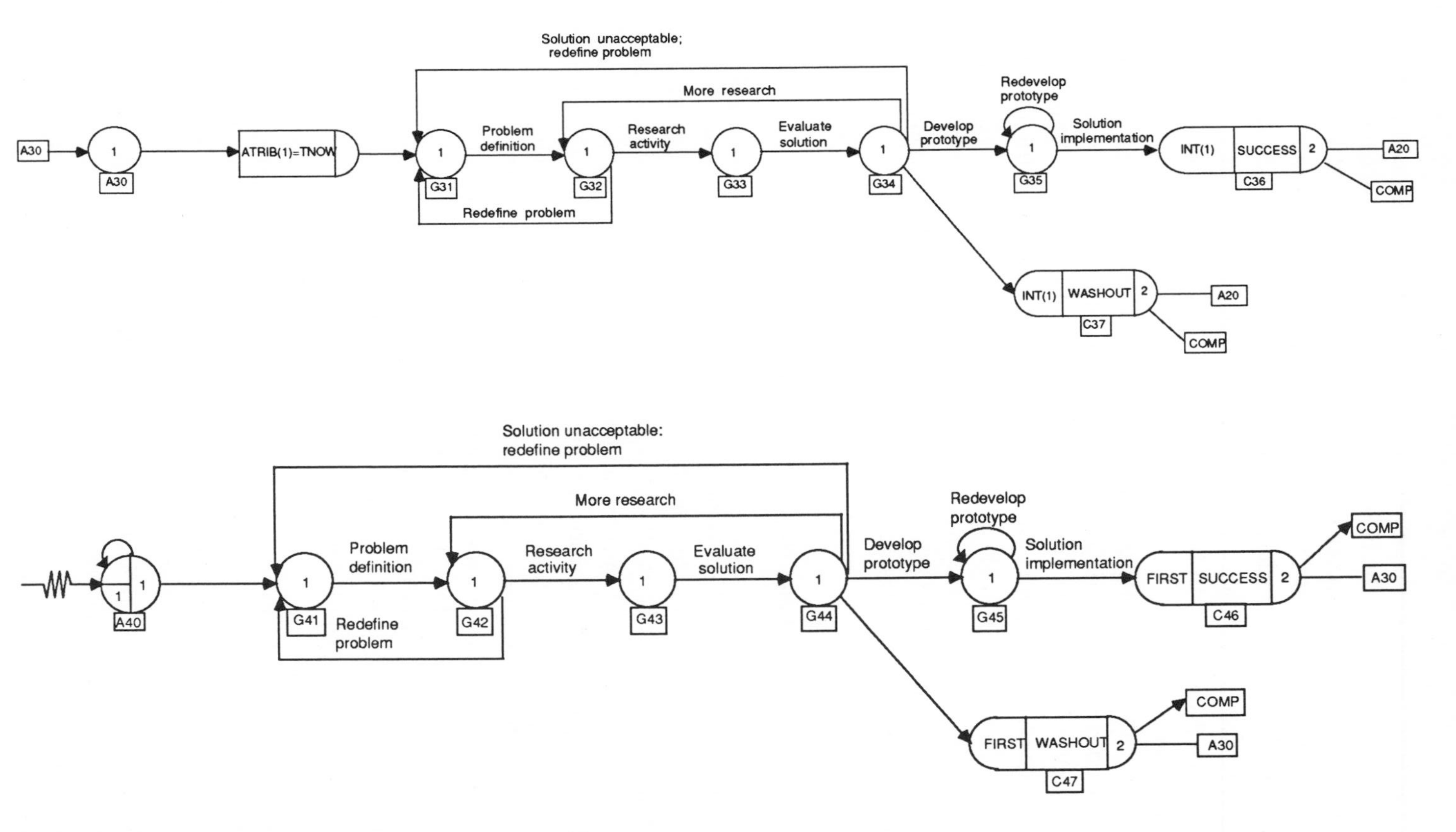

Figure 10-9 SLAM II network of multiteam, multiproject team R&D process (concluded).

The network shows that one research team starts on project 1 (CREATE node A10) at the same time the other team starts on project 4 (CREATE node A40). When the first team finishes project 1, either by failure at COLCT node C17 or successful completion at COLCT node C16, it proceeds to project 2 (ACCUMULATE node A20). The same procedure describes the second team's movement from project 4 to project 3 (ACCUMULATE node A30). In addition, a signal from each project is sent to ACCUMULATE node COMP to indicate a project completion.

Once either ACCUMLUATE node A20 or A30 has been released, representing a start up for project 2 or 3, these nodes cannot be released a second time (as signified by the infinite subsequent release requirement). When either team terminates from its second project, it moves to a potential third project to begin work on that project if it has not been started. All network activity is terminated when four projects have been completed, which is modeled by signaling ACCUMULATE node COMP with each project completion event and specifying that four entities are required to release the node.

10.3.3 Model Output and Use

Moore and Taylor [9] provide an analysis of this type of R&D project. The intent of this section is to review the model output in order to illustrate the types of responses that can be made to the managerial concerns discussed previously.

Estimates of project success and failure for each of the four projects are obtained from the proportion of the runs that different COLCT nodes are released. Total network performance is based on reaching ACCUMULATE node COMP, and statistics are obtained for this event at COLCT node C50. At ASSIGN nodes, XX(I) is set to the beginning time of a project I, where I is the project number. At COLCT nodes C26 and C36 values of [TNOW-ATRIB(1)] are used to collect statistics on the individual performance times of projects 2 and 3. Tables 10-1 and 10-2 give a summary of time and cost statistics for the individual projects and the total network. Project success and failure probabilities are also given.

In Table 10-1, the expected completion time of the network is estimated at 982 days with a standard deviation of 252 days. The associated expected cost is given in Table 10-2 as $892,600, with a standard deviation of $180,600. Using these results, the R&D firm based their proposal estimates on a total duration of 1300 days at a cost of $1,100,00. These values correspond to a 90% chance of being within the time limitation and an 88% chance of being within the cost limitation, assuming that normal distributions pertain. Additionally, the firm computed a three standard deviation level (99.7%), which is 1738 days and

Table 10-1 Summary of Time Statistics for Project Network

Project Times	Prob.	Time (days)			
		E(t)	σ_t	Min. t	Max. t
Project 1					
Successful completion	0.745	419	125	277	1514
Washout	0.255	182	76	108	676
Overall project	1.000	358	154	108	1514
Project 2					
Successful completion	0.954	277	96	131	757
Washout	0.046	173	93	78	487
Overall project	1.000	272	99	78	757
Project 3					
Successful completion	0.638	717	207	453	1831
Washout	0.362	376	165	163	1218
Overall project	1.000	593	253	163	1831
Project 4					
Successful completion	0.970	371	120	208	1118
Washout	0.030	142	56	84	297
Overall project	1.000	364	125	84	1118
All projects					
Successful completion	0.427	1096	224	758	2210
Overall network	1.000	982	252	481	2210

Source [9]

$1,434,400. These values are within break-even points and this reduced the perceived risk of the projects.

An area of concern to the firm is the relatively low probability (0.427) of overall success for all four projects. Although the firm would not experience a substantial financial loss on a project failure, each failure affects their return and their reputation. When this is the case, it is desirable to perform a sensitivity analysis of the results. The impact on the overall probability of success was assessed for different probabilities associated with the stages of individual projects. There is, of course, a cost associated with increasing a stage probability of success (SLAM II could be used to model the R&D project necessary to improve the probabilities). Additional probabilistic calculations revealed project 3 as the most risky in terms of variation in project time and cost.

The statistical information also enabled the firm to evaluate potential personnel, equipment, and capital needs. Specifically, bottlenecks at the stages are a concern when there is a high probability of repeating a set of activities.

Table 10-2 Summary of Cost Statistics for Project Network.

		Cost (thousands of dollars)			
Project Costs	Prob.	E(c)	σ_c	Min. c	Max. c
Project 1					
Successful completion	0.745	473.0	128.5	316.5	1147.9
Washout	0.255	195.1	72.1	129.9	663.4
Overall project	1.000	402.1	168.3	129.9	1147.9
Project 2					
Successful completion	0.954	290.8	92.1	103.5	759.1
Washout	0.046	185.6	95.5	86.0	515.7
Overall project	1.000	287.9	95.0	86.0	759.1
Project 3					
Successful completion	0.638	564.6	149.2	287.2	1247.4
Washout	0.362	329.5	119.0	129.1	827.0
Overall project	1.000	480.7	179.8	129.1	1247.4
Project 4					
Successful completion	0.970	411.5	129.4	231.9	1142.1
Washout	0.030	162.7	59.7	105.9	329.1
Overall project	1.000	404.0	134.7	105.9	1142.1
All projects					
Successful completion	0.427	1008.7	148.5	726.6	1658.0
Overall network	1.000	892.6	180.6	468.4	1658.0

Source [9]

This is particularly the case for the relatively high probabilities of failure on project 1 (0.255) and project 3 (0.362). Additional resources to alleviate this potential problem were considered. However, after checking the original probabilistic branching estimates, it was determined that the high failure probabilities are a function of the nature of these individual projects. Thus, additional resources would not substantially reduce these probabilities. As a result, no extra resources were allocated. This is considered an important use of the model for future planning episodes and demonstrates the communication capabilities of the network procedure.

Sensitivity analyses performed when data estimates are subjective allow the effect of data inaccuracies to be quantified. In performing a sensitivity analysis, key activity durations and node probabilities are identified and adjusted to observe the overall effect on the outputs. Changing activity durations has only a moderate effect. As expected [3, 9], changes in node probabilities have a pronounced effect, especially on the feedback branches from nodes G14, G24, G34, and G44.

In this example, the networking approach was used to evaluate different team strategies, prepare overall time and cost estimates, provide inputs to contract negotiations, plan and schedule personnel, equipment, and capital, and identify bottlenecks. The model presented above is for only one configuration, employing two research teams currently in use by the firm. With the model, alternatives were tested to determine the effect of adding additional research teams and several additional projects into the planning horizon of the firm.

10.4 TECHNOLOGY FORECASTING: EVALUATING THE EFFECTIVENESS OF DIFFERENT FUNDING LEVELS IN NUCLEAR FUSION RESEARCH

This study, performed by Vanston and others[13, 14], demonstrates a network approach to evaluate different funding alternatives of an engineering research project. This study differs from the one presented in the previous section in that detailed descriptions of the activities in the project were developed, including data-collection procedures. Vanston refers to the total project approach, including the model networking, as PAF for partitive analytical forecasting. Important network concepts employed in the model are the chance of failure in planned research, the repetition of unsuccessful tasks, and the alteration of planned activities as a result of new developments in related research efforts. The outputs from the model provide estimates of the probability of successful project completion for each funding alternative, as well as time estimates for the realization of research goals. With such probability and time information for each funding alternative, decision makers can rationally evaluate and select a course of action that best meets the research objectives.

The following section gives a brief historical summary of nuclear power developments. A model of a particular project within fusion research is then described. A discussion is included on how data were obtained to estimate probabilities and time durations. Finally, the use of the model in evaluating funding alternatives is explained. Throughout the section, a heavy reliance is made on the papers and reports of Vanston and others [13, 14].

10.4.1 Historical Background and Management Concerns [14]

Fusion reactors are expected to represent the third generation of nuclear power reactors. The first two generations of reactors are of the fission type, which produce heat by splitting large atoms into pairs of lighter atoms. All U.S. nuclear plants are currently powered by first-generation reactors, also called thermal reactors. Fuel for these reactors is being used at an increasingly rapid rate. The second-generation reactor is the fast breeder reactor, which produces more fuel atoms than it burns. Because they increase the recoverable energy from natural uranium by about 60 times, fast breeder reactors are a prime hope for extending the world's finite fuel supplies.

The third generation of reactors will utilize the energy released when light atoms fuse together to form heavier ones. Since the early 1950s, when the uncontrolled, massive release of fusion energy in hydrogen bombs became a reality, scientists and engineers have sought to develop controlled fusion as a source of power. Although still in a developmental stage, research has made steady progress, and many researchers in the field hope to see fusion become a commercial reality by the end of the century. Although eventual success is not certain, the nature of both the obstacles and the potential solutions has become increasingly clear.

The primary fuels for fusion are deuterium and tritium, both hydrogen isotopes, that is, forms of the element with higher atomic weights than ordinary hydrogen. Deuterium can be extracted from seawater. The reactor produces tritium from the metal lithium. Controlled fusion is important because it not only promises a potentially infinite energy source, but also because fewer problems related to reactor operation and radioactive waste disposal are involved.

The principal obstacle to the development of fusion power is that the interacting nuclei are positively charged and strongly repel each other. Thus, to achieve fusion, the deuterium and tritium nuclei must be made to collide at velocities high enough to overcome this electrostatic repulsion. To attain this velocity, the nuclei must be raised to temperatures in the vicinity of 100 million degrees Celsius. At this temperature, the atoms are fully ionized and exist in the form of a plasma of positive and negative ions. Furthermore, to release a net output of energy, the fuel isotopes must be confined long enough to allow a large number of fusion reactions to occur.

To demonstrate that fusion power is scientifically feasible, researchers must be able to meet temperature and containment requirements simultaneously. Researchers are now working to produce the strong magnetic fields necessary to confine plasma while heating it to the high ignition temperature that will start the fusion process.

The Tokamak (an acronym for the Russian words for torus, room, and magnet) device is one of several potential methods for achieving magnetic confinement in a fusion reactor. Researchers differ on the exact sequence of events that might lead to a commercially successful Tokamak reactor. However, there is enough consensus to project and refine one probable scenario. Vanston, Nichols, and Soland [14] modeled the Tokamak program to estimate the probability of success and the time to the completion of the first commercial fusion reactor.

10.4.2 SLAM II Model Concepts

The Tokamak development project can be viewed in three successive stages: (1) the demonstration of scientific feasibility, (2) the demonstration of engineering feasibility, and (3) the completion of the first commercial reactor. Demonstrating scientific feasibility involves showing that plasma temperature and confinement requirements can be achieved simultaneously in an experimental device. The demonstration of engineering feasibility requires the development of the necessary hardware for a continuously operated reactor. The final stage of the project involves the construction of the first demonstration plant designed to test the economic or commercial feasibility of fusion power.

A hierarchical approach is used to model this project to illustrate the iterative aspects of model building [12]. This process is important to the maintenance of an overall perspective with respect to modeling objectives. In addition, communication among model builders and between model builders and decision makers is facilitated when one progresses from aggregate models to more detailed ones.

The process starts with a network that models the three stages, as shown in Figure 10-10. Demonstrating scientific feasibility, stage 1, requires the successful completion of plasma-heating experiments, as well as the testing of either a turbulent or nonturbulent heating method. The breakdown of stage 1 is modeled in Figure 10-11. The first (1) and subsequent (∞) release requirements of ACCUMULATE node A16 specify that only the first activity to finish (either nonturbulent or turbulent heating testing) will release the node. The output activity from node A16 signifies that successful completion of either activity results in successful completion of stage 1. At ACCUMULATE node A15, both nonturbulent and turbulent approaches are required to fail before assessing a project (stage 1) failure. Plasma-heating research is modeled as a single activity that can result in success or failure. Typically, this activity would be disaggregated and repeated experiments would be modeled.

To illustrate further the hierarchical approach, stages 2 and 3 are described in slightly more detail in Figure 10-12. Stage 2, demonstration of engineering feasibility, includes two parallel approaches: the deuterium-tritium (DT) process and the deuterium-deuterium (DD) process. The first of these is a fusion process using both tritium and deuterium atoms, whereas the second requires only deuterium atoms. The engineering problems of the former are easier to overcome. Since either process can lead to pilot plant design, stage 3, an "or" situation exists so that only one input requirement is needed at ACCUMULATE node A8. Stage 3 is modeled by the following three sequential activities: designing a pilot plant, constructing the plant, and testing the full-scale operation. Two points of project failure are identified, one in the construction and one in

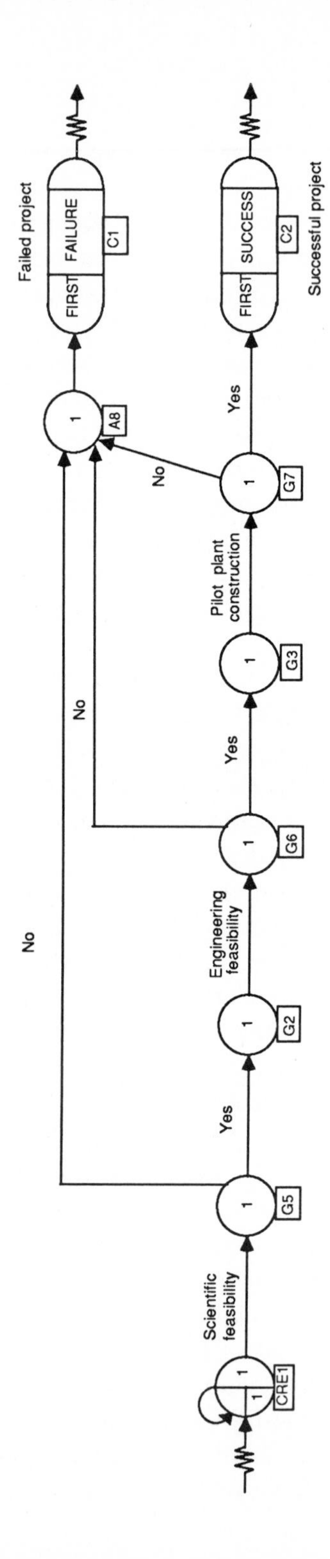

Figure 10-10 Aggregate model Tokamak project.

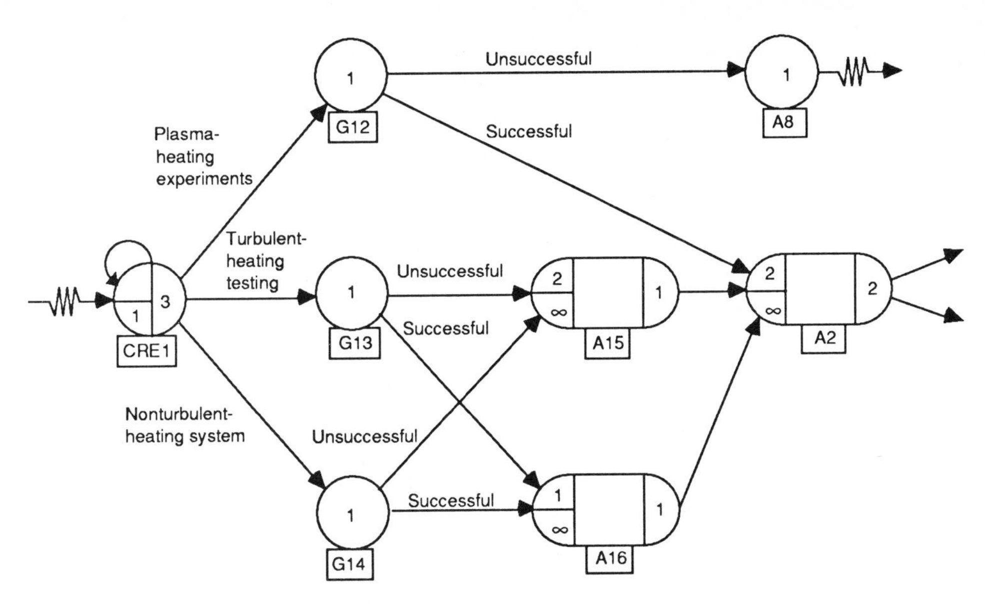

Figure 10-11 Breakdown of Stage 1 activity description.

testing. Testing has three possible outcomes: success, failure, or a requirement for retesting. For purposes of this discussion, the model need not be disaggregated further. It is clear that each process or design would consist of many subactivities. The interested reader should refer to Vanston [13], where a 300-node, 600-activity network is described. The discussion here focuses on data collection, model outputs, and their use in decision making.

10.4.3 Data Collection

To gather the activity time information and node probability data, Vanston conducted interviews with knowledgeable researchers and administrators associated with the Tokamak project. A systematic sampling of informed judgments was made using written questionnaires. This approach is similar to the Delphi family of techniques [4, 5]. The interviewer carefully defined the information desired and permitted the participant to qualify answers in any manner deemed appropriate.

Vanston employed the following procedures in the interviews. The interviewer presented the participant with a copy of the overall network and the specific subnetwork in the area of the participant's expertise. The use of the network put the data request in an integrated context and greatly facilitated communication. The general nature of the interview was explained, together

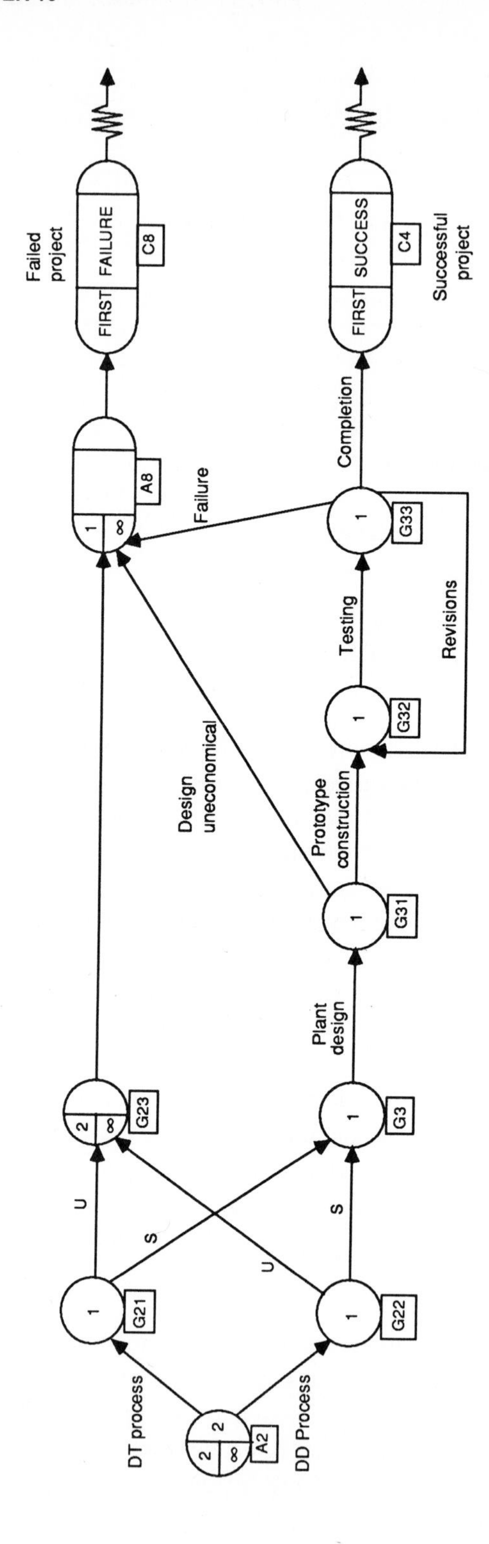

Figure 10-12 Breakdown of stages 2 and 3 of Tokamak project.

with an outline of how the information was to be used. Assurances of anonymity, if desired, were formally stated. For each activity to be estimated, the participant's experience was requested and rated on a scale of 1 to 3. This self-rating was later used to give added weights in favor of qualified experience. The opportunity to qualify experience levels in each task area also reduced a participant's reluctance to assign estimates in areas where their competence was not extensive.

A major advantage of the structured interview technique is that it provides for individualized challenges, that is, the interviewer is able to match the participant's subjective responses against previous estimates. The interviewer is required to be thoroughly familiar with the network interrelationships and to be alert to any inconsistencies in a participant's estimates. If a participant's estimate differs significantly from those of others, this fact is disclosed and a request made to explain possible reasons for the difference. The participant could change an estimate, if desired, but the original estimates are also recorded for possible future analysis.

During an interview, the participant is asked to give three types of estimates: (1) the likelihood that each activity would be completed, (2) the probable time that would be required to complete each activity, and (3) the costs associated with activity completions. For each node with probabilistic branching, the participant is asked either to estimate the likelihood of an event's occurrence as a numerical probability or to choose the most appropriate of seven adjectival statements, such as an event "very probably will occur."

For each activity, the participant is asked to give a minimum and a maximum practical time for completion, together with an indication of when the activity is most likely to be completed. This is done initially for the whole subnetwork under the assumption of a certain funding level. Later, similar estimates are requested based on different funding levels. The interviewer then compares the duration estimates for each activity for the different funding levels and brings apparent discrepancies to the attention of the participant. This technique adds a new element of self-challenge to Delphi-type procedures.

10.4.4 Model Output and Use

In this project, there are three major milestones:

1. Demonstration of scientific feasibility
2. Completion of prototype (plasma test) reactor
3. Completion of first commercial reactor

The output of the network model may be used directly to portray the probability of success for each of these milestones. Figures 10-13 and 10-14 are extracted from Vanston's report [13] and portray the probability of reaching the first milestone under two funding levels and the probability of successfully developing a commercial reactor as a function of time, respectively. From Figure 10-14, it is seen that increasing the funding level does not significantly increase the probability of success during the early years. The impact of increased funding occurs in the middle years (1983-1988). A decision maker must evaluate the increased cost due to higher funding against the gain of approximately five years in achieving milestone 1. Note that the increased funding results in only a slight improvement from 0.717 to 0.758 in the probability of successfully demonstrating scientific feasibility.

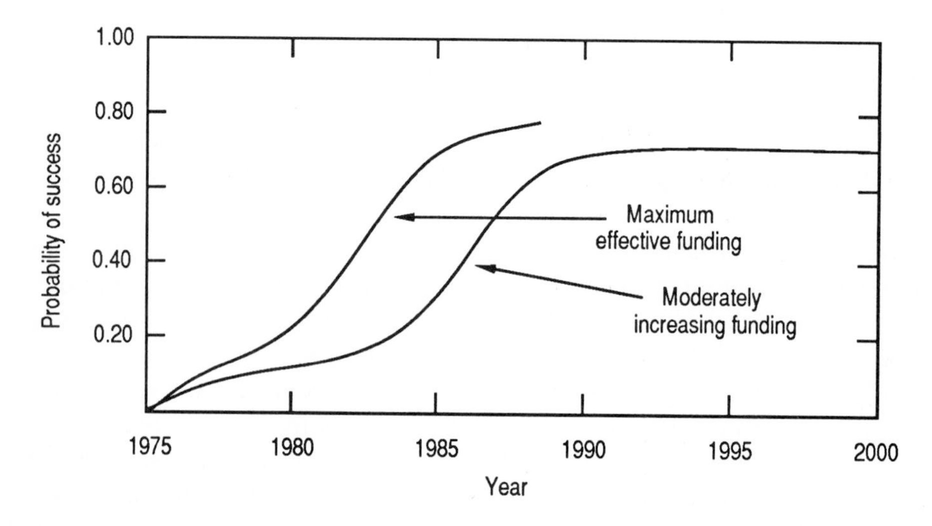

Figure 10-13 Probability of successfully demonstrating scientific feasibility under moderately increasing and maximum effective funding. Source [13]

Figure 10-14 presents a different picture of the effect of increased funding. Here there is a large disparity in the times to complete a first commercial reactor. The difference in mean times under the two funding levels is approximately 18.8 years. In this situation, the decision maker must evaluate the investment in increased funding versus the potential payoff resulting from the more rapid availability of commercial reactors. This decision is complicated by the estimate that there is only a 30% to 35% chance that the project will result in the completion of a first commercial reactor. Results like those presented in Figures 10-13 and 10-14 clearly identify the impact of budget allocations.

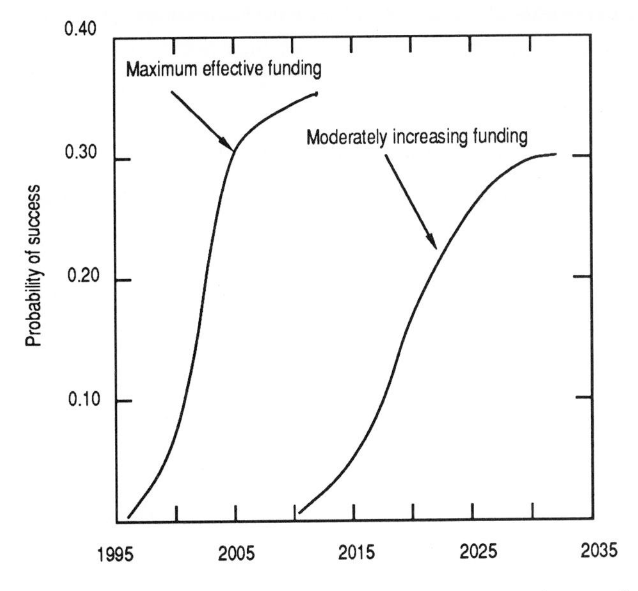

Figure 10-14 Likelihood of first commercial reactor under moderately increasing and maximum effective funding. Source [13]

In addition to modeling the Tokamak project with two funding levels, other funding strategies were devised and the network used to determine the probability of success and mean time to success for the three milestones indicated [13]. Table 10-3 presents the results of these analyses. Soland, Vanston, and Nichols [12] extended the research described by performing optimization studies for the project described. The optimization study employed branch-and-bound techniques and attempted to maximize the likelihood that a successful demonstration reactor is completed before the end of the year 2010, subject to an expenditure limit of $11 billion.

10.5 SUMMARY

Three examples of the use of network modeling for project planning are presented in this chapter. The examples demonstrate that network models can be built at different levels, ranging from the very aggregate to the very detailed. The examples demonstrate that, in project planning, there is a need for probabilistic and conditional branching, variability in the time description for project activities, and acyclic modeling capabilities. The examples illustrate how the outputs from a simulation analysis can assist in project planning and decision making.

Table 10-3 Expected Tmes Required to Reach Key Milestones with Their Associated Probabilities of Success[a]

	Demonstration of Scientific Feasibility		Completion of Prototype (Plasma Test) Reactor		Completion of First Commercial Reactor	
Funding Strategy	Mean Time in Years	Probability	Mean Time in Years	Probability	Mean Time in Years	Probability
MIF (Moderately Increasing Funding)	13.2	0.727	26.9	0.580	48.8	0.303
MEF (Maximum Effective Funding)	9.0	0.758	17.5	0.661	30.0	0.337
CPF (Continue Present Funding)	18.6	0.638	32.9	0.502	54.4	0.259
MIF with MEF of first wall R&D	13.2	0.716	24.0	0.590	45.8	0.290
MIF with MEF of blanket R&D	13.3	0.714	26.7	0.687	49.0	0.290
MIF with MEF of CMS R&D	12.6	0.742	26.7	0.666	46.5	0.314
MIF with MEF of CMS after DSF	13.2	0.732	26.9	0.652	46.9	0.325
MIF with MEF of first wall and blanket R&D	13.1	0.767	23.8	0.629	45.8	0.286
MIF with MEF of all supporting technology R&D	12.6	0.733	23.0	0.638	42.5	0.303
MIF with MEF after DSF	13.1	0.743	23.2	0.651	35.7	0.305

[a] Source [13]

10.6 EXERCISES

Chapter 10 exercises are to be performed for the situation described below [7].

The National Center for Drug Analysis (NCDA) of the Food and Drug Administration (FDA) is an analytical laboratory specializing in the analysis of unit dosage forms of various pharmaceutical products. A primary function of the center is to develop and test automated methods to be used in the analytical programs. After a decision is made to investigate a particular category of drugs, the research director begins the planning and scheduling process by reviewing the schedule of drugs and noting the dosage levels, different product forms, specific chemical and physical properties, and similarities to products previously analyzed. A flowchart of the research procedure used in NCDA is in Figure 10-15 The flowchart provides guidelines to be used by the research chemists in performing projects related to drug testing. All the steps, up to and including the Preparation of Reports of NCDA Methodology, must be completed before the research can be considered complete. A list of activities and time estimates is provided in Table 10-4.

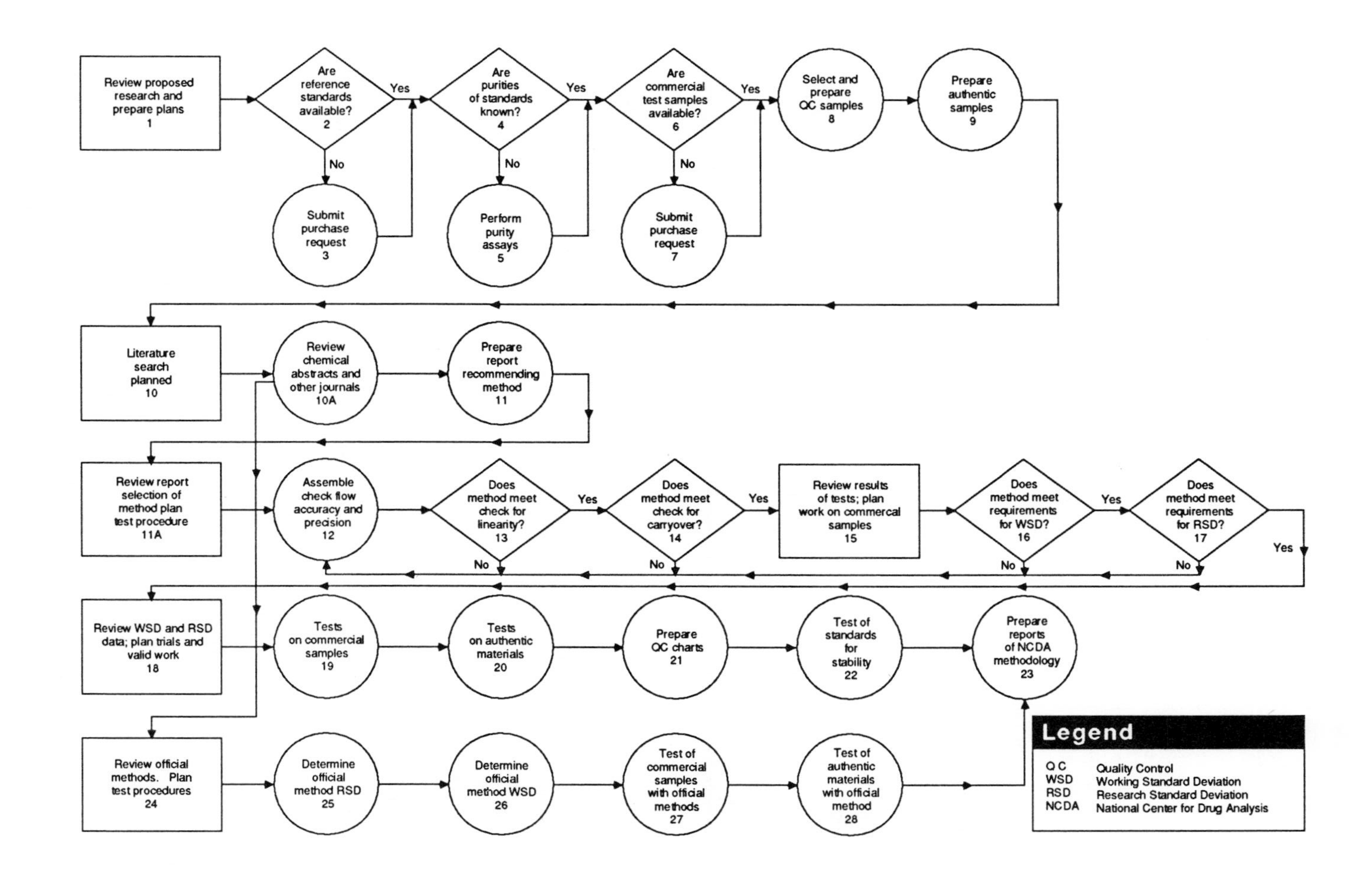

Figure 10-15 Flowchart for drug testing procedures. Source [7]

Table 10-4 Activity List for Drug Testing Procedures

Activity	Predecessor Activities	Time Estimates (days) Optimistic	Most likely	Pessimistic
1. Review and prepare plans	-	1	1	1
2. Review reference standards	1	1	2	3
3. Set reference standards	2	10	21	90
4. Review purity standards	2,3	2	4	6
5. Do purity assays	4	1	4	10
6. Determine availability of samples	1	1	2	3
7. Get commercial samples	6	5	10	20
8. Prepare QC samples	6,7	0.25	0.5	3
9. Prepare authentic samples	4,5	0.5	1	3
10. Perform literature search	1	1	4	15
11. Develop NCDA method	10	0.5	1	5
12. Assemble system	8,9,11	1	2	4
13. Linearity tests	11	1	3	10
14. Carryover tests	11	1	2	10
15. Review tests and plan work	13,14	3	4	5
16. WSD met	11	10	12	15
17. RSD met	11	0.5	1.0	2.0
18. Review WSD and RSD data	16,17	3	12	13
19. Tests on commercial samples	8,16	2	4	6
20. Validation tests	9,17,19	1	2	3
21. Prepare QC chart	20	0.2	0.5	1
22. Tests of standards	21	10	14	21
23. Prepare reports	22,27,28	3	5	10
24. Review methods	11	10	15	20
25. Determine RSD	24	0.5	1	2
26. Determine WSD	24	5	12	15
27. Tests on commercial samples	25	2	4	6
28. Validation samples	26	1	2	3

10-1. (a) Without reference to the list of activities, convert the flowchart to a PERT diagram. List all assumptions required.
(b) Prepare a PERT diagram from the list of activities.
(c) Perform an analysis on the PERT networks developed in parts (a) and (b).

10-2. (a) Develop a SLAM II network from the flowchart for drug testing procedures. List any additional data requirements for the SLAM II model.
(b) Develop a SLAM II model from the list of activities.
(c) Compare the SLAM II models developed in parts (a) and (b).
(d) Develop procedures to obtain estimates for the additional information required for the SLAM II network. Assume values for the required additional data and perform a SLAM II analysis.

10-3. Specify to which activities you would allocate additional personnel or dollars in order to shorten the project duration. Perform a SLAM II analysis assuming that the activities selected have been shortened. Select activity times so that there are two or more paths that have approximately the same expected completion time. Perform a SLAM II analysis to estimate the criticality indexes for each activity.

10.7 REFERENCES

1. Bellas, C. J., and A. C. Samli, "Improving New Product Planning with GERT Simulation," *California Management Review*, Vol. 15, No. 4, Summer 1973, pp. 14-21.
2. Bird, M. M., E. R. Clayton, and J. Moore, "Industrial Buying: 2 Method of Planning for Contract Negotiations," *Journal of Economics and Business*, Vol. 26, 1974, pp. 209-213.
3. Clayton, E. R., and L. J. Moore, "GERT vs. PERT, "*Journal of Systems Management*, Vol. 22, No. 2, 1972, pp. 11-19.
4. Dalkey, N., and O. Helmer, "An Experimental Application of the Delphi Method to the Use of Experts," *Management Science*, Vol. 9, No. 3, 1963.
5. Eisner, H., *Computer Aided Systems Engineering*, Prentice-Hall, Englewood Cliffs, NJ, 1988.
6. Hill, T. W., "System Improvement: A Sensitivity Approach Using GERT," Master's Engineering Report, Arizona State University, 1966.
7. Kwak, N. K., and L. Jones, "An Application of PERT to R&D Scheduling," *Information Processing and Management*, Vol. 14, 1978, pp. 121-131.
8. Moore, L. J., and E. R. Clayton, *Introduction to Systems Analysis with GERT Modeling and Simulation*, Petrocelli Books, New York, 1976.
9. Moore, L. J., and B. W. Taylor III, "Multiteam, Multiproject Research and Development Planning with GERT," *Management Science*, Vol. 24, No. 4, December 1977, pp. 401-410.
10. Raju, G. V. S., "Sensitivity Analysis of GERT Networks," *AIIE Transactions*, Vol. 3, No. 2, 1971, pp. 133-141.
11. Samli, A. C., and C. Bellas, "The Use of GERT in the Planning and Control of Marketing Research," *Journal of Marketing Research*, Vol. 8, August 1971, pp. 335-339.
12. Soland, R. M., J. H. Vanston, Jr., and S. P. Nichols, "Optimal Resource Allocation in the Nuclear Fusion Development Program," *ORSA/TIMS National Meeting*, November 1975.
13. Vanston, J. H., Jr., "Use of the Partitive Analytical Forecasting (PAF) Technique for Analyzing of the Effects of Various Funding and Administrative Strategies on Nuclear Fusion Power Plant Development," University of Texas, TR ESL-15 Energy Systems Laboratory, 1974.
14. Vanston, J. H., Jr., S. P. Nichols, and R. M. Soland, "PAF-A New Probabilistic, Computer-based Technique for Technology Forecasting," *Technological Forecasting and Social Change*, Vol. 10, 1977, pp. 239-258.

15. Vanston, J. H., Jr., *Technology Forecasting: An Aid to Effective Technology Management*, Technology Futures, Inc., Austin, TX, 1987.
16. Wilson, J.R., and others, "Analysis of Space Shuttle Ground Operations", *Simulation*, June 1982, pp. 187-203.

11 INVENTORY CONTROL

11.1 INTRODUCTION

Inventory systems have been analyzed for many years and continue to attract attention from practitioners and theoreticians in the industrial engineering and management science fields. This interest is justified because of the potential cost savings that can be achieved through improved decision making in inventory situations. The number of models of inventory situations is extensive and the analysis of these models has produced cost-saving solutions [2, 3, 4].

All inventory systems involve the storage of items for future sale or use. Demand is imposed on an inventory by customers seeking supplies of the items. A company must establish an inventory policy that specifies (1) when an order for additional units should be placed, and (2) how many of the items should be ordered at the time each order is placed. The answers to these two questions depend on the revenues and costs associated with the inventory situation. Revenue is a function of the number of units sold and the selling price per unit. Costs are more complex and include ordering expenses, inventory carrying charges (which may include an opportunity cost due to loss of the use of capital tied up in inventory), and stock-out costs. Stock-out costs can be considered either as the cost associated with a ***backorder*** (an order that is satisfied when inventory becomes available) or the cost associated with a ***lost sale***.

Inventory theory deals with the determination of the best inventory policy. Equations have been developed for setting parameter values in specific situations. These equations, however, are based on restrictive assumptions in order to make an analysis tractable. Through network simulation, such assumptions can be avoided. In this chapter, the common concerns in inventory analysis are discussed, and methods for applying network modeling to the analysis of inventory situations are described.

11.2 TERMINOLOGY

To introduce the terminology employed in inventory analysis, consider a single-commodity inventory system. The units of a commodity physically held in storage are collectively referred to as ***stock on hand***, or ***inventory on hand.*** The level of inventory on hand decreases when a customer's request, or ***demand***, for stock is satisfied. Note that it is the demand that is important in inventory situations and not just a customer's arrival. Inventory on hand increases when a shipment of goods (an order) arrives. An example of the dynamic nature of inventory on hand over time is illustrated by the graph in Figure 11-1. Initially, there are three units in stock. The arrow at time 1 indicates that a request for 1 unit of stock is made at time 1. This demand is satisfied since there is ample supply in stock. Hence, stock on hand decreases to 2 at time 1. Another demand at time 2 causes a second decrease in stock. At this point, an order is placed for 5 new units as indicated by the letter O. The shipment is not received until time 5. This 3-time unit delay between the time an order is placed and the time that it is received is called the ***lead time***. The lead time is important because stock outages can occur during this interval. For example, at time 3 another demand depletes the stock. At time 4, a demand occurs, but it cannot be satisfied. This creates a stock-out situation. Assuming that the customer agrees to wait until stock arrives, the request is recorded as a ***backorder***. If the customer does not agree to wait, a ***lost sale*** would result. In Figure 11-1, the stock on hand does not change when a backorder occurs. When the order is received at time 5, the backorder is satisfied and the number of units sent to inventory is 4. In general, at the time of receipt of an order, the stock level increases by the order

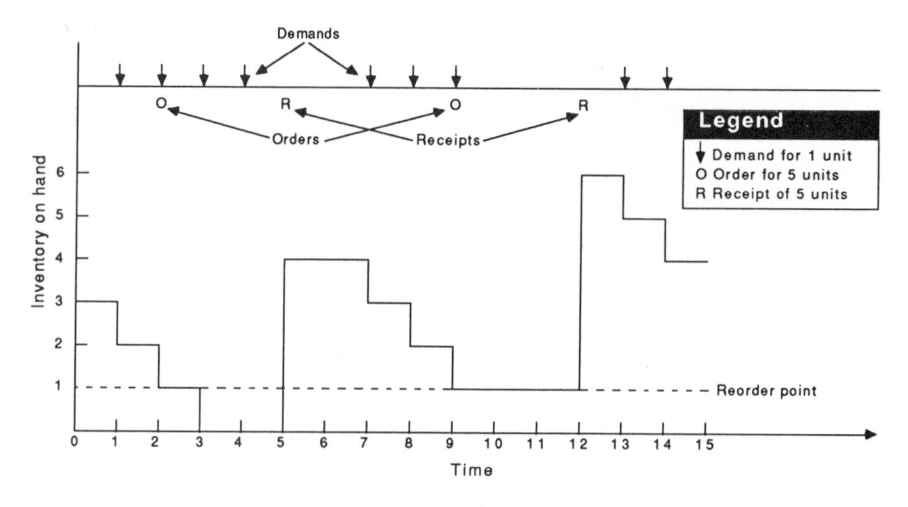

Figure 11-1 Example of inventory-on-hand changes.

size minus the number of current backorders. Note in Figure 11-1 that no demand occurs at time 5 or 6. Demand does occur, however, at times 7, 8, and 9. When the inventory on hand reaches 1 unit at time 9, another order for 5 units is placed. Again, the order arrives 3 time units later and the stock level is raised to 6 units.

The amount of inventory on hand at the time just prior to the addition of a newly arrived order is referred to as ***safety stock***. The term safety is used because this quantity reveals how "safe" the company is from an undesirable stock-out situation. Safety stock at time 5 is zero and at time 12 it is 1.

Another important system variable is inventory position. ***Inventory position.*** is equal to the inventory on hand plus the number of units on order (the number due in), minus the number of units that are on backorder (the number due out). While inventory on hand represents the number of units physically available to be sold, inventory position represents the number of units that are or will be available to the company for potential sale. Inventory decisions regarding the timing for placing orders and the quantity to order for each order are normally based on the inventory position. The inventory position for the example in Figure 11-1 is shown in Figure 11-2. At times 1 and 2, there are no backorders or stock on order. Therefore, the inventory position equals the inventory on hand. When an order is placed at time 2, the inventory position increases to 6, reflecting the fact that there are 5 units on order and 1 unit in stock. A review of only the inventory on hand at time 4 might lead to the decision to order more units because inventory on hand is zero at this point. However, by checking the inventory position at time 4, it is known that units are on order and a new order need not be placed. Consequently, inventory position represents the number of

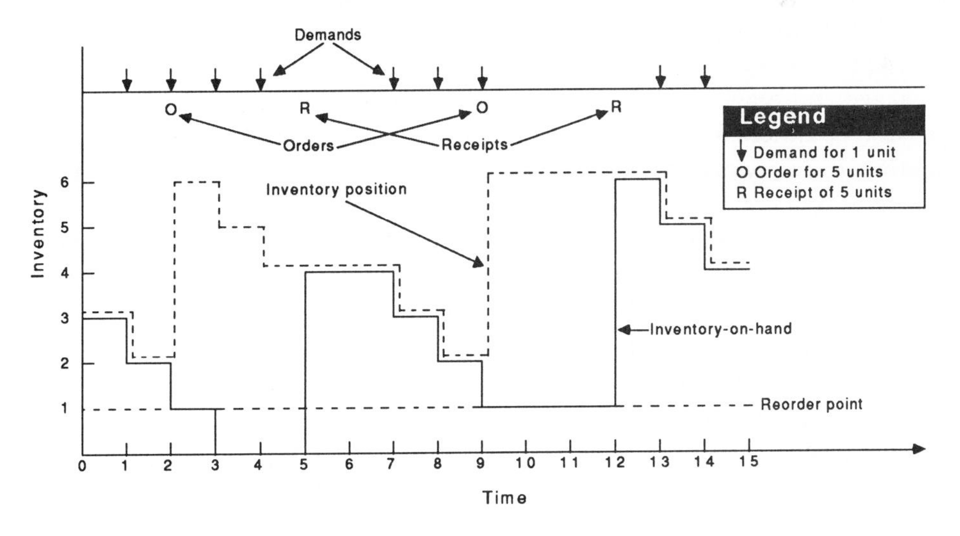

Figure 11-2 Graph of inventory-on-hand and inventory position.

units that are expected to be available to meet demand and, therefore, ordering decisions are based on inventory position rather than just the inventory on hand. Whenever a demand occurs, inventory position is reduced. Hence, inventory position is reduced by 1 at time 4 to reflect the backorder. When the order is received, inventory position is not changed, as it is increased when the order is placed.

In the preceding discussion, the process for deciding to place an order at times 2 and 9 was not specifically described. Orders for more units occur at points in time when the inventory position is reviewed. Two types of review policies are common in industry, ***periodic*** review and ***continuous*** review. Periodic review measures the inventory position every TBR time units, where TBR is a fixed time between reviews. Inventory position measurement may require a physical inventory count and an examination of both the outstanding order and backorder files. A continuous inventory review policy examines the inventory position at each demand occurrence. The continuous inventory review policy is also referred to as transaction reporting or as an on-demand review policy.

11.3 INVENTORY COSTS AND SYSTEM PERFORMANCE MEASURES

Revenues and costs are associated with inventory situations and are influenced by management policies and decisions. Through an examination of the costs, performance measures are defined that are important to the analysis of inventory systems. The costs associated with inventory include holding costs, stock-out costs, and the cost of management procedures, such as inventory review and ordering.

The holding of inventory has associated with it the cost of insurance, taxes, breakage, pilferage, warehousing, handling equipment, and operational expenses. Furthermore, since capital invested in inventory is not available, holding inventory results in a lost opportunity cost. Typically, holding costs are expressed as a percentage of the total cost invested in inventory per unit of time. For example, a company might estimate that each dollar invested in inventory for a year incurs a holding cost of 30 cents. If an average inventory stock level valued at $10,000 is maintained for one year, then the annual holding cost would be $3000. This cost often includes interest incurred on loans related to inventory and the foregone income that might have accrued had the money been invested in something other than inventory (usually calculated at a company's internal rate of return).

To generalize, let HC represent the holding cost per inventory dollar per unit of time. Let CPU represent the cost for each unit in inventory, and let ASV represent the average inventory on hand for a given time interval T. Then the total holding cost incurred for time interval T is given by HC*CPU*ASV*T. HC, CPU, and T are normally known, or given values, and therefore, given an estimate of the average level of inventory on hand for time period T, a total holding cost can be computed.

Although it costs money to keep goods in inventory, there are lost profits associated with being without inventory. Obviously, lost profits due to lost sales because of lack of inventory are rather straightforward and usually identifiable. However, it is difficult to measure lost profits due to such factors as the loss of customer goodwill. If it is assumed that estimates can be obtained for the intangible costs associated with inventory, then a cost of lost sales can be computed. Let CSL equal the cost of losing a sale and let TLS equal the total number of lost sales over a given time interval because of a lack of inventory. Then, stock-out cost due to lost sales would equal CSL*TLS. If backorders are incurred, the cost due to backorders may be expressed as CPB*TB, where CPB is the estimated cost per backorder and TB is the total number of backorders per period of time. To include a cost for the time a customer is "backordered," a cost per backorder per unit of time the backorder exists, CPBT, is used, and ATB would be a time weighted average of the number of backorders. Over a given time interval, T, the total backorder cost would be CPB*TB+CBPT*ATB*T.

One method to attempt to avoid stock-out situations is to review inventory and to place inventory replenishment orders frequently. The degree to which this is advantageous is limited by the added holding costs incurred for any excess inventory and the additional costs due to the review procedure and order processing. Let CPO and CPR represent the cost per order and the cost per review, respectively. If TO represents the total number of orders and TR represents the total number of reviews during a specified time interval, then CPO*TO is the cost of ordering and CPR*TR equals the cost of reviews.

The purpose of incurring these costs is to maintain an inventory of goods that results in profits. Let TS equal the total sales, CPU represent the cost per unit, and PPU equal the sale price per unit. The gross profit from sales is TS*(PPU-CPU). By combining all costs and revenues, the profit, P, for an inventory policy for T time units can be defined as

P=TS*((PPU-CPU)-(CPO*TO)+(CPR*TR)+(HC*CPU*ASV*T)+(CSL*TLS)
+(CPB*TB)+(CBPT*ATB*T)))

where TS = total sales in period T
PPU = price per unit
CPU = cost per unit
CPO = cost per order
TO = total orders in period T
CPR = cost per review

TR = total reviews in period T
HC = holding cost per inventory dollar per time unit
ASV = average inventory on hand in period T
CSL = cost per lost sale
TLS = total lost sales in period T
CPB = cost per backorder
CBPT = cost per backorder per time unit
TB = number of backorders in period T
ATB = average number of backorders in period T
T = time interval

It is assumed that knowledge of the coefficients PPU, CPU, CPO, CPR, HC, CSL, CPB, and CPBT is available or may be collected when analyzing an inventory system. Inventory analysis is used to estimate the following performance variables in order to complete the calculation of profit:

1. Level of inventory on hand over time at receipt times (safety stock)
2. Number of backorders
3. Number of sales (satisfied demand)
4. Number of orders
5. Number of reviews
6. Number of lost sales

The profit equation developed above is just one method for evaluating inventory policies by combining performance values. Sometimes it is not necessary to combine the measures. For example, a corporation could be most concerned about the loss of customers and use safety stock as the prime measure of performance, even though safety stock is not a direct part of the profit or cost calculations. In all cases, it is important to recognize that system performance measures need to be analyzed if an inventory policy evaluation is to be performed. Later in this chapter, SLAM II output is related to each of the performance variables.

11.4 THE CENTRAL QUESTIONS: WHEN TO ORDER? HOW MUCH TO ORDER?

To avoid stock-out situations, large inventory levels should be maintained. However, large inventory in storage results in high holding costs. If this situation were countered by ordering small amounts, frequent orders would be required to avoid stock-out costs. But frequent orders are undesirable because of the costs associated with placing a large number of orders. Other factors, such as quantity discounts, also make small order quantities disadvantageous. Thus, there is a cost trade-off between placing frequent small orders and not so frequent larger orders.

Inventory policies specify how often to order and how much to order. Typically, inventory policies are specified in one of the following two ways:

1. If the inventory position is less than a specified ***reorder point***, R, order Q units.
2. If the inventory position is less than a specified reorder point, R, order up to a ***stock control level*** of SCL units.

Both of these policies use an ordering point based on inventory position and specify how much to order. Neither policy, however, specifies the actual time of ordering, which depends on when the inventory position is reviewed. As discussed in Section 11.2, two types of review procedures are common. A ***periodic review*** procedure specifies that a review is to be performed at equally spaced points in time, for example, once at the end of every month. The ***continuous review*** procedure, also called a transaction reporting or on-demand review procedure, involves examining the inventory position everytime it decreases and requires maintaining a running record of the inventory position (known as a perpetual inventory system).

Networks can be used to model the dynamic behavior of inventory situations to obtain estimates of the performance measures discussed above. Such models incorporate inventory policies as a part of the network, so the various strategies can be evaluated. Consistent with the approach to problem solving presented in the text, inventory analysis begins with a basic SLAM II network model, and then embellishments are added to include more complex inventory situations.

11.5 EVALUATING A CONTINUOUS REVIEW POLICY FOR AN INVENTORY SYSTEM HAVING BACKORDERS

The inventory system to be considered involves a single commodity in which backorders occur. No sales are lost and a continuous review procedure is used to monitor inventory position. When the inventory position decreases to a specified reorder point, R, an order for Q units is placed.

11.5.1 The SLAM II Model

The SLAM II model of the inventory system is shown in Figure 11-3. The demand for goods is modeled by the CREATE node, where interarrival times correspond to times between demands. When a demand arrives in the model at ASSIGN node ASN1, the variable POS, representing the inventory position, is

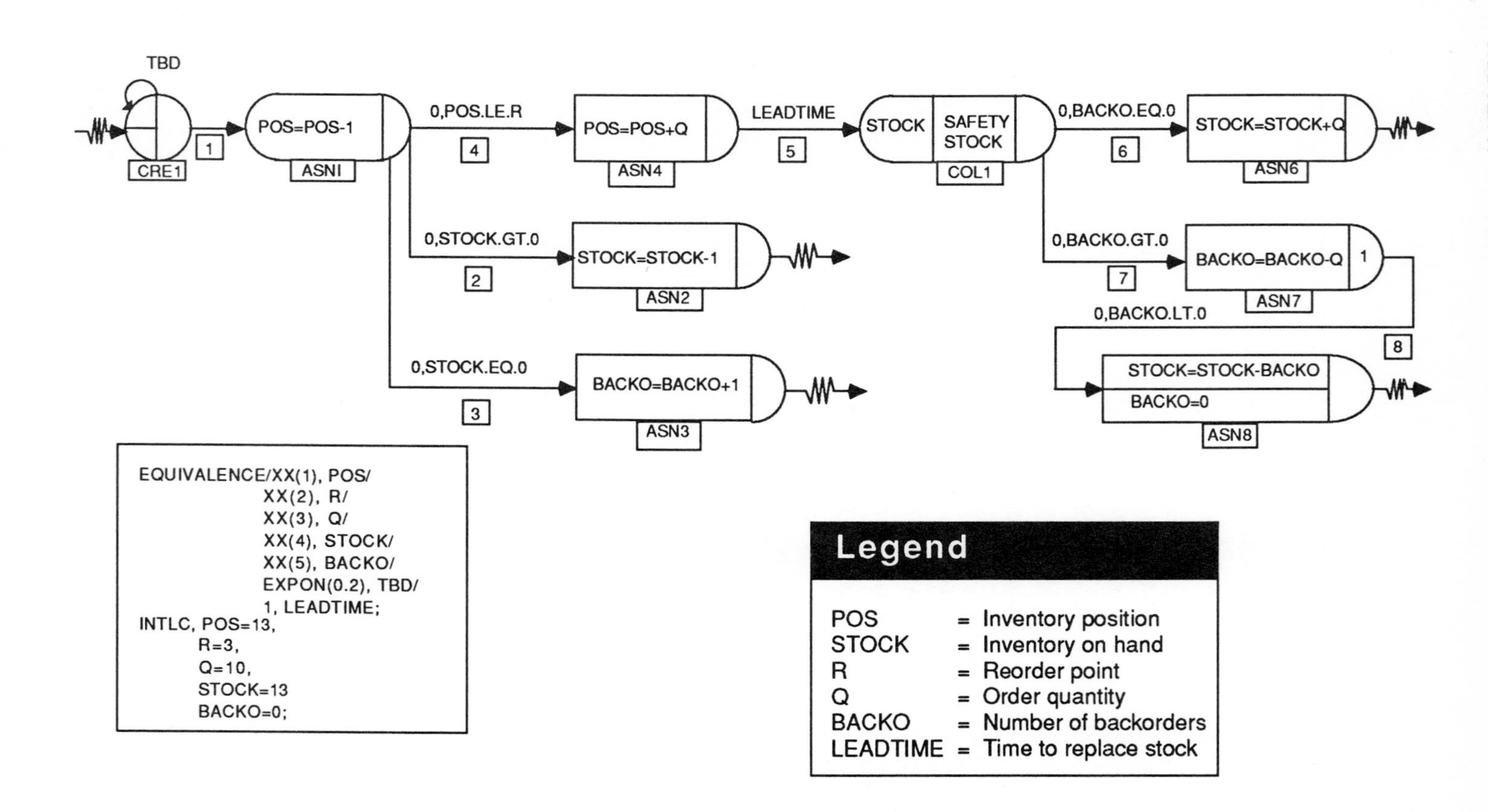

Figure 11-3 SLAM II model of an inventory situation with backorders and continuous review.

reduced by 1. Two of the three branches may be taken from node ASN1, depending on the inventory position relative to the reorder point and the number of units in stock. For example, the variable STOCK, representing the level of stock on hand, is always checked when a demand is made on the system. If STOCK is greater than zero, items are on hand and 1 unit is provided to the customer. This is accomplished at node ASN2, where STOCK is decreased by 1. If STOCK is equal zero, the demand is converted to a backorder by increasing BACKO, representing outstanding backorders, by 1 at node ASN3.

The inventory position is also checked after POS is decreased by 1 at node ASN1 for each demand made on the system. If POS is greater than the reorder point R, no further action is required. However, should the demand reduce POS to R, the reorder process is initiated. An order is placed by first increasing the inventory position (POS) by the number of units ordered (Q) at node ASN4 and then starting activity 5, which represents the lead time before the order arrives. The order arrives at COLCT node COL1 at the end of the delay for the lead time. Statistics are collected at node COL1 on the number of units in STOCK before placing the new units in the inventory, thus collecting statistics on safety stock. The backorder level is then checked to determine how many of the new units are needed to satisfy backorders, with the rest of the order becoming part of inventory.

The conditional branching stipulated at node COL1 places all the Q units ordered into STOCK at node ASN6 if there are no backorders (BACKO.EQ.0). If BACKO is greater than zero, then the backorder level is reduced by Q units at node ASN7. Should BACKO remain positive or zero after this reduction, there are insufficient units in the arriving order to satisfy the existing backorders and no change is made to STOCK. Consequently, no further action is taken until the next order arrives. However, a negative value for BACKO at node ASN7 indicates that all backorders have been satisfied and that there are units in the arriving order that are to be placed into inventory. This is accomplished at node ASN8, where STOCK is increased from zero to the current value of BACKO (with a change in sign), and the variable BACKO is reset to zero.

For the model illustrated in Figure 11-3, an EQUIVALENCE statement is employed in order to use generic names for the variables instead of SLAM II variable names, functions and constants. In addition, initial values for the variables are input using the INTLC statement. At the start of a simulation, inventory position, POS, is initialized to 13 units, stock on hand, STOCK, to 13 units, and the number of BACKO to 0. The reorder point, R, is 3 units, and the quantity ordered, Q, is 10 units. Statistics and histograms could be automatically obtained for each variable by including COLCT nodes in the model. For the model in Figure 11-3, the only statistics collected are at node COL1 to obtain an estimate for the level of safety stock.

11.5.2 Summary of Modeling Concepts

This inventory example introduces most of the general SLAM II concepts that are necessary to model inventory systems. The following list summarizes the modeling concepts used in the example and generalizes them where appropriate.

1. Demands are modeled as system arrivals. The arrival pattern can be varied to represent different demand processes.
2. Inventory on hand can be modeled as a global variable or a resource. When a demand arrives and inventory on hand is available, a sale results. When a demand arrives and inventory on hand is not available, a backorder or lost sale results.
3. Backorders can be maintained as a global variable. Since backorders are demands that wait, backorders can also be modeled using AWAIT or QUEUE nodes.
4. A receipt of an order results in satisfying backorders and then increasing the inventory on hand.
5. The lead time is represented by an activity time.
6. Inventory position can be maintained as a global variable or a resource. Inventory position decreases when a demand occurs and increases by the order quantity when an order is placed.
7. A continuous inventory review procedure is modeled by decreasing inventory position by the demand amount at an ASSIGN node. Orders are created by using routing logic or attribute assignment. The order quantity may be a constant amount or determined as a function of system status at an ASSIGN node.
8. Safety stock is the inventory on hand at the time of the receipt of an order. Inventory on hand values collected at COLCT nodes are used to obtain estimates of safety-stock values.
9. Total sales, total reviews, and total orders equal the number of entities that pass through activities associated with these functions. Statistics can be obtained on these values by inserting COLCT nodes into the network. In subroutine OTPUT, values for these quantities can be obtained using standard SLAM II functions.

11.5.3 SLAM II Assessment of Performance Measures, Cost, and Profits

The inventory performance measures discussed in Section 11-3 are directly obtainable from the outputs of SLAM II. The translation of these performance measures into inventory costs, revenues, and profits is accomplished using subroutine OTPUT. In Table 11-1, each inventory performance measure is listed, together with the relevant SLAM II concept, its manifestation in the example of Section 11.5.1, and its associated SLAM II output.

Table 11-1 Inventory Performance Measures for Backorder and Continuous Review Model of Figure 11-3

Performance Measure	SLAM II Concept	Item in Example	SLAM II Output
Inventory on hand	Global variable	Variable STOCK	Time-weighted statistics
Number of backorders	Global variable	Variable BACKO	Time-weighted statistics
Number of sales	Number of entities and global variable	Activity 1 and variable BACKO	NNCNT(1) - BACKO
Number of orders	Number of entities	Activity 4	NNCNT(4)
Number of reviews	Number of entities	Activity 1	NNCNT(1)
Safety stock	COLCT node statistics and histogram	Variable STOCK prior to receipt of an order	Statistics collected at node COL1

The average inventory on hand corresponds to the time-weighted average of the value of the variable STOCK. The SLAM II processor automatically computes this time-weighted average at the end of each simulation run when directed by a TIMST statement for the time-persistent variable STOCK. The SLAM II Summary Report includes this time-weighted average as the estimate of the mean value for the variable and also includes its standard deviation, minimum value, maximum value, time interval, and current value. Statistics are also obtained for the variables POS (inventory position) and BACKO (backorders) by including TIMST statements for these variables as part of the SLAM II input. Histograms of the fraction of time a variable has a set of values can be obtained as part of the summary report for each of these variables by specifying histogram parameters on the TIMST statement.

In the SLAM II inventory model illustrated in Figure 11-3, there is a correspondence between specific activity completions and the number of sales, orders, backorders, and inventory position reviews. The SLAM II processor maintains values and computes statistics for each of these variables, which can be accessed in user-written subroutines through the subprograms discussed in Chapter 5. For example, the number of entities that have completed an activity is obtained through the SLAM II function NNCNT(NACT), where NACT is the activity number. In Figure 11-3, the total number of demands is the number of entities that have completed activity 1, NNCNT(1). Consequently, total sales can be calculated as TS = NNCNT(1) - BACKO. Since the user written subroutine OTPUT is called automatically at the end of each simulation, additional performance measures can be computed and output in this subroutine.

The SLAM II statement model is shown in Figure 11-4. The STAT statements are used to collect statistics on average and total profit and the total number of sales, orders, and inventory reviews for 100 runs. The collection of these statistics is accomplished in subroutine OTPUT, where calls are made to subroutine COLCT for each variable.

```
GEN,PRITSKER,INVENTORY,4/26/88,100,,N,,N,Y/100;
LIM,,,10;
EQUIVALENCE / XX(1),POS / XX(2),R / XX(3),Q / XX(4),STOCK /
              XX(5),BACKO / EXPON(0.2),TBD / 1,LEADTIME;
TIMST,POS,INVPOS;
TIMST,STOCK,AVSTOCK;
TIMST,BACKO,AVBACKO;
STAT,1,AVERAGE PROFIT;
STAT,2,TOTAL PROFIT;
STAT,3,TOTAL SALES;
STAT,4,TOTAL ORDERS;
STAT,5,TOTAL REVIEWS;
STAT,6,AVERAGE INV;
STAT,7,AVERAGE BO;
STAT,8,AVERAGE SAFET,10/0/.03;
INTLC,POS=13,R=3,Q=10,STOCK=13,BACKO=0;
       NETWORK;
CRE1   CREATE,TBD;
       ACT/1;
ASN1   ASSIGN,POS=POS-1;
       ACT/2,0,STOCK.GT.0,ASN2;
       ACT/3,0,STOCK.EQ.0,ASN3;
       ACT/4,0,POS.LE.R,ASN4;
;
ASN2   ASSIGN,STOCK=STOCK-1;
```

Figure 11-4 SLAM II statement model of an inventory situation with backorders and continuous review.

```
        TERM;
ASN3    ASSIGN,BACKO=BACKO+1;
        TERM;
;
ASN4    ASSIGN,POS=POS+Q;
        ACT/5,LEADTIME;
COL1    COLCT,STOCK,SAFETY STOCK,,1;
        ACT/6,0,BACKO.EQ.0,ASN6;
        ACT/7,0,BACKO.GT.0,ASN7;
;
ASN6    ASSIGN,STOCK=STOCK+Q;
        TERM;
;
ASN7    ASSIGN,BACKO=BACKO-Q;
        ACT/8,0,BACKO.LT.0,ASN8;
        ACT,,,TRM;
;
ASN8    ASSIGN,STOCK=STOCK-BACKO,BACKO=0;
TRM     TERM;
        END;
INIT,0,312,NO/9;
FIN;
```

Figure 11-4 SLAM II statement model of an inventory situation with backorders and continuous review (concluded).

Subroutine OTPUT for the basic inventory system is listed in Figure 11-5. By combining the estimates for the performance measures obtained during a simulation run, an average profit for the run can also be computed. DATA statements are used to set the cost and price coefficients of the profit equation. The total sales are computed, as described above, the number of orders TO is NNCNT(4), and the number of inventory reviews TR is NNCNT(1). The time-integrated average value of the number of backorders ATB is TTAVG(3), where the index 3 results from BACKO being the third variable for which TIMST statistics are requested on the input statements. The time-integrated average for inventory on hand is obtained through the function TTAVG(2). The estimated total profit, P, is computed using the equation developed in Section 11.3 with the assumption of no cost per backorder (CPB). A profit per unit of time is computed by dividing P by the simulation time at the end of the run, which is TNOW, the time at which subroutine OTPUT is called. This assumes a start time of zero. The SLAM II summary report for 100 runs of the inventory model is shown in Figure 11-6. A histogram of the average safety stock on each of the 100 runs is given in Figure 11-7.

```
      SUBROUTINE OTPUT
C
      COMMON /SCOM1/ ATRIB(100),DD(100),DDL(100),DTNOW,II,MFA,MSTOP,
     +              NCLNR,NCRDR,NPRNT,NNRUN,NNSET,NTAPE,SS(100),
     +              SSL(100),TNEXT,TNOW,XX(100)
      EQUIVALENCE (XX(1),POS), (XX(2),R),  (XX(3),Q), (XX(4),STOCK),
     +           (XX(5),BACKO)
      DATA PPU,CPU,CPBT,HC,CPO,CPR / 77., 40., 10., .004, 25., 2. /
C       *** PPR = SELLING PRICE/UNIT,   CPU = COST/UNIT         ***
C       *** CPBT = COST/BACKORDER/TIME, CPO = COST/ORDER        ***
C       *** HC = HOLDING COST/INVENTORY $/TIME UNIT             ***
C       *** CPR = COST/INVENTORY REVIEW                         ***
       TS = NNCNT(1) - BACKO
       TO = NNCNT(4)
       TR = NNCNT(1)
       ASV = TTAVG(2)
       ATB = TTAVG(3)
       P = TS * (PPU - CPU) - (CPO*TO + TR*CPR + HC*CPU*ASV + CBPT*ATB)
       AVP = P/TNOW
       CALL COLCT (P,2)
       CALL COLCT (TS,3)
       CALL COLCT (TO,4)
       CALL COLCT (TR,5)
       CALL COLCT (AVP,1)
       CALL COLCT (ASV,6)
       CALL COLCT (ATB,7)
       SAFET = CCAVG(9)
       CALL COLCT (SAFET,8)
       RETURN
       END
```

Figure 11-5 Subroutine OTPUT for basic inventory model.

```
                       S L A M   I I   S U M M A R Y   R E P O R T

             SIMULATION PROJECT INVENTORY                   BY PRITSKER

             DATE  4/26/1988                                RUN NUMBER  100 OF  100

                      CURRENT TIME   0.3120E+03
                      STATISTICAL ARRAYS CLEARED AT TIME  0.0000E+00

                      **STATISTICS FOR VARIABLES BASED ON OBSERVATION**
```

	MEAN VALUE	STANDARD DEVIATION	MINIMUM VALUE	MAXIMUM VALUE	NUMBER OF OBSERVATIONS
AVERAGE PROFIT	0.1624E+03	0.3916E+01	0.1532E+03	0.1714E+03	100
TOTAL PROFIT	0.5068E+05	0.1222E+04	0.4781E+05	0.5347E+05	100
TOTAL SALES	0.1559E+04	0.3757E+02	0.1471E+04	0.1645E+04	100
TOTAL ORDERS	0.1555E+03	0.3738E+01	0.1470E+03	0.1640E+03	100
TOTAL REVIEWS	0.1559E+04	0.3765E+02	0.1471E+04	0.1645E+04	100
AVERAGE INV	0.3847E+01	0.4226E-01	0.3536E+01	0.3869E+01	100
AVERAGE BO	0.3316E+00	0.1163E-01	0.3247E+00	0.3976E+00	100
AVERAGE SAFET	0.1739E+00	0.4120E-01	0.6790E-01	0.3041E+00	100

Figure 11-6 SLAM II summary report for basic inventory model.

```
                              **HISTOGRAM NUMBER   8**

                                  AVERAGE SAFET

OBSV    RELA     CUML       UPPER
FREQ    FREQ     FREQ     CELL LIMIT      0         20         40         60         80        100
                                          +    +    +    +    +    +    +    +    +    +    +
  0    0.000    0.000    0.0000E+00       +                                                   +
  0    0.000    0.000    0.3000E-01       +                                                   +
  0    0.000    0.000    0.6000E-01       +                                                   +
  2    0.020    0.020    0.9000E-01       +*                                                  +
  6    0.060    0.080    0.1200E+00       +***C                                               +
 20    0.200    0.280    0.1500E+00       +**********    C                                    +
 29    0.290    0.570    0.1800E+00       +**************               C                     +
 24    0.240    0.810    0.2100E+00       +************                             C         +
 14    0.140    0.950    0.2400E+00       +*******                                          C +
  3    0.030    0.980    0.2700E+00       +**                                                 C+
  1    0.010    0.990    0.3000E+00       +*                                                  C+
  1    0.010    1.000       INF           +*                                                   C
  -                                       +    +    +    +    +    +    +    +    +    +    +
100                                       0         20         40         60         80        100
```

Figure 11-7 Histogram of average safety stock level for basic inventory model.

11.6 MODELING A PERIODIC REVIEW PROCEDURE AND A STOCK CONTROL LEVEL ORDERING POLICY

A SLAM II network model of an inventory situation in which a periodic review procedure is employed and a policy to order up to a stock control level is presented in Figure 11-8. The network logic is similar to the model given in Figure 11-3, except that an inventory review does not occur after each demand created at node CRE1. Instead, an entity representing a time delay between reviews (TBR) is created, and the reorder process is modeled by the disjoint network beginning with CREATE node CRE2. In this model, the subnetwork following node CRE1 is used only to decrease the inventory position (POS) by 1 for each demand and to adjust the inventory level (STOCK) or backorder level (BACKO) accordingly. The subnetwork starting with CREATE node CRE2 represents the inventory review, the ordering process, and the restocking process when units are received. Node CRE2 creates the first entity at the time of the first review (TFREV) and subsequent reviews after TBR time units. When the reorder point R is reached, the condition for activity 4 is satisfied and the order quantity Q is calculated at ASSIGN node ASN4.

The order quantity Q is the stock control level (SCL) minus the inventory position (POS). For this policy, the order quantity is determined to increase the inventory position up to a prescribed level; that is, POS = SCL. The values of Q and POS are determined at node ASN4. The lead time and restocking process are accomplished in the same manner as previously described for the model in Figure 11-3.

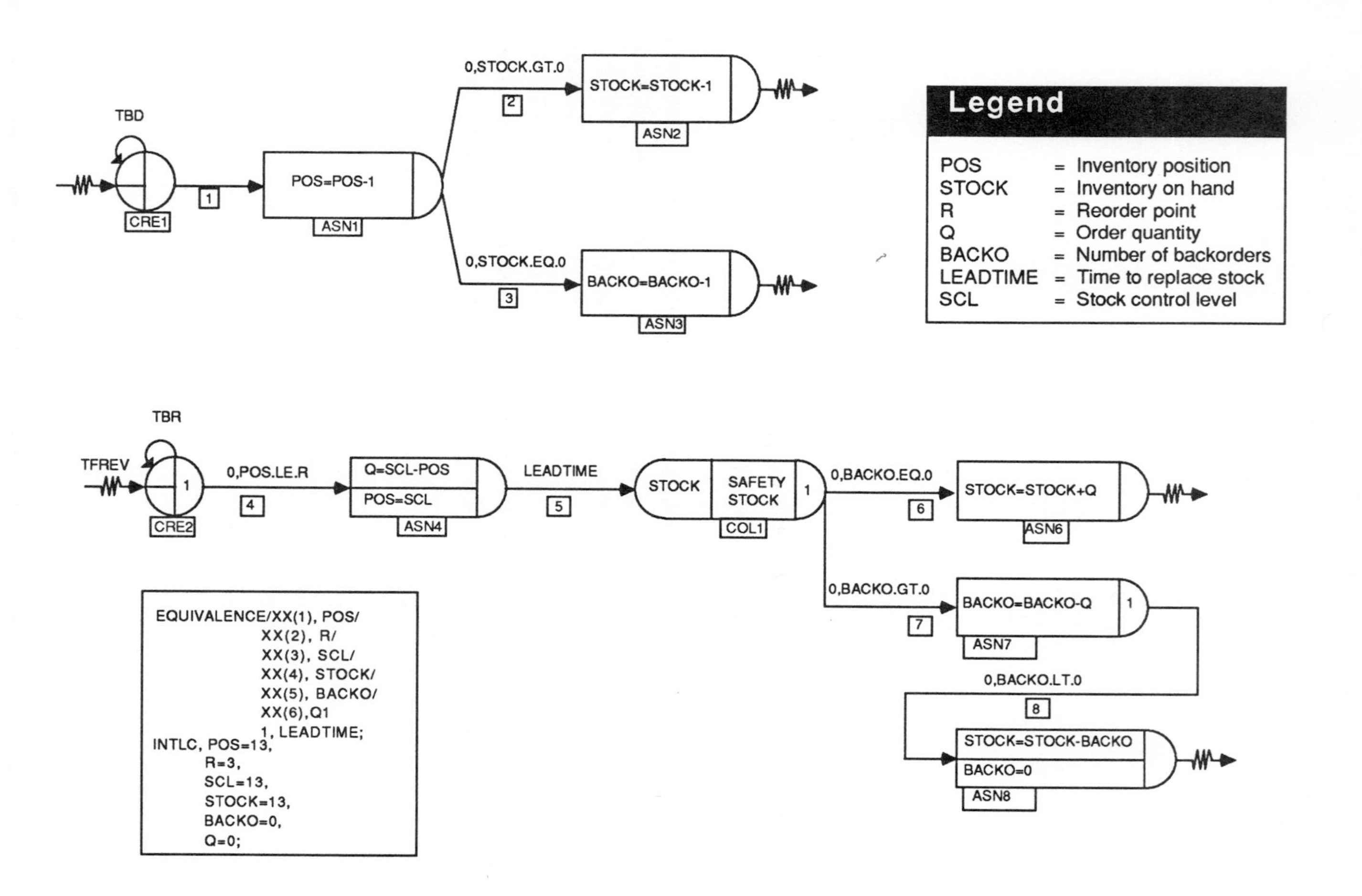

Figure 11-8 SLAM II model of an inventory system with backorders, periodic review, and a stock control level.

Note in the previous example that the SLAM II global variable XX(3) was made equivalent to Q on the EQUIVALENCE statement and that the input value for Q on the INTLC statement was 10. For the model in Figure 11-5, XX(3) is made equivalent to SCL and the initial value for SCL is 13.

In the two inventory situations presented, standard rules for carrying out inventory policies are used. SLAM II models can be adapted to incorporate more advanced decision-making concepts. One such possibility is to set the reorder point as an inverse function of the safety stock. Thus, if the average safety stock is large, a reduced value of the reorder point is established. Another possibility is to make the amount ordered a function of the time since the last order was placed, the average number of backorders, and/or the average inventory on hand. The complexity incorporated into a SLAM II model for establishing the decision variables is limited only by the analyst's ability to conceptualize new strategies.

11.7 MODELING LOST SALES

In inventory systems, there are situations where a sale will not be made if the demand cannot be satisfied immediately. This occurs when backorders are not taken or when a customer decides to go elsewhere for the item. In such cases, a sale is considered as lost.

The model for the lost-sale situation simply requires that all variables, activities, and nodes that are used to track backorders be removed from the networks previously presented. Usually, it is desired to track the number of lost sales. To do this, a variable (LSALE) is defined, and an ASSIGN node is used to increment lost sales when there is a demand and the stock level is zero. Note that a lost-sale does not change inventory position. A lost-sales inventory model is shown in Figure 11-9. The number of lost sales is increased by 1 at node ASN3 when there is a demand and STOCK equals 0.

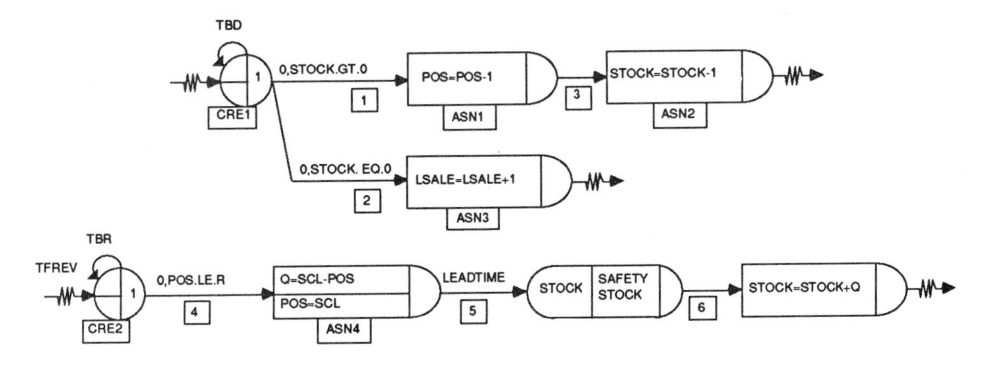

Figure 11-9 Network model for lost sales and periodic review.

A combination of the backorder and lost-sales situation is possible. Assume that four backorders are allowed to accumulate before a lost sales occurs. Customer demands and inventory status changes for this situation are modeled in Figure 11–10. The number of backorders is computed as before, however, they are not allowed to exceed 4 units. Lost sales (LSALE) are incremented by 1 when the variable BACKO equals 4 and there is a demand.

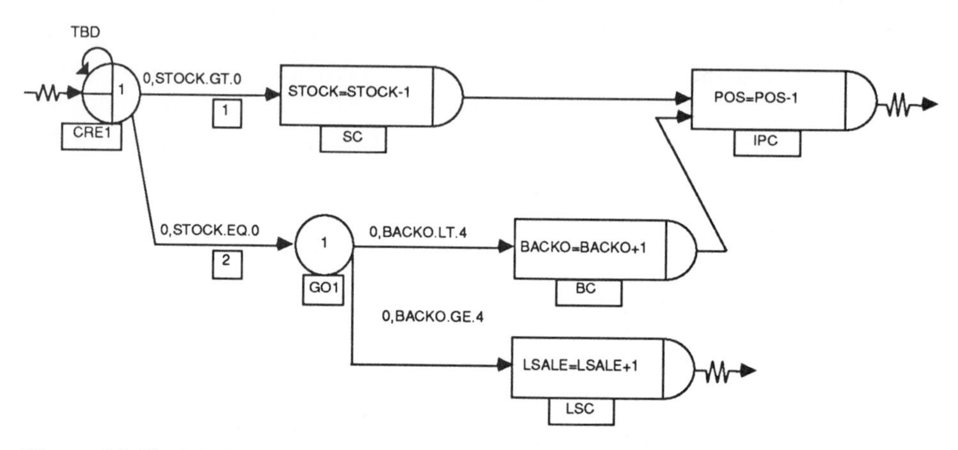

Figure 11-10 Model of a backorder and lost-sales situation.

Another method that may be used to specify lost sales is to introduce a probability that a customer does not backorder. Suppose that 10% of the time a customer decides not to backorder, which results in a lost sale. The network segment in Figure 11-11 represents this situation under the assumption of unlimited backorders.

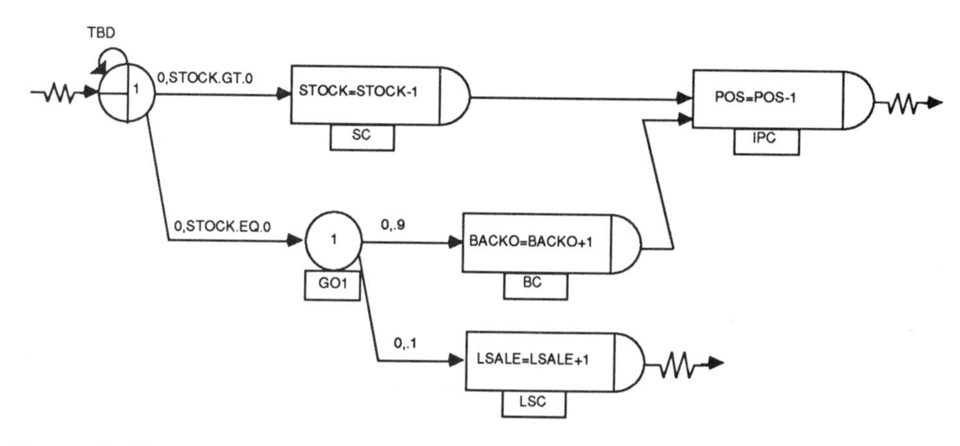

Figure 11-11 Lost sales based on probability of not backordering.

Sometimes a sale is lost because a customer, for whatever reason, decides not to accept a backordered unit when it arrives. This situation is modeled in Figure 11-12. In this model, a lost sale occurs when the current number of backorders exceeds 4 and also because 10% of the customers that have units on backorder refuse the item when it comes in. This situation is accounted for in the model after a replenishment order arrives. As before, statistics are collected at node COL1 for the level of safety stock.

At node ASN6, the variable STOCK is incremented by Q units, which is the order quantity. Conditional branching from node GO2 is based on the status of backorders in the system. If there are no backorders (BACKO,EQ.O) or no inventory on hand (STOCK.EQ.0), then no further action is required. If there are backorders, the branch to node GO3 is taken. If neither of these conditions holds, no further action is required.

At node GO3, a test is performed to determine if a customer still wants the item on backorder. Ninety percent of the customers accept the unit, and activity 7 to node ASN7 is taken. At node ASN7, both STOCK and BACKO are decreased by 1, confirming the sale, with no change in inventory position (POS). The backorder evaluation is continued by returning to node GO2. Ten percent of the customers refuse to accept a unit, and activity 8 routes the customer entity to node ASN8. At node ASN8, lost sales (LSALE) are increased by 1, backorders (BACKO) are decreased by 1, and the inventory position (POS) is increased by 1 to reflect that a backordered unit is to be retained as stock. A return to node GO2 is then made to repeat the process.

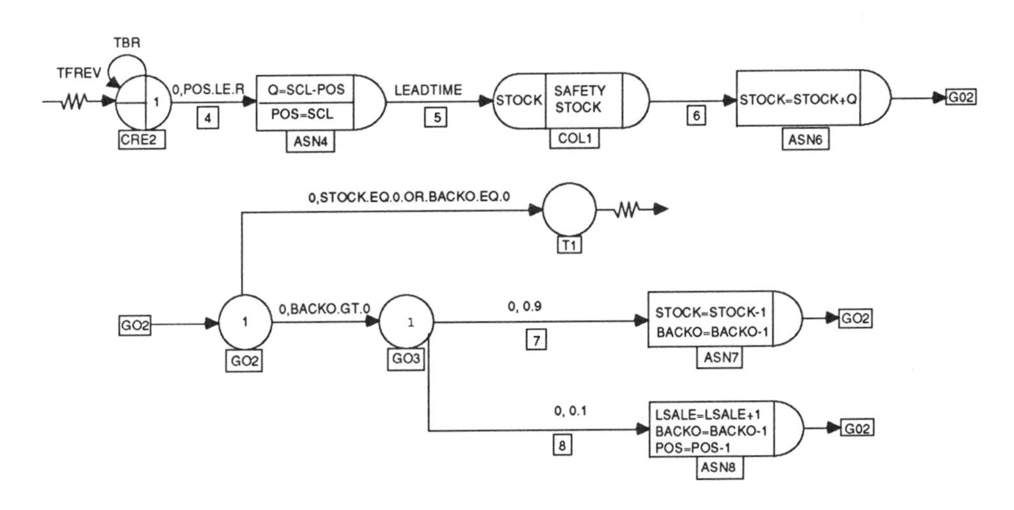

Figure 11-12 Lost sales due to backorder refusal.

In this model, it is assumed that the reneging of backordered customers is not known until an order is received. That is, the decision is not made by the customer until the customer has been informed that a unit is available to be purchased.

11.8 USING SIMULATION TO IMPROVE INVENTORY CONTROL

Once an inventory network model is constructed, the effect of changes to system performance due to changes in the component parts of the network may be measured. Both procedural changes in inventory policies and changes in the business environment are candidates for evaluation. Common changes in the business environment are:

1. A change in the demand for goods
2. A change in the cost of operation
3. A change in lead times

Although these types of changes may not be controllable, their impact on operations and profits can be calculated. It may be critical, for example, to know the impact on inventory *before* a forecasted increase or decrease in demand occurs. Similarly, expected changes in operating costs and product lead times need to be assessed. A network approach provides a method for analyzing these types of changes through embellishments to the SLAM II model. Making a change in a demand process may be as simple as new data input for the time between demands at a CREATE node, or it can be as complex as adding nodes and branches to create special demand conditions. Examples of such changes are presented in Chapter 12. Cost changes are accounted for by adjusting the cost variables used in subroutine OTPUT. Lead-time changes are controlled by activity branch specifications. For a particular business operation, the procedural aspects of inventory policy may be evaluated and controlled by analyzing new values for the quantity ordered, the reorder point, and the inventory review methods that are used.

11.9 SUMMARY

Diverse inventory situations have been modeled in terms of SLAM II networks in this chapter. The modeling of inventory control procedures involves the allocation of units of inventory to meet demand levels and the decision logic to determine when inventory is to be replenished. Complex situations are shown to be extensions of simple ones. Thus, the SLAM II approach of combining elements into network models of systems facilitates the modeling and analysis of complex inventory situations. This chapter demonstrates how SLAM II models of procedures are derived from control policy specifications.

11.10 EXERCISES

11-1. Discuss the following statements:
(a) Orders to replenish inventory should be based on inventory position and not inventory on hand.
(b) Safety stock is a measure of how well the reorder point is set.
(c) Order quantity is more important than reorder point in determining the cost of keeping goods in inventory.
(d) Relate the lead-time demand (the demand that occurs during the lead time) to the reorder point and safety stock values.

11-2. Discuss the differences between the continuous review and periodic review procedures with respect to the setting of order quantities, reorder point, back-orders, and stock control levels.

11-3. Perform a SLAM II analysis for the model presented in Section 11.5. Attempt to increase the profit per week by resetting the values for the decision variables.

11-4. Develop a SLAM II model in which an economic order quantity (the order quantity that minimizes total cost) is used and price discounts are given for ordering in multiples of 100.

11-5. For the inventory model presented in Figure 11-9, collect statistics on the lead-time demand.

11-6. Show how the inventory model presented in Section 11.5 can be driven by demand data stored in a database. Use an available database (instructor prepared) or create a database and run the inventory model for diverse policies.

11-7. Give reasons why the histogram of Figure 11-7 should have a bell-shaped form.

11-8. For the inventory model of Figures 11-4 and 11-5, determine the values of reorder point and reorder quantity that maximize average profit.

11.11 REFERENCES

1. Banks, J., and J. S. Carson, II, *Discrete-Event System Simulation*, Prentice-Hall, Englewood Cliffs, NJ, 1984.
2. Hadley, G., and T. M. Whitin, *Analysis of Inventory Systems*, Prentice-Hall, Englewood Cliffs, NJ, 1963.
3. Peterson, R., and E. A. Silver, *Decision Systems for Inventory Management and Production Planning*, Wiley, New York, 1979.
4. Wemmerlov, U., "Inventory Management and Control", *Handbook of Industrial Engineering*, G. Salvendy, ed., Wiley, New York, 1982.

12 MORE ON INVENTORY CONTROL

12.1 INTRODUCTION

The major focus of this chapter is the modeling of detailed demand processes for multiple types of items stored in inventory. Methods are described to specify the time between demands on a SLAM II network when the number of demands per unit of time is known or when explicit demand sequences are specified. The modeling approach for seasonal and bulk demands, including procedures for differentiation among different types of demands, is also discussed. The last section in this chapter presents a SLAM II model for a two-commodity inventory system.

12.2 MODELING DEMAND PROCESSES

The modeling of the demand for a product or service is a complex subject. In this book, the discussion is limited to the characterization of demand data. The underlying processes that cause demands to occur are not investigated. Demand data can be input to a model directly or can be characterized by a distribution function. When the data are used directly, a tacit assumption is made that a historic record is sufficiently representative of future demand patterns to be used in the evaluation of new policies. By fitting a distribution function to the data, an attempt is made to model all representative demand time histories. Techniques for building models of demand data are described in this section.

Many factors influence demand such as the time of day, the season of the year, and the service provided to past customers. A good modeler determines which factors are important when characterizing demands. A list of basic constructs relating to demand modeling is given on the next page.

1. If the time between demands is characterized by different distribution types, the next demand time is selected on a probabilistic basis. This may be necessary if demands arise from multiple sources.
2. If demand times are known, that is, a schedule of demands exists, then the demand times should be used directly.
3. If a histogram is available that describes the time between demands, then a discrete probability distribution should be created and function DPROBN should be used to obtain the interdemand time.
4. If the demand distribution varies over time (seasonal variations for example), or if demands for goods arrive in batches, or if demands have priorities, different models (subnetworks) to capture the functional specifications for these special features are required.

12.3 CHARACTERIZING THE INTERDEMAND TIME FROM DATA ON THE NUMBER OF DEMANDS

Frequently, demand data are in the form of a specified number of units demanded per time interval. For example, sales data may reveal that in any given week that the number of units demanded is Poisson distributed with a mean of λ. When simulating, the time a demand occurs is required. This necessitates the conversion of the distribution of the number of units per time interval to a distribution on the time between demands. In this discussion, demand occurrences will be referred to as events. In addition, the distribution of the number of events occurring in a given time interval will be referred to as the distribution of counts.

For the case where the distribution of counts is Poisson, converting to a distribution of the time between events is direct, because the time between the events of a Poisson process with a mean number per unit time λ is exponentially distributed with a mean time of $1/\lambda$. With other distributions of counts, there is no direct transformation from the distribution of counts to the distribution of time between events. That is, more information than count data is required to model the time between events. If it is not feasible to obtain additional data, the modeler has several possibilities for making use of available data. For example, there is a version of the central limit theorem [2, 3] that states that the distribution of the number of counts N_t in time interval t approaches the normal distribution with an expectation of

$$E[N_t] = \frac{t}{\mu}$$

and

$$VAR[N_t] = \frac{t\sigma^2}{\mu^3}$$

where μ is the mean time between events and σ is the standard deviation of the time between events.

Pritsker [5] discusses how this result can be used to verify assumptions about the distribution of the time between events, given data on N_t. The intent here is to use this result for another purpose, which is to estimate μ and σ given data describing $E[N_t]$ and $VAR[N_t]$. Although this is a convenient way to estimate μ and σ, a statement still cannot be made regarding the distribution of the time between events.

Nevertheless, this is a method for approximating parameters for an assumed distribution. The behavior of the assumed distribution can be tested using the network shown in Figure 12-1. The distribution of the time between events is specified as the interarrival time for the CREATE node. The network is simulated for t time units, and n runs are made. Statistics are collected at the end of each run at COLCT node COL2 to observe the number of counts on activity 1 for each run of the network. For each run, NNCNT(1) represents one observation of N_t. The statistical array is not cleared at the end of each run for node COL2, as specified on the INIT statement, and a histogram is requested for the observations at this node. In this way, an observation is made of the number of demands on a run, and a histogram is obtained of these observations over multiple runs. This information can then be compared to count distribution data using standard goodness of fit techniques. The between-statistics collected at node COL1 provide experimental data on the distribution of the time between demands specified for the CREATE node.

Another option is to assume a constant time between demands on a run, but that the demand over all runs has the distribution N_t. This is modeled by drawing a sample of N_t for a run and dividing the sample value by the length of a run. Since, on each run, a sample of N_t is drawn, the distribution on N_t over all runs is modeled. This scheme is illustrated for a run length of 1000 in the network

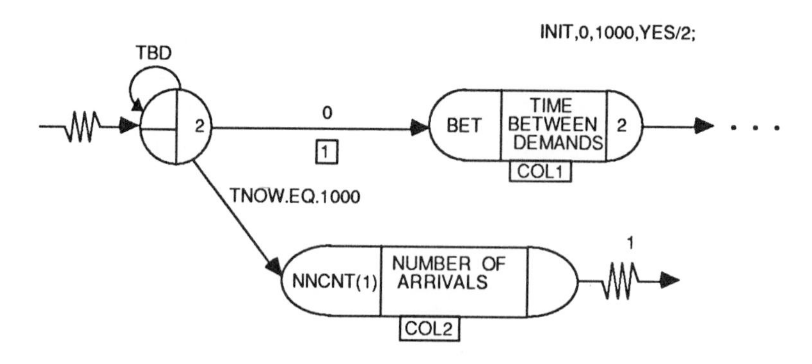

Figure 12-1 Network for testing arrival distribution assumptions.

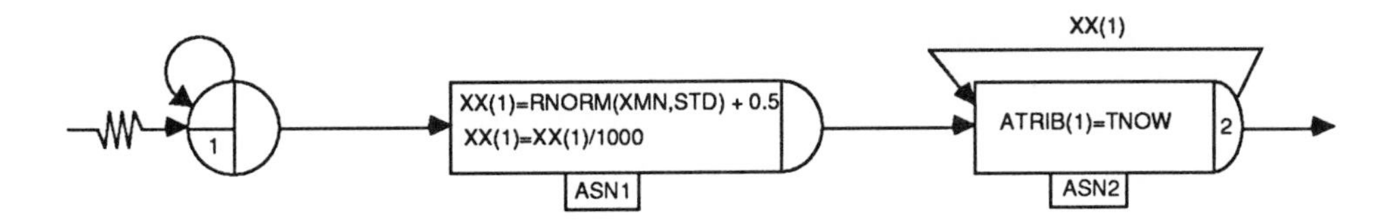

Figure 12-2 Network for specifying time between demands by sampling the number of counts, N_t.

segment of Figure 12-2. A sample value is obtained at node ASN1 and assigned to the SLAM II global variable XX(1). For the illustration, a normally distributed number of counts is assumed, and the sample is obtained from function RNORM(XMN,STD,IS). XX(1) is then divided by 1000 at node ASN1 to obtain the constant time between demands. Demands are then created at node ASN2, where the time of creation for each entity is marked on attribute 1. Subsequent demands are created every XX(1) time units by the self-loop activity at node ASN2. Since the network is simulated for TTFIN time units, the number of demands entering the system will be the original sample value. Over n runs, the number of arrivals to the system will be characterized by the distribution for N_t. Since the distribution of N_t is not really dependent on the distribution employed for the time between demands, this approach indicates the nonuniqueness of the interdemand time distribution for a specified distribution of the number of counts.

12.4 PREDETERMINED DEMAND SEQUENCES

As an alternative to characterizing demand data in functional form, it may be desirable to use a predetermined sequence of demands abstracted directly from purchasing records. This can be done using the network segment and code shown in Figure 12-3. The CREATE node generates one entity. The activity emanating from the CREATE node has a duration taken from function USERF shown below. The time of the first demand, TIME, and the attributes of the first demand, ATRIB, are read. The delay time until this demand is placed in the network is set equal to the time of the demand minus the current simulated time; that is, USERF = TIME - TNOW. If the order is late in starting, so that this delay time is negative, USERF is set to zero.

Upon arrival to GOON node G1, branching causes two entities to proceed into the model: one represents the current demand and the other signals the CREATE node to read the next demand event time and attributes. When an end of file is detected by the READ statement in function USERF, the value of ATRIB(1)

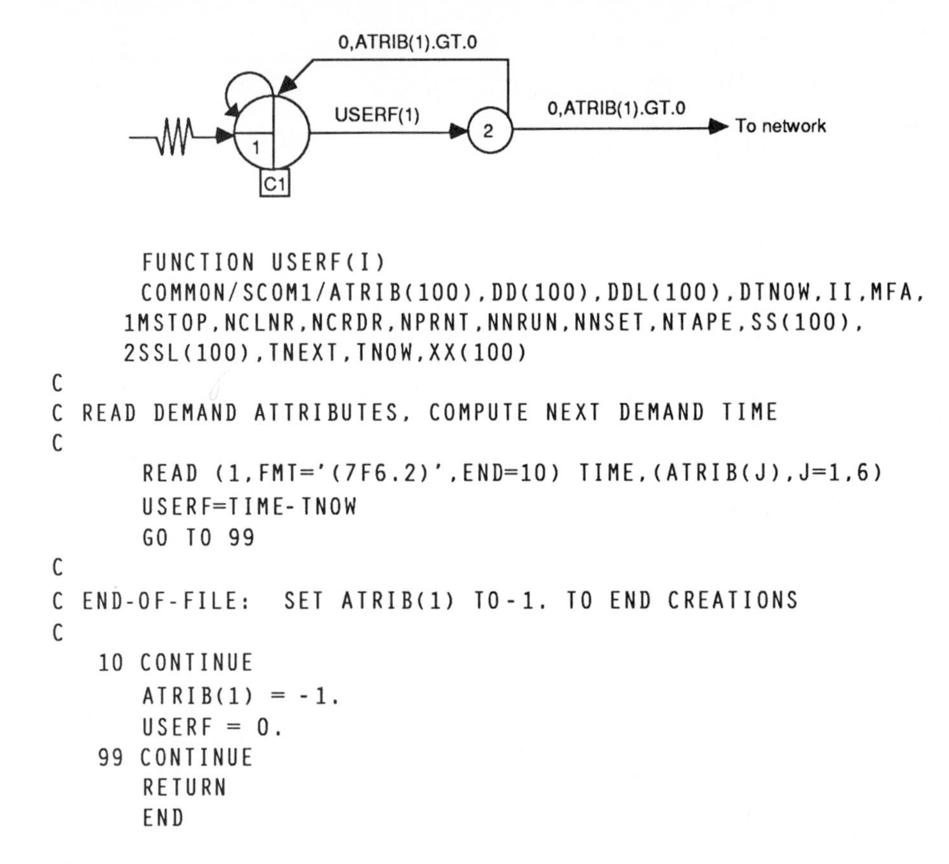


```
      FUNCTION USERF(I)
      COMMON/SCOM1/ATRIB(100),DD(100),DDL(100),DTNOW,II,MFA,
     1MSTOP,NCLNR,NCRDR,NPRNT,NNRUN,NNSET,NTAPE,SS(100),
     2SSL(100),TNEXT,TNOW,XX(100)
C
C READ DEMAND ATTRIBUTES, COMPUTE NEXT DEMAND TIME
C
      READ (1,FMT='(7F6.2)',END=10) TIME,(ATRIB(J),J=1,6)
      USERF=TIME-TNOW
      GO TO 99
C
C END-OF-FILE:  SET ATRIB(1) TO-1. TO END CREATIONS
C
   10 CONTINUE
      ATRIB(1) = -1.
      USERF = 0.
   99 CONTINUE
      RETURN
      END
```

Figure 12-3 Model for predetermined demand sequences.

is set to zero to stop entities from leaving the GOON node. This example illustrates a do-it-yourself CREATE node that not only inserts entities into a model, but also defines their attribute values. From function USERF, any source, including a company database, can be accessed.

12.5 INTERDEMAND TIME DISTRIBUTIONS SPECIFIED BY A USER-DEFINED HISTOGRAM

In some cases, interdemand time data are characterized in the form of a histogram. If it is not desirable or convenient to characterize this distribution by one of the standard distribution functions, the analyst may sample from the histogram directly using an activity duration specification of DPROBN (IC, IV, IS). The arguments to DPROBN are a row number IC of ARRAY, where the cumulative probability is defined for obtaining a corresponding value given in row number IV of ARRAY. IS is a random number stream and is an optional argument on the network.

DPROBN may be used as a duration or in an ASSIGN node. The values of the required two rows of ARRAY are set using an ARRAY statement. As an example of the use of DPROBN, consider the probability mass function, PROB, and its associated cumulative distribution function, CP, shown in Table 12-1.

Table 12-1 Values for Use with DPROBN

I	PROB(I)	CP(I)	VAL(I)
1	0.10	0.10	10
2	0.05	0.15	15
3	0.20	0.35	25
4	0.30	0.65	50
5	0.20	0.85	75
6	0.15	1.00	100

To obtain a sample time for activity 7 from this distribution employing random number stream 4, the following statements are used:

```
ARRAY(1,6) / 0.10, 0.15, 0.35, 0.65, 0.85, 1.00;
ARRAY(2,6) /   10,   15,   25,   50,   75,  100;
ACT/7, DPROBN(1, 2, 4);
```

With the use of this statement, 10% of the durations of activity 7 are expected to be 10, 5% are expected to be 15, 20% to be 25, and so on. The SLAM II function DPROB is available to obtain samples from a discrete probability function in user-written code.

12.6 SEASONAL DEMAND

Sometimes demand fluctuates over time according to seasons. To model time-varying arrival patterns, conditional branching in conjunction with a disjoint network that models seasons may be used. To illustrate, consider the network segments shown in Figure 12-4. Activities 1, 2, and 3 represent a different interdemand time specification. Together they can be viewed as representing three different patterns based on a season. The pattern that is used depends on which activity branch is take from GOON node GO1. The seasonal branch to

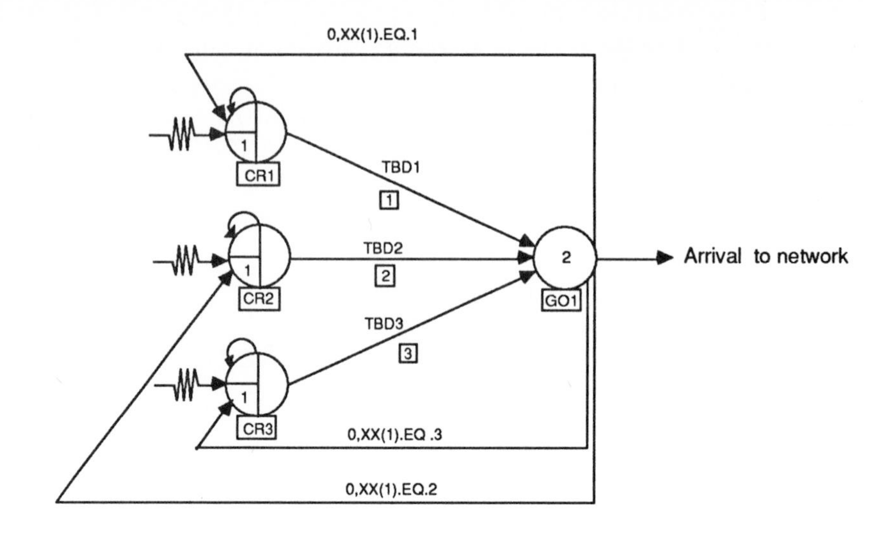

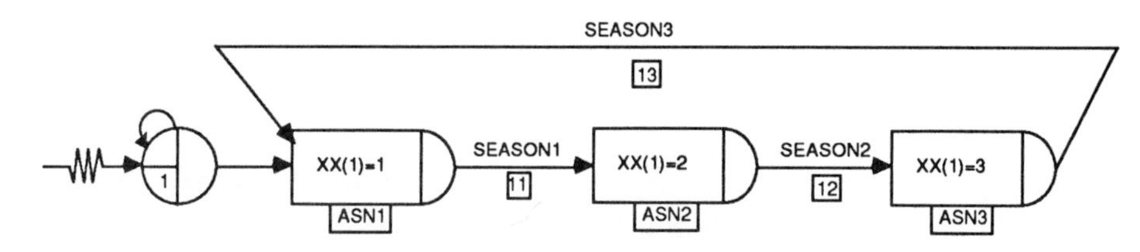

Figure 12-4 Network segments modeling seasonal demand.

be taken from node GO1 is controlled by the value of the global variable XX(1).

The value of XX(1) is established in the disjoint network at ASSIGN node ASN1, where XX(1) is set equal to 1, representing season 1. The time delay on activity 11 represents the length of season 1. Likewise, node ASN2 and activity 12 cause the demand pattern for season 2 to be in effect, and node ASN3 and activity 13 represent season 3.

The length of a season could also be modeled where conditional branching is dependent on activity completions using function NNCNT(NACT). This would enable the modeling of complex time-dependent demand patterns that could be incorporated into both the seasons and the interdemand networks.

12.7 DIFFERENTIATING AMONG DEMANDS

To differentiate among demands for the same item, attribute values are used. In this way, demands can be ranked in QUEUE nodes using attribute-based priority rules. Different types of demands may be routed to different network segments by using their attribute values in conditional statements on activities. Complex routing at SELECT nodes is another possibility, where queue selection rules are written to be based on attributes. In addition, resources allocated to demands at AWAIT nodes can be performed according to a specific priority.

12.8 BULK DEMANDS

When entities are used to represent bulk or batch demands, an attribute can be used to specify the number of units demanded by the entity. Satisfying bulk demands then follows the procedures described for multiple arrivals associated with a single entity. The procedures for aggregating and disaggregating bulk demands using BATCH, UNBATCH, and ACCUMULATE nodes may also be used.

12.9 MODELING MULTICOMMODITY INVENTORY SYSTEMS

The procedures for modeling multiple item or multicommodity inventory situations are a direct extension of the single-item models presented in Chapter 11. A two-commodity model is presented in Figure 12-5. The subnetwork for the upper half of the figure represents the demand, backorder, and stock components of the system. The subnetwork in the lower half of the figure represents a periodic review process with stock control levels for each item.

Consider the upper subnetwork first. The time between demands for a unit of commodity type 1 is generated at CREATE node CRE1, and the time between demands for a unit of commodity type 2 is created at CREATE node CRE2. Following each of these CREATE nodes is an ASSIGN node that sets COMTYPE to 1 or 2, depending on the type of inventory unit demanded. In the EQUIVALENCE statement, COMTYPE is made equivalent to ATRIB(1). Inventory position, stock, and backorders are then updated for each type of unit demanded. Note that these inventory variables are equivalenced to elements of ARRAY. This allows COMTYPE to specify the column of ARRAY to update, that is, column 1 is used for commodity type 1 variables and column 2 is used for commodity type 2 variables.

The inventory review subnetwork in the middle of Figure 12-5 starts the review process at time TFREV as stipulated for CREATE node CRE3. Subsequent reviews occur every TBR time unit as specified for node CRE3. At each review, two entities are created: one with COMTYPE set at 1 and one with COMTYPE set at 2. The processing of both these review entities proceeds as described in Section 11.6.

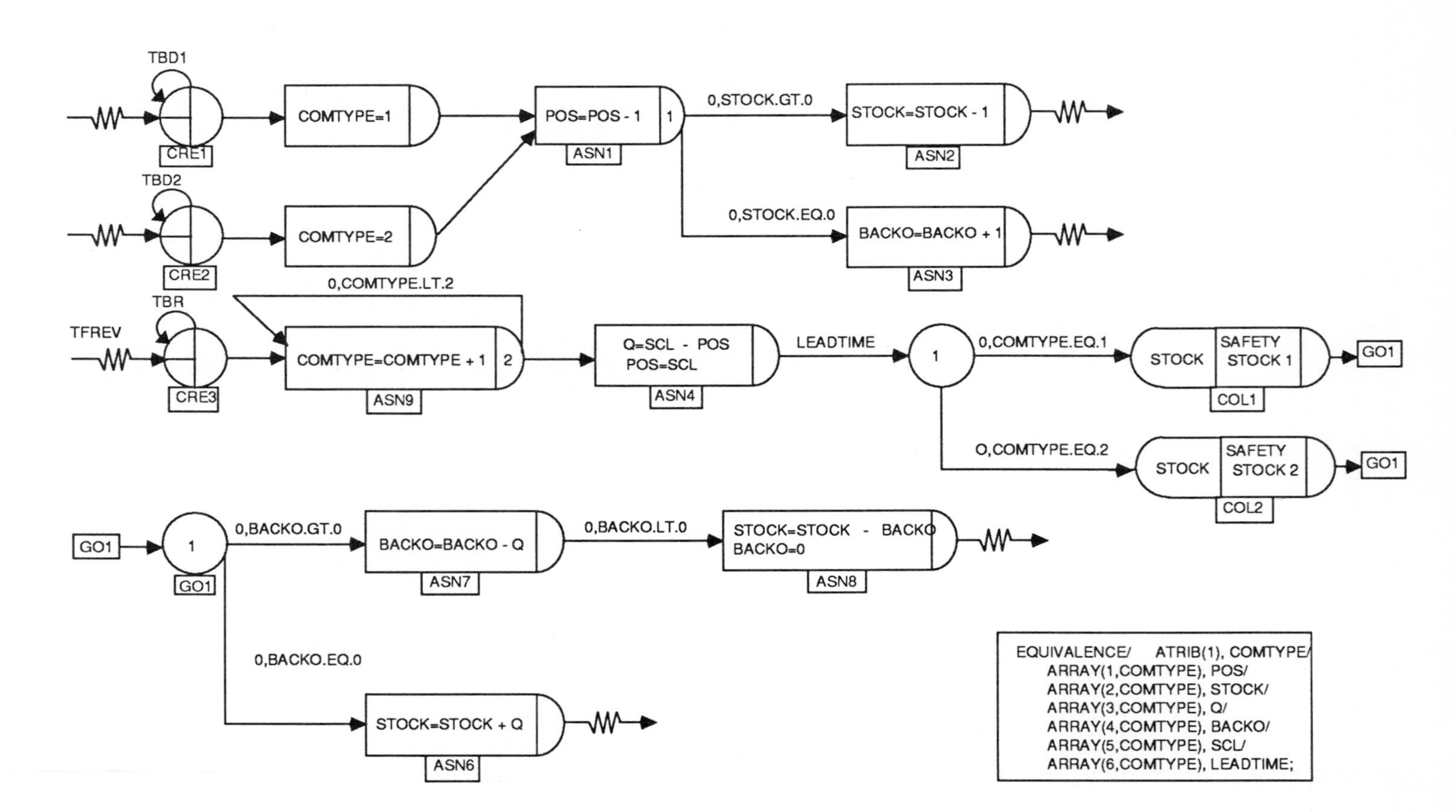

Figure 12-5 SLAM II model for multicommodity inventory situation.

The two-commodity model can be easily generalized to the N-commodity case. Furthermore, the logic described in Chapter 11 for lost sales and continuous review procedures and those previously described in this chapter for complex demand processes can be incorporated for any or all of the item types. Profit calculations can also be programmed in subroutine OTPUT to account for individual item types in addition to any aggregate values that might be desired.

12.10 SUMMARY

The procedures for modeling inventory demand processes are similar to those for modeling job arrivals for production systems. The use of probabilistic and conditional branching, together with the use of SLAM II global variables, facilitates the modeling of complex demand processes. A SLAM II network model for a two-commodity situation is presented, which is not significantly different from the single-commodity model developed in Chapter 11.

12.11 EXERCISES

12-1. Discuss the assumptions involved in using the following distributions as the interdemand time distribution: exponential, normal, lognormal, Erlang ($k = 2$), triangular, and uniform.

12-2. Give three reasons for selecting a distribution type to represent interdemand times rather than using a histogram of the actual demand times experienced over the past year.

12-3. Modify the network for specifying the time between demands (Figure 12-3 in the multicommodity inventory model) so that the average obtained is used as the mean for an exponential distribution.

12-4. Use the concepts presented in Section 12.6 relating to seasonal demand to build a network model describing the sale of baseball bats.

12-5. Modify the inventory model presented in Section 11.6 to the situation in which the number of demands per demand instant is Poisson distributed with a mean of 3. Assume that if a value of zero is obtained as the Poisson sample the customer does not place an order.

12-6. Build a general SLAM II model for an inventory situation in which the demand distribution is characterized by a generalized Poisson distribution, a compound Poisson distribution, and a stuttering Poisson distribution. Provide operational definitions for these variants to the Poisson distribution.

12-7. Run the multicommodity inventory system as modeled in Section 12.9 and develop strategies for relating the inventory policies for the two-item system.

12-8. Build a model of an inventory policy in which inventory position is monitored with a DETECT node.

12.12 REFERENCES

1. Banks, J., and J. S. Carson, II, *Discrete-Event System Simulation*, Prentice-Hall, Englewood Cliffs, NJ, 1984.
2. Feller, W., *An Introduction to Probability Theory and Its Applications*, Vol. 1, Wiley, New York, 1957.
3. Feller, W., *An Introduction to Probability Theory and Its Applications*, Vol. 2, Wiley, New York, 1972.
4. Lewis, C. D., *Demand Analysis and Inventory Control*, Saxon House, Westmead, Farnborough, England, 1975.
5. Pritsker, A. A. B., *Introduction to Simulation and SLAM II*, *3rd* ed., Wiley and Systems Publishing, New York and West Lafayette, IN, 1986.
6. Hadley, G., and T. M. Whitin, *Analysis of Inventory Systems*, Prentice-Hall, Englewood Cliffs, NJ, 1963.

13 RELIABILITY, QUALITY CONTROL, AND EQUIPMENT REPLACEMENT

13.1 INTRODUCTION

The reliability and quality of a system have become extremely important factors in the competitive marketplace. The prime example of this is in the automotive industry, where an increasing percentage of the market has been acquired by foreign companies due to the improved reliability and quality of their products. In this chapter, models are developed for assessing reliability and quality control plans. Equipment replacement models are also included in this chapter because the procedures for modernizing a plant have an impact on reliability and quality. The procedure for evaluating equipment replacement plans corresponds to finding the shortest path through a network. A general description of the stochastic shortest-route problem is included in this chapter. This use of networks is illustrated for assessing equipment replacement plans and safeguard designs at nuclear facilities. This latter problem involves an assessment of the reliability of a system in the face of a knowing adversary. First, a description of reliability and its assessment is provided with an emphasis on the use of network modeling procedures.

13.2 RELIABILITY AND ITS ASSESSMENT

When system components are placed in series, the failure of any component causes the failure of the system. To increase the working time of the system, components are placed in parallel, with any of the components being capable of performing the unit's operation. In this way, the failure of a component does not result in the failure of the system. A reliability assessment requires the analysis of series, parallel, and combinations of series and parallel subsystems.

Definitions of reliability vary in the literature. One reason for this is that reliability can be considered in either a static sense or in a dynamic sense. For a fixed time period, the reliability of a component can be considered as the probability that the component fails in the time period. This probability is the static reliability for the component performing satisfactorily during the lifetime of the system or the mission time for the system. Alternatively, reliability can be defined in terms of the mean time to failure, considering the repair of a component that has failed while a parallel component continues to perform the required operation. In this case, the reliability of the system is determined on a dynamic basis. Therefore, the reliability of a system depends on both the failure distribution of the components and the repair distribution for failed components.

In reliability evaluations, a system is decomposed into its component parts, an analysis is performed on each component part to determine part reliability, and then the part reliabilities are combined in a mathematical and logical fashion to obtain system reliability predictions. Networks, and SLAM II in particular, also follow this systems analysis approach by providing elements (nodes and branches) for describing system components about which data are collected. The data are then transformed into system performance measures through the use of a network analyzer. Thus, it is natural that networks would be used in reliability analysis.

Many authors have used the GERT network simulation language for analyzing reliability situations. R. Skeith and M. Skinner have developed analytical procedures that employ Mellin transforms on GERT exclusive-or networks for special types of reliability problems. G. Whitehouse presents several approaches to reliability analysis using GERT [23, 24]. The use of GERT IIIZ for reliability analysis has also been demonstrated [1, 7].

Since reliability problems occur frequently, a specialized GERT IIIZ program was developed for analyzing the reliability of systems that includes both failure and repair. The program, called GRASP, was initially developed by R. Shulaker and D. Phillips and was finalized by J. Polito and C. Petersen [10]. The GRASP program has been used for complex reliability analysis for the Navy and at several steel corporations.

13.3 POWER STATION MAINTENANCE REPAIR EVALUATION

As an example of reliability modeling using SLAM II, consider a power station that requires three generators to be on line at all times. Since all three generators must be operative to prevent a power system failure, company policy is to have one spare generator that can replace any generator that fails. In addition, it is company policy to start repair work immediately on a generator that has failed. The statistical characteristics describing the failure time t_f for each generator are assumed to be identical. However, the spare generator is not of the same quality as the on-line generators. Its failure time is described by t_{fs}. Let the time to repair a generator be given by the random variable t_r.

The SLAM II network model of the power station generators† to obtain statistics on time to system failure due to the simultaneous failure of two generators is shown in Figure 13-1. The three generators are created at the CREATE node CR1 and placed into operation on activity 1. The time to failure for each generator is represented by the activity time for activity 1. When a generator fails, time between failure statistics are collected at COLCT node C1. Up to three activities are started when a generator fails. Activity 2 represents the repair of the generator. If less than two generators are operating, that is, NNACT(1).LT.2, the power station goes down and a signal is sent to COLCT node C2. When a generator fails, an entity is routed to AWAIT node A1 to seize the spare generator and put it on line. The time for failure for the spare generator, t_{fs}, is prescribed for activity 3. The spare generator is modeled as a resource to allow it to be preempted when the regular generator is repaired. If the spare fails before the regular generator is repaired, the power station goes down and a system failure occurs as indicated by activity 3 leading to COLCT node C2.

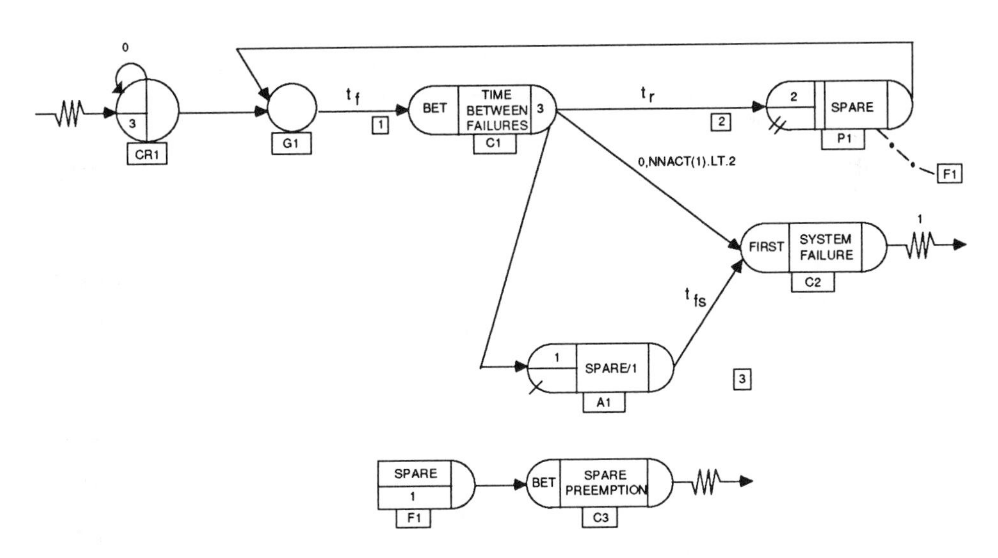

Figure 13-1 SLAM II model of a power station generator system.

† This example was developed in conjunction with Robert Trent when he was associated with the Construction Evaluation and Research Laboratory (CERL) of the U.S. Army.

Following the repair of a regular generator, the spare generator is taken off line by preempting it at PREEMPT node P1. The spare generator is routed to FREE node F1 using the *send node* field of the PREEMPT node, where one unit of the resource SPARE is freed. The time between shutdowns of the spare generator is collected at node C3. The regular generator is put back on line by routing it to node G1 from PREEMPT node P1, which initiates activity 1. Statistics collected in this example are the following:

1. Time between on-line generator failures at node C1
2. Time of failure for the power station at node C2
3. Utilization of on-line generators on activity 1
4. Utilization of the repair person on activity 2
5. Utilization of spare generator on activity 3
6. Time between spare preemptions at node C3

These statistics can be used by management to answer the following types of questions:

1. Is power station reliability sufficient to meet customer requirements?
2. Should a better spare generator be acquired?
3. How many repairers are needed to maintain failed generators?

The model presented can easily be modified to represent more generators and repair personnel.

13.4 MODELING RELIABILITY USING SLAM II

This section formalizes the concepts introduced in the preceding example. First, a general approach to modeling reliability problems in SLAM II is presented.

13.4.1 Static and Dynamic Reliability

To evaluate the static reliability of systems with SLAM II is straightforward. A model for determining the probability of the system working during a fixed mission time for three units in series and three units in parallel is shown in Figure 13-2. ACCUMULATE node ACM1 represents system success, and the probability of reaching this node reflects system reliability. When components are in series, all three components must work during the entire mission; hence, the

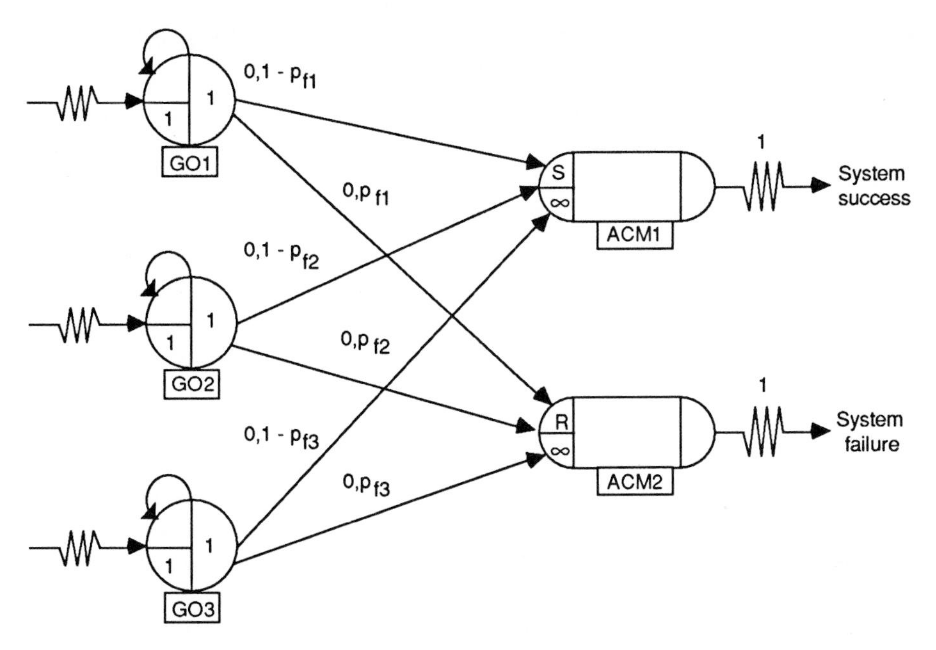

Figure 13-2 Static reliability for components in series and parallel. For components in series: S = 3, R = 1; for components in parallel: S = 1, R = 3; p_{fi} = probability that component i fails during the mission time.

three branches incident to node ACM1 must be completed in order to have system success. Thus, the first release requirement, S, for node ACM1 is 3. If any of the components fail, node ACM2 would be released, and a system failure would occur since the first release requirement, R, at this node would be prescribed as 1. When components are in parallel, the values of S and R are reversed. Consequently, it would take only one activity incident to node ACM1 to achieve system success, but all three components must fail in order for node ACM2 to be released. For subsystems in series, the SLAM II model in Figure 13-1 would be repeated for each subsystem with appropriate values of S and R inserted.

Now consider dynamic reliability assessment. In Figure 13-3, a SLAM II model of three units in series is shown. Each unit is represented by an activity, with the time to failure of a unit being the activity time. Since the units are in series, failure of any unit results in the failure of the system. Thus, the activities are shown in parallel in the SLAM II network, and

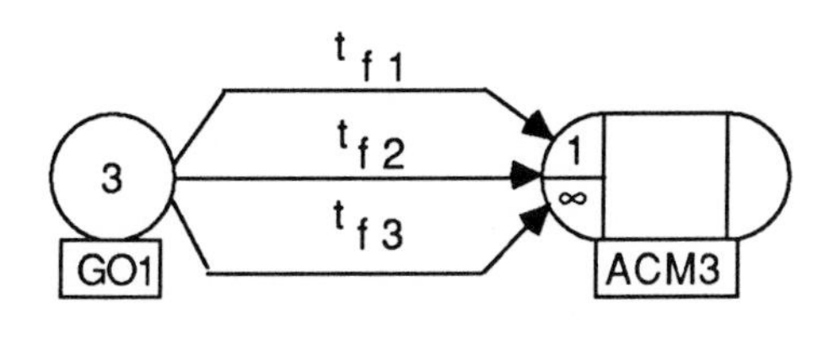

Figure 13-3 Three units in series; t_{fi} = time to failure for component i.

the time to system failure is the time of the first component failure. Therefore, only one entity is required to reach ACCUMULATE node ACM3 for system failure to occur. Note that in modeling reliability in a dynamic fashion it is not a question of whether the system will fail, but when it will fail. A histogram of the times of release of node ACM3 provides an estimate of the distribution of the time to system failure. The probability of successful system operation up to time t is estimated by the fraction of runs on which the system time to failure is greater than t. Thus, the complementary value of the cumulative frequency is used to estimate system reliability.

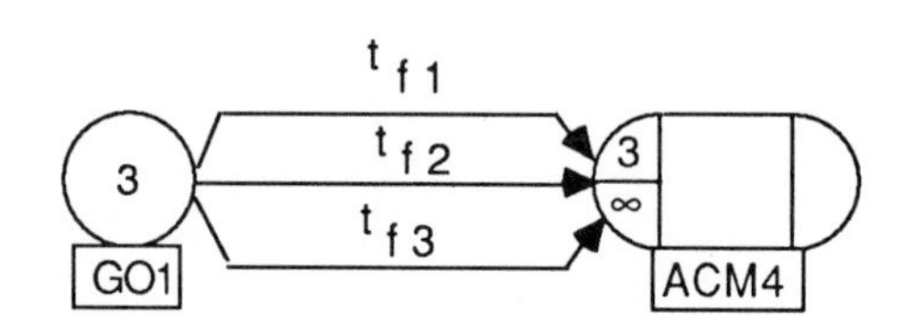

Figure 13-4 Three units in parallel; t_{fi} = time to failure for component i.

In Figure 13-4, the SLAM II network for measuring time to failure for three units in parallel is given. This network is similar to the model for units in series presented in Figure 13-3. The only difference is the requirement for three activity completions at ACCUMULATE node ACM4. This specifies that all units must fail before the system fails. With this minor change, the network has been converted from a series system model to a parallel system model.

13.4.2 Reliability of Combined Parallel-Series Systems [1, 11]

To demonstrate the use of SLAM II to model combined series and parallel units, consider the system configuration shown in Figure 13-5. In this system, unit A is in series with a subsystem consisting of units B, C, D, E, and F. For the subsystem, unit B is in series with unit D, and unit C is in series with both unit D and a subsystem consisting of units E and F in parallel.

The SLAM II model to determine the time to system failure without component repair is shown in Figure 13-6. GOON node GO6 represents system failure, which occurs when activity 1 is completed or when activity 7 is completed. Activity 7 is completed when the subsystem that is in series with unit A fails. The modeling of the subsystem is more complex in that it represents three parallel paths whose failure is modeled by nodes GO3, GO4, and GO5. Node GO3 is released if either unit B or D fails, node GO4 is released if unit D or C fails, and node GO5 is released if unit C fails or both units E and F fail. Figure 13-6 illustrates how SLAM II portrays these potential failure paths, and demonstrates the parallelism between estimating system reliability from component reliabilities and estimating system failures based on component failure times.

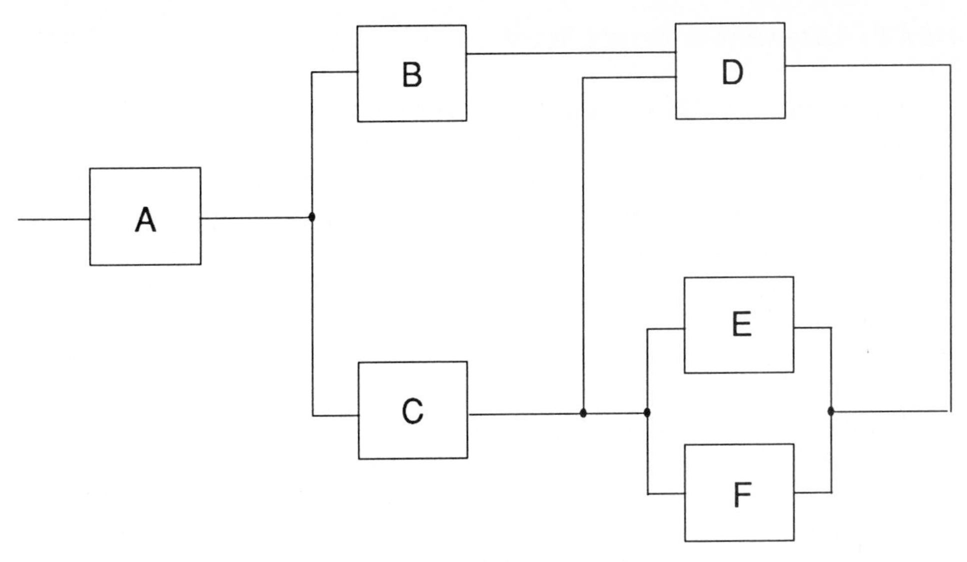

Figure 13-5 System consisting of series and parallel units.

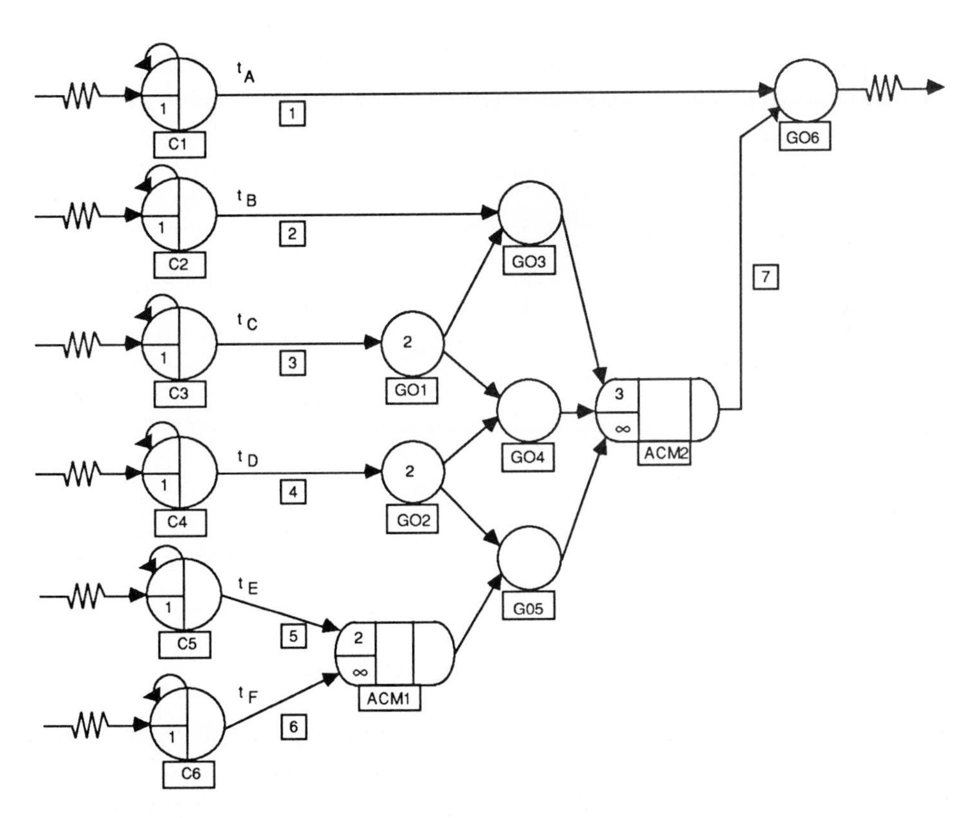

Figure 13-6 SLAM II model of a combination of series-parallel units.

13.4.3 System Reliability with Repair

If components are in series and every component is required for the system to function, reliability assessment does not change due to the ability to repair the component, because as soon as it fails the system fails. Therefore, the discussion relative to the repair of components is restricted to situations in which there are parallel components.

When repair is possible, a unit can be in one of two states: operative or in the process of being repaired (nonoperative). Each state of a unit is modeled by an activity so that two activities are associated with a unit. Since the units are in parallel, it is necessary to indicate a system failure only when all units have a failed status; that is, all units are in the "being repaired" activity.

As a first model, the number of units working is counted using the SLAM II global variable XX(1). This modeling approach to the reliability problem for two units in parallel is given in Figure 13-7. Activity 1 represents the time to failure for unit 1, and activity 3 represents the repair time for unit 1. Activities 2 and 4 represent analogous states for unit 2. The system is turned on at the start of a simulation by creating an entity at each of the CREATE nodes and initiating activities 1 and 2 to start the delay for the time to failure for each component. When either component fails (activity 1 or activity 2 is completed), the SLAM II global variable XX(1) is decreased by 1 at the following ASSIGN node, and the time to repair the failed unit (activity 3 for unit 1 or activity 4

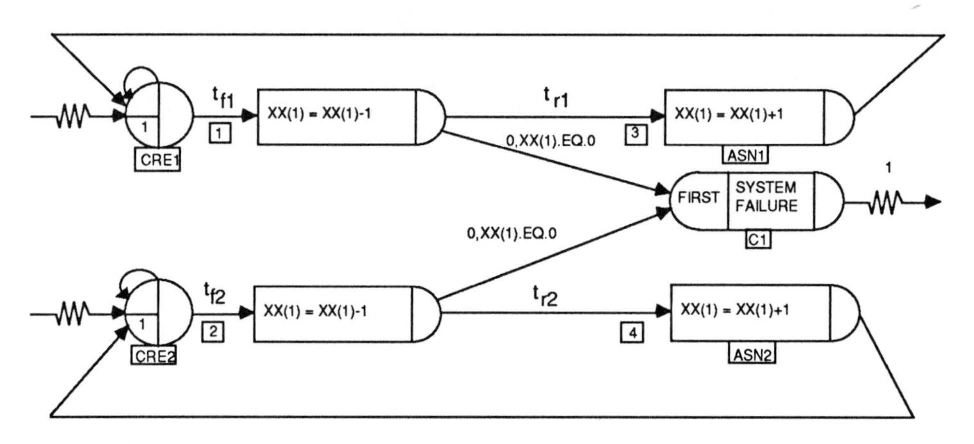

Figure 13-7 Model to evaluate the reliability of two units in parallel with repair capability. XX(1) > 0 indicates at least one parallel unit is working. XX(1) = 0 indicates both units have failed: t_{fi}, time to failure of component i; t_{ri}, time to repair component i.

for unit 2) is started. When the repair time is completed, XX(1) is increased by 1 at the following ASSIGN node, and the entity is rerouted to activity 1, where the unit is operational and the activity duration represents a time to the next failure. The CREATE node in this model acts as a GOON node. If at any time both units are in the repair cycle (failed), XX(1) will equal 0, indicating that both units are not working. In this case, an entity is routed from the ASSIGN node of the last failed unit to the COLCT node C1 to indicate that the system has failed.

Although the repair process is shown by a single activity for each unit, it should be clear that each activity could be replaced by a complex SLAM II network, which involves the unit waiting for a repairer and other types of resources if required.

The SLAM II global variable XX(1) is used to count the number of parallel units that are operating. Three alternate techniques to model the system reliability problem have also been developed. These employ user-written functions, resources, and balking concepts. However, before discussing these techniques, a brief history of the evolution of network modeling approaches to the system reliability problem is presented.

A network simulation model for the system reliability problem was first reported by Hammesfahr, Rakes, and Clayton in 1978 [7]. This model included several QUEUE nodes, a MATCH node, and a disjoint network. Using this model as a basis, personnel at Pritsker & Associates developed an improved model consisting of two nodes, two branches, two user functions for each component in parallel, and a single node to represent system failure. In the latter model, use was made of subroutine ENTER to place entities into the network. While reviewing this model, Ken Musselman of Pritsker & Associates developed an approach to the problem that used resource-modeling concepts. Based on this resource approach, Pritsker and Sigal developed an approach based on QUEUE nodes and service activities that represent system failure through balking.

The essence of this discussion is that models evolve over time and that they become simpler after they have gone through several stages of development [15, 16]. Similar observations have been made with regard to mathematical programming formulations [14]. Therefore, to highlight this model-evolution process, all three of the above network modeling approaches to the system reliability problem are presented.

13.4.4 Multiple Components

The network required to model a system consisting of N parallel units can be developed from the model shown in Figure 13-7. Rather than presenting such a model, a user function approach is presented in Figure 13-8. In this model, N entities are created at a CREATE node. (A value for N must be prescribed.) A unit number, UNIT, is assigned as attribute 1 by counting the number of entities entering the network, that is, completing activity 3. Activity 1 represents units that are working. The working time is the time until the next failure. It is prescribed by user function 1, where the next failure time for a unit is determined based on its unit number. Activity 2 represents the repair time, which is computed in USER(2). System failure occurs when there are no entities in activity 1, which is detected by a DETECT node monitoring NNACT(1) to see if it crosses 0. Note that statistics on activity 1 describe the number of units working and statistics on activity 2 describe the number of units being repaired. With networks and program inserts, complex failure and repair operations can be modeled to assess reliability measures.

13.4.5 A Resource Approach

A resource approach to modeling system reliability when units are repairable involves defining operational units as a resource. Each time a unit fails and

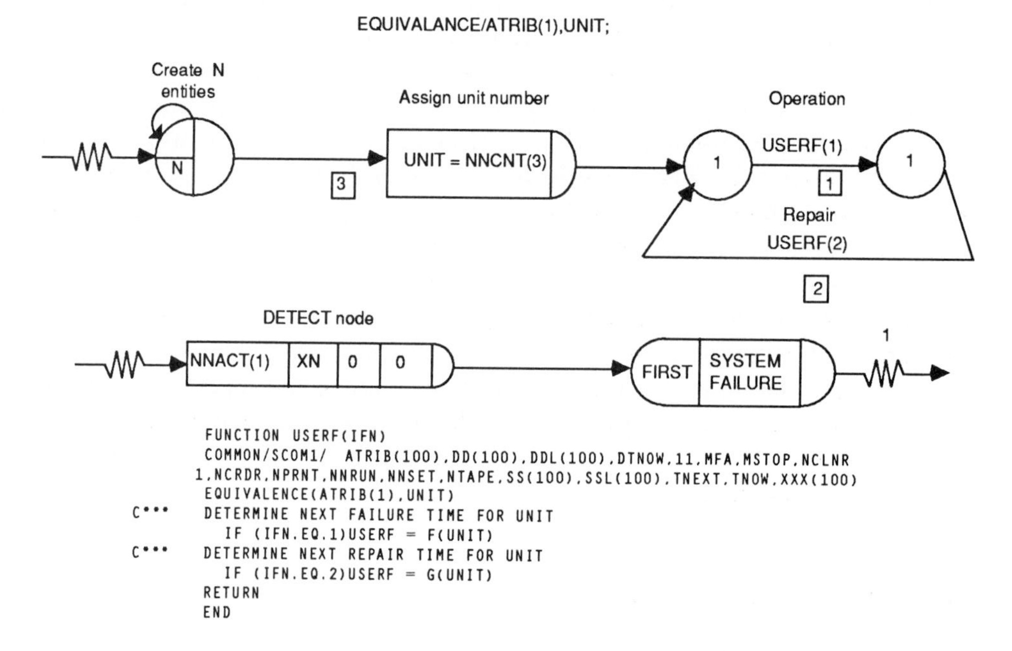


```
      FUNCTION USERF(IFN)
      COMMON/SCOM1/  ATRIB(100),DD(100),DDL(100),DTNOW,II,MFA,MSTOP,NCLNR
     1,NCRDR,NPRNT,NNRUN,NNSET,NTAPE,SS(100),SSL(100),TNEXT,TNOW,XXX(100)
      EQUIVALENCE(ATRIB(1),UNIT)
C***  DETERMINE NEXT FAILURE TIME FOR UNIT
        IF (IFN.EQ.1)USERF = F(UNIT)
C***  DETERMINE NEXT REPAIR TIME FOR UNIT
        IF (IFN.EQ.2)USERF = G(UNIT)
      RETURN
      END
```

Figure 13-8 Model and user function for estimating system time to failure when repairs are included.

begins repair, the capacity of the resource is decreased by 1. Following the failure of a unit, a check is made to see if the capacity of the resource has been reduced to zero. If this occurs, all units are in the failed state and a system failure has occurred.

The SLAM II model of this situation is shown in Figure 13-9. In the model, N parallel units are inserted into the network at the CREATE node with a zero time delay. As each entity, representing a unit, enters the ASSIGN node, the SLAM II global variable, II, is set equal to the SLAM II status variable NNCNT(1), which counts the number of entities that have completed activity 1. Therefore, the global variable II identifies each unit as it enters the network. The mean time to failure for each unit is stored in ARRAY(1,II), and the mean time to repair each unit is stored in ARRAY(2,II). At the ASSIGN node, the variable MFAIL is set to the mean time to failure for each unit, that is, ARRAY(1,II). Likewise, the variable MREPAIR is set to the mean time to repair each failed unit, that is, ARRAY(2,II). The values for each of these variables are stored in attributes 1 and 2, respectively, for each entity through the use of an EQUIVALENCE statement.

Each entity is then routed to ALTER node ALT1, where the capacity of resource NWORK, initially set at zero on the RESOURCE block, is incremented by 1. Each unit is then turned on at activity 2 for a period of time that is specified by a random sample drawn from an exponential distribution with a mean of MFAIL. The capacity of the resource NWORK is decreased by 1 at node ALT2 when a unit fails. Conditional branching from this node starts the repair on the unit on activity 3. The time to repair a unit is randomly drawn from an exponential distribution with a mean of MREPAIR. When the repair cycle is complete, the entity is rerouted to node ALT1, where the capacity of NWORK is increased by 1, and the unit starts working again. When the capacity of

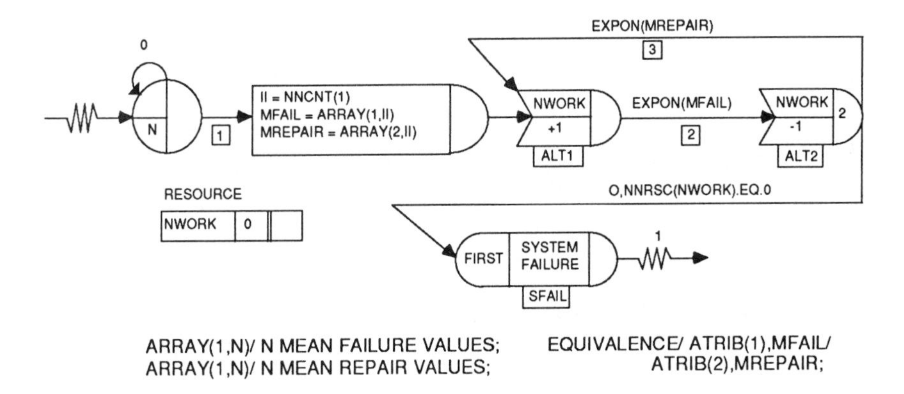

Figure 13-9 SLAM II model using resources for system reliability assessment.

resource NWORK equals zero, indicating there are no units in operation, the branch from node ALT2 to the COLCT node is taken to record a system failure.

Within the framework of this model, the availability of the resource NWORK represents the number of units in an operative state. Thus, statistics on the availability of this resource provide information on the average number of units working. This model illustrates an interesting use of resources because they are never allocated.

13.4.6 System Failure as a Balking Entity

A third model for the system reliability problem is presented in Figure13-10. The creation of the N components is accomplished at the CREATE node, and the variable assignments are made at the ASSIGN node as before. However, the ALTER nodes are replaced by QUEUE nodes Q2 and Q3. In addition, there are N parallel servers for units in operation on service activity 2, one for each unit in the system, and N-1 parallel servers for the repair service represented by activity 3. Thus, if a unit fails and there are N-1 repairs ongoing, it is the N*th* failure and a system failure has occurred. This is detected on the network by balking from node Q3 to COLCT node SFAIL when a unit cannot gain access to activity 3. Therefore, the time that an entity arrives at node SFAIL represents the time of a system failure. The expected number of components operating is estimated from statistics on activity 2.

This presentation of three alternative methods of modeling the same problem illustrates the versatility of network modeling. There is no one best way to model a system. Consequently, having the flexibility to model a system from different viewpoints is one of the great advantages of SLAM II network-modeling concepts.

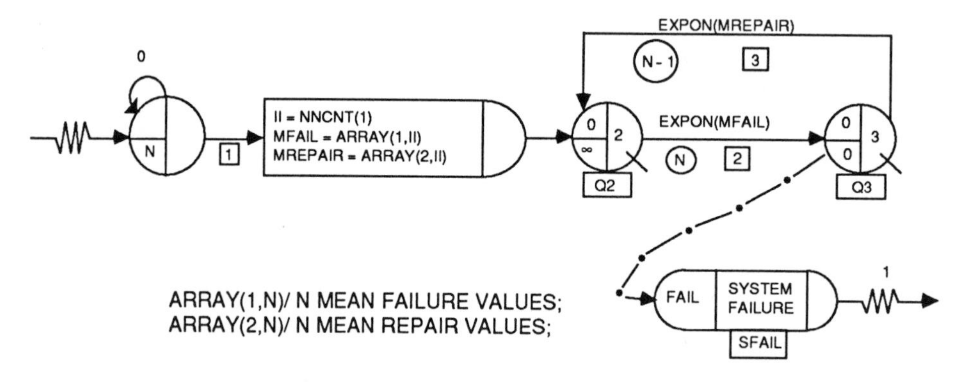

Figure 13-10 SLAM II model using balking for system reliability assessment.

13.5 MODELING QUALITY CONTROL SAMPLING PLANS

Quality control sampling plans provide procedures for evaluating the quality of manufacturing lots. A properly designed quality control plan results in actions and methods improvement that can reduce the losses resulting from rejections, scrap, and rework. The investment in terms of time and cost for inspection typically yields high returns due to the avoidance of such losses and the gains obtained from increased customer satisfaction with a quality product. Network models have been used to assess sampling plans. For example, Whitehouse [23] reports on a study by J. Fry, G. Powell, and C. Mullin for the DODGE Continuous Sampling Plan, CSP-1, using the GERT simulation language [18]. Other diverse quality control systems have been modeled with GERT by Whitehouse and Hsuan [25]. In addition, both Whitehouse [23] and Ron Skeith have developed network models of Military Standard 105D which is a complex sampling plan.

Acceptance sampling plans are an integral part of quality control. The purpose of acceptance sampling is to determine a course of action, rather than to estimate lot quality. The plan prescribes that a sample of size n be taken from the lot. If there are c or fewer defects in the sample, the lot should be accepted. From the acceptance sampling plan, the risk of accepting lots of a given quality can be estimated. In other words, acceptance sampling yields quality assurance.

To model acceptance sampling plans in SLAM II, probabilistic branching is employed to specify whether an item is defective or not. A count is kept to determine whether the number of defects detected has reached the acceptance number. A count is also kept on the number of items inspected. If the count of defective items increases above the acceptance number before the count for the sample reaches the total sample size, the lot is rejected. Otherwise, the lot is accepted. Separate nodes represent the rejection and acceptance of the lot. By making multiple runs, estimates of the probabilities of accepting or rejecting a lot based on the sampling plan are obtained. In the next section, an example that demonstrates the use of SLAM II to model a double sampling plan is given.

13.6 SLAM II MODEL OF A DOUBLE SAMPLING PLAN

In a double sampling plan, a sample of size N1 is taken first. If C1 or fewer defective parts are detected, the lot is accepted based on this one sample size. If more than C2 defective parts are detected, the lot is rejected. If more than C1 but less than or equal to C2 defective parts are discovered, a second sample of size N2 is examined. If more than C3 parts are found to be defective, including those from the first sample, the lot is rejected; otherwise, it is accepted.

To evaluate the performance of this plan for various values of N1, N2, C1, C2, C3, and a given probability of a defective item† P1, the SLAM II network given in Figure 13-11 is employed. The SLAM II statement model is given in Figure 13-12. In this example, testing times are omitted, and only the number of parts tested is assessed. If times are added to the model, the total testing time required for the sampling plan can be determined. Similarly, the costs of these activities may be evaluated.

A create node is used to insert an entity into the network starting at time 1 and continuing at intervals of one time unit. For this model, time is advanced by one unit and can be interpreted as the number of parts tested. In this way, the output provides estimates of the number of parts sampled before acceptance or rejection. The CREATE node is the only location in the entire network where a duration or delay is included. Activity 1 represents testing the part. At GOON node GO1, the part is classified as nondefective with a probability of 1-P1 by activity 2 and as defective with a probability of P1 by activity 3. At any point in time, the number of entities that have completed activity 2, NNCNT(2), is the number of parts that have been classified as nondefective. Similarly, NNCNT(3) is the number of parts that have been classified as defective. The number of parts that have been tested is NNCNT(1). These quantities may be used directly to determine the disposition of the lot in accordance with the double sampling plan.

Following GOON node GO2, at most one branch is taken in accordance with the sampling plan. Activity 4 accepts the lot on the first sample if the number of parts tested, NNCNT(1), is equal to N1 and the number of defective parts, NNCNT(3), is less than or equal to C1. Activity 5 accepts the lot on the second sample if the number of parts tested is greater than or equal to N1 + N2. Activity 6 rejects the lot on the first sample if the number of parts tested, NNCNT(1), is less than or equal to N1 and the number of defective parts, NNCNT(3), is greater than C2. Activity 7 rejects the lot on the second sample, which occurs when the number of defective parts, NNCNT(3), is greater than C3. Six collect nodes are used to estimate statistics on the number of parts tested when the lot is accepted on the first sample, accepted on the second sample, rejected on the first sample, rejected on the second sample, accepted on either sample, or rejected on either sample.

Times could be added to the network to represent the testing time on activity 1 and the classification or inspection times for activities 2 and 3. These times could be made a function of how a part is classified. If these times are added to the network, the total testing time required for a sampling plan could be

† If a lot fraction defective is estimated, the probability of classifying a part as defective will change slightly each time a part is tested. This would be included in the SLAM II model using probabilistic branching based on attribute values.

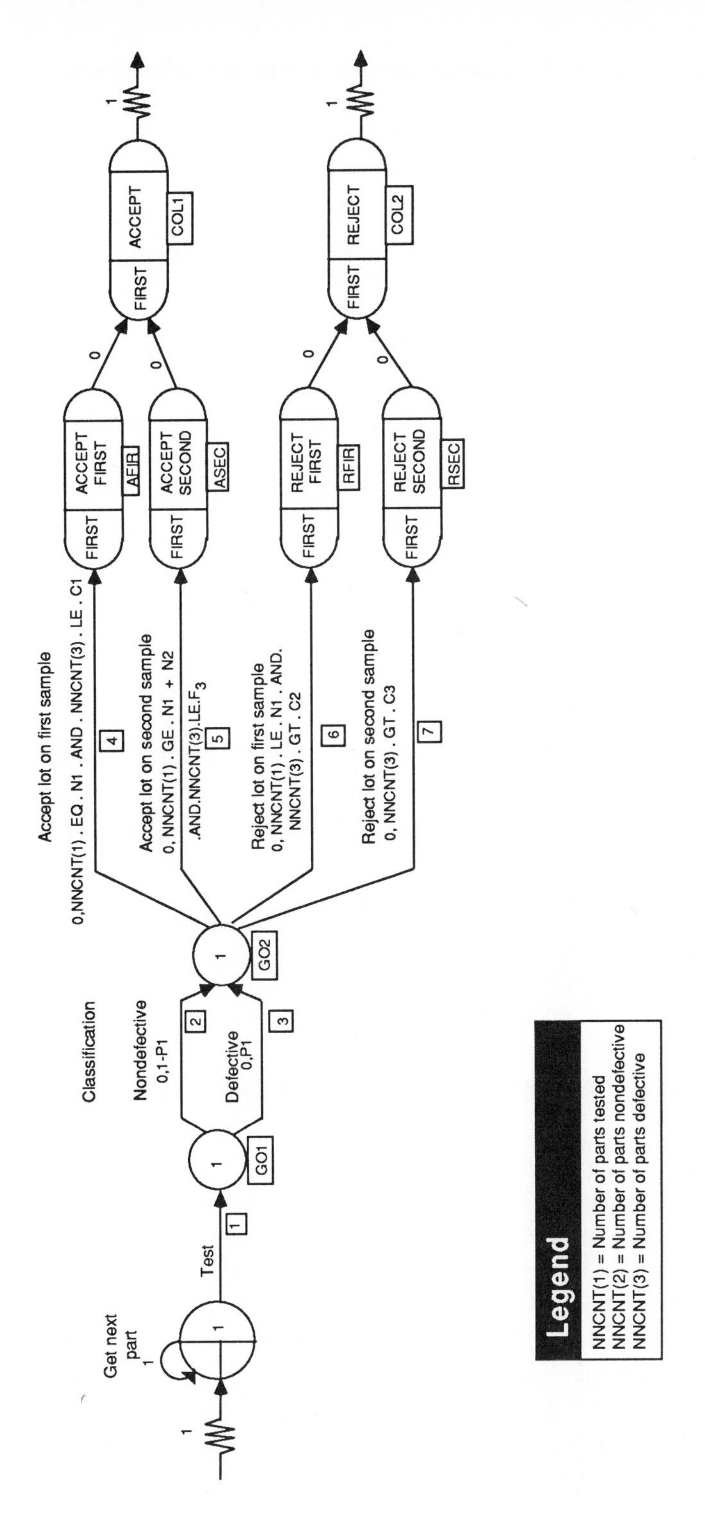

Figure 13-11 SLAM II network of a double sampling plan.

```
GEN,PRITSKER,DOUBLE SAMPLING,3/7/88,1000,,N,,N,Y/1000;
LIMITS,,,6;
EQUIVALENCE/ 2,C1/   6,C2/ 6,C3/
            50,N1/ 100,N2/.08,P1;
NETWORK;
      CREATE,1,1;
      ACT/1;                 TEST PARTS
G01   GOON,1;
      ACT/2,0,1-P1,G02;   NONDEFECTIVE PARTS
      ACT/3,0,P1,G02;     DEFECTIVE PARTS
G02   GOON,1;
      ACT/4,0,NNCNT(1).EQ.N1 .AND. NNCNT(3).LE.C1,AFIR;
      ACT/5,0,NNCNT(1).GE.N1+N2 .AND. NNCNT(3).LE.C3,ASEC;
      ACT/6,0,NNCNT(1).LE.N1 .AND. NNCNT(3).GT.C2,RFIR;
      ACT/7,0,NNCNT(3).GT.C3,RSEC;
      ACT,,,T1;
AFIR  COLCT,FIRST,ACCEPT FIRST;
      ACT,,,COL1;
ASEC  COLCT,FIRST,ACCEPT SECOND;
      ACT,,,COL1;
RFIR  COLCT,FIRST,REJECT FIRST;
      ACT,,,COL2;
RSEC  COLCT,FIRST,REJECT SECOND;
      ACT,,,COL2;
T1    TERM;
COL1  COLCT,FIRST,ACCEPT;
      TERM,1;
COL2  COLCT,FIRST,REJECT;
      TERM,1;
      END;
INIT,,,NO;
FIN;
```

Figure 13-12 SLAM II statement model for evaluating a double sampling plan.

determined. The cost of applying a sampling plan could then be evaluated and compared with the benefits that are derived from the sampling plan to determine the desirability of implementing the plan.

13.7 RESULTS FOR THE DOUBLE SAMPLING PLAN

Table 13-1 provides the values for a double sampling plan, which is referred to as a strategy. The strategy was analyzed with the SLAM II model by simulating it for 1000 lots. The resulting SLAM II Summary Report is shown in Figure 13-13. From this output, the probability of acceptance (node COL1) is estimated at 0.233 and the probability of rejecting the lot (node COL2) at 0.767. The results observed from QC curves [4] for this double sampling plan are 0.23 for the probability of acceptance and 0.77 for the probability of rejection. The SLAM II model provides a close estimate for these quantities (the standard deviation for a binomial variable with a mean of 0.77 is approximately 0.0133).

Table 13-1 Strategy for Double Sampling Plan

Definition	Variable	Value
Acceptance number		
First sample	C1	2
Second sample	C3	6
Rejection number		
First sample	C2	6
Sample size		
First sample	N1	50
Second sample	N2	100
Part fraction defective	P1	0.08

S L A M I I S U M M A R Y R E P O R T

SIMULATION PROJECT DOUBLE SAMPLING BY PRITSKER

DATE 3/ 7/1988 RUN NUMBER 1000 OF 1000

CURRENT TIME 0.5000E+02
STATISTICAL ARRAYS CLEARED AT TIME 0.0000E+00

STATISTICS FOR VARIABLES BASED ON OBSERVATION

	MEAN VALUE	STANDARD DEVIATION	MINIMUM VALUE	MAXIMUM VALUE	NUMBER OF OBSERVATIONS
ACCEPT FIRST	0.5000E+02	0.0000E+00	0.5000E+02	0.5000E+02	219
ACCEPT SECOND	0.1500E+03	0.0000E+00	0.1500E+03	0.1500E+03	14
REJECT FIRST	0.4257E+02	0.5959E+01	0.2700E+02	0.5000E+02	110
REJECT SECOND	0.8247E+02	0.2074E+02	0.5100E+02	0.1500E+03	657
ACCEPT	0.5601E+02	0.2382E+02	0.5000E+02	0.1500E+03	233
REJECT	0.7675E+02	0.2386E+02	0.2700E+02	0.1500E+03	767

Figure 13-13 SLAM II outputs for double sampling plan.

13.8 INCORPORATING MISCLASSIFICATION ERRORS IN THE SAMPLING PLAN

In an inspection process for maintaining high-quality levels of manufactured parts, there is always the possibility of rejecting good units (Type I error) and/or accepting bad ones (Type II error). These considerations may be incorporated into the SLAM II model.

To incorporate misclassification errors into the double sampling plan illustrated in Figure 13-11, GOON nodes G1, G2, G3, and G4 are added to the network as shown in Figure 13-14. Decisions are based on how parts are

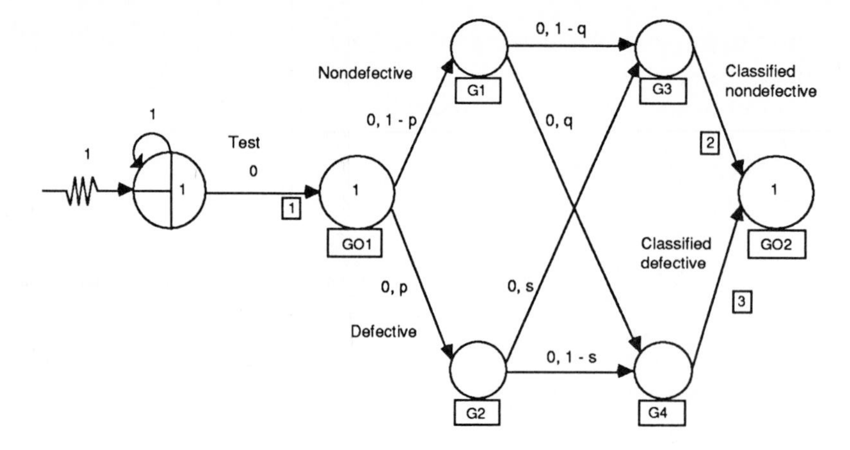

Figure 13-14 Adding part misclassification to the sampling plan network: p, probability a part is defective; q, probability of classifying a good part as defective; s, probability of classifying a defective part as good.

classified, including the testing errors, so that the number of parts completing activities 2 and 3 is used in the acceptance or rejection decision.

In the network segment illustrated in Figure 13-14, there is the probability s that a defective part is classified as nondefective, and a nondefective part has the probability of q of being classified as defective. Correspondingly, the probability that a nondefective part is classified as nondefective is 1-q and the probability that a defective part is classified as defective is 1-s. Thus, misclassification is easily incorporated into a sampling plan network model.

13.9 STOCHASTIC SHORTEST-ROUTE ANALYSIS

A special type of decision problem that has received considerable attention in the literature is the determination of the shortest route in a network. Many management science, operations research, and industrial engineering problems have been shown to be equivalent to defining the shortest path in a network [6].

In this chapter, two illustrations of shortest-route analysis are presented. An equipment replacement example describes a procedure for evaluating a least cost replacement strategy where the decision variable is the year at which replacement should be made. The second example involves an assessment of a security system. Such an assessment is related to the reliability of a system when an adversary is attempting to defeat or destroy an operating system. The shortest path in the network can be found by finding the shortest path from the CREATE node to an intermediate node and then finding the shortest path from the intermediate node to the TERMINATE node. By doing this for all intermediate nodes, an algorithm for finding the shortest-path results [2, 3].

When activity times are not deterministic, there is a probability associated with a path being the shortest or optimal. These probabilities have been defined as optimality indexes by Sigal who has done the pioneering research in stochastic shortest-route problems [19, 20, 21]. Areas of application for stochastic shortest-route analysis are equipment replacement, reliability, security and safeguards, stochastic maximal flow, and dynamic programming problems. In the following section, SLAM II methods are demonstrated for the analysis of stochastic shortest-route problems. Next, decision problems related to equipment replacement and security procedures are formulated as shortest-route problems.

13.10 SLAM II APPROACH TO SHORTEST-ROUTE PROBLEMS

There are many ways to model shortest-route problems with SLAM II. In this section, networks are constructed using nodes to represent decisions points and deterministic branching to allow all decisions to be selected. In this way, all activities emanating from a node are started, and all possible paths from the node are initiated. On a given run, only the first activity to be completed that is incident to a node is on the shortest path, hence, all other activities incident to a node that has been released need not be considered as candidates for the shortest path. However, if more than one activity is completed at the same time, they may be on alternative shortest paths.

By collecting the time of releasing the end node of the network and averaging over all runs, an estimate is obtained of the shortest time to traverse the network. If deterministic activity times are employed, a single run yields the paths that are shortest. Should the activity times be characterized by random variables, then different runs may have shortest paths. To automatically obtain the probability that a given path is shortest (the optimality index), user code is employed. Because of the possibility of ties, multiple paths on a run may be optimal, and the sum of the optimality indexes may not be 1.00.

13.11 FINDING THE SHORTEST PATH AND ESTIMATING PATH OPTIMALITY INDEXES

For the shortest-path problem, a network is constructed such that each activity is only started once. Each entity traversing an activity represents an alternate path through the network. Each node is released by an entity arriving to it. As an entity traverses the network, the number of nodes it passes through is established as ATRIB(1), with the node numbers maintained starting in ATRIB(2). A path is described by this sequence of node numbers as stored in ATRIB(2) through ATRIB(K), where K=ATRIB(1)+1. ASSIGN nodes are used following each activity to update the attributes of the entity arriving to the ASSIGN node. All activity durations are set at the first ASSIGN node, which is given the label N1 and is referred to as node 1. A CREATE node is used to generate one entity, which is routed to ASSIGN node N1. Multiple activities may be taken from each ASSIGN node, and the general characteristics of an activity and ASSIGN node combination are shown in Figure 13-15. Activity N with duration XX(N) is incident to ASSIGN node NK. The attributes of the entity are updated by increasing ATRIB(1) by 1 and setting the value K in the next available attribute for storing the path over which this entity is traversing. After the last ASSIGN node in the network, a COLCT node is used to record the time or cost associated with the first entity arrival to the COLCT node. In EVENT node E1, an entity's path is determined from its attributes, and the number of times that this path has been optimal is then increased by 1.

If the time of arrival of an entity to the COLCT node is greater than the shortest time, the run is terminated. Note that multiple entities could arrive at the same time. In subroutine EVENT, it is necessary to record which paths are optimal, as it is not known before the simulation which of the paths may be optimal. At the end of all runs, the paths that were optimal are printed along with estimates of their optimality indexes in subroutine OTPUT.

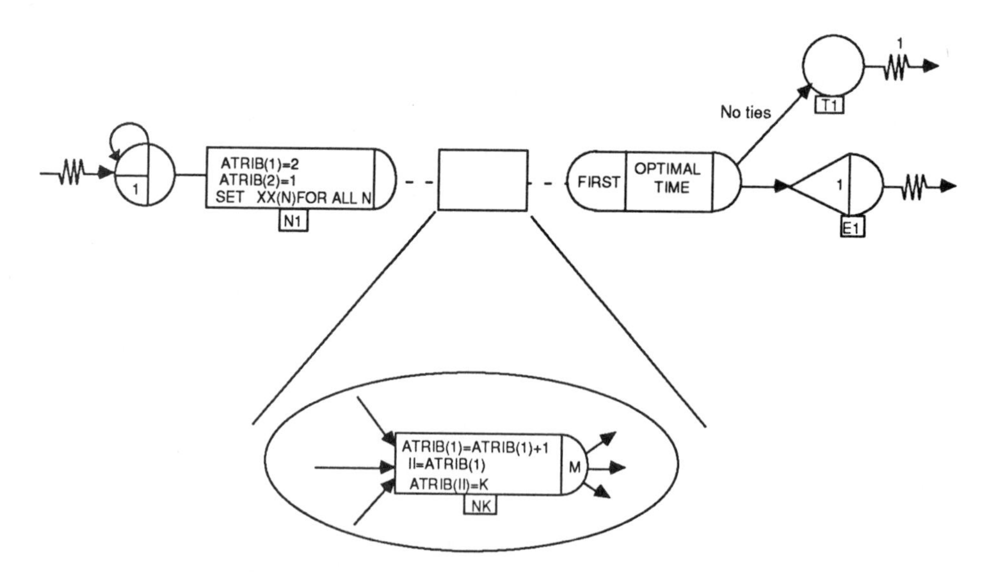

Figure 13-15 Network sketch for estimating shortest path and optimality indexes.

13.12 STOCHASTIC SHORTEST-ROUTE ANALYSIS FOR EQUIPMENT REPLACEMENT

As equipment ages, maintenance and operation costs increase. Deciding how often to replace expensive equipment is an important management concern. Costs may be reduced by frequently replacing existing equipment with new equipment, but each replacement increases capital expenditures. Equipment replacement policies may be modeled as a network, where each path represents the total cost of maintenance, operation, and capital outlay for a given replacement strategy. Finding the shortest route through the network corresponds to selecting the policy that minimizes total cost.

As an illustration, consider a company planning its equipment replacement during the next four years [18]. Let c_{ij} represent the cost of purchasing the equipment at the start of year i plus the cost of operating and maintaining the equipment until the start of year j. The network in Figure 13-16 represents the replacement options over the four-year planning horizon. Node i represents the start of year i. The arc ij represents the costs, c_{ij} incurred from year i to year j and is the difference between the purchase price and the salvage value plus the sum of the operational and maintenance costs. By this construction, each path in the network represents a strategy of equipment replacement over the four-year planning horizon. For example, one option is to replace the equipment at the start of every year at a total cost of C12 + C23 + C34 + C45.

When the c_{ij} are known, a shortest-path algorithm can be used to find the minimum cost solution. For this example, the c_{ij} are assumed to be independent random variables with known probability distributions, and the procedures described in the preceding section are used to analyze this stochastic shortest-route problem.

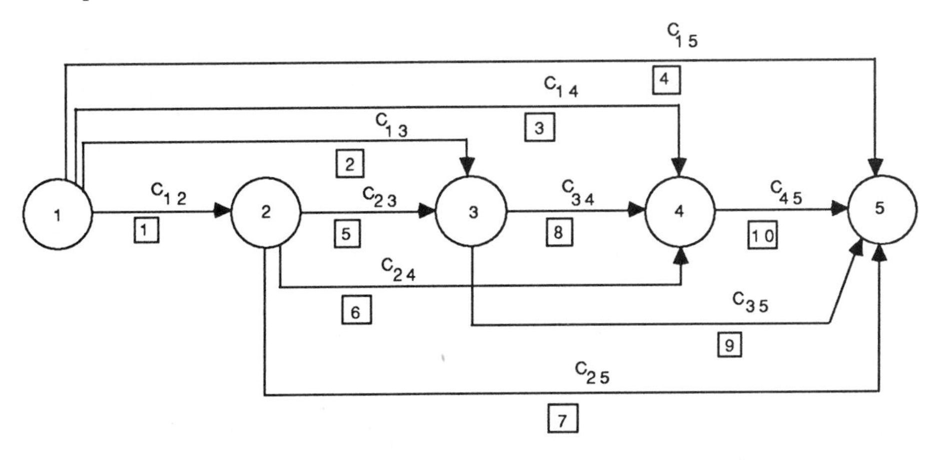

Figure 13-16 Network representing equipment replacement strategies.

The cost distributions for the example are given in Table 13-2. Although all the distributions used here are discrete, this is not a restriction, since any of the SLAM II distribution types could have been used. The objective of this stochastic version of equipment-replacement problems is to select the replacement strategy that has the greatest probability of being the least-cost option.

Table 13-2 Cost Distributions for the Network in Figure 13-16

Activity Number	Cost Variable	Distribution		
1	c_{12}	(10, 1)		
2	c_{13}	(20, 1)		
3	c_{14}	(30, 1)		
4	c_{15}	(40, 1)		
5	c_{23}	(10, 0.5)	(12, 0.5)	
6	c_{24}	(18, 0.3)	(20, 0.3)	(22, 0.4)
7	c_{25}	(26, 0.2)	(35, 0.8)	
8	c_{34}	(9, 0.5)	(12, 0.5)	
9	c_{35}	(18, 0.2)	(20, 0.4)	(22,0.4)
10	c_{45}	(9, 0.5)	(13, 0.5)	

The SLAM II network is similar to the network presented in Figure 13-16 with ASSIGN nodes as described for Figure 13-15 replacing the milestone nodes. The SLAM II statement model is presented in Figure 13-17. An N has been prefixed to each node number to form the node labels. In the following discussion, node numbers and node labels are used interchangeably. A CREATE node generates one entity from which all entities traversing the paths of the network flow. At node N1, ATRIB(2) is set to 1 to indicate that node N1 is on the optimal path. ATRIB(1) is set to 2 to indicate that the last node on the path is stored as the second attribute. The values of XX(5) through XX(10) are set equal to the sample costs of activities 5 through 10. Activities 1, 2, 3, and 4 have constant durations. XX(20) is used to store the shortest path and is initially set to a large number so that the first optimal path does not terminate a run of the network. In EVENT 1, XX(20) is reset to the time of the first entity arrival at EVENT node E1. Following the collection of the optimal cost, a test of TNOW being greater than XX(20) is made and, if it is, no further ties for the optimal path exist and the run is terminated. At the COLCT node, a histogram is collected to provide an estimate of the distribution of the optimal cost. The statement model indicates that 10,000 runs are to be made.

```
GEN,PRITSKER,EQUIP REPLACEMENT,1/11/88,10000,N,N,,N,Y/10000;
LIMITS,,6,10;
ARRAY(1,2) /.5,1.0/
     (2,3) /.3,.6,1.0/
     (3,2) /.2,1.0/
     (4,3) /.2,.6,1.0/
     (5,2) /10,12/
     (6,3) /18,20,22/
     (7,2) /26,35/
     (8,2) /9,12/
     (10,2)/9,13;
;
;  ATRIB(1) IS NUMBER OF NODES ON PATH
;  ATRIB(N) IS NTH NODE ON PATH
;
NETWORK;
       CREATE,,,,1;                START TIME TO EQUIPMENT REPLACEMENT
N1     ASSIGN,ATRIB(1)=2,ATRIB(2)=1,XX(20)=9999.,
                XX(5)=DPROBN(1,5),XX(6)=DPROBN(2,6),XX(7)=DPROBN(3,7),
                XX(8)=DPROBN(1,8),XX(9)=DPROBN(4,6),XX(10)=DPROBN(1,10),4;
       ACT/1,10,,N2;               REPLACE AT END OF 1ST YEAR
       ACT/2,20,,N3;               REPLACE AT END OF 2ND YEAR
       ACT/3,30,,N4;               REPLACE AT END OF 3RD YEAR
       ACT/4,40,,N5;               REPLACE AT END OF 4TH YEAR
;
N2     ASSIGN,ATRIB(1)=ATRIB(1)+1,II=ATRIB(1),ATRIB(II)=2,3;
       ACT/5,XX(5),,N3;            REPLACE AT END OF YEARS 1 & 2
       ACT/6,XX(6),,N4;            REPLACE AT END OF YEARS 1 & 3
       ACT/7,XX(7),,N5;            REPLACE AT END OF YEARS 1 & 4
;
N3     ASSIGN,ATRIB(1)=ATRIB(1)+1,II=ATRIB(1),ATRIB(II)=3,2;
       ACT/8,XX(8),,N4;            REPLACE AT END OF YEARS 1, 2 & 3
       ACT/9,XX(9),,N5;            REPLACE AT END OF YEARS 2 & 4
;
N4     ASSIGN,ATRIB(1)=ATRIB(1)+1,II=ATRIB(1),ATRIB(II)=4,1;
       ACT/10,XX(10),,N5;          REPLACE AT END OF YEARS 1, 2, 3, & 4
;
N5     ASSIGN,ATRIB(1)=ATRIB(1)+1,II=ATRIB(1),ATRIB(II)=5;
       COLCT,FIRST,OPTIMAL COST,4/36/1,1;
       ACT,,TNOW.GT.XX(20),TERM; TERMINATE SIMULATION, NO MORE TIES
       ACT;
;
;      RECORD OPTIMAL ACTIVITIES AND COST
;
E1     EVENT,1;
       TERM;
;
TERM   TERM,1;
       END;
INIT,0,,N;
```

Figure 13-17 SLAM II statement model of equipment replacement strategies.

When an entity reaches EVENT node E1, an optimal path has been traversed. In subroutine EVENT shown in Figure 13-18, a count on the number of times a path is optimal is maintained. On the first run of the network, the variable COUNT(I) is set to zero, where I is a path number and it is assumed that there are 20 or fewer optimal paths; that is, I is less than or equal to 20. The value of XX(20) is then set equal to TNOW to provide the optimal cost for this run. The attributes of the entity arriving to the EVENT node are placed into the vector IPATH. If the number of paths, NPATH, is greater than zero, then the current path is tested to see if it has been optimal on a previous run. The previous optimal paths are maintained in the array JPATH(NP,N), where NP is the path number. In JPATH(NP,1), the number of nodes on the path is stored, which provides a quick check on whether the current optimal path is equivalent to path NP. If JPATH(NP,N)=IPATH(N) for all N, then path NP has been optimal one more time

```
      SUBROUTINE EVENT(IEVT)
      COMMON/SCOM1/ATRIB(100),DD(100),DDL(100),DTNOW,II,MFA,MSTOP,NCLNR
     1,NCRDR,NPRNT,NNRUN,NNSET,NTAPE,SS(100),SSL(100),TNEXT,TNOW,XX(100)
      COMMON/UCOM1/IPATH(6),COUNT(20),NPATH,JPATH(20,6)
      COMMON QSET(1)
      DIMENSION NSET(1)
      EQUIVALENCE (NSET(1),QSET(1))
C     *** INITIALIZE COUNT AND NPATH ON FIRST RUN ***
      IF (NNRUN .EQ. 1) THEN
         DO 20 I = 1,20
            COUNT(I) = 0.
 20      CONTINUE
         NPATH = 0
      ENDIF
C     *** SET XX(20) TO TIME OF SHORTEST PATH ***
      XX(20) = TNOW
      NUMNOD = ATRIB(1)
      DO 10 N = 1,NUMNOD
         IPATH(N) = ATRIB(N)
 10   CONTINUE
      IF (NPATH .GT. 0) THEN
C        *** TEST TO SEE IF PATH WAS PREVIOUSLY SHORTEST ***
         DO 30 NP = 1,NPATH
            DO 40 N = 1,NUMNOD
               IF (IPATH(N) .NE. JPATH(NP,N)) GO TO 30
 40         CONTINUE
           COUNT(NP) = COUNT(NP) + 1
            RETURN
 30      CONTINUE
      ENDIF
C     *** ESTABLISH NEW PATH ***
      NPATH = NPATH + 1
      DO 60 N = 1,NUMNOD
         JPATH(NPATH,N) = IPATH(N)
 60   CONTINUE
      COUNT(NPATH) = COUNT(NPATH) + 1
      RETURN
      END
```

Figure 13-18 Subroutine EVENT for equipment replacement model.

and the variable COUNT(NP) is increased by 1. If there is no match, then the number of different optimal paths, NPATH, is increased by 1 and the current path is stored in row NPATH of JPATH. The count for this new path is then increased by 1. The conditional branching after the COLCT node causes subroutine EVENT to be called for any path whose cost is equal to XX(20), but not for paths whose cost is greater.

In subroutine OTPUT shown in Figure 13-19, a report of all optimal paths and their optimality index is printed. These results are shown in Figure 13-20. From Figure 13-20, the path with the highest optimality index (0.3183) contains nodes 1 and 5. By subtracting the node numbers, the replacement strategy and years between replacement are identified; in this case, it is replacement after four years. The strategy of replacing after two years identified by the path from nodes 1 to 3 to 5 has the second highest optimality index of 0.2936. If the replacement strategy is based solely on optimality index, then replacement after every four years would be chosen. Replacement every four years involves a cost of 40 units. Replacement every two years involves a cost of 38 units 20% of the time, 40 units 40% of the time, and 42 units 40% of the time. For these two paths, selection based on the optimality index is consistent with the choice based on lower expected value and lower variability.

The histogram of the optimal cost is shown in Figure 13-21. The average of the optimal cost is 38.2 units and varies between 36 and 40 units. It is seen from Figure 13-21 that an optimal cost of 36 units is estimated to be possible 20% of the time. For this example, the network is small enough that comparisons between the paths and a complete enumeration of all paths is possible.

```
      SUBROUTINE OTPUT
      COMMON/SCOM1/ATRIB(100),DD(100),DDL(100),DTNOW,II,MFA,MSTOP,NCLNR
     1,NCRDR,NPRNT,NNRUN,NNSET,NTAPE,SS(100),SSL(100),TNEXT,TNOW,XX(100)
      COMMON/UCOM1/IPATH(6),COUNT(20),NPATH,JPATH(20,6)
      DATA NRUNS/10000/
C     *** CHECK IF TOTAL NUMBER OF RUNS HAS BEEN COMPLETED ***
      IF (NNRUN .GE. NRUNS) THEN
C        *** COMPUTE AND PRINT OPTIMALITY INDEXES ***
         WRITE(NPRNT,100)
         DO 90 NP = 1,NPATH
            XINDX = COUNT(NP)/FLOAT(NRUNS)
            NUMNOD = JPATH(NP,1)
            WRITE(NPRNT,200) NP, XINDX, (JPATH(NP,N),N=2,NUMNOD)
   90    CONTINUE
      ENDIF
      RETURN
  100 FORMAT(2X,'PATH',10X,'OPTIMALITY'10X,'NODES',/,2X'INDEX',11X,
     +       'INDEX',12X,'ON PATH',//,50('-'))
  200 FORMAT(1X,I4,11X,F10.7,5X,6I4)
      END
```

Figure 13-19 Subroutine OTPUT for equipment replacement model.

PATH INDEX	OPTIMALITY INDEX	NODES ON PATH
1	0.2001000	1 2 5
2	0.3183000	1 5
3	0.2936000	1 3 5
4	0.1479000	1 2 3 5
5	0.1693000	1 2 4 5
6	0.1159000	1 4 5
7	0.1412000	1 3 4 5
8	0.0714000	1 2 3 4 5

Figure 13-20 Optimal path definitions and optimality index for equipment replacement model.

```
                                **HISTOGRAM NUMBER  1**

                                     OPTIMAL COST

OBSV     RELA     CUML       UPPER
FREQ     FREQ     FREQ     CELL LIMIT    0         20        40        60        80        100
                                         +    +    +    +    +    +    +    +    +    +    +
2001     0.200    0.200    0.3600E+02    +**********                                      +
1197     0.120    0.320    0.3700E+02    +******         C                                +
2460     0.246    0.566    0.3800E+02    +************               C                    +
1159     0.116    0.682    0.3900E+02    +******                           C              +
3183     0.318    1.000    0.4000E+02    +****************                                C
   0     0.000    1.000       INF        +                                                C
   -                                   +    +    +    +    +    +    +    +    +    +    +
10000                                    0         20        40        60        80        100

                      **STATISTICS FOR VARIABLES BASED ON OBSERVATION**

                     MEAN          STANDARD       MINIMUM       MAXIMUM       NUMBER OF
                     VALUE         DEVIATION      VALUE         VALUE         OBSERVATIONS

  OPTIMAL COST     0.3823E+02     0.1502E+01     0.3600E+02    0.4000E+02      10000
```

Figure 13-21 Optimal cost statistics and histogram for equipment replacement model.

13.13 SHORTEST-ROUTE ANALYSIS FOR SAFEGUARDS SYSTEM DESIGN

Researchers at Sandia Laboratories and Pritsker & Associates have applied shortest-route analysis to security procedures for nuclear facilities [9, 19]. The network model consists of possible routes a thief could use to penetrate a site, access a target, and escape. The network arcs represent the time involved in traversing distances, removing obstacles, and performing other tasks involved in theft and escape. The primary research objective is to compare the thief's minimum time with the security force's response time. A secondary objective is the identification of optimal routes of penetration, which provides information on policies to recommend regarding the nature and allocation of security procedures and expenditures.

To illustrate the general nature of the approach to safeguards modeling, a simplified example is presented. Figure 13-22 represents a schematic drawing of potential routes of penetration by a thief entering a protected area. Walls of the building are shown by solid lines. Theft routes are shown by dashed lines. The target of the theft is located at point N6. The thief, however, is not sure of the exact location of internal walls relative to point N6. The thief's initial entry is at point N1. Four options are available to penetrate the first barrier:

1. Travel to one side of the structure where a door is located, break the lock, and proceed to G1 via N21.
2. Bore through the wall and arrive at G1 via N22.
3. Scale the wall to arrive at G1 via N23.
4. Move to another location to scale the wall and arrive at point N3.

It is assumed that travel times after breaking in to point G1 are zero. From point G1, the thief might travel directly to point N4 or first go to N3 and then decide to try point N4. At point N4, the thief bores through the wall to reach point N6.

When arriving at point N3, the thief might decide to move to point N4 or to bore through the wall at N3 to reach N5. After boring through the wall at N3, it is necessary to bore through another wall to reach N6. In this model, the thief does not retrace a path because it increases the probability of detection.

The problem is to compute the probability that each route is the shortest. Also of interest in this problem is the distribution of the minimum time to reach the target. This distribution provides information on the requirements for a safeguards system response time once a break-in has started.

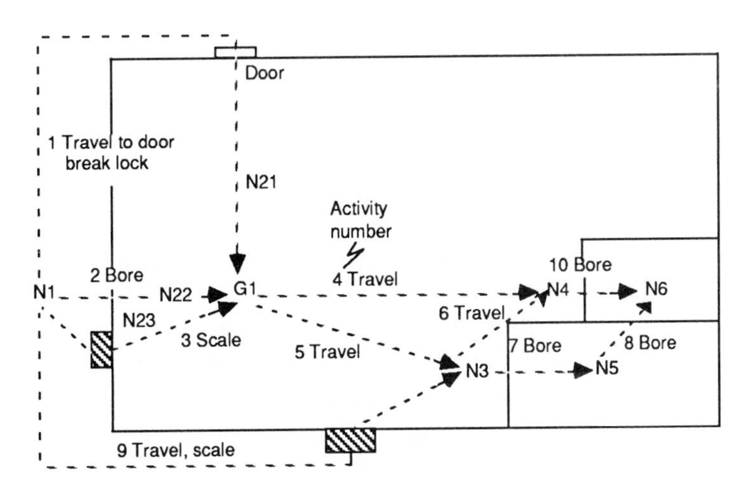

Figure 13-22 Schematic of possible penetration routes.

The SLAM II network for this example is shown in Figure 13-23 and the statement model in Figure 13-24. Estimates of the probability distributions of the activity times are given in Table 13-3. The method for obtaining samples of the activity times obtained in subroutine INTLC is shown in Figure 13-25. Statistics collected at node N6 provide an estimate of the minimum time distribution. Subroutines EVENT and OTPUT, presented in Figures 13-18 and 13-19 for the equipment replacement example, are also employed in this illustration to identify paths and to estimate their optimality indexes.

Table 13-3 Probability Distributions of Theft Activity Times

Activity	Distribution		
1	(5, 0.2)	(7, 0.2)	(15,0.6)
2	(15, 0.9)	(20, 0.1)	
3	(10, 0.5)	(20, 0.5)	
4	(5, 0.5)	(6, 0.5)	
5	(3, 0.3)	(4, 0.7)	
6	(4, 0.5)	(6, 0.5)	
7	(10, 0.4)	(20, 0.3)	(30,0.3)
8	(10, 0.5)	(15, 0.5)	
9	(10, 0.5)	(15, 0.5)	
10	(20, 0.3)	(25, 0.6)	(30, 0.1)

Figure 13-26 presents the results of 10,000 runs of the network. It is seen that any one of seven paths may be optimal, with five of the paths having a probability greater than 0.1 of being the shortest. The range of minimum times is from 28 to 51, which indicates a wide variability in the theft process.

13.14 SUMMARY

Reliability and quality control are fundamental concepts of systems analysis. This chapter illustrates that SLAM II networks can incorporate reliability and quality control considerations into systems models. A network approach based on building system models from network elements parallels the reliability systems analysis approach of computing system reliability from component reliabilities. The concept of a stochastic shortest route is presented, and examples are given of SLAM II models to evaluate equipment replacement plans and safeguards for theft or sabotage deterrence.

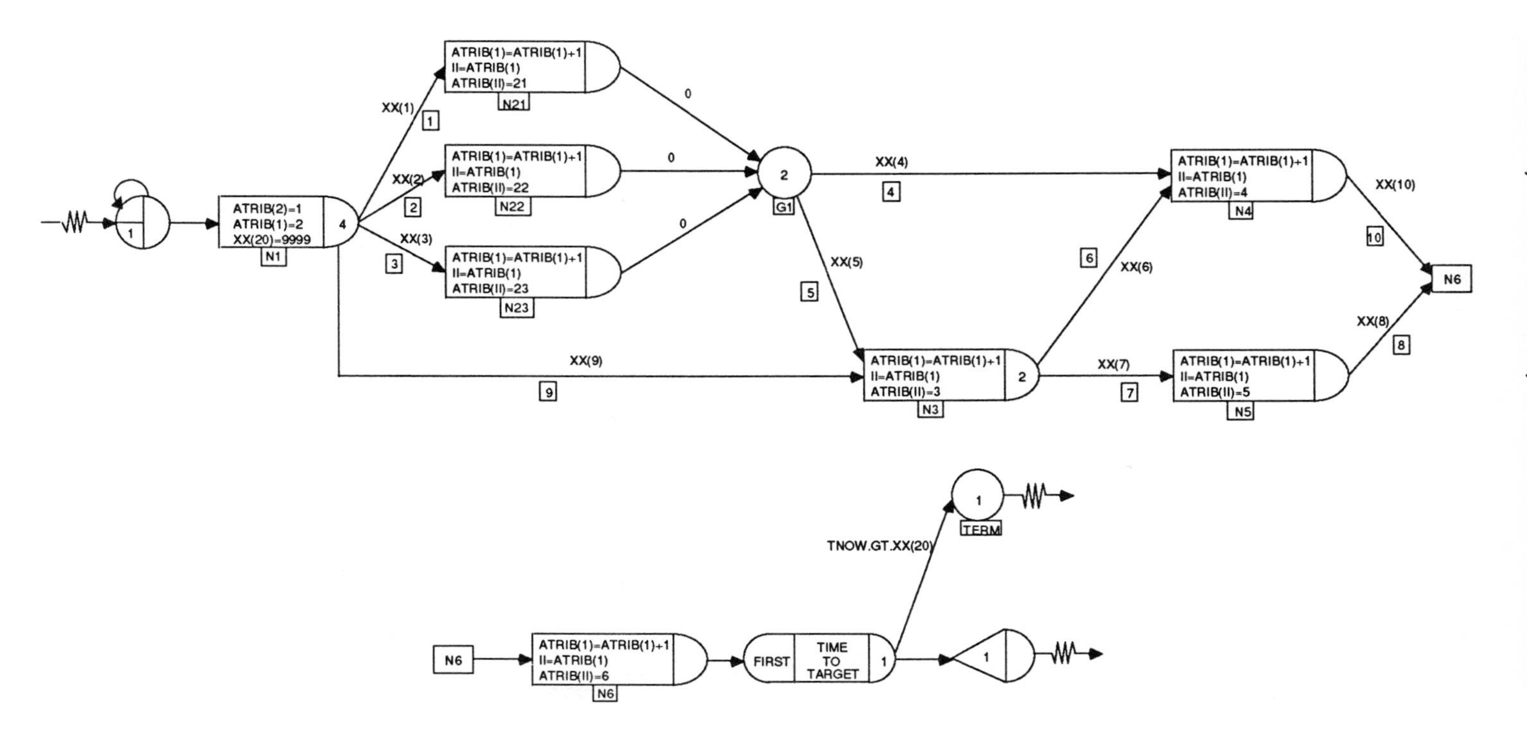

Figure 13-23 SLAM II network for theft time evaluation.

```
GEN,PRITSKER,THEFT,4/18/88,10000,N,N,,N,Y/10000;
LIMITS,,6,20;
;
;  ATRIB(1) IS NUMBER OF NODES ON PATH
;  ATRIB(N) IS NTH NODE ON PATH
;
NETWORK;
      CREATE,,,,1;                  START BREAK-IN
N1    ASSIGN,ATRIB(1)=2,ATRIB(2)=1,XX(20)=9999.;
      ACT/1,XX(1),,N21;             ENTER THROUGH DOOR AND GO TO B
      ACT/2,XX(2),,N22;             BORE HOLE IN WALL AND GO TO B
      ACT/3,XX(3),,N23;             ENTER THROUGH WINDOW AND GO TO B
      ACT/9,XX(9),,N3;              SCALE WALL AND GO TO C
;
N21   ASSIGN,ATRIB(1)=ATRIB(1)+1,II=ATRIB(1),ATRIB(II)=21,1;
      ACT,,,G1;
N22   ASSIGN,ATRIB(1)=ATRIB(1)+1,II=ATRIB(1),ATRIB(II)=22,1;
      ACT,,,G1;
N23   ASSIGN,ATRIB(1)=ATRIB(1)+1,II=ATRIB(1),ATRIB(II)=23,1;
      ACT,,,G1;
;
G1    GOON,2;
      ACT/4,XX(4),,N4;              TRAVEL FROM B TO D
      ACT/5,XX(5),,N3;              TRAVEL FROM B TO C
;
N3    ASSIGN,ATRIB(1)=ATRIB(1)+1,II=ATRIB(1),ATRIB(II)=3,2;
      ACT/6,XX(6),,N4;              TRAVEL FROM C TO D
      ACT/7,XX(7),,N5;              TRAVEL FROM C TO E
;
N4    ASSIGN,ATRIB(1)=ATRIB(1)+1,II=ATRIB(1),ATRIB(II)=4,1;
      ACT/10,XX(10),,N6;            BORE INTO POINT F FROM D
;
N5    ASSIGN,ATRIB(1)=ATRIB(1)+1,II=ATRIB(1),ATRIB(II)=5,1;
      ACT/8,XX(8),,N6;              BORE INTO POINT F FROM E
;
N6    ASSIGN,ATRIB(1)=ATRIB(1)+1,II=ATRIB(1),ATRIB(II)=6;
      COLCT,FIRST,TIME TO TARGET,24/28/1,1;
      ACT,,TNOW.GT.XX(20),TERM; TERMINATE SIMULATION, NO MORE TIES
      ACT;                          IGNORE ENTITY
      EVENT,1;
      TERM;
;
TERM  TERM,1;
      END;
;
INIT,0,,N;
FIN;
```

Figure 13-24 SLAM II statement model for theft time evaluation.

```
      SUBROUTINE INTLC
      COMMON/SCOM1/ATRIB(100),DD(100),DDL(100),DTNOW,II,MFA,MSTOP,NCLNR
     1,NCRDR,NPRNT,NNRUN,NNSET,NTAPE,SS(100),SSL(100),TNEXT,TNOW,XX(100)
      DIMENSION CPROB(3,10),VALUE(3,10),NVAL(10)
      DATA ((CPROB(I,J),I=1,3),J=1,10) /
     +  .2,  .4, 1.0,     .9, 1.0, 0.0,     .5, 1.0, 0.0,
     +  .5, 1.0, 0.0,     .3, 1.0, 0.0,     .5, 1.0, 0.0,
     +  .4,  .7, 1.0,     .5, 1.0, 0.0,     .5, 1.0, 0.0,
     +  .3,  .9, 1.0 /
      DATA ((VALUE(I,J),I=1,3),J=1,10) /
     + 5.,  7., 15.,    15., 20., 0.,    10., 20., 0.,
     + 5.,  6.,  0.,     3.,  4., 0.,     4.,  6., 0.,
     + 10., 20., 30.,   10., 15., 0.,    10., 15., 0.,
     + 20., 25., 30. /
      DATA (NVAL(J),J=1,10) / 3, 2, 2, 2, 2, 2, 3, 2, 2, 3 /
      DO 100 J = 1,10
         XX(J) = DPROB (CPROB(1,J),VALUE(1,J),NVAL(J),1)
  100 CONTINUE
      RETURN
      END
```

Figure 13-25 Obtaining activity times for theft model.

PATH INDE	OPTIMALITY INDEX	NODES ON PATH
1	0.1194000	1 21 3 5 6
2	0.1356000	1 23 4 6
3	0.2381000	1 3 5 6
4	0.0529000	1 23 3 5 6
5	0.2144000	1 3 4 6
6	0.3002000	1 21 4 6
7	0.0397000	1 22 4 6

```
                                                      **HISTOGRAM NUMBER  1**
                                                            OPTIMAL COST
 OBSV    RELA    CUML       UPPER
 FREQ    FREQ    FREQ     CELL LIMIT      0        20        40        60        80       100
                                          +    +    +    +    +    +    +    +    +    +    +
   137   0.014   0.014   0.2800E+02      +*                                                 +
   280   0.028   0.042   0.2900E+02      +*C                                                +
  1118   0.112   0.154   0.3000E+02      +****** C                                          +
   393   0.039   0.193   0.3100E+02      +**       C                                        +
   232   0.023   0.216   0.3200E+02      +*         C                                       +
   381   0.038   0.254   0.3300E+02      +**          C                                     +
   773   0.077   0.331   0.3400E+02      +****            C                                 +
  1611   0.161   0.493   0.3500E+02      +********                C                         +
   854   0.085   0.578   0.3600E+02      +****                        C                     +
   347   0.035   0.613   0.3700E+02      +**                            C                   +
   428   0.043   0.655   0.3800E+02      +**                              C                 +
   879   0.088   0.743   0.3900E+02      +****                                C             +
   890   0.089   0.832   0.4000E+02      +****                                     C        +
   694   0.069   0.902   0.4100E+02      +***                                          C    +
    51   0.005   0.907   0.4200E+02      +                                             C    +
    58   0.006   0.913   0.4300E+02      +                                              C   +
   351   0.035   0.948   0.4400E+02      +**                                             C  +
   264   0.026   0.974   0.4500E+02      +*                                                C+
   184   0.018   0.993   0.4600E+02      +*                                                 C
     0   0.000   0.993   0.4700E+02      +                                                  C
     0   0.000   0.993   0.4800E+02      +                                                  C
    35   0.004   0.996   0.4900E+02      +                                                  C
    26   0.003   0.999   0.5000E+02      +                                                  C
    14   0.001   1.000   0.5100E+02      +                                                  C
     0   0.000   1.000   0.5200E+02      +                                                  C
     0   0.000   1.000      INF          +                                                  C
  _                                       +    +    +    +    +    +    +    +    +    +    +
 10000                                    0        20        40        60        80       100

                       MEAN        STANDARD      MINIMUM       MAXIMUM      NUMBER OF
                       VALUE       DEVIATIO      VALUE         VALUE        OBSERVATIONS

  OPTIMAL COST      0.3627E+02    0.4543E+0    0.2800E+02    0.5100E+02      10000
```

Figure 13-26 Results for 10,000 runs of the theft model.

13.15 EXERCISES

13-1. Develop a SLAM II model to estimate the reliability of the following system configurations.

(a)

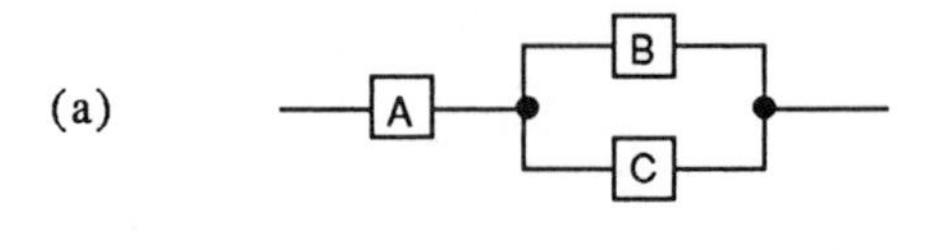

(b)

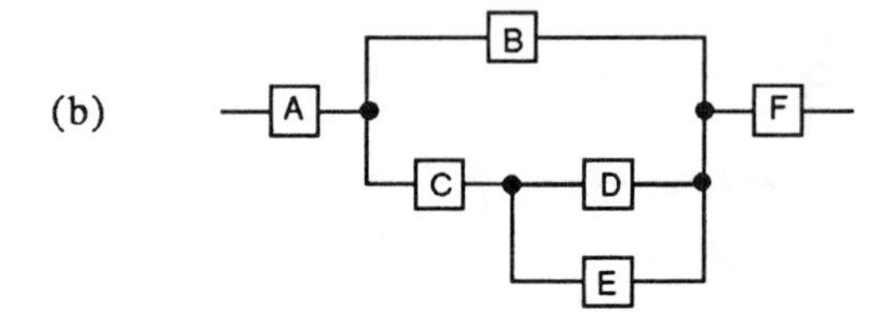

13-2. Given three components in series each of which has an exponentially distributed time to failure with a mean rate of λ, determine the distribution of the system time to failure.

13-3. Given three components in parallel each of which has en exponentially distributed time to failure with a mean rate of λ, determine the distribution of the time to failure.

13-4. In the model of the reliability of two units in parallel with repair capabilities (Figure 13-1), assume that the failure and repair times are exponentially distributed with a failure rate of 5 and repair rate of 3. Using a table of random numbers, perform 10 runs manually to obtain the time to system failure. *Embellishment*: For the system described in this example, compute the theoretical time to failure. (*Hint:* This is a three-state Markov process, with the states being zero failed systems, one failed system, and two failed systems.)

13-5. Discuss the advantages and disadvantages of the alternative models for assessing system reliability presented in Section 13.4.

13-6. Change the model of the power station maintenance and repair to include ten generators on line, three spares, and two repair personnel.

13-7. Build a SLAM II model for assessing the worth of MIL Standard 105D [24].

13-8. Describe in words the quality control principles involved in a double sampling plan to a manager using the SLAM II network of Figure 13-11.

13-9. Build a SLAM II model of a single sampling plan with which you are familiar.

13-10. Incorporate the double sampling plan of Figure 13-11 into a production system model and cause the production line to be shut down if two consecutive lots are rejected.

13-11. The Soviet government is planning to transport waste from a nuclear plant from Vladivostok to Rostov [19, 23]. The network shown here represents different routes between the two cities. There is an expense associated with each route due to the fact that precautionary action must be taken along each leg of the journey to avoid accidents or sabotage. This expense is not known with certainty and can only be estimated. The estimated expenses are shown in the accompanying table. Build a SLAM II model to select the route that has the greatest probability of incurring the least cost.

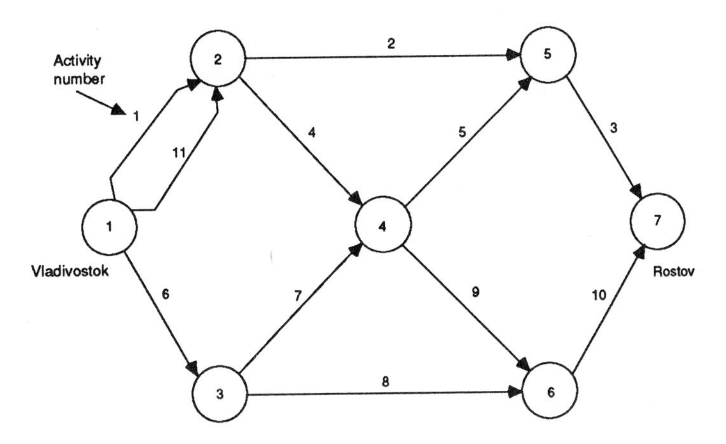

Branch	Distribution of Cost		
1	(10, 0.3)	(20, 0.3)	(30, 0.4)
2	(20, 0.6)	(40, 0.2)	(60, 0.2)
3	(30, 0.1)	(50, 0.9)	
4	(15, 0.5)	(20, 0.5)	
5	(40, 0.4)	(55, 0.6)	
6	(30, 0.9)	(40, 0.05)	
7	(10, 0.15)	(15, 0.75)	(20, 0.1)
8	(40, 0.5)	(60, 0.5)	
9	(30, 0.7)	(40, 0.3)	
10	(20, 0.75)	(40, 0.25)	
11	(15, 0.35)	(25, 0.65)	

13-12. For the equipment replacement model, show that the theoretical optimality indexes for the paths as listed in Figure 13-20 are 0.2, 0.32, 0.296, 0.148, 0.168, 0.112, 0.140, and 0.07.

13-13. For the equipment replacement model, show that the lowest expected cost for a path is 40 units.

13-14. For the equipment replacement strategies problem, build a model to select the path with the lowest expected cost.

13-15. Convert the cost distributions for the activities in the equipment replacement model so that the path with the highest optimality index does not have the lowest expected cost.

13-16. For the theft model, show that the theoretical optimality indexes for the paths as listed in Figure 13-26 are 0.1171, 0.1389, 0.2324, 0.0532, 0.2171, 0.3002, and 0.0380. Show that the probability that the minimum time to target being equal to 35 is 0.1599.

13.16 REFERENCES

1. Case, K. E., and K. R. Morrison, "A Simulation of System Reliability Using GERTS III," *Virginia Academy of Science Meeting*, May 14, 1971.
2. Dreyfus, S. E., "An Appraisal of Some Shortest Path Algorithms," *Operations Research*, Vol. 17, No. 3, May-June 1969, pp. 395-412.
3. Dijkstra, E. W., "A Note on Two Problems in Connection with Graphs," *Numerische Mathematik*, Vol. 1, 1959, pp. 269-271.
4. Duncan, A. J., *Quality Control and Industrial Statistics*, Irwin, Homewood, IL, 1965.
5. Eisner, H., *Computer-aided Systems Engineering*, Prentice-Hall, Englewood Cliffs, NJ, 1987.
6. Elmaghraby, S. E., *Network Models in Management Science*, Lecture Series on Operations Research, Springer-Verlag, New York, 1970.
7. Hammesfahr, R. D. J., T. R. Rakes, and E. R. Clayton, "An Application of Q-GERT to the System Reliability Problem," *Proceedings, S.E. TIMS Conference*, October 1978, pp. 237-243.
8. Henley, E. J., and K. Kumamotoh, *Reliabilty, Engineering and Risk Assessment*, Prentice-Hall, Englewood Cliffs, NJ, 1981.
9. Hulme, B. L., "Graph Theoretic Models of Theft Problems, I. The Basic Theft Model," Sandia Laboratories, Albuquerque, NM, SAND 75-0595, November 1975.
10. Polito, J., Jr., and C. C. Petersen, *User's Manual for GRASP*, Purdue Laboratory for Applied Industrial Control, Report Number 75, April 1976.
11. Pritsker, A. A. B., *Introduction to Simulation and SLAM II*, 3*rd* ed., Wiley and Systems Publishing, New York and West Lafayette, IN, 1986.
12. Pritsker, A. A. B., and C. E. Sigal, *The GERT IIIZ User's Manual*, Pritsker & Associates, West Lafayette, IN, 1974.
13. Pritsker, A. A. B. and C. E. Sigal, *Management Decision Making: A Network Simulation Approach*, Prentice-Hall, Englewood Cliffs, NJ, 1983.
14. Pritsker, A. A. B., L. J. Watters, and P. M. Wolfe, "Mathematical Forumlation: A Problem in Design," *Proceedings 19th AIIE Conference*, May 1968, pp. 205-210.
15. Pritsker, A. A. B., "Model Evolution I: A Rotary Index Table Case History," *Proceedings, 1986 Winter Simulation Conference*, 1986, pp. 703-707.
16. Pritsker, A. A. B., "Model Evolution II: An FMS Design Problem," *Proceedings, 1987 Winter Simulation Conference*, 1987, pp. 567-574.
17. Pritsker, A. A. B., *Modeling and Analysis Using Q-GERT Networks*, 2*nd* ed., Wiley, New York, 1979.
18. Ravindrin, A., D. T. Phillips, and J. Solberg, *Introduction to Operations Research*, 2*nd* ed., Wiley, New York, 1987.
19. Sigal, C. E., "Stochastic Shortest Route Problems," Ph.D. dissertation, Purdue University, December 1977.
20. Sigal, C. E., A. A. B. Pritsker and J. J. Solberg, "Cutsets in Monte Carlo Analysis of Stochastic Networks," *Mathematics and Computers in Simulation*, Vol. 21, No. 4, December 1979, pp. 376-384.

21. Sigal, C. E., A. A. B. Pritsker, and J. J. Solberg, "The Stochastic Shortest Route Problem," *Operations Research*, Vol. 28, No. 5, September-October 1980, pp. 1122-1129.
22. Wagner, H., *Principles of Operations Research with Applications to Managerial Decisions*, 2nd ed., Prentice-Hall, Englewood Cliffs, NJ, 1975.
23. Whitehouse, G. E., "GERT, A Useful Technique for Analyzing Reliability Problems," *Technometrics*, February 1970.
24. Whitehouse, G. E., *Systems Analysis and Design Using Network Techniques*, Prentice-Hall, Englewood Cliffs, NJ, 1973.
25. Whitehouse, G. E., and E. C. Hsuan, "The Application of GERT to Quality Control: A Feasibility Study`," (NASA Contract NAS-12-2079) Department of Industrial Engineering, Lehigh University, Bethlehem, PA.

14 LOGISTICS SYSTEMS ANALYSIS

14.1 INTRODUCTION

A logistics system consists of a collection of personnel and machines organized to procure, store, and transport material or people. A logistics system may exist in one facility, such as an airport terminal, or it may be a collection of facilities, such as the network of Air Force bases. In this chapter, the basic concerns in the analysis of logistics systems are introduced through simple examples that describe a network approach to logistics system analysis. Sections are included that define common performance measures and the SLAM II calculation procedures to obtain these measures. In addition, special SLAM II topics in logistics are identified and SLAM II applications of logistics analysis are presented.

A logistics system has characteristics similar to a production system where the processing operations are replaced by the transport of goods from one location to another. However, when modeling logistics processes, consideration is usually given to the effects of reliability, maintainability, maintenance planning, support, and test equipment in addition to the normal elements of supply and transportation. In fact, integrated logistics support (ILS) not only includes the elements listed above, but defines facilities, personnel and training, technical data, funding, and management information as part of the logistics system. A logistics analysis includes the modeling of any of the elements of an ILS. For example, a model presented in this chapter involves the receipt of material from many sources and organizing it to support a battle engagement. A characteristic of logistics systems is the arrival of both the suppliers of and the demanders for material.

A prime advocate of the analysis of logistics systems is Mortenson, and the material presented in the next section is based on his application [17].

14.2 EXAMPLE OF A LOGISTICS PROCESS

Figure 14-1 depicts a simplified support process for an air transportation system. When a piece of equipment fails on an aircraft, a diagnostic process is initiated that results in the removal of the component that caused the equipment to fail (the faulty box). The faulty box becomes an input to a repair process that may involve a transport or move activity. The box eventually arrives at a repair facility to be fixed. To fix the box requires tools, test equipment, and a repairer. The unavailability of any of these resources may cause the box to wait. When all the required resources are available, the box is fixed. Following another transport activity, it becomes a spare. When another box of the same type fails, the repaired box that became a spare is installed, and a logistics cycle is completed.

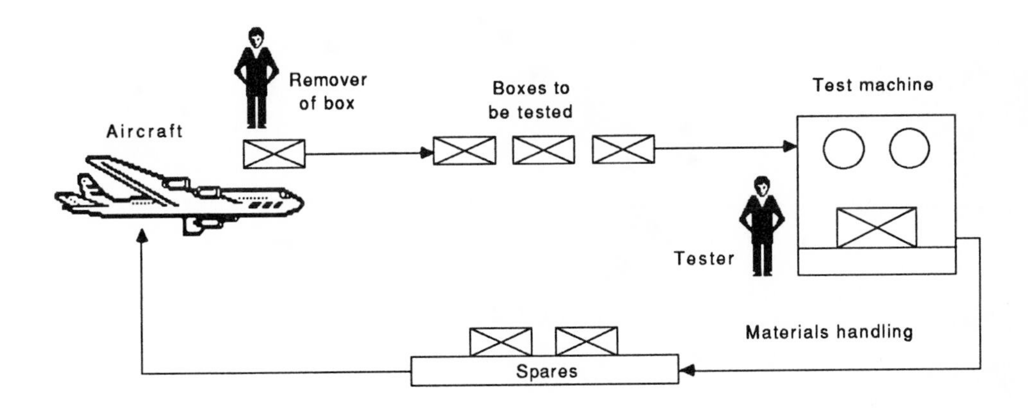

Figure 14-1 Simplified aircraft support sytem.

Many performance aspects are associated with the simplified support process described above. Operational questions relate to the aircraft and its readiness. Some of the factors that affect operational readiness (OR) are the number of spares available, the cycle time for repairing a spare, the failure rate associated with the equipment, and the level of maintenance support on the flight line. These latter quantities are a function of how the logistics support system operates and how its performance is measured. Support systems performance can be considered in terms of its cost, efficiency, and utilization.

To introduce logistics systems analysis, a baseline SLAM II model of the aircraft support system presented in Figure 14-1 is developed. The SLAM II model is shown in Figure 14-2. The time for the first entity to be created (TF)

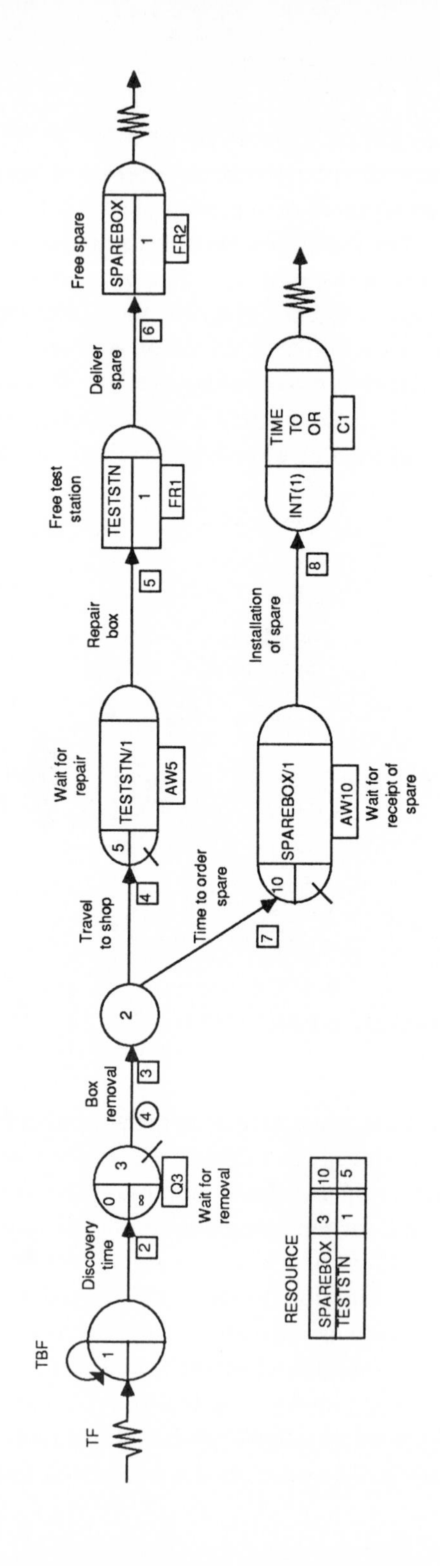

Figure 14-2 Baseline SLAM II model of a logistics system.

in the network represents the time to the first failure. At the CREATE node, the time between equipment failures is specified. The branch from the CREATE node to node Q3 represents the time to discover the box that failed. After discovery, the aircraft waits at QUEUE node Q3 for a specialist to remove the failed box. In this example, the removal of the box, activity 3, may be performed by any of four support personnel. After removal, two activities occur: the box is moved to the repair shop, and the aircraft waits for a spare box. Therefore, a GOON node is used to route two entities. The entity traveling to the shop, activity 4, represents the faulty box. The entity initiating an order for a spare, activity 7, represents the aircraft.

When the box entity arrives at the shop, it waits for a test station resource (TESTSTN) at node AW5. When a test station is available, the box is repaired with a repair time that is associated with activity 5. At FREE node FR1, the test station resource is made available for the repair of other faulty boxes. Activity 6 represents the time required to deliver the repaired box as a spare. The spare, modeled as a resource, is made available for reissue at FREE node FR2, which increases the number of spare resources by 1. The RESOURCE block in Figure 14-2 indicates that three spare boxes are initially available and that they are allocated (issued) at AWAIT node AW10.

An aircraft waits for a spare at node AW10. When one becomes available, the installation of the spare begins, which is modeled by the activity from node AW10 to COLCT node C1 (activity 8). At node C1, the aircraft becomes operational. The time that the aircraft is out of commission is collected at node C1 using INTERVAL statistics, which in this case is the time from when a failure was detected until the time the aircraft is operationally ready (OR). Other performance measures associated with the model in Figure 14-2 include the availability of spares (resource 1), the utilization of the test station (resource 2), the utilization of support personnel to remove failed boxes (activity 3), and the various waiting times associated with the resources and servers of the model.

The model illustrated in Figure 14-2, although simple, represents a baseline model for understanding logistics systems. In the following sections, changes are made to illustrate how other functions associated with an integrated logistics support system may be modeled.

14.3 RELIABILITY

In the baseline model, failure of boxes is represented by one activity, and a combined failure rate for boxes is used. This assumes that a box is a generic piece of equipment for which there is a set of spares. To include the reliability aspects of different subsystems, nodes and activities representing multiple-

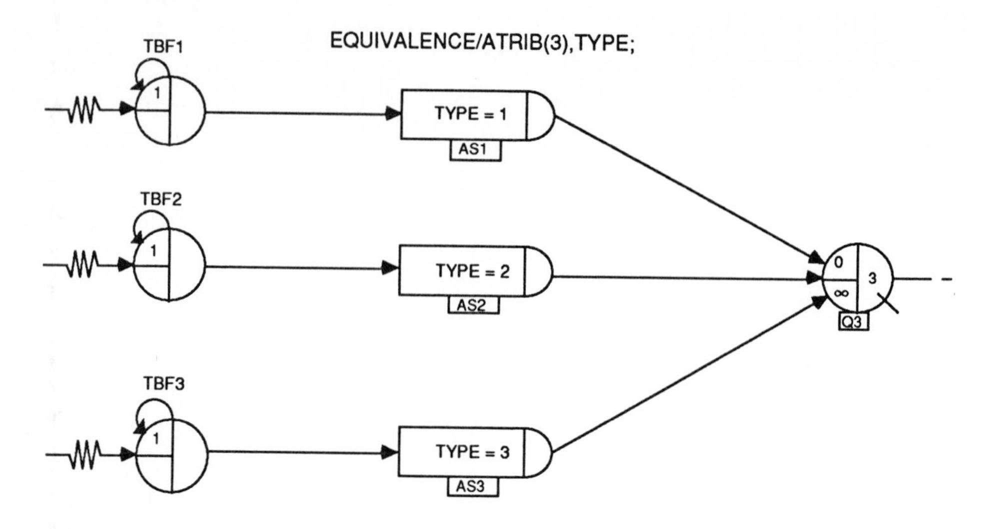

Figure 14-3 Modeling failures for three equipment types.

failure generation and repair are incorporated in the model. Modeling failure generation for three boxes is illustrated in Figure 14-3. This model has three nodes and branches in parallel to model-failure generation for three types of boxes. An attribute, TYPE, assigned at ASSIGN nodes AS1, AS2, and AS3, is used to identify the failed box type. The failed box entity is then routed to node Q3.

For this extended model, a resource block for each type of spare is also required. Although additional nodes, branches, and resources are required to model the reliability of subsystems, the structure of the baseline SLAM II model is unchanged. SLAM II networks that include these reliability concepts have been built at the Air Force Test and Evaluation Center (AFTEC) to study the combined reliability of the B-52 bomber and Air Launch Cruise Missiles (ALCM).

14.4 MAINTAINABILITY

Maintainability relates to the detection, discovery, removal, and replacement operations associated with failures. Each of these activities can be modeled in as much detail as necessary to portray the operation of the system under study. In the baseline model, the activities associated with maintainability are activities 2, 3, 5, and 8. These operations are modeled at an aggregate level in Figure 14-2. The detailed tasks associated with each of these operations may be inserted directly in the network. The level of detail depends both on the use to be made of the model and the extent to which independence may be assumed among maintenance activities. If independence holds, separate analysis for each operation using a detailed SLAM II network of tasks could be made to obtain the distribution of the time to perform an operation. The desirability and feasibility of such a decomposition of the problem is normally contingent on the dependence between the waiting time required and the failure time associated with the aircraft equipment.

Maintenance resources for detection and discovery operations are expanded when a false alarm is issued. False alarms are common when built-in test capabilities are designed into a system. From a modeling point of view, an additional input node needs to be included to represent this situation. Figure 14-4 illustrates inclusion of false alarms into the logistics network model. In Figure 14-4, detection of false alarms is assumed after a box is removed. A more complex model would involve the identification of false alarms after other activities in the network using probabilistic branching. Probabilistic branching can also be used to model incorrect detection and categorization, that is, classifying a failure as a false alarm, and vice versa.

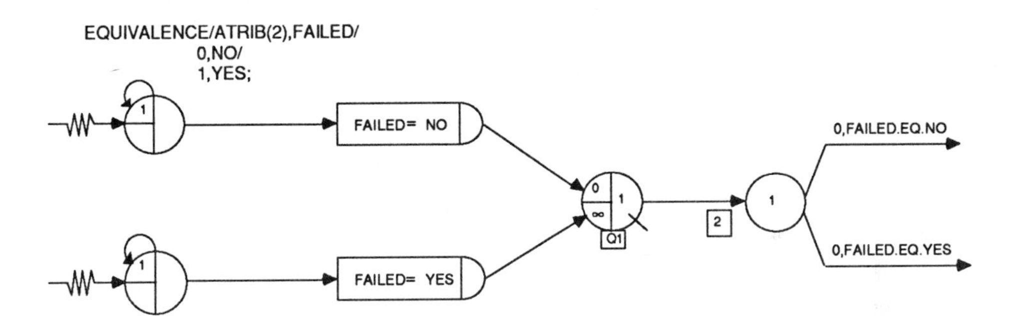

Figure 14-4 Modeling false alarms.

14.5 MAINTENANCE PLANNING

Maintenance planning involves decisions regarding the type of organizational structure for maintenance (in place, shop, or depot repair), the number of personnel associated with maintenance activities, and the inventory of spares. Although maintenance planning relies heavily on production and scheduling procedures, the discussion of modeling of these procedures is delayed until Chapters 16 and 17.

The network segment shown in Figure 14-5 may be used to portray organizational questions associated with maintenance planning. In this figure, fail-

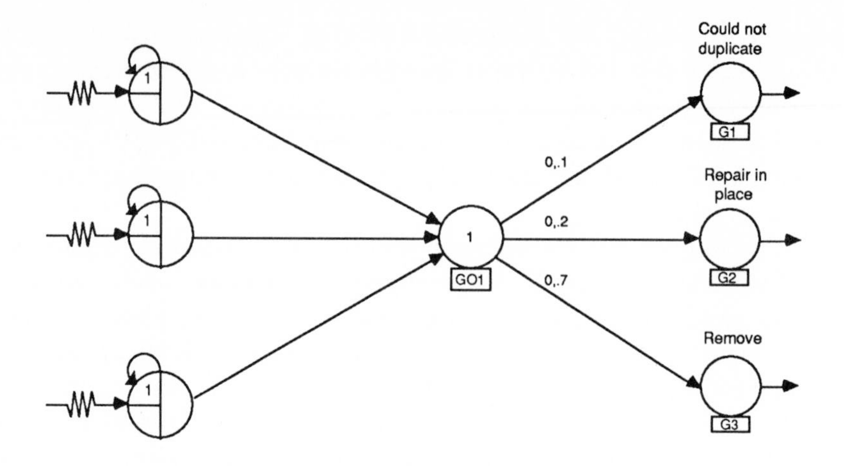

Figure 14-5 Model to investigate maintenance organizational structures.

ures are generated for different boxes. Classification of the type of repair action that is required is based on probabilistic branching from the GOON node GO1. In Figure 14-5, 10% of the failures are classified as CND, that is, could not duplicate. For CNDs, the only maintenance actions required are associated with the discovery activities. Twenty percent of the indicated failures are to be repaired in place, which requires personnel to be available on the flight line, but eliminates the need for supply or transportation operations. In 70% of the indicated failures, a box is removed and the operations described in Figure 14-2 are performed.

By modeling the flow of the entities subsequent to this network classification segment, the effect of personnel availability may be assessed as a function of the probabilities associated with failure classification. For example, the model in Figure 14-6 depicts shop repair actions following the removal and transportation of a box. The activities after removal involve the classification of the failed box with regard to its potential disposition. Probabilistic branching at GOON node GO2 results in the following possible courses of action:

1. The failed unit checks out satisfactorily and the unit is deemed acceptable as a spare without further action (bench check OK).
2. The unit is not good enough to repair and it is condemned. This action reduces the number of spares available in the system or could institute a reorder for more units of this type (see Section 6.6).
3. The unit is classified as NRTS, not repairable this station, and a routing of the unit to another station is made.
4. The unit is routed through the repair operations performed at this station.

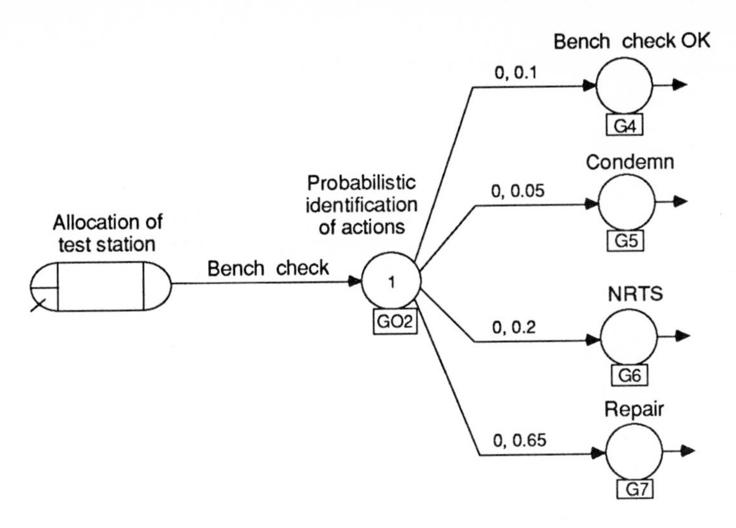

Figure 14-6 Model to assess effects of bench checks.

Each of these outcomes results in a failed box entity being routed to another network segment. For the repair decision, a model of repair operations would show staff loading at the shop level to be evaluated.

The effects of adding or deleting spares to the baseline model (Figure 14-2) can be investigated directly by changing the number of resources available on the RESOURCE block. Another way of capturing the effect of spares is to include an inventory model within the logistics system model. Thus, as items are condemned or classified as NRTS, levels could be checked and additional spares introduced into the system. Delays incurred for ordering, producing, and transporting spares would be modeled as activities.

Inventory or wholesale supply is covered in detail in Chapter 11, and it is an integral part of maintenance planning. Figure 14-7 illustrates the procedures for including a depot inventory control system within the logistics model. The significant feature of this model is the allocation of spare parts at AWAIT node PREQ and the receipt of new parts at FREE node PREC. The reorder point is set at node PORD, and when five failures occur, node PORD is released and an order for five new spares is initiated. Activity 9 represents the time delay to acquire or produce the five spare parts.

Another procedure for providing spares production in the model is to include a production subnetwork in which the operations associated with producing spares is included, that is, a breakout of activity 9. A model of this type has been developed for the Tennessee Valley Authority in which the production of equipment for nuclear facilities is modeled in order to determine the construction time for new nuclear reactor facilities.

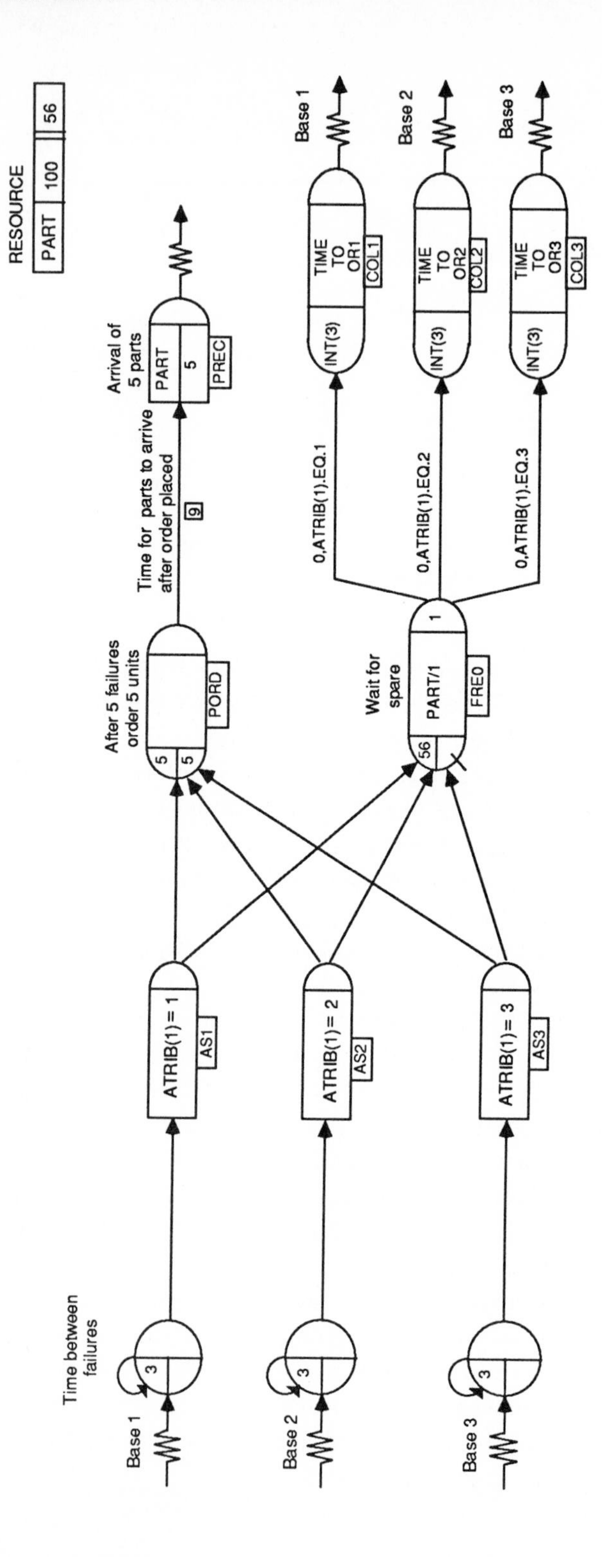

Figure 14-7 Spares reordering subnetwork.

14.6 TRANSPORTATION

A transportation activity associated with logistics support may be modeled in a simple form as an activity with a transportation duration. An activity may represent different types of transportation such as the flying of an aircraft, the transportation of spare parts or failed units, and the movement of maintenance personnel. In some models, it has been found that the movement of maintenance personnel to a work site is significant. For example, the time to transport personnel to a remote site that houses communications equipment or intercontinental ballistic missiles can be larger than the actual repair time. The concepts described in Chapters 18 and 19 on the modeling of materials handling systems are sometimes used for modeling transportation delays.

14.7 PERSONNEL TRAINING

As was discussed in the baseline model, crew size is modeled as the number of parallel servers or as the number of resources available. Separate resources are used to represent different skill levels of people and different types of workers. By requiring multiple resources to do a job, different crew makeups are modeled. Three ways to model multiple-resource requirements are by a series of AWAIT nodes, by a GWAIT node, or by using a SELECT node with the assembly (ASM) selection rule with QUEUE nodes. For complex multiple-resource requirements, ALLOC(I) is specified in place of a resource type at an AWAIT node, and the complex resource allocation procedure is coded in subroutine ALLOC(I,IFLAG).

The training of personnel is included in a model by a network segment that produces new repair personnel by increasing the capacity of the repair personnel resource at an ALTER node. Learning effects are included by making activity durations a function of the current time and the number of repairs completed. An interesting use of SLAM II has been proposed for studying job performance aids. A job performance aid, such as a maintenance manual or directory, adds tasks to a maintenance person's job. However, by performing such tasks as looking up or checking procedures, the actual time to perform the maintenance task is reduced.

14.8 SUPPORT AND TEST EQUIPMENT

Detailed models that incorporate activities relating to support and test equipment have been developed [8, 21]. Specifically, two models of the EF-111A test stations have been completed. These models include activities representing independent failures of line replaceable units (LRU), statistics on spare LRU usage, and failures of the test stations. An advanced model includes the capacity to change the configuration of individual test stations by defining separate test bays as resources. An analysis was performed to determine the impact of different test station configurations at the individual bays. The evaluation was made in terms of both projected spare LRU usage and the time the various resources were in use. A model of the F-16 automated intermediate shop (AIS) included an organizational-level maintenance action breakout. The F-16 AIS model was used to study the effects of scheduling rules associated with station outputs.

14.9 LOGISTICS FACILITIES ANALYSIS

Aerial ports receive cargo from trucks, rail, and aircraft. The cargo is unloaded, inspected, documented, sorted into aircraft loads, packaged, and moved through the terminal. The packaged cargo is then loaded onto aircraft and flown to specified destinations. During this processing, the cargo is frequently set aside in temporary storage areas. Thus, the system may be viewed as a large queueing network where cargo arrives by different methods and is routed by various material handling systems through different storage queues. This system has multiple resource constraints in that special crews, forklifts, conveyors, towline carts, and other equipment are needed for the service operations.

Network models have been used extensively to analyze the logistics involved in Air Force base operations [1, 2, 13, 14]. McNamee and Lee [16] were first to note the need for a network analysis in the design of port facilities. They showed how simulation could be used to determine design requirements for new facilities, as well as to evaluate suggestions for the redesign of existing facilities. Auterio [1] defined the role for network analysis within the Military Airlift Command as a means for managers of the airlift system to measure the productive capacity and effectiveness of aerial port cargo processing. He further identified the need to determine the effect of fluctuating demands for cargo on airlift system performance and posed the following common managerial questions:

1. Is it worthwhile to introduce new material handling equipment at a given port? If so, what are the desirable equipment specifications?
2. How many aircraft can the air terminal handle simultaneously for different demand loads?
3. During contingencies, when demand activity is increased, what is the

maximum performance that can be attained? What kind of additional resources would be necessary to improve performance?

In the next section, a model abstracted from projects performed by Duket, Wortman, and Auterio for the Military Airlift Command addressing these types of questions is presented [1, 9].

14.10 ANALYZING THE LOGISTICS SYSTEM AT DOVER AIR FORCE BASE

Figure 14-8 depicts a schematic of the Dover Air Terminal.† Trucks arrive with cargo to one of three unloading docks. If a dock is not available, a queue of trucks forms. An unloading crew utilizes forklifts to unload truck cargo, sort it by destination, and place the sorted cargo units onto carts. A cart towline system then transports carts of cargo units to one of two storage areas (based on destination), where cargo units are accumulated into pallet loads. In this example, it is assumed that cargo for only two destinations is generated and that segregation of cargo in the storage areas is not required. When an entire pallet load is accumulated, the cargo is palletized and inspected at the palletization station. Inspection may result in minor adjustments and rework for the pallet formation. A material handling machine, called an elevating transfer vehicle (ETV), transports pallets from the palletization stations to an outbound storage

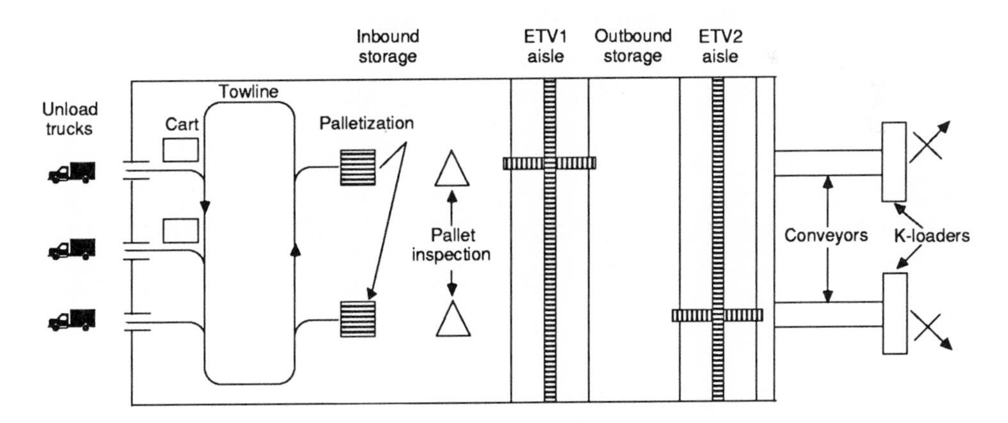

Figure 14-8 Schematic of air terminal.

† In the interest of confidentiality and ease of presentation, hypothetical data and a simplified version of the actual model are employed. In particular, document handling, which is a large part of the logistics system, is not included. Also, the features of logistics analysis previously discussed in this chapter are not included, since examples of network modeling of such operations have been demonstrated.

area. At different points in time, a decision manager, called a load planner, determines if an aircraft load should be formed. When an aircraft load is available, a second ETV transfers the individual pallets making up the aircraft load to one of the two conveyor lines leading to a transport vehicle known as a K-loader. Ramp crews utilize the K-loaders to load an aircraft, which departs the system after it is loaded.

Different types of aircraft arrive at the base and wait for cargo loading by the K-loaders. In this example, only two aircraft types are considered. These types are large and small. Large aircraft are capable of carrying 36 pallet loads from the base to destination 2; small aircraft have a capacity of 18 pallet loads and travel to destination 1.

The actual study considered further complexities for carriers and cargo, which are briefly mentioned here, but are not included in the example. In the actual system, cargo is also received by rail and air. Aircraft are unloaded, and the same resources are utilized to transport the cargo from arriving planes to the inbound storage areas through a depalletization process. Cargo can also be moved out of the system by truck or rail. In addition, cargo types defined as explosive cargo, special handling cargo, and outsized cargo were included in the original study but omitted here. Each of these three cargo types involved the use of special facilities beyond competing for the resources described above. Documentation processing, an important part of the system, is also omitted in this abstracted example. Details of the actual study are included in the literature [9].

The managers of the aerial port were interested in port performance under different cargo flow conditions. Port performance is measured in terms of throughput, equipment and labor utilization, and the time aircraft are delayed on the ground.

By varying cargo input and measuring throughput, the model provides a means for determining the peak port throughput, that is, port capacity. The model also permits the identification of system bottlenecks and suggested alternatives to improve port operations, which would be evaluated by an analysis of a revised model.

14.10.1 The SLAM II Model

Figure 14-9 shows the SLAM II model of the system described above. CREATE nodes SMALL and LARGE along with activities 7 and 8 model the arrival of small and large aircraft, respectively. Attribute 3 of an aircraft entity is marked to indicate its time of arrival, and ASSIGN nodes ASN5 and ASN6 are used to set attribute 2 to the number of pallets that can be carried by each

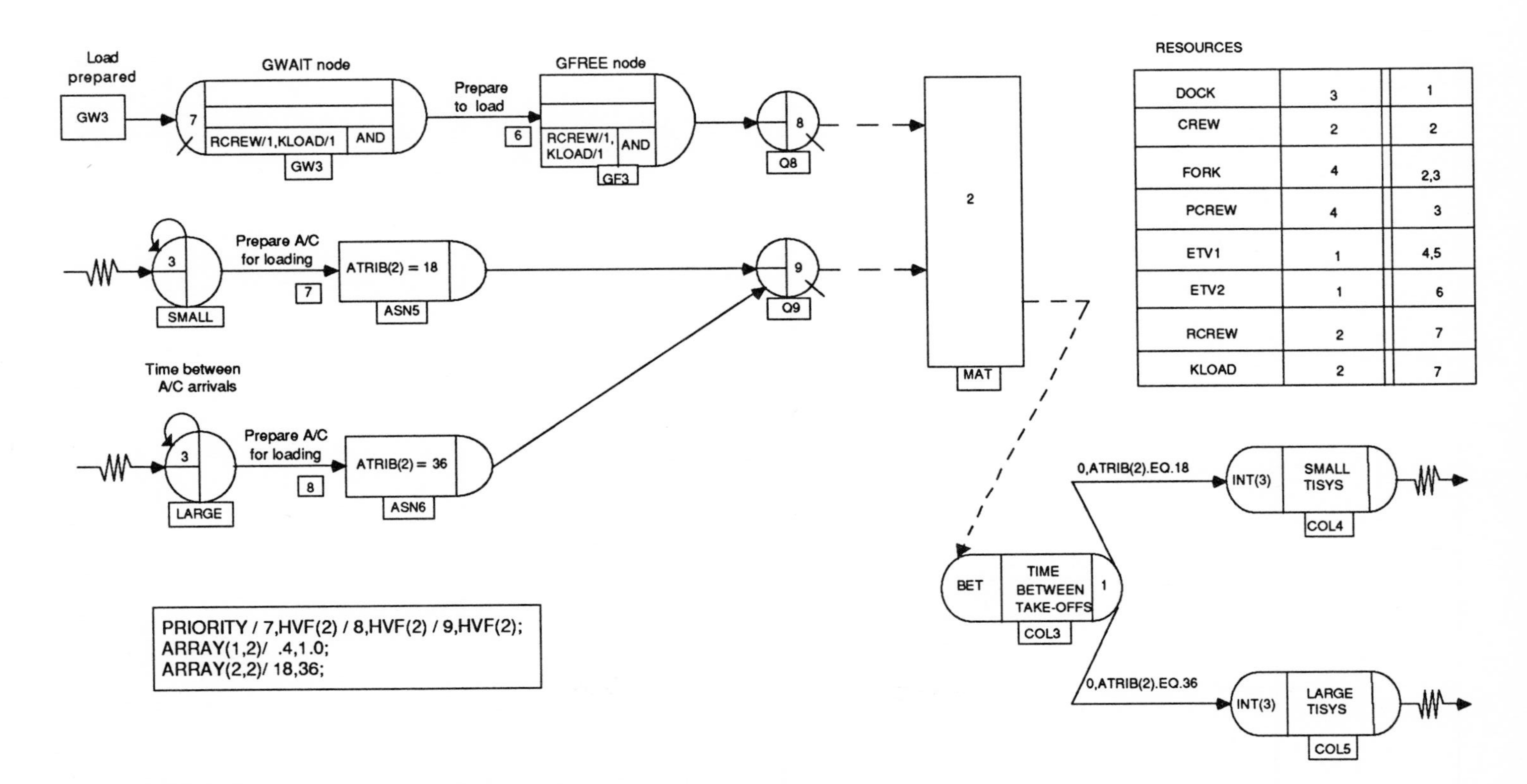

Figure 14-9 SLAM II model of an aircraft terminal.

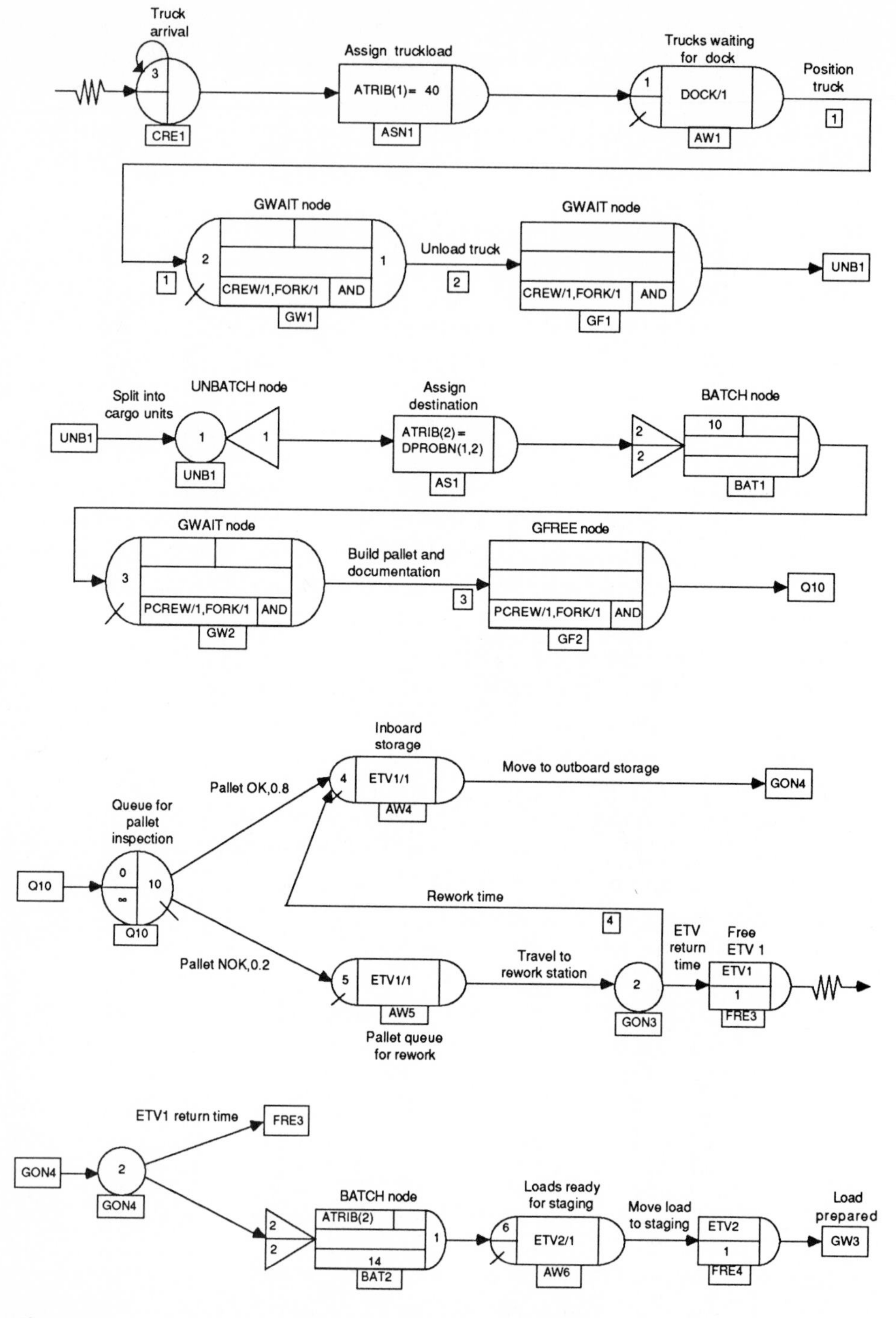

Figure 14-9 SLAM II model of an air terminal (concluded).

aircraft type. Small aircraft can carry 18 pallets and large aircraft can carry 36 pallets. ATRIB(2) is set to 18 or 36 depending on the size of the aircraft. Arriving aircraft are placed in QUEUE node Q9, with large aircraft placed at the head of the file by setting the priority for file 9 to the high value of attribute 2, HVF(2). Aircraft waiting in node Q9 are matched with cargo entities, which are stored in QUEUE node Q8.

The processing of material arriving to the air terminal is modeled starting at node CRE1. ASSIGN node ASN1 sets the value of ATRIB(1) 1 to 40, indicating the number of cargo units on an arriving truck. (Note that this is a simplified representation of the actual model, which differentiated types of trucks and truckloads.) Trucks wait for a dock at node AW1. When a dock is assigned, a time to position the truck is modeled as activity 1. The unloading of a truck does not begin until both an unloading crew and a forklift are available. In Chapter 19, the Material Handling Extension to SLAM II is presented, which includes the generalized await node, GWAIT. In this model, GWAIT nodes are used to require two resources to be available before an entity can proceed in the network. GWAIT node GW1 causes trucks to wait in file 2 for both the crew and forklift. When both the crew CREW and forklift FORK are available, they are allocated to the truck, and activity 2 is initiated which unloads the truck. At GFREE node GF1, described in Chapter 19, both the CREW and FORK resources are freed for their next assignment. This use of GWAIT and GFREE nodes represents the most elementary operations of these nodes to require multiple resources at the same time.

Following unloading, the entity is routed to UNBATCH node UNB1, where ATRIB(1) entities are established to represent the 40 loads that arrived on the truck. For each load, an aircraft type is designated for attribute 2 by assigning a value of 18 or 36 with probability 0.4 and 0.6, respectively, at ASSIGN node AS1. This is accomplished by using the discrete probability function DPROBN (1,2), which references the cumulative probabilities of 0.4 and 1.0 in row 1 of ARRAY with the associated values 18 and 36 in row 2 of ARRAY. Each load entity is then sent to BATCH node BAT1, where batches of loads of size 10 are made. Two batches are established at the node based on the value of attribute 2 for small (18) and large (36) aircraft.

When ten loads are batched, a pallet entity is routed to GWAIT node GW2 to wait for both a pallet crew PCREW and a forklift FORK. When both of these resources are available, a pallet of ten loads is built and documentation is established at activity 3. The pallet crew and the forklift are freed at GFREE node GF2.

Next, the pallet is inspected to determine if it is built in accordance with requirements (pallet OK) for loading onto an aircraft. Pallets waiting for inspection are placed in file 10 at node Q10. Pallets passing inspection are routed to AWAIT node AW4, where an accepted pallet is placed in the inboard storage area awaiting the elevating transfer vehicle (ETV1). A rejected pallet (pallet NOK) is also moved by resource ETV1 to a rework station. After the travel to the rework station, the ETV1 is returned. The pallet is reworked, and it is assumed that all pallets are acceptable after they are reworked. It is further assumed that the rework area is close enough to the inboard storage area so that resource EVT1 can access it directly for movement to the outbound storage area. The rework time is shown on activity 4, which connects GOON node GON3 to AWAIT node AW4.

Following the move of the pallet by resource EVT1, resource EVT1 has a return trip before it is freed at node FRE3. The pallet is sent to BATCH node BAT2, where a batch of 18 pallets for small aircraft is made and a batch of 36 pallets for large aircraft is made. The threshold specification at node BAT2 is ATRIB(2), which is either 18 or 36. In this way, the first load of a batch defines the threshold. Attribute 2 is also used to differentiate between the two batches accumulated at node BAT2. Thus, all pallets with ATRIB(2) equal to 18 are placed in the same batch until the required number of pallets is available to create a planeload. In this model, ATRIB(2) serves as the size of the pallet and the type of pallet. When sufficient pallet entities arrive to node BAT2, a batched entity representing a planeload is routed to AWAIT node AW6, where it waits for resource EVT2 to move it to staging.† After the move to staging, resource EVT2 is freed at node FRE4. The planeload is then routed to GWAIT node GW3, where both a ramp crew, RCREW, and a K-loader, KLOAD, are required to prepare the load for the aircraft. Activity 6 models the load preparation task, and node GF3 releases the ramp crew and K-loader. The prepared aircraft load is then placed in file 8 to await a match with an aircraft that can transport the planeload of pallets created. The matching of the cargo and an aircraft is done at node MAT and is based on attribute 2. When a match occurs, the aircraft entity is routed to COLCT node COL3, where the time between take-offs is collected. The cargo entity is not routed from the MATCH node. COLCT nodes COL4 and COL5 are used to obtain the time in the system for the small aircraft and the large aircraft, respectively.

14.10.2 Model Output and Use

For the actual air terminal studied, extensive statistical summaries were obtained. Table 14-1 lists the complete set of output variables obtained from

† Load planning decisions at a terminal are varied and complex. In one of the applications of this model, a user function was employed to test the type of aircraft currently on the ground. In addition, it is possible to schedule an aircraft to land, given that a load is available. The reader should note that modeling aircraft as resources facilitates adding these embellishments.

the actual application of the model. The footnoted entries signify that the output variable was not included in the abstracted version of the model that has been presented in this chapter. Basically, performance measures related to documentation processing that must accompany any material processing have been left out of the abstracted model.

Table 14-1 Performance Outputs from Air Terminal Models

1. Quantity and frequency of input to the facility
   - 1.1 Truck and aircraft arrivals
   - 1.2 Cargo pieces and shipments received
   - 1.3 Transshipments[a]
   - 1.4 Missions scheduled
   - 1.5 Advance notification documents[a]
2. Quantity and frequency of facility production
   - 2.1 Total inbound aircraft loads (including documentation) handled
   - 2.2 Total outbound aircraft loads (including documentation) handled
   - 2.3 Pallet loads produced
   - 2.4 Pallet positions of unpalletizable cargo[a]
   - 2.5 Reports to local authorities[a]
   - 2.6 Reports to headquarters and airlift authorities
3. Personnel/equipment requirements and busy times
   - 3.1 Receipt card runner[a]
   - 3.2 Load planning runner[a]
   - 3.3 Off loading crews[a]
   - 3.4 On loading crews[a]
   - 3.5 Pallet buildup crews
   - 3.6 Forklifts for 15 locations[a]
   - 3.7 K-loaders/operators
   - 3.8 Customs inspectors[a]
   - 3.9 Documentation manual edits[a]
4. Terminal facility bottleneck indications
   - 4.1 Tally collections
   - 4.2 All cargo transfers
   - 4.3 Machine room preparation of advance receipt documents
   - 4.4 Load planning document preparation
   - 4.5 Ramp loading/unloading
   - 4.6 Truck dock unloading
   - 4.7 Pallet buildup/breakdown stations
   - 4.8 Load pulling and staging operations

[a]Outputs not available from abstracted example.

One of the most important quantities associated with a cargo terminal is referred to as ***port saturation***, which is defined as the total output level for which no additional output is obtained when there is an increase in inputs. Saturation values were obtained from the network model by increasing cargo input rates and observing output rates from successive simulation runs.

Additional runs were made to determine the ability of the port to respond to emergency conditions. Of interest in this study was the amount of output that could be obtained in a 30-day period following the initiation of a decision to use the port at saturation levels. For this study, different initial conditions associated with inputs, pallet loads, and aircraft availability were specified, and the output as a function of increased input levels to the port was investigated.

Resource studies were also made using the port model. At one of the terminals, it was determined that the primary constraint involved documentation processing prior to load-planning activities. When these constraints were removed in the model, saturation capacity increased by an average of eight aircraft loads per day. A secondary constraint then became active, which involved the distribution procedures for forklift trucks. The removal of the forklift constraint resulted in an additional one and a half aircraft loads being processed per day [1].

14.11 SUMMARY

A framework for studying logistics systems problems using network models has been presented in this chapter. In particular, the concepts of reliability, maintainability, maintenance planning, supply, transportation, personnel and training, and support and test equipment are discussed in terms of SLAM II modeling capabilities. An example showing how SLAM II can be used for logistics facilities analysis is presented, which illustrates the activities and logic involved in moving material through Dover Air Force Base.

14.12 EXERCISES

14-1. For the baseline model of a logistics system, develop qualitative relationships between the following performance measures and system elements:

(a) Time to replenish a spare and the time to order a spare.

(b) Time to replenish a spare and number of support personnel available to remove a box.

(c) Number of aircraft waiting for repair and the time to install a spare.

(d) Time an aircraft is out of commission to the number of support personnel available.

14-2. In the modeling of false alarms shown in Figure 14-4, embellish the network to model the misclassification of failures and false alarms.

14-3. Build a SLAM II model to represent a system consisting of three parallel units that fails only when all three units fail.

14-4. Combine the various submodels presented in this chapter into an integrated model of a logistics system. Discuss the impacts of each submodel on total system performance and how a study could be performed to identify the critical subsystems in a total logistics system.

14-5. Discuss how regression analysis could be used to build a model that relates the SLAM II output statistics to SLAM II input values. Discuss the advantages and disadvantages of having a regression equation that models the SLAM II model.

14-6. Cargo arrives at an air terminal in unit loads at the rate of two unit loads per minute. At the freight terminal there is no fixed schedule, and planes take off as soon as they can be loaded to capacity. Two types of planes are available for transporting cargo. There are three planes with a capacity of 80-unit loads and two planes that have a capacity of 140-unit loads. The round-trip time for any plane is normally distributed with a mean of 3 hours, a standard deviation of 1 hour, and a minimum and maximum of 2 and 4 hours, respectively The loading policy of the terminal manager is to employ smaller planes whenever possible. Only when 140-unit loads are available will a plane of type 2 be employed. Develop a SLAM II network to model this system to estimate the waiting time of unit loads and the utilization of the two types of planes over a 100-hour period.

Embellishments

(a) Model failures in aircraft that occur with probability 0.1 after each flight. Repair time is 1.5 hours if a spare is available and 8 hours if no spare is available. Assume that spares are available 90% of the time.

(b) Modify the model for embellishment (a) by incorporting an inventory model for spare parts that keeps only one spare on line and that orders a spare everytime one is used. The time to obtain a spare is 32 hours. Increase the simulation period for this model to 320 hours and evaluate system performance.

14.13 REFERENCES

1. Auterio, V. J., "A-GERT Simulation of Air Terminal Cargo Facilities," *Proceedings, Pittsburgh Modeling and Simulation Conference*, Vol. 5, 1974, pp. 1181-1186.
2. Auterio, V. J., and S. D. Draper, "Aerial Refueling Military Airlift Forces: An Economic Analysis Based on Q-GERT Simulation," Material Airlift Command, *Chicago ORSA Conference*, 1974.
3. Blanchard, B. S., *Logistics Engineering and Management*, Prentice-Hall, Englewood Cliffs, NJ, 1974.
4. Blanchard, B. S., and W. J. Fabrycky, *Systems Engineering and Analysis*, Prentice-Hall, Englewood Cliffs, NJ, 1981.
5. Bowersox, D. J., *Logistical Management*, Macmillan, New York, 1974.

6. Castillo, D., and J. K. Cochran, "A Microcomputer Approach for Simulating Truck Haulage Systems in Open Pit Mining," *Computers in Industry*, Vol. 5, No. 1, 1987, pp. 37-47.
7. Doring, B., and A. Knauper, "A Simulation Study with a Combined Network and Production Systems Model of Pilot Behavior on an ILS-Approach," *Automatica*, Vol. 19, 1983, pp. 741-748.
8. Drezner, S. H., and A. A. B. Pritsker, "Network Analysis of Countdown," RAND Corporation, RM-4976-NASA, March 1966.
9. Duket, S., and D. B. Wortman, "Q-GERT Model of the Dover Air Force Base Port Cargo Facilities," MACRO Task Force, Material Airlift Command, Scott Air Force Base, IL, 1976.
10. Eisner, H., *Computer-Aided Systems Engineering*, Prentice-Hall, Englewood Cliffs, NJ, 1987.
11. Erdbruegger, D. D., W. G. Parmelee, and D. W. Starks, "SLAM II Model of the Rose Bowl Staffing Plans," *Proceedings, 1982 Winter Simulation Conference*, 1982, pp. 127-135.
12. Hoffman, S. E., M. M. Crawford, and J. R. Wilson, "An Integrated Model of Drilling Vessel Operations," *Proceedings, 1983 Winter Simulation Conference*, 1983, pp. 45-53.
13. Lee, C., and L. P. McNamee, "A Stochastic Network Model for Air Cargo Terminals," *Ninth Annual Allerton Conference on Circuit and Systems Theory*, 1971, pp. 1140-1150.
14. Maggard, M. J., W. G. Lesso, and others, "GERTS IIIQR: A Multiple Resource Constrained Network Simulation Model," *Management Datamatics*, Vol. 5, No. 1, 1976, pp. 5-14.
15. McCallum, J. N., and B. B. Nickey, "Simulation Models for Logistics Managers," *Logistics Spectrum*, Vol. 18, No. 4, Winter 1984, p.10.
16. McNamee, L. P., and C. Lee, "Development of a Standard Data Base and Computer Simulation Model for an Air Cargo Terminal," U.S. Army, CERL, Champaign, IL, 1973.
17. Mortenson, R. E., "R, M, and Logistics Simulations Using Q-GERT," *Proceedings, 1980 Annual Reliability and Maintainability Symposium*, pp. 1-5.
18. Ratcliffe, L. L., B. Vinod, and F. T. Sparrow, "Optimal Prepositioning of Empty Freight Cars", *Simulation*, June 1984, pp. 269-275.
19. Standridge, C. R., and J. R. Phillips, "Using SLAM and SDL to Assess Space Shuttle Experiments", *Simulation*, July 1983, pp. 25-35.
20. U.S.A.F. R&M 2000 Process, Office of the Special Assistant for Reliability and Maintainability, Department of Defense, October 1987.
21. Watters, L. J., and M. J. Vasilik, "A Stochastic Network Approach to Test and Checkout," *Proceedings, Fourth Conference of Application of Simulation*, 1970, pp. 113-123.
22. Yu, J., W. E. Wilhelm, and S. A. Akhand, "GASP Simulation of Terminal Air Traffic," *Transportation Engineering Journal*, Vol. 100, 1974, pp. 593-609.

15 PRODUCTION SYSTEMS

15.1 INTRODUCTION

A ***production system*** is an organized collection of personnel, materials, and machines that operates within a well-defined work space. Inputs to the system include raw materials and orders for manufactured goods, and outputs are manufactured items. The diagram in Figure 15-1 depicts a production system. Work space, labor, machines, and materials are referred to as system resources. As will be discussed in Chapters 18 and 19, material handling equipment is often considered separately from other machines. Scheduling procedures regulate the flow of the process and provide managers with a means to modify the system output.

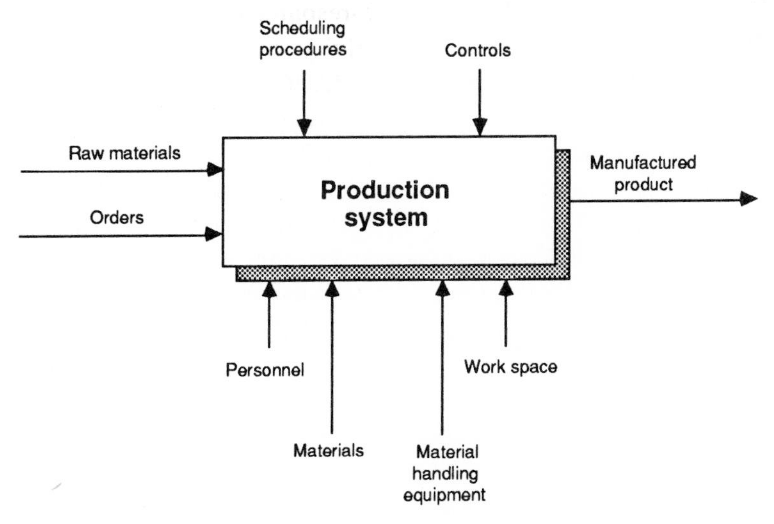

Figure 15-1 Diagram of a production system.

In this chapter, system performance measures for production systems are discussed, and procedures for estimating these performance measures using SLAM II networks are described. Examples of the use of SLAM II concepts for modeling production systems are presented with an emphasis on the close correspondence between elements of the model and components of the production system.

15.1.1 Introductory Concepts

To introduce the use of SLAM II concepts in production planning, a model of a manufacturing process is illustrated. The system produces a finished product that is processed by one machine using a single piece of raw material. The raw material, which is assumed to be an unfinished part, is delivered to the machine area. An operation is performed by the machine and the finished part leaves the machine area. The challenge posed by this simple system, as with more complex production systems, is to meet the demands of production in a timely fashion with efficient use of space and machines.

To model a production system, interest is focused on the following three aspects of the system:

1. Arrival of unfinished parts to the machine area
2. Buildup of unfinished parts awaiting the service operation
3. Service operation

The process can be viewed as a single-resource queueing system, where the parts to be processed arrive to a service activity, that is, the machining operation. The buildup of parts awaiting service is modeled as a queue. A pictorial diagram of this one-machine work center is given in Figure 15-2, and a SLAM II network model of this system is shown in Figure 15-3.

The service operation is a machining activity. If the service activity is ongoing, the machine is busy and arriving parts must wait at QUEUE node QUE. When a part is machined, another part, if waiting, is removed from the queue and service on it begins. If a part arrives when the machine is busy, it waits by joining the queue. The arrival of parts at the work center is modeled by CREATE node ARI.

When service is completed, the part exits the system through COLCT node DPART. This node collects observations on the interval of time that the part has been in the system. This time interval for a part in the network represents the product flowtime. At the end of a simulation run, the number of parts processed, the average flowtime, and the standard deviation of the flowtime for all

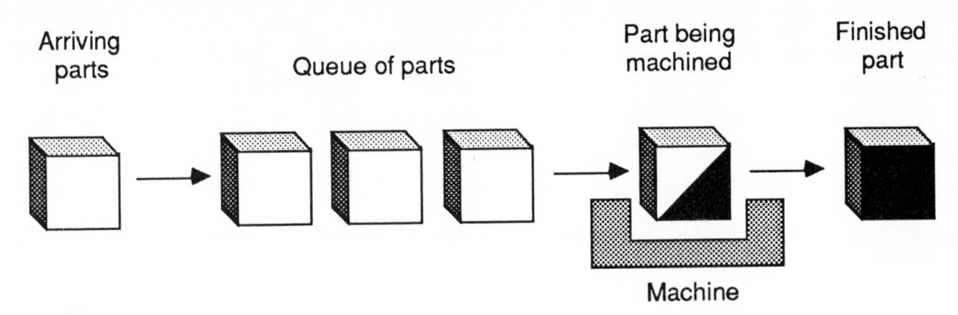

Figure 15-2 Diagram of a one-machine work center.

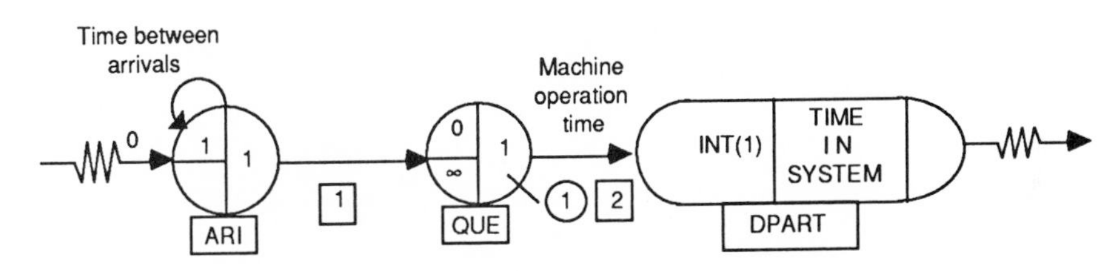

Figure 15-3 SLAM II model of a one-machine work center.

parts are reported as a part of the simulation output. Other outputs from the model are the average time the machine is busy, the average time a part waits for the machine, and the average number of parts in the queue.

More complexity is now introduced into the machining example to illustrate the ease of changing a SLAM II model to reflect a change in system structure or operating policy. Suppose that there are three identical machines. The number of parallel servers is then specified as 3. Often machines are not identical. For example, each machine may have a different operation speed, and one of the available machines must be selected. In this case, three separate branches are used, one to represent each machine, and a SELECT node is introduced in front of the servers. With multiple machines, multiple queues may form. The SELECT node provides the capability to choose between sets of parallel queues and servers. To illustrate, consider the situation depicted in Figure 15-4. Here there are two arrival processes, two queues, and three servers. There are two types of decisions to be made in this situation. First, when a machine finishes processing a part, should the next part come from queue 1 or queue 2? Second, when a part arrives and more that one machine is available, which machine should be selected to process the part? Such operational procedures can affect the total performance of the system. A SLAM II model of this manufacturing situation is shown in Figure 15-5.

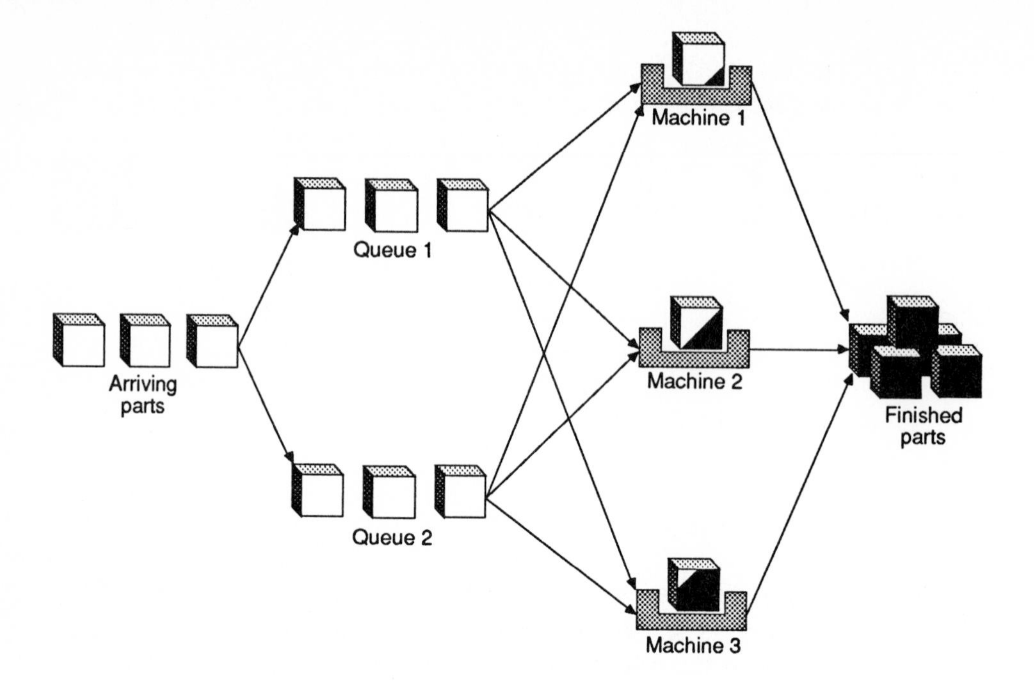

Figure 15-4 Diagram of a multiple-machine work center.

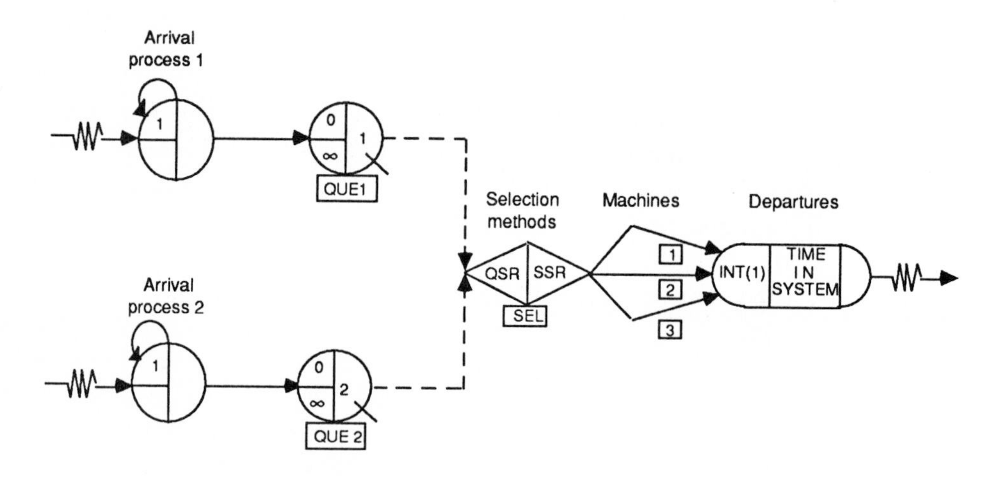

Figure 15-5 SLAM II model of a multiple-machine work center.

CREATE nodes model the arrival processes. QUEUE nodes QUE1 and QUE2 represent the queues. Service activities 1, 2, and 3 represent the three machines, and the time specifications (not shown) for each branch represent the operation characteristics of each machine.

Queue and server selection procedures are modeled through the specification of selection rules for the SELECT node SEL. If a part arrives at either queue when one or more machines are available, the SELECT node schedules the part for service by selecting one of the available machines according to the specified SSR rule. When a machine completes service, the SELECT node takes a part from one of the queues in accordance with the QSR rule and routes the part to the machine. All parts leave the system at the COLCT node. The SLAM II processor, by simulating the network, computes waiting times, time in the system, server utilization, and queue length statistics.

SLAM II resource concepts provide additional constructs for modeling service operations. With resources, complicated starting conditions for service operations may be specified. Resource and service activity models are similar. The definition of service in the resource model is easily expanded to include a set of branches between the AWAIT and FREE nodes. In addition, a SERVER resource may be augmented or removed during the operation of the system using ALTER and PREEMPT nodes. The service activity orientation allows complex queue selection and server selection procedures to be modeled through the use of the SELECT node.

SLAM II symbols serve as powerful building blocks that may be assembled to model systems that contain many structural and procedural complexities. Additional examples of network features for modeling production systems are presented later in this chapter and in the chapters that follow. First, performance measures in production planning are discussed and then related to SLAM II network outputs.

15.1.2 System Performance Measures

The performance of a production system is measured by its effectiveness in converting raw material into finished product. Performance measures can be grouped into four categories:

1. Measures of throughput
2. Measures of ability to meet deadlines
3. Measures of resource utilization
4. Measures of in-process inventory

Throughput is the output produced in a given period of time. Another name for this measure is the ***production rate***. However, the term rate is often avoided, as it implies continuous output, which is not always the case. System capacity is often defined as the maximum throughput that can be obtained.

The ability to meet deadlines is measured by product lateness, tardiness, or flowtime. ***Lateness*** is the time between when a job is completed and when it was scheduled to be completed. ***Tardiness*** is the lateness of a job only if it fails to meet its due date; otherwise, it is zero. ***Flowtime*** is the amount of time a job spends in the system. In some cases, the total time it takes to complete all jobs is of importance. This time is referred to as the ***makespan***. All these measures are indications of the effectiveness and efficiency of the system, including its scheduling procedures, in satisfying customer orders.

System resources include personnel, machines, materials, and work space. The utilization of these resources, as measured by the fraction of time they are productive, is another measure of system effectiveness. Measures of resource utilization relate to the degree to which a system is operating at capacity.

The final category of performance measures is concerned with the buildup of raw materials and unfinished parts during production. The buildup of unfinished parts is called in-process inventory or work in process (WIP). In-process inventory is usually due to parts waiting for available resources. Since inventory requires storage space and ties up capital, it is of great importance in production planning.

This discussion has dealt with system performance measures without reference to system objectives. The purpose has been to focus on the evaluation of the performance of systems in light of specified objectives that are determined by management for a particular situation. Objectives are satisfied when system performance measures reach prescribed levels. The measures outlined provide the basis for evaluating diverse objectives. By concentrating on performance measures, we bypass the question of identifying the objectives of production planning, which are, by necessity, situation dependent. For example, it has been frequently noted that the objectives of a production system are highly dependent on production volume.

In many situations, objectives are established in terms of either cost effectiveness or system profitability. The measures of performance described provide input values for computing cost effectiveness and system profitability. The approach presented in this book is to present methods for obtaining values to insert into profit and cost equations.

In the system diagram of Figure 15-1, scheduling procedures are viewed as a system input. It is a special input in that the procedures are under the designer's and scheduler's control. By varying scheduling procedures, the performance of the system may be improved at little or no cost. System performance is also affected by changes in the structure of the system. For example, the purchase of new machines, the expansion of the work space, or an adjustment in the labor force changes the system and modifies system performance. Production planning involves the design and evaluation of new facilities, machinery, and operating procedures to meet future demand.

15.2 SLAM II ESTIMATION OF PERFORMANCE MEASURES IN PRODUCTION PLANNING

In this section, methods are presented to generate outputs from SLAM II models to estimate throughput, lateness in meeting deadlines, resource utilization, and in-process inventory.

15.2.1 Throughput

Throughput is the number of units produced during a time interval. It is an important performance measure for determining whether a system can meet expected demand. To estimate throughput, information is needed on the number of entities that became finished products and the time span of the simulation. SLAM II maintains the variable CCNUM(ICLCT) as the number of entities that pass through COLCT node number ICLCT. The current simulation time in SLAM II is maintained in the variable TNOW. Therefore, at the end of a simulation run, the throughput for COLCT node number ICLCT is estimated by TPUT = CCNUM(ICLCT)/TNOW. If the COLCT node is placed to observe all completed parts, the variable TPUT represents system throughput.

The value for TPUT for each simulation run is an observation of a random variable. Its computation would be made in the user-written SUBROUTINE OTPUT, which is called at the end of each simulation run. Consequently, statistics may be computed over multiple network runs for throughput with the statement CALL COLCT(TPUT,IC), where IC is a user-defined index number specified on a STAT statement to collect the throughput values over multiple runs. By specifying that STAT variable IC not be cleared at the end of each run, SLAM II automatically computes statistics on TPUT over the number of runs requested. A histogram of system throughput values over multiple runs aids in estimating the percentage of time that throughput is within a specified range.

Sometimes it is of interest to measure the time between departures from the system, which is the reciprocal of system throughput. To do this, BETWEEN statistics are collected at a COLCT node. The time between departure is useful if the output of the production system serves as the input to another system.

In the case where an entity represents a batch of items to be processed, the size or volume of the batch is modeled as an attribute of the entity, for example, ATRIB(2). The number of units produced in this case is not the number of entities passing through a COLCT node, but the sum of the ATRIB(2) values that represent a batch size. To calculate the system's throughput, a running total of units processed is maintained as a SLAM II variable. An ASSIGN node, placed before the COLCT node, may be used to make the computation. For example, the relationship XX(1) = XX(1) + ATRIB(2) could be computed at an ASSIGN node, where XX(1) is the total number of parts produced. Then, in subroutine OTPUT, the unit throughput is calculated by stipulating that UNITP = XX(1)/TNOW. This value is then used in the same manner as TPUT in the above paragraphs. Alternatively, a BATCH node may be used to sum an attribute value for all arriving entities, or an UNBATCH node may be used to create individual parts from the entity representing a batch of parts.

15.2.2 Meeting Due Dates

Production managers strive to satisfy customer demand. One measure of a production system's performance is the length of time a part spends in the system and is referred to as flowtime. Interval statistics are used to compute flowtime.

Time-in-subsystem statistics and frequency histograms may be obtained for any segment of a network by marking an attribute at an ASSIGN node and collecting INTERVAL statistics at a COLCT node placed in the network.

Often a job is assigned a specified due date. The lateness of a job is the time difference between when the job is finished and when it was due. Let's assume that the due date for a job is ten time units from the time it arrives. In this model, a constant two time unit delay is required for the job to move to the service area, that is, from the CREATE node to the QUEUE node. The machine service time follows a normal distribution, with a mean of four time units and a standard deviation of one time unit. The network model is shown in Figure 15-6.

Each job is assigned a DUEDATE equal to its creation time plus 10 at an ASSIGN node. DUEDATE is made equivalent to ATRIB(2) so that the DUEDATE is a characteristic of the entity as it travels through the network. A time delay of 2 is prescribed for activity 1, which is a travel time to the machine. Service activity 2 has a machining time that, for each entity, is a sample from a normal distribution having a mean of 4 and a standard deviation of 1. The lateness status of a completed job is determined by comparing its DUEDATE with TNOW. If TNOW is larger, the job is tardy. If TNOW is smaller, the job is early. Statistics are then collected on early and tardy jobs at COLCT nodes EARLY and TARDY.

Attribute 1 is used in this example to mark the time that a job entered the

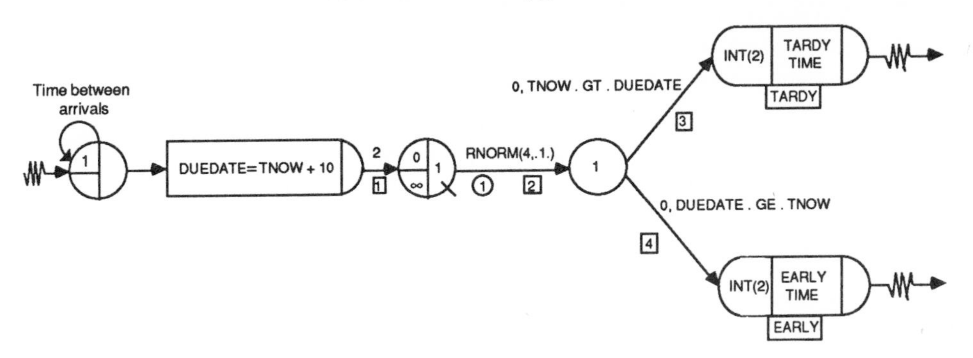

Figure 15-6 SLAM II model of jobs with due dates.

model at the CREATE node. Therefore, all the statistics relating to throughput that were presented previously in this section may also be computed. In the ASSIGN node where DUEDATE is established, a user function could be employed, for example, DUEDATE = USERF(1). In this case, the due date can be made a function of the current status of the network elements, such as queue length, NNQ(IFILE), and the number of entities being processed by a server, NNACT(NACT). Consequently, a wide variety of due date specification policies may be modeled and evaluated.

15.2.3 Makespan

Makespan is the time to complete a specified number of jobs. Suppose that makespan statistics are desired for the first 100 jobs processed by the single-server system and for every 100 jobs after the first 100 jobs. These statistics may be collected by adding three nodes at the end of a network as shown in Figure 15-7. The first added node ACUM is an ACCUMULATE node. Its purpose is to combine entities and, in this model, 100 entities must reach node ACUM before it releases two entities (M=2): one entity to COLCT node MAKE and one to COLCT node AMAKE. FIRST statistics obtained at COLCT node MAKE provide the time to complete the first 100 jobs, that is, the makespan for 100 jobs. The average, standard deviation, minimum, and maximum for makespan is computed over the remainder of the jobs by collecting BETWEEN statistics at COLCT node AMAKE. A histogram at node AMAKE would characterize the makespan density function. When collecting statistics by specifying BET at a COLCT node, the first arriving entity is used as a reference point. Therefore, by adjusting the first release requirement at node ACUM, a different number of jobs can be eliminated from the average makespan computations.

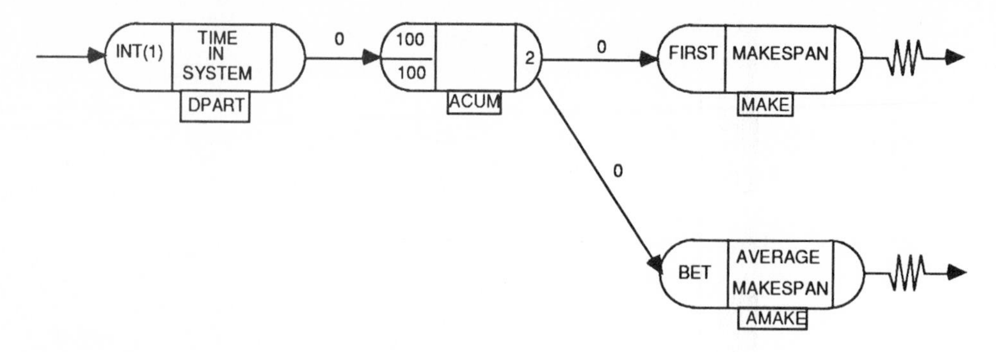

Figure 15-7 SLAM II model to compute makespan.

15.2.4 Server and Resource Utilization

Service activities and resources restrict entity flow through a network. When a service activity completes a job, it attempts to reschedule its server based on entities waiting in QUEUE nodes preceding the activity. A resource, on the other hand, can be allocated at more than one node in the network. Therefore, in manufacturing terms, service activities are passive and resources are active. Resources are allocated to entities waiting in AWAIT and PREEMPT nodes in the order prescribed on a RESOURCE block. Also defined at the RESOURCE block is the initial capacity of the resource type. The utilization of these physical resources is a measure of system performance. Observations on the utilization of a system's resources are collected during each simulation run of a network. The SLAM II output processor then automatically generates statistical reports on utilization for all service activities and resources. The resource statistics output report provides information on both the utilization and availability of SLAM II resources relative to the maximum available units for each resource type. Statistics are compiled on the number of resources in use (allocated to entities) and available (not allocated to entities).

15.2.5 In-process Inventory Requirements

As parts move through a production system, they queue up at different points in the system. This buildup may occur in machine areas, in temporary storage areas, and in buffers for material handling equipment. The aggregate amount of raw material and unfinished parts in the system is known as in-process inventory. It is costly to have goods in inventory because inventory represents a nonproductive use of capital. Therefore, the production manager is continually faced with the trade-off of maintaining proper inventory levels to assure sufficient units in the system to maintain production and the need to keep inventory cost under control.

Another problem associated with in-process inventory relates to storage space, which is usually limited. This problem raises the following questions:

1. How much space will be needed for in-process inventory if the production rate is increased, existing facilities are expanded, or a new facility is designed?
2. How do changes in operating procedures affect in-process inventories?

The importance of measuring and controlling in-process inventories relates to the requirements for capital, space, and lost production due to blocking and the nonavailability of parts.

In a SLAM II model, the buildup of parts may occur at QUEUE, AWAIT, ACCUMULATE, BATCH, and PREEMPT nodes. For those nodes with files, the average number of entities waiting is obtained by averaging over time the number of entities that are in the file. Figure 15-8 illustrates the computation of this time-integrated average.

For each run of a network, the average number, standard deviation, and maximum number of entities in the file are printed along with the current number in the file at the end of the run. The average waiting time of entities in the file is also printed on the summary output report. In addition, statistics relating to the file for the current events calendar are reported.

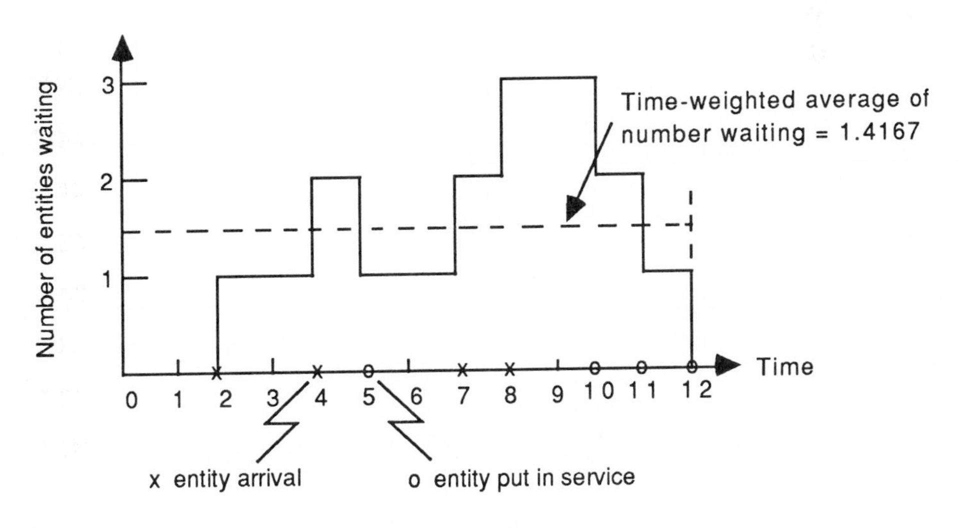

Figure 15-8 Number of entities in a file and calculation of average number in the file.

Estimates of queue length statistics are an invaluable aid in gauging the storage space requirements of a proposed system or in the evaluation of proposed changes to an existing system. Queue lengths are indicative of the space requirements for each storage area and, when taken collectively, are a measure of the total in-process inventory for the system. Furthermore, knowledge of where excessive queue lengths might occur in a proposed system assists in the prevention of bottlenecks. It is important to note that queue length is a random variable, and variability in space requirements must be considered in a production plan. Therefore, estimates of average queue length, the standard deviation, and the maximum number in a queue are necessary for proper planning.

Storage areas often consist of a finite space. In SLAM II, this space limitation corresponds to a queue's maximum capacity (QC). Three events could occur when an entity arrives at a QUEUE or an AWAIT node that is at capacity. First, if an item arrives from a service activity and blocking has been specified, the entity waits until it can enter the node; second, the item can balk to another node; finally, the item, having no where to go, is terminated from the network. In this last instance, SLAM II provides a message that the item has been destroyed. In most production systems, items that cannot be placed into the in-process inventory queue are stored in another area. Balking is the technique that SLAM II employs to model this situation. Through balking, an item that cannot be placed into a queue because of storage limitations is routed to another node. The network segment following this node then determines the disposition of the balked item. Balking is a consequence of space limitations and is, therefore, directly related to in-process inventory requirements. The number of balkers from a queue may be collected by routing a balked item through a COLCT node. The type of statistics to be collected are specified by the analyst.

Some of these performance measures are related. For example, production managers know that, all else being equal, in-process inventory increases if the time in the system for a job increases. This relationship is known as Little's formula [29, 33], which specifies that

$$N = R * T$$

where N is the average number in the system, R is the average rate at which items arrive, and T is the average time in the system.† It can be shown that if R is held constant an increase in T will increase N. Another relationship is that the expected number of departures equals the expected number of arrivals in a stationary system. Relationships among system variables are important, but

† The equation assumes that the time in the system for an item not completely processed is the ending time of a run minus the item's arrival time.

they are often obscure and system specific. One of the responsibilities of the analyst is to ascertain the significant relationships between system inputs and outputs. SLAM II aids in the discovery of these relationships.

15.3 SPECIFYING PRODUCTION TIMES

The time duration of a production activity is modeled like any other activity duration in SLAM II. Because of the importance of production activities, a discussion of commonly used methods for defining production times is included here. A sample value may be drawn from any of the SLAM II distributions to represent the production time for an entity. A constant service time may be specified or read in as an input. Furthermore, it is possible to look up a value in a predefined table, specify a time through a user function, or establish the production time as a function of the current status of SLAM II variables. SLAM II provides extensive flexibility to model production times.

Probabilistic branching is also used in specifying a production time. For example, if 30% of the parts to be processed require ten time units on the machine, 45% of the parts require fifteen time units, and 25% of the parts would need twenty time units. This situation may be modeled as in Figure 15-9. The use of probabilistic branching to model service activities is restricted to a set of parallel, identical servers that follow a QUEUE node. The routing of entities to different nodes is, however, allowed. A generalization of the network segment of Figure 15-9 can be made by employing the function DPROBN to model a large number of class frequencies.

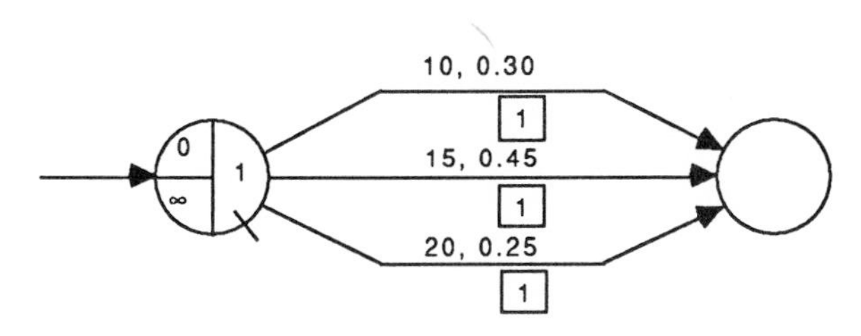

Figure 15-9 Probabilistic service times.

Frequently, a machine can perform different operations on parts being processed. The operation to be performed depends on the characteristics of the entity entering service. The service time information for this situation is usually modeled by assigning values to a set of attributes associated with the entity. Consider the case where two job types enter a machine area. Job type 1 requires an operation with a service time that is exponentially distributed with a mean of two time units. Job type 2 requires an operation that has a normally distributed service time with a mean of 5 and a standard deviation of 2. This situation may be modeled as in Figure 15-10.

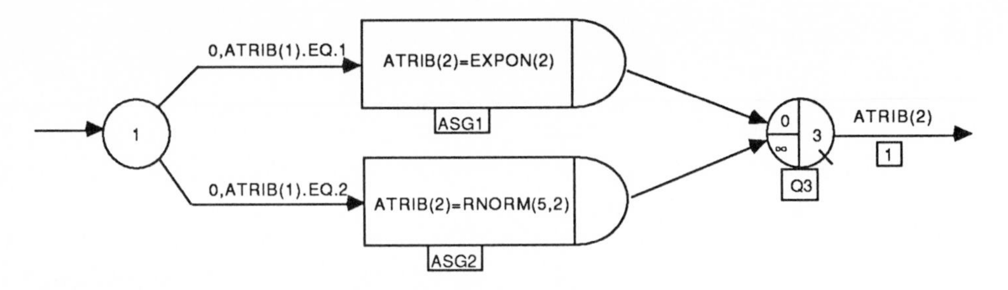

Figure 15-10 Model for multiple job types.

Attribute 2 is assigned the desired machine processing time. Assume attribute 1 designates the type of job that is to be processed. After identifying the type of job, attribute 2 is set to the machine processing time, which is made the duration for activity 1. If the values assigned to attribute 2 are considered as estimates of the production time, and the actual production time is plus or minus one time unit, then the error of the estimate may be added to the estimate directly on the service activity as shown in Figure 15-11.

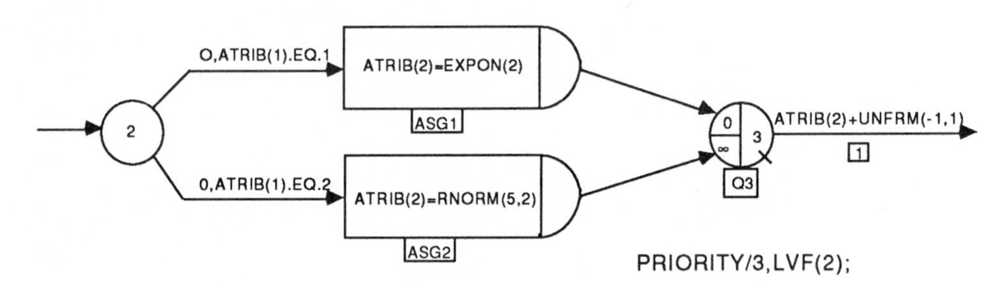

Figure 15-11 Model of job sequencing based on shortest processing time.

In Figure 15-11, the statement

PRIORITY/3,LVF(2);

gives a priority to entities in the queue (file 3) with the shortest estimated processing time (the value of attribute 2). An actual processing time is established when an item is removed from the queue and placed into service. The actual processing time is computed as the estimated processing time, ATRIB(2), plus an estimation error that is uniformly distributed, with a lower limit of -1 and an upper limit of +1 time units, that is, UNFRM(-1,1).

15.3.1 Modeling Setup Operations

If service involves both a setup phase and a processing phase, then the activity duration is specified as the sum of the setup time and the processing time. When prescribing a service activity duration in this form, it is presumed that the server is busy during both setup and processing. Furthermore, it is assumed that the next setup cannot be performed until the processing of the current job is completed. If a setup may be performed while processing is ongoing, the setup and processing operation is modeled as a sequence of service activities. If only one setup may be performed during processing, then a QUEUE node is used to block the setup and processing operations, as shown in Figure 15-12.

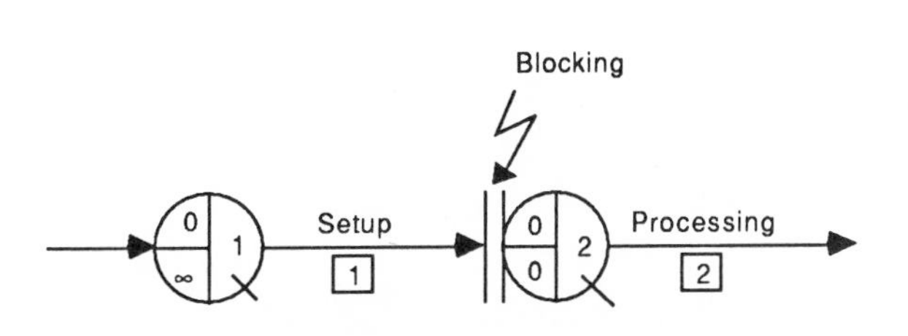

Figure 15-12 Blocking model

If it is desired to model the setup and processing operations as separate activities, but to prohibit the start of another setup operation until processing is completed, then the resource concepts shown in Figure 15-13 are employed. In this network segment, jobs wait in AWAIT node AW1 for a server to become available for the setup and processing operations. When a server is free, an entity is removed from the AWAIT node and the setup operation is initiated. The processing operation is started as soon as the setup is completed. When processing is finished, the part exits through the FREE node, which releases the server to begin another setup operation. The RESOURCE statement allocates one unit of resource SERVER to AWAIT node AW1, whose file number is 12.

The choice of modeling technique for setup operations depends on the problem being modeled. Single and sequential service activities and resource allocations are used. No general guidelines, other that those illustrated above, are available.

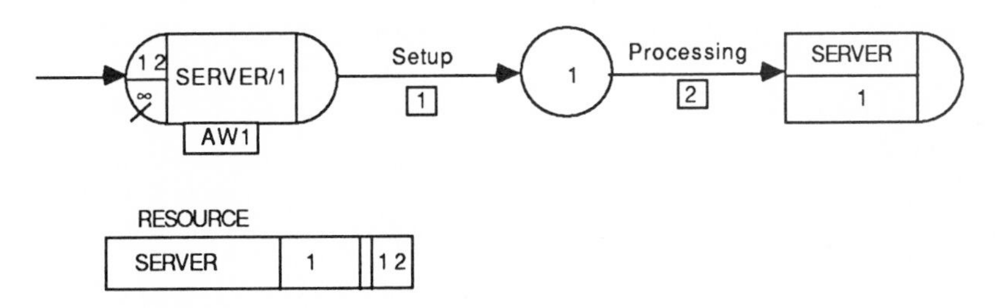

Figure 15-13 SLAM II network segment illustrating resource concepts.

15.3.2 User Functions for Machine Times

When service time is computed in a user function, the modeler has access to information regarding an entity's attributes, the current simulation time, the status of servers and resources, the length of queues, and other network information. In this way, any functional relationship may be prescribed for the service time.

Assume, for example, that a SELECT node is used to determine which of four servers is to be assigned a job, as illustrated in Figure 15-14.

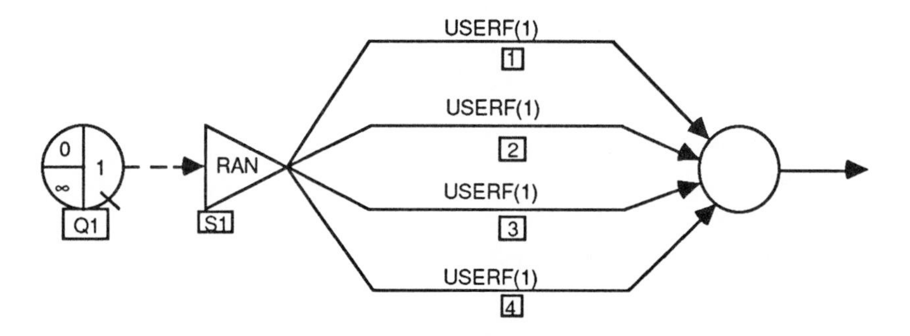

Figure 15-14 Random selection of servers at a SELECT node.

In this network segment, each of the four servers associated with the SELECT node has a user-defined service time, USERF(1). When USERF(1) is called, an activity has been selected by the SELECT node. The service time for the selected activity can be made a function of the following five network variables:

1. Job type as indicated by the value of an attribute
2. The server selected, NNUMB(IDUM)
3. Number of ongoing service activities,
   NNACT(1) + NNACT(2) + NNACT(3) + NNACT(4)
4. Current number of entities waiting in QUEUE node Q1, NNQ(1)
5. Batch size for the job as indicated by the value of an attribute

The reader is referred to Chapter 5 for more details on the use of these functions and variables in USERF. One note of caution: if multiple entities are routed from a node at the same time, then any change to a SLAM II variable made in USERF applies to any other entity routed from the node. In particular, if the ATRIB vector is changed for one entity, the new value is used for any entity routed after the change.

15.4 ANALYZING COMPUTER INTEGRATED MANUFACTURING SYSTEMS

More and more industries are considering the automation of their manufacturing processes. A computer-integrated manufacturing (CIM) system consists of machine tools, material handling equipment, and storage facilities that are integrated and computer controlled. Among the potential benefits of a CIM are lower labor costs, less setup time, decreased in-process inventory, increased machining accuracy, increased production rates, and greater information about the operation of the system. The capital expenditures involved are great, however, and for this reason it is essential that the system be designed and operated efficiently. SLAM II simulation models have been used to analyze these systems and have been useful in their evaluation and design [15, 20, 27, 31, 39]. This section presents a SLAM II model that illustrates the important concepts required to analyze a CIM without encumbering the discussion with extensive details of a large system's operation.

15.4.1 The System and Management Concerns

Figure 15-15 illustrates a schematic diagram for a proposed CIM. Parts to be processed arrive at a load/unload area. After loading, all parts travel through

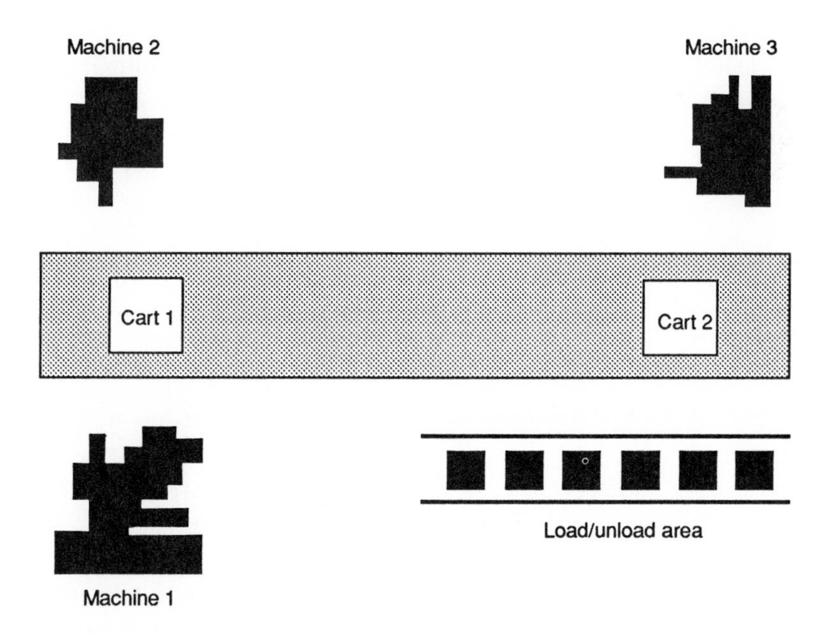

Figure 15-15 Schematic diagram of CIM system.

the system on special pallets. Pallets are a constraining resource of the system since parts must wait if all pallets are in use. Six pallets are included in the proposed design. The parts are loaded manually onto the pallets. Manual assistance is always considered available in both loading and unloading, that is, loading and unloading are not considered a limiting resource. Once loaded, the pallets are ready to travel to one of three machines. This transport is accomplished automatically by a material handling system consisting of two computer-controlled carts. It is assumed that all parts can be machined by any of three machines. Following each machine operation, a pallet must be transported to an area for part inspection. After inspection, the pallet requires a cart for transport back to the load/unload area. After unloading, the part leaves the system and the pallet is available for reuse. An important operational consideration is that a machine cannot service the next pallet until one of the carts has removed the current pallet from the machining area. The same constraint applies to the inspection station. Thus, a machine may be busy, idle, or blocked.

The types of questions that are addressed in the analysis of the operation of computer-integrated manufacturing systems are as follows:

1. What is the production rate and throughput of the system?
2. What is the utilization of each machine?
3. How often is each machine prevented (blocked) from working because a cart is unavailable or because a part is unavailable (idle time)?
4. What is the utilization of each cart?
5. How many parts wait for pallets, how many pallets wait for carts, and how many pallets wait for machines?

Answers to these questions provide a basis for evaluation of the effectiveness of the CIM system.

In addition, questions regarding the capability of the CIM to accommodate changes in system configuration and operational procedures are posed. The following are examples of such questions regarding potential ways to improve the CIM:

1. What effect does adding more carts have on the production rate or machine utilization?
2. How does the system perform under different part arrival patterns?
3. What is the effect of increasing the number of pallets?
4. What is the productive capacity of the system?
5. Can new dispatching methods increase the production rate and, if they can, what method is best?
6. What is the effect of employing a different material handling system?

The following section describes a SLAM II model designed to represent the CIM described above and to address managerial concerns. The model is abstracted from studies dealing with a Sundstrand CIM in operation at Caterpillar [31] and a Kearney and Trecher CIM in use at Ingersol-Rand [6].

15.4.2 The SLAM II Model

Figure 15-16 contains a SLAM II model for the CIM presented in Figure 15-15 The arrival of parts at the system is represented by the CREATE node with attribute 1 marked so that interval statistics can be obtained when parts leave the system. If a pallet is not available when a part arrives, it waits in AWAIT node AWT4. Pallets and carts are modeled as resources. The RESOURCE block indicates that resource 1 represents pallets, that there are six pallets available, and that pallets are allocated to entities waiting in file 4 at node AWT4; and resource 2 represents carts, that there are two carts available, and that carts are allocated to entities waiting in files 8, 6, and 5, which are associated with nodes AWT8, AWT6, and AWT5, respectively.

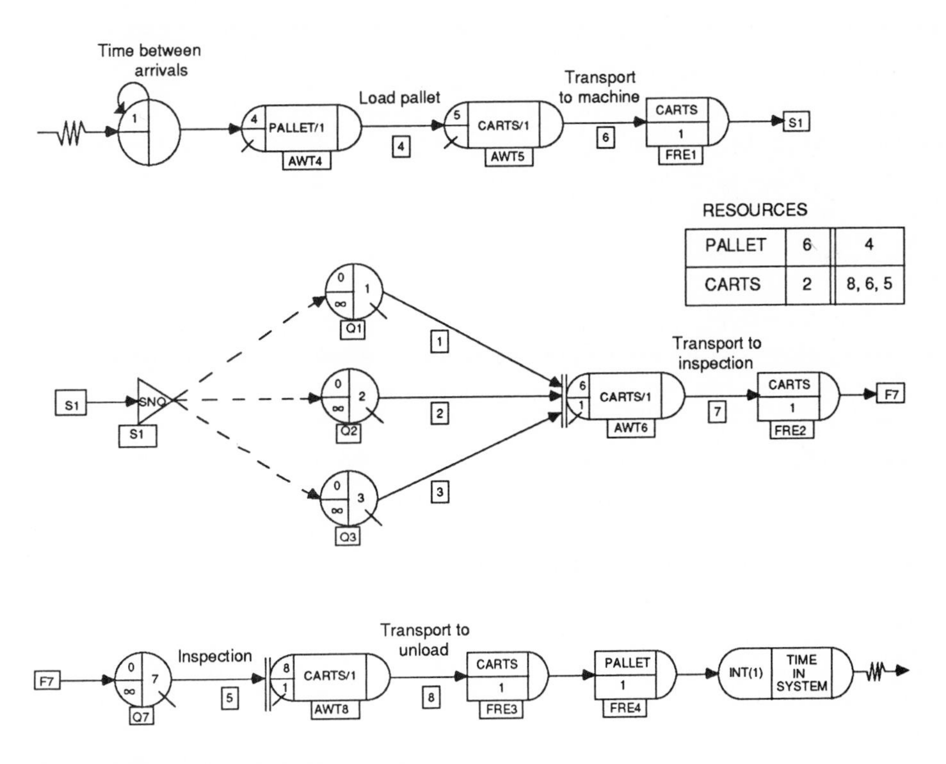

Figure 15-16 SLAM II model of a CIM system.

When a part is allocated a pallet, the part entity is loaded onto the pallet, which is modeled by activity 4. After loading the part on the pallet, it waits for a cart at AWAIT node AWT5 to transport it to a machine. For this model, priority for cart transport is given first to entities leaving the inspection area, then to parts that have just been machined, and finally to parts arriving for machine processing. After a part entity is transported to its next station, FREE node FRE1, FRE2, or FRE3 frees the cart for the next transport activity. If an entity is not waiting at any of the three AWAIT nodes, the cart resource is made idle.

Activity 6 represents the movement of a loaded pallet from the holding area at node AWT5 to the machining area. Following transport, the cart is freed at node FRE1 and is available for reallocation. The part is routed to the QUEUE node that has the smallest number of parts in it by SELECT node S1 using the smallest number in the queue (SNQ) selection rule. Machines 1, 2, and 3 are represented by activities 1, 2, and 3, and the queues for these machines are represented by QUEUE nodes Q1, Q2, and Q3. Following a machine operation, it is presumed that only one pallet and part from any of the machines may wait on the track for a cart. Thus, the logic is established to block any machine that completes its operation when a part is still waiting of be transported to inspection. This is modeled at node AWT6 by specifying a capacity of 1 and the blocking of all machines when node AWT6 is at this capacity. When a cart is available and allocated to the movement of parts from machines to the inspection station, the entity in node AWT6 is routed to node FRE2, where the cart is freed, and the part entity is placed into QUEUE node Q7 to wait for inspection. Following inspection, the part waits for a cart at node AWT8. Again, only one part may wait for a cart at node AWT8. Therefore, should this node have a loaded pallet waiting for a cart, the just inspected part blocks the inspection station, activity 5, until a cart transports a part from the inspection area. The movement of an inspected part to the unload area is accomplished by activity 8, after which the cart is freed at node FRE3. Next, the pallet is freed at node FRE4, and it is made available for parts waiting at node AWT4. The part entity leaves the system through the COLCT node, where interval statistics are collected. This provides information on the time in the system for each part. By dividing the number of observations at the COLCT node by the total simulated time period, an estimate of the production rate can be obtained. Following a discussion of model output and potential uses of this model, procedures for embellishing the model are presented.

15.4.3 Model Output and Use

Table 15-1 summarizes the SLAM II output statistics that correspond to the

Table 15-1 Summary of CIM System Performance Measures

Performance Measures	SLAM II Statistic
Time parts spend in the system	INTERVAL statistics, COLCT node
Production rate	Number of observations per unit time
Machine utilization	Satistics for servers 1, 2, and 3
Machine blocking, no cart	Blocking statistics, servers 1, 2, and 3
Use of carts	Busy statistics, resource 2
In-process inventory	
Machine area	File statistics, nodes Q1, Q2, and Q3
Load/unload area	File statistics, nodes AWT4, AWT5, AWT6, and AWT8
Inspection station	File statistics, node Q7

desired performance measures. The time parts spend in the system is collected at the COLCT node through the specification of INTERVAL statistics. Machine utilization estimates are derived from the utilization outputs for servers 1, 2, and 3. Blocking statistics for these servers represent the fraction of time the machines were neither busy or idle, but were delayed from continuing work because a cart was unavailable. The utilization of resource 2 represents the usage of the material handling system. The in-process inventory estimates are obtained from statistics on the number of entities in nodes Q1, Q2, Q3, AWT4, AWT5, AWT6, Q7, and AWT8.

Runner and Leimkuhler [31] used a network simulation model to investigate several operation and design parameters for a Sunstrand system installed at Caterpillar. One of these involved determining the relationship between production rate and the speed of the material handling system. In the model in Figure 15-16, cart speed is related to the time for transport on activities 6, 7, and 8. By varying this speed for several runs, a graph like the one depicted in Figure 15-17 may be obtained. From such experiments, the trade-off between a high-speed transport system and a desired level of production can be made. For this model, a diminishing rate of return is observed at high speeds.

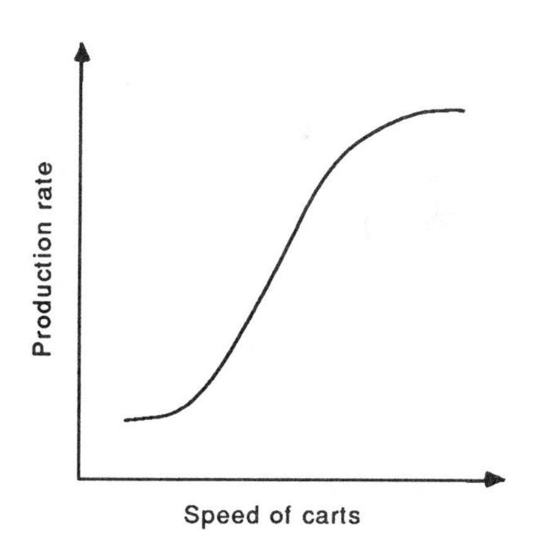

Figure 15-17 Relationship between material handling speed and production rate.

Similarly, a relationship between the number of transport pallets and production rate can be portrayed. In the SLAM II model, this involves simply changing the capacity of resource type 1 (PALLET). Figure 15-18 characterizes the output observed for this potential design change.

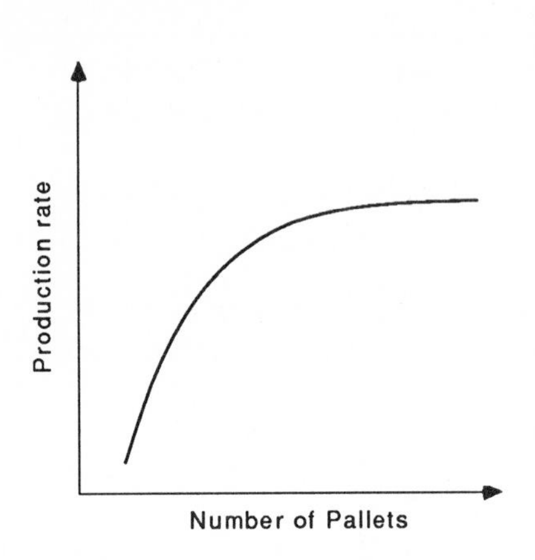

Figure 15-18 Relationship between number of pallets and production rate.

Other operational procedures can be readily assessed. For example, an increase in the number of machines involves associating additional QUEUE nodes with SELECT node S1. By changing the file sequencing order on the RESOURCE block, the effects of dispatching procedures for the carts can be investigated.

To add more carts to the system, the capacity of resource 2 is increased. In fact, a useful way to determine the number of carts needed to avoid delays is to assign a large number of carts in one run. SLAM II will then automatically compute the maximum number of units of a resource in use, which corresponds to the maximum number of carts needed to avoid all delays due to material handling requirements. Another use of the model was to perform a sensitivity analysis of system performance due to variation in the arrival pattern. This permitted evaluation of the system under different forecasted needs and allowed estimates of the capacity of the system to be calculated.

The order in which the files are listed on the RESOURCE block for CARTS establishes the priority of the allocation of carts to inspected parts, then machined parts, and finally to arriving parts. It may be desirable to prevent an excessive buildup of parts waiting for carts in node AWT5 of Figure 15-16 which could starve the system for inputs. One method to accomplish a priority modification is to specify ALLOC(1) at AWAIT node AWT6 as shown in Figure 15-19. By replacing the cart resource allocation at AWAIT node AWT6 in Figure 15-15 with the ALLOC(1) specification, the allocation of carts could be assigned according to any priority structure designed for transporting parts to the inspection station. System design logic is translated to the priority structure coded in subroutine ALLOC. In this case, if NNQ(5) is greater than a prescribed number, then no allocation (IFLAG=0) should be made by ALLOC. SLAM II then polls file 5, and a reallocation of a cart is made to an arriving part. This flexibility inherent in SLAM II through the availability of user subroutines enables the modeling of complex systems to obtain accurate estimates of system performance.

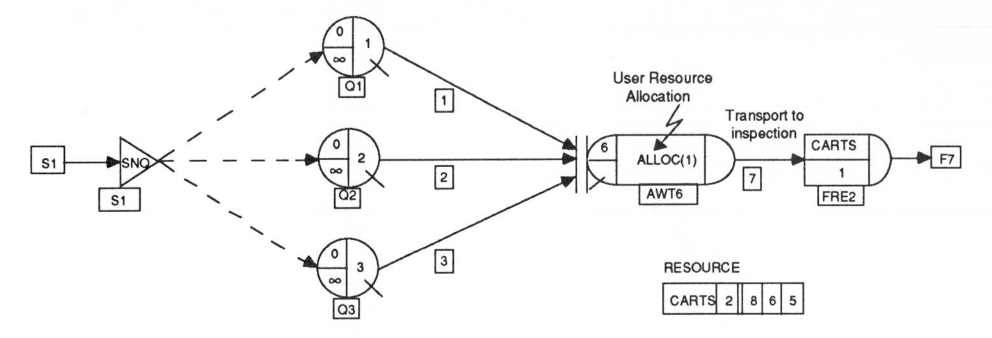

Figure 15-19 Model illustrating user allocation of resources.

15.5 MODELING JUST-IN-TIME PROCEDURES WITH KANBANS

The control of costs associated with in-process inventory in production systems has long been a major concern of production managers. One technique, first introduced successfully in Japan, to control these costs is the just-in-time (JIT) procedure with Kanban (card) controls. A basic premise of JIT is that significant savings can be achieved if in-process inventory is reduced to a minimum. Manufacturers and industrial engineers worldwide are designing and implementing JIT into production systems to reduce manufacturing costs [4, 19].

The significant savings through a reduction of in-process inventory associated with JIT requires evaluation in light of potential increases in other manufacturing costs. Reducing in-process inventory at each stage of production has the potential for increasing setup costs. Variation in production times could cause idle time for machines and increase overtime costs to meet production schedules.

Kanbans is a demand ***pull*** system that requires production scheduling integration from the final product inventory back through each stage of production to the input of raw material. Careful consideration for such a planning process is needed to establish the system's capacity and the potential for a bottleneck at each production stage. One observation made by many managers is that the design and implementation of a JIT system is a complex and lengthy process requiring, for example, ten years in one Japanese firm [11].

The basic modeling structure for the development of a SLAM II model for JIT with Kanbans is adapted from a simulation model developed by Huang, Rees and Taylor [8]. Their paper evaluated overtime requirements for changes in the number of Kanbans included in a JIT system, processing time variance, and demand levels. In the next sections, the use of SLAM II concepts to model JIT with Kanban is described. First, a discussion of JIT procedures is given based on the summary presented by Huang, Rees, and Taylor.

15.5.1 Just-in-Time Procedure

Traditional production techniques use a ***push*** concept in which a forecast for end product demand is made and, based on lead times, translated into a forecast for each stage of production. This forecast establishes in-process inventory levels, which include an allowance for safety stock. Considerable labor, material, storage, and handling operations are invested in this in-process inventory. In a push-type production environment, inventory levels are difficult to control because inflated safety stock levels and fluctuations in demand are amplified at each stage in the system.

In contrast, production systems using JIT pull products through the stages and attempt to minimize in-process inventory. Additional benefits attributed to JIT are simplified inventory control procedures, reduced fluctuations in demand from stage to stage, increased shop floor control, and a reduction in defects. The objective of JIT is to reduce the level of in-process inventory at each stage of production to one unit or batch. To accomplish this, each stage is driven by a demand from a following stage in the production process. That is, the final stage of production will not produce the next end product until there is a demand for the last completed unit. The preceding stage(s) to the final stage operates in a similar fashion and does not produce component parts required for final product assembly until the final stage has shipped its last completed unit, placed waiting parts into processing, and demanded the parts for another final assembly. The resulting demand is passed by each stage in reverse succession to all preceding stages of production down to the source of the raw material.

A Kanban system controls the timing of production at each stage in JIT systems. The typical flow of Kanbans and unit production through succeeding stages of production is shown in Figure 15-20. This illustration demonstrates the use of two Kanbans. One is the production Kanban that is routed internally within a stage to authorize the production of the next unit. The other Kanban is used to authorize the withdrawal of a processed unit from a stage to a following stage. In this example, assume that there is one completed unit at the output side and one unit awaiting processing at the input side of each stage, but there is no ongoing processing at any stage. When a demand is placed on the system, it is reflected through the system by removing a completed unit at the top stage and simultaneously passing a production Kanban back to the input side of the top stage, which causes the waiting input to be moved into processing. When this unit is moved into processing, its withdrawal Kanban is routed back to the previous stage to authorize the movement of the output from this middle stage to the input side of the top stage with the withdrawal Kanban attached. The production Kanban that was attached to this unit is then routed to the input side of the middle stage to authorize the processing of the unit waiting. This

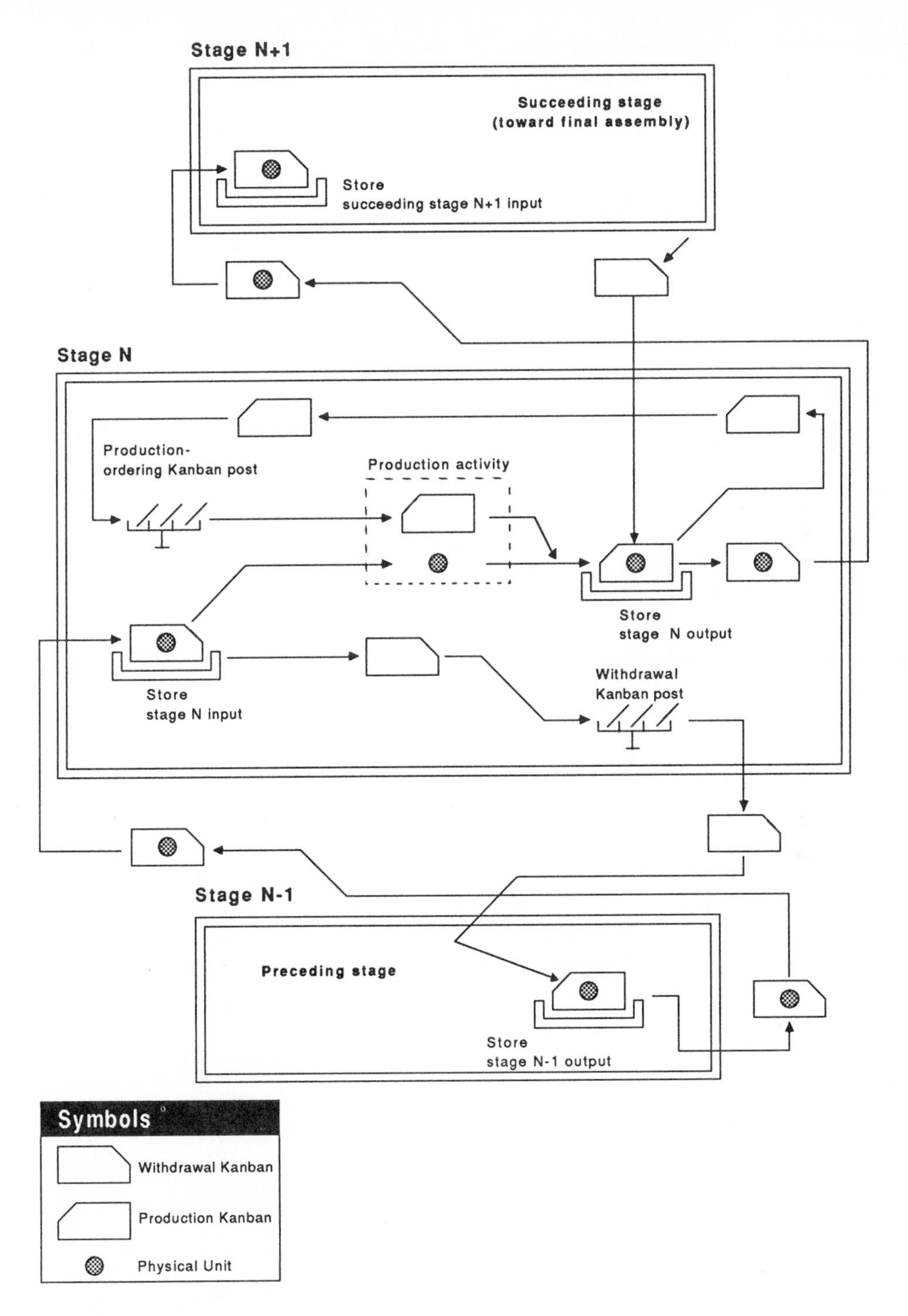

Figure 15-20 Flow of two Kanbans in a pull JIT system. Source [8]

releases the withdrawal Kanban for the bottom stage to move the next unit to the middle stage. The process continues down the system to the raw material source.

15.5.2 SLAM II Model for a JIT Production Stage

A SLAM II network for a JIT stage is illustrated in Figure 15-21. In this network segment, nodes Q2 and Q4 have an initial number in queue of 1 so that when the first pull demand is made, the production process can begin. A demand is placed when an entity representing a withdrawal Kanban arrives from the following stage to node Q5. SELECT node ASM2 assembles the units from nodes Q4 and Q5 and routes a unit to GOON node GO2, where two entities are released. One entity is routed to the following stage representing a processed unit. The second entity represents the production Kanban and is routed to node Q1 to initiate the processing of the input unit to the stage waiting at node Q2.

When a unit is in node Q1 and the production Kanban is in node Q2, they are assembled at node ASM1 and an entity is routed to node GO1. From node GO1, an entity is routed to the preceding stage representing the withdrawal Kanban to that stage. The second entity represents the unit to be processed by server 1, the production activity of the stage. Upon completion of service, the entity is placed into node Q4 to await the next demand for the stage, which occurs when the next entity arrives to node Q5.

If the stage is balanced, the queue length for QUEUE nodes Q1, Q3, and Q5 should be zero. In addition, the maximum number of entities in QUEUE nodes Q2 and Q4 will never exceed 1, and the wait time in these queues should be minimal. For a well-balanced stage, the service activity should be busy most of the time. An example illustrating the use of SLAM II modeling to estimate performance measures of a multiline, multistage production process using JIT procedures with Kanbans is presented in the next section.

15.6 MULTILINE, MULTISTAGE JIT SYSTEM WITH KANBANS

A model for a multiline, multistage JIT system with Kanbans is extracted from a network model developed by Huang, Rees, and Taylor [8]. The objective of the model is to evaluate the effects of processing time variability and the number of Kanbans on the overtime cost to meet given demand levels. Other planned uses of the model were to investigate improvements in systems design, the procurement of new equipment, the expansion of current facilities, and the effect of fluctuating demand.

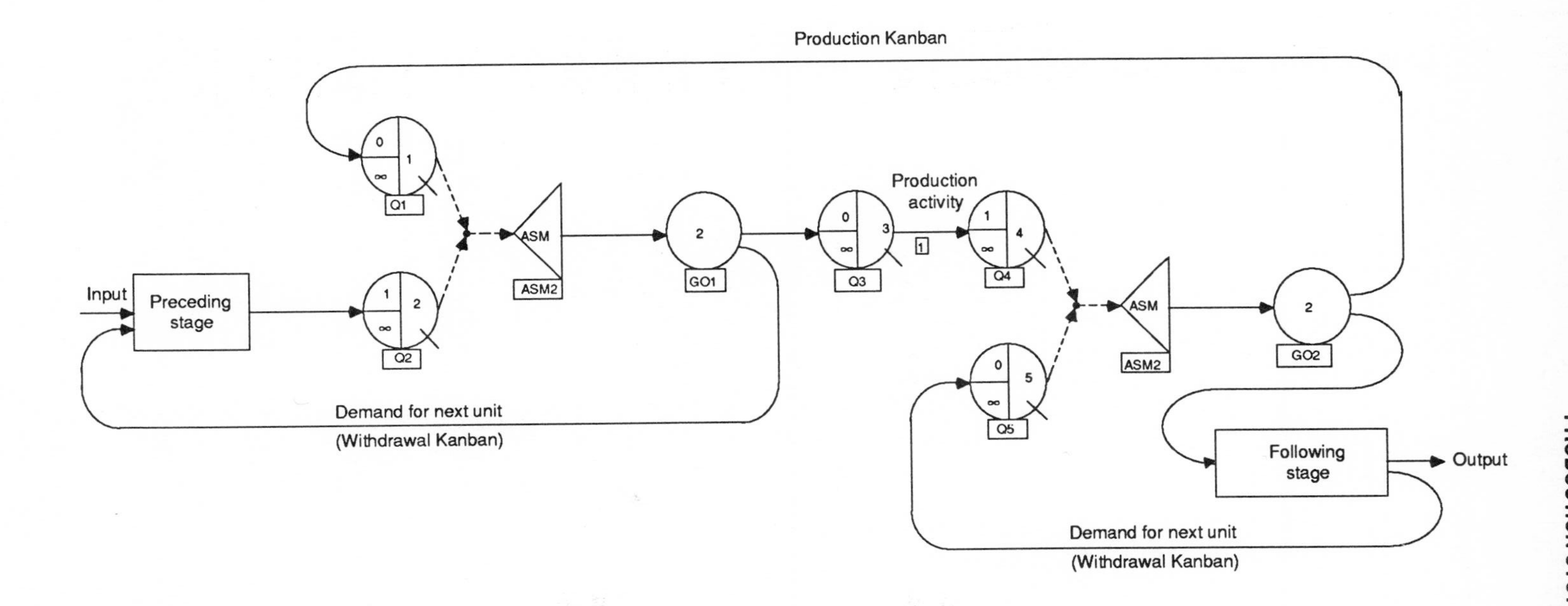

Figure 15-21 SLAM II model segment for Kanban controlled pull JIT system.

The three-line, four-stage production process shown in Figure 15-22 is to be modeled. The system has a withdrawal Kanban to pull a unit from a preceding stage and a production Kanban to authorize unit processing within a stage. An external demand placed on the system at the final assembly stage results in the movement of a finished product at that stage. This finished product output results in an assembly authorization for three subunits at the final stage and the movement of three subassemblies from the output stages of their respective production lines to the final stage. Each production line for subassemblies then pulls units from its preceding stages and ultimately from the source for raw materials.

15.6.1 SLAM II Model for the Final Assembly Stage

The SLAM II network segment representing each stage in this production system follows the SLAM II network segment given in Figure 15-20. The SLAM II network segment for the final assembly stage is shown in Figure 15-23. This network is similar to the stage segment previously presented with the addition of the arrival of demands at CREATE node CRE1. Also, it illustrates a method for accumulating subassemblies. The SLAM II network for the overall JIT system combines segments from Figure 15-21 and 15-23.

In Figure 15-23, demand for the system is modeled by the creation of entities at the CREATE node CRE1, with the first demand occurring at time TF. Each demand entity is routed to QUEUE node Q12 and results in the movement of a final product from QUEUE node Q11 through SELECT node ASM4. At GOON node GO4, the final product exits the system, and a production Kanban is routed to QUEUE node Q6. This production Kanban entity completes the assembly requirement for SELECT node ASM5 (nodes Q7, Q8, and Q9 are initialized with one entity). Four entities then emanate from node GO3. Three of these entities represent withdrawal Kanbans for the preceding three production lines, and the fourth entity represents the three subassemblies that are to be combined into the final product by service activity 2. The three withdrawal Kanbans effect the release of a subassembly from each of the preceding lines, which enter the final assembly stage at nodes Q7, Q8, and Q9 to wait for the next production Kanban to arrive at node Q6. (A batch of ten units is the output from stages prior to the final stage. The batch is converted back to units at the final stage.) When service is completed on activity 2, an entity representing the finished product is routed to node Q11. If the next demand for a finished product is waiting in node Q12, the entity is immediately passed through node ASM4 to node GO4 and the cycle continues. Otherwise, the entity waits in node Q11 until the next demand occurs.

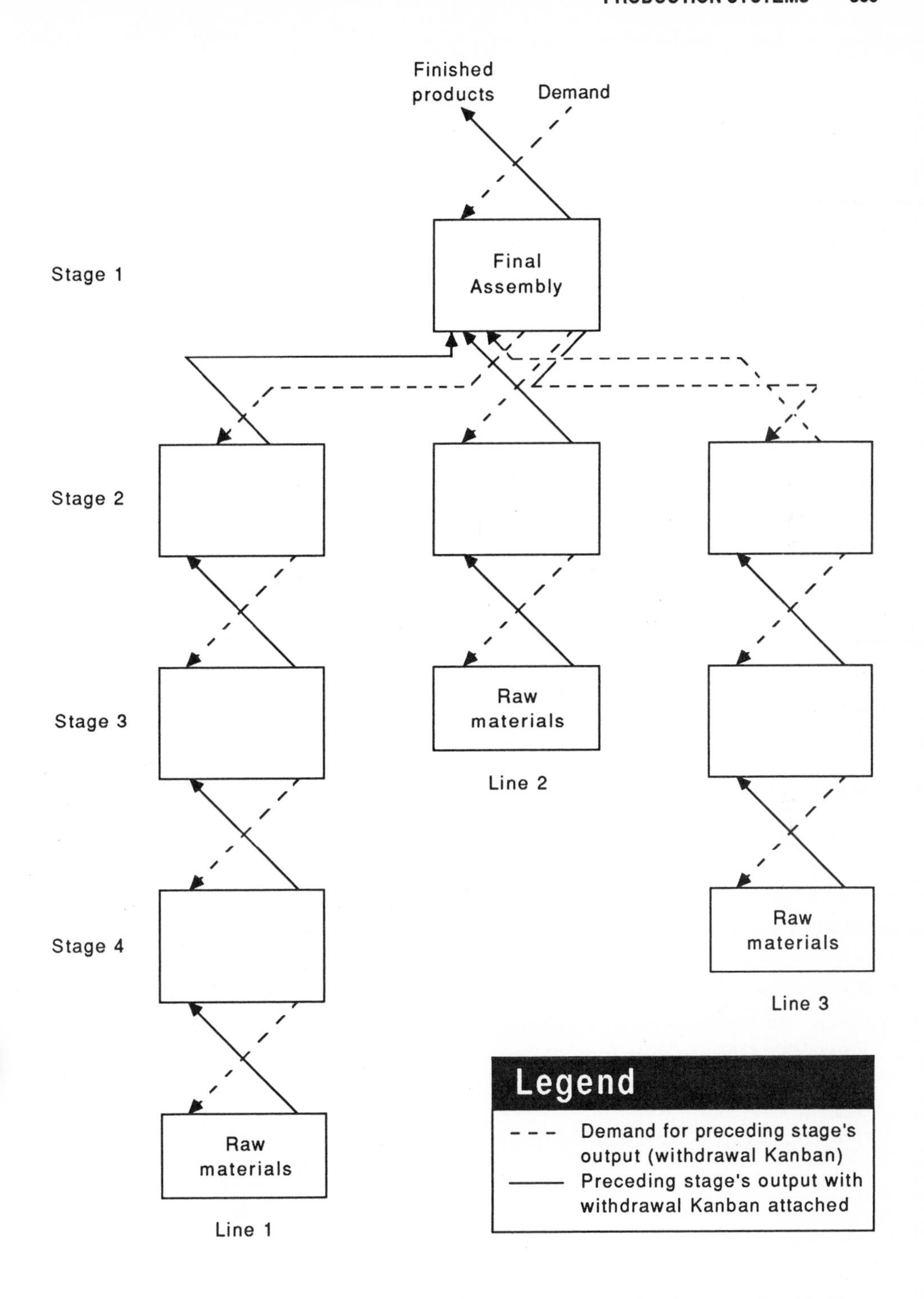

Figure 15-22 A multiline, multistage production process using Kanbans. Source [8]

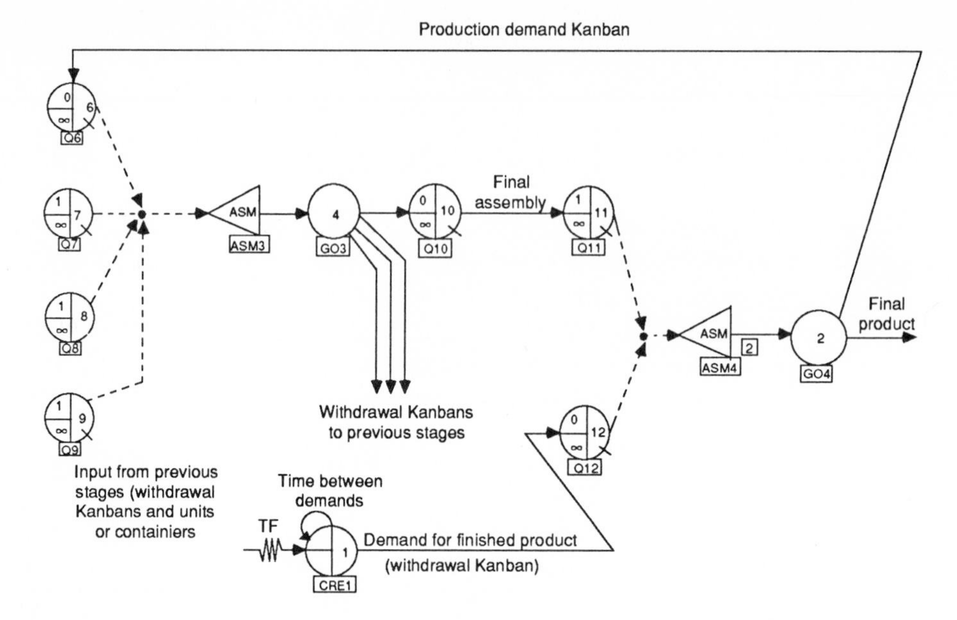

Figure 15-23 SLAM II model segment for final assembly stage in a Kanban controlled pull JIT production system.

15.6.2 JIT System Simulation and Outputs

Huang, Rees, and Taylor [8] reported on four different experiments with the JIT/Kanbans model, only one of which is discussed here. The objective of the experiment is to determine the effect of different processing time distributions on system performance when buffered in-process inventory is designed into the system. This design is evaluated for a constant demand of 1000 finished units per day for a normal 480-minute shift. The primary measure of system performance is the average overtime (in minutes) required per day to meet the daily demand of 1000 units. Statistics are also obtained for the final stage of production on the average number of units per day in preproduction inventory (nodes Q7, Q8, and Q9) and postproduction inventory (node Q11). Simulation runs were performed under each of the following conditions:

1. Constant activity service time of 48 minutes
2. Exponential activity service with a mean of 48 minutes
3. Normal activity service time with a mean of 48 minutes and a standard deviation of 4.8 minutes (small variance)
4. Normal activity service time with a mean of 48 minutes and a standard deviation of 24 minutes (large variance)

5. Each simulation is repeated for each of the activity distributions, with one and two Kanbans allowed in the system to observe the effects of buffered inventory in the system

A simulation run for a constant service time with one and two Kanbans established the standard output of the system at 1000 units per day with no overtime required. The simulation is then repeated for the eight other situations. Each run is continued until the 1000 daily demand quota is satisfied. The amount of overtime required to meet the daily demand is then recorded.

Figure 15-24 shows the effect of stochastic service times on the amount of overtime to meet the specified level of demand. The plots quantitatively illustrate that overtime requirements increase with greater production service time variability and smaller in-process inventory (fewer Kanbans).

The effect of increasing the number of Kanbans on overtime is illustrated in Figure 15-25. For the larger service time variance, overtime can be reduced an average of 50 minutes per day if two Kanbans are allowed in the system. The introduction of additional Kanbans does not achieve significant savings as shown in Figure 15-25.

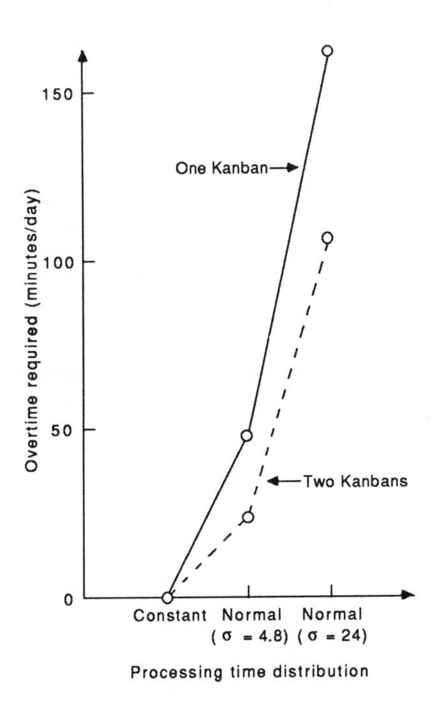

Figure 15-24 The effect of stochastic processing time on overtime required with constant demand. Source [8]

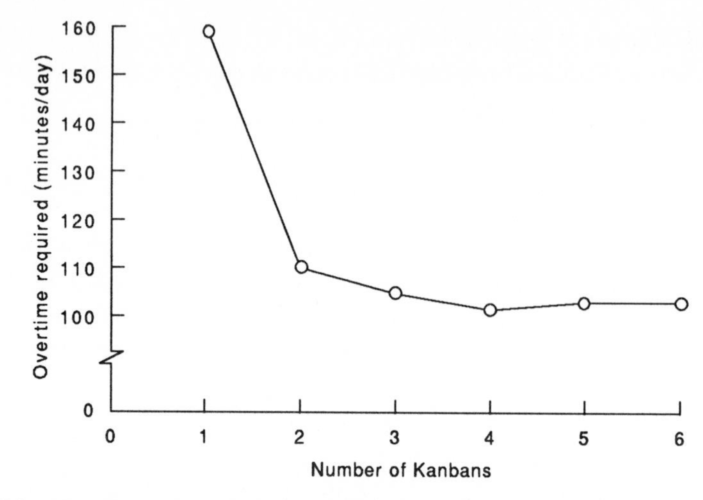

Figure 15-25 The effect of the number of Kanbans on overtime required with normal processing time. Source [8]

An interesting trade-off involves the cost of reducing service time through training and experience, improvements in processing, or more efficient equipment and machinery, compared to inventory carrying costs savings for different control strategies. For this discussion, the number of Kanbans is the variable that represents a control strategy. Figure 15-26 depicts overtime cost curves for

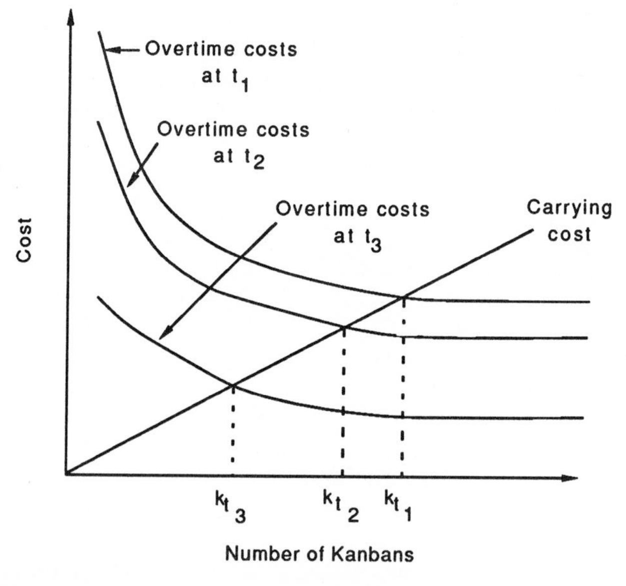

Figure 15-26 The overtime-inventory tradeoff. Source [8]

three service activity times, $t_1 > t_2 > t_3$, relative to the number of Kanbans in the system. Carrying costs increase linearly with the number of Kanbans. From basic inventory theory, the optimum occurs where the component costs are equal. This provides a means to set the number of Kanbans for each design and to obtain a cost to compare with the implementation expense required to obtain a reduced processing time. Trade-off analysis of this type combines design issues and manufacturing control strategies. Such analyses are required to introduce computer-integrated manufacturing effectively.

15.7 SUMMARY

System performance for production planning is categorized into measures for throughput, ability to meet deadlines, resource utilization, and in-process inventory. The procedures for obtaining these measures of performance from network models are described. An evaluation of a computer-integrated manufacturing system is presented. A model of just-in-time procedures using Kanban for a production system is developed, and the use of simulation to analyze the design of control procedures is illustrated.

15.8 EXERCISES

15-1. Build network models of the following situations:

(a) Two products are stored in separate buffer areas to be processed by one of three machines. Product 1 can be processed using machine 1 or machine 2; product 2 can be processed using machine 2 or machine 3. When machine 2 is available, the operating policy is to select a type 1 product in preference to a type 2 product.

(b) Two products are stored in separate buffer areas to be processed by one of three machines. Product 1 can be processed by machines 1 and 2, and product 2 can be served by machines 1, 2, and 3. Machine 3 is to be used to serve product 2 only if machines 1 and 2 are not busy.

(c) The situation is the same as described in part (a) except that machine 2 should process product 2 before product 1 if the number of type 2 products waiting exceeds 5.

15-2. Discuss methods by which jobs are assigned due dates in practice. Build SLAM II networks for assigning due dates. Discuss methods by which due date assignment procedures may be included in an overall model of a production facility.

15-3. Discuss situations in which makespan is an important measure of a production system's performance.

15-4. Discuss how performance measures of a production system influence management decision making regarding costs of alternatives, designs for system capacity, and the determination of operational requirements.

15-5. Design a user function to incorporate learning, fatigue, and other service characteristics so that these factors are included in the duration of the service time associated with a human-machine system.

15-6. Develop a specification for a SLAM II model of a crane system that has two cranes on a single runway that serves 3 machines. Develop a procedure for resolving interference conflicts between the cranes.

15-7. Modify the model of the computerized manufacturing system given in Figure 15-16 so that the movement of the cart to the machine involves bringing a new part to the machine and removing a processed part. The only time a cart will load or unload a machine is when it cannot perform both operations for lack of a part or machine availability.

15-8. Given the data on the following 20 jobs, compute the following performance measures for a one-machine situation: throughput, lateness, tardiness, production rate, and makespan.

Order Book for Twenty Jobs

Job Number	Processing Time	Arrival Time	Due Date
1	7	0	10
2	4	0	10
3	10	0	15
4	6	3	25
5	5	3	30
6	7	3	35
7	3	6	40
8	10	6	45
9	2	6	50
10	5	12	55
11	9	12	60
12	1	12	65
13	6	16	70
14	3	16	75
15	6	16	80
16	2	20	85
17	4	20	90
18	3	20	95
19	10	25	100
20	7	25	100

15-9. Explain the qualitative relationship depicted by the curves in Figures 15-17 and 15-18 between production rate and the independent variables material handling speed and number of pallets.

15-10. For each of the performance measures listed in Table 15-1, specify the anticipated changes due to the following elemental modifications to the CIM system model presented in Figure 15-16. Develop a qualitative relationship between the output measures (where appropriate), assuming each elemental change is made independent of other changes.

(a) An increase in the input rate by decreasing the time between arrivals.
(b) Introduction of a probability of rejection from inspection which causes pallets to be sent back for machining.
(c) An increase in the time to transport to the unload area.
(d) An increase in the length of activities 1, 2, and 3.
(e) An increase in the number of unit entities that can be handled concurrently by a cart.
(f) A limit on the number of pallets at nodes Q1, Q2, and Q3.

15-11. A chemical plant has been designed to produce two grades of liquid product [25]. The location of the plant dictates the use of rail traffic as the primary method of product distribution, and business considerations require a highly reliable distribution system. The company distributes to the following five consumers: two packaging units, an export terminal, a redistribution terminal, and an outside customer. The physical facilities required at the production site are storage tanks and a tank car fleet for each product grade, a tank car loading rack, and a marshaling yard. New storage and unloading rack facilities are also required at the export terminal and the redistribution terminal. In addition, product receiving and storage facilities at the two packaging units may require upgrading to be capable of handling the new product flows. The major activities of the distribution network to be modeled are described next.

The production unit makes two product grades, which are stored in separate tanks. Upon arrival of an empty tank car, product is removed from the appropriate tank and loaded into the empty car. A consumer demand causes a train of full cars to be assembled and shipped. The train travels to its destination, is disassembled, and the cars are unloaded into a storage tank that supplies the consuming unit. The empty cars are assembled into a train and returned to the production unit for reuse.

Build an aggregate network model to determine qualitatively the size of storage and rack facilities and the tank car fleet required to meet the project objectives. The primary measures of effectiveness of the distribution system are the frequency and the duration of both the stock-outs at the consumer locations and shutdowns due to high inventory at the production unit. Additional statistics to be collected are rack utilization, the number of surplus cars, the required marshaling yard capacity, and the delays at the consumers' unloading facilities. Discuss how different alternative designs would be represented.

Embellishment: Hypothesize data values and evaluate alternative design strategies.

15-12. Consider the Just-In-Time with Kanbans production system of Figure 15-22.

(a) Develop the SLAM II statement model for the final-stage network given in Figure 15-23 and simulate this stage for a constant output of 1000 units per day. Assume that the input from all previous stages is adequate to meet the required output.

(b) Add the required SLAM II statements to complete the model as shown in Figure 15-22 using the network illustrated in Figure 15-21 and rerun the simulation. Assume there is sufficient raw material at the input to each stage of the simulation. Develop the service time and incorporate any characteristics necessary to produce 1000 units of output per day at the final stage.

(c) Increase the in-process inventory allowed at each stage by adding one additional Kanban to each stage and rerun the simulation. What effect does this have on overtime hours required to meet the required output of the system? Develop a cost structure for the system to evaluate the introduction of additional in-process inventory. Use the simulation ourputs to evaluate the design of alternative system procedures.

15-13. The simulation model developed in Exercise 15-12 uses a combination of QUEUE and SELECT nodes with the ASM option to assemble units into the next higher assembly and to authorize movement into the system with Kanbans. Delays in the flow of entities in a SLAM II network can also be modeled with resources, gates, and AWAIT nodes. In addition, entities can be combined using ACCUMULATE, BATCH, and QBATCH nodes. Develop a SLAM II simulation model of the network illustrated in Figure 15-22 using these modeling options. Simulate the network using the same assumptions given for Exercise 15-12b and compare the results with those obtained for the previous simulation.

15-14. Develop a model that controls in-process inventory with a pull system that uses information about the needs of two future stages. Under what conditions is it necessary to use information beyond the next stage of a process?

Embellishment: Develop a procedure that schedules production based on demand. List the scheduling rules necessary to smooth production requirements.

15.9 REFERENCES

1. Bandy, D. B., and S. D. Duket, "Q-GERT Model of a Midwest Crude Supply System," *Milwaukee ORSA/TIMS Joint National Meeting*, October 1979.
2. Compton, W. D., ed., *Design and Analysis of Integrated Manufacturing Systems*, National Academy Press, Washington, DC, 1988.
3. Dessouky, M., H. Grant, and D. Gauthier, "Simulation of an Injector Plunger Production Line," *Proceedings, Winter Simulation Conference*, 1985, pp. 303-307.
4. Evans, J. R., and others, *Applied Production and Operations Management*, 2nd ed., West Publishing, St. Paul, MN, 1987.
5. Godziela, R., "Simulation of a Flexible Manufacturing Cell," *Proceedings, Winter Simulation Conference*, 1986, pp. 641-648.
6. Herald, M. J., and S. Y. Nof, "Modeling Analysis and Design Issues in a CMS with a Closed Loop Conveyor," School of Industrial Engineering, Purdue University, June 1978.
7. Hira, D. S., and P. C. Pandey, "A Computer Simulation Study of Manual Flow

Lines," *Journal of Manufacturing Systems*, Vol. 2, No. 2, 1983, p. 117.
8. Huang, P. Y., L. P. Rees, and B. W. Taylor III, "A Simulation Analysis of the Japanese Just-in-Time Technique (with Kanbans) for a Multiline, Multistage Production System," *Decision Sciences*, Vol. 14, 1983, pp. 326-344.
9. Huang, P. Y., L. P. Rees, and B. W. Taylor III, "Integrating the MRP-based Control Level and the Multi-stage Shop Level of a Manufacturing System via Network Simulation," *International Journal of Production Research*, Vol. 23, No. 6, 1985, pp. 1217-1231.
10. Juran, J. M., "Japanese and Western Quality-A Contrast," *Quality Progress*, 1978, Vol. 11, No. 12, pp. 10-18.
11. Kimura, O., and H. Terada, "Design and Analysis of Pull System, a Method of Multi-stage Production Control," *International Journal of Production Research*, Vol. 19, 1981, pp. 241-253.
12. Lilegdon, W. R., C. H. Kimpel, and D. H. Turner, "Application of Simulation and Zero-One Programming for Analysis of Numerically Controlled Machining Operations in the Aerospace Industry," *Proceedings, 1982 Winter Simulation Conference*, 1982, pp. 281-289.
13. Lin, L., and J. K. Cochran, "Optimization of a Complex Flow Line for Printed Circuit Board Fabrication by Computer Simulation," *Journal of Manufacturing Systems*, Vol. 6, No. 1, pp. 47-57.
14. Martin, D. L., "Simulation Analysis of an FMS During Implementation," *Proceedings, Winter Simulation Conference*, 1986, pp. 628-632.
15. Martin, D. L., "Gaining Insights into Manufacturing Processes Using Simulation", Society of Manufacturing Engineers, Technical Paper MS86-1009, 1986.
16. Martin, D. L., "Gaining Insights Into Manufacturing Processes Using Simulation," *Proceedings, Ultratech-Simulation Conference*, Long Beach, CA, 1986.
17. Mills, M. C., "Using Group Technology, Simulation and Analytic Modeling in the Design of a Cellular Manufacturing Facility," *Proceedings, Winter Simulation Conference*, 1986, pp. 657-660.
18. Miner, R. J., D. B. Wortman, and D. Cascio, "Improving the Throughput of a Chemical Plant," *Simulation*, Vol. 35, 1980, pp. 125-132.
19. Monden, Y., "Adaptable Kanban System Helps Toyota Maintain Just-in-Time Production, *Industrial Engineering*, 1981, Vol. 13, No. 5, pp. 29-46.
20. Musselman, K. J., "Computer Simulation: A Design Tool for FMS," *Manufacturing Engineering*, Vol. 93, 1984, pp. 115-120.
21. Philipoom, R. R., and others, "An Investigation of the Factors Influencing the Number of Kanbans Required in the Implementation of the JIT Technique with Kanbans," *Int'l. Journal of Production Research*, Vol. 25, 1987, pp. 457-472.
22. Pritsker, A. A. B., "Three Simulation Approaches to Queueing Studies Using GASP IV," *Computers and Industrial Engineering*, Vol. 1, 1976, pp. 57-65.
23. Pritsker, A. A. B., "Applications of SLAM," *IIE Transactions*, Vol. 14, 1982, pp. 70-77.
24. Pritsker, A. A. B., "Applications of Simulation," *Proceedings, IFORS*, J. P. Brans, ed., Elsevier Science Publishers, 1984, pp. 908-920.
25. Pritsker, A. A. B., *Introduction to Simulation and SLAM II*, *3rd* ed., Wiley and Systems Publishing, New York and West Lafayette, IN, 1986.

26. Pritsker, A. A. B., "Model Evolution: A Rotary Index Table Case History," *Proceedings, Winter Simulation Conference*, 1986, pp. 703-707.
27. Pritsker, A. A. B., "Model Evolution II: An FMS Design Problem", *Proceedings, Winter Simulation Conference, 1987, pp. 567-574.*
28. Polito, J., and A. A. B. Pritsker, "Computer Simulation and Job Analysis", in *Job Analysis Handbook*, S. Gael, ed., Wiley, New York, 1986.
29. Reitman, J., *Computer Simulation Applications,* Wiley, New York, 1971.
30. Rolston, L. J., and R. J. Miner, *MAP/1: Manufacturing Analysis Program Using Simulation*, Pritsker & Associates, West Lafayette, IN, 1988.
31. Runner, J. S., and F. F. Leimkuhler, "CAMSAM: A Simulation Analysis Model for Computer-aided Manufacturing Systems," *Proceedings, Summer Computer Simulation Conference*, 1978, Newport Beach, CA, July 1978.
32. Schooley, R. V., "Simulation in the Design of a Corn Syrup Refinery," *Proceedings, Winter Simulation Conference,* 1975, pp. 197-204.
33. Stidham, S., Jr., "A Last Word on $L = \lambda W$", *Operations Research*, Vol. 22, No. 2, 1974, pp. 415-421.
34. Sugimori, Y., and others, "Toyota Production System and Kanban System - Materialization of Just-in-Time Respect-for-Human Systems," *International Journal of Production Research*, Vol. 15, 1977, pp. 553-564.
35. Taha, H. A., *Operations Research,* Macmillan , New York, 1971.
36. Taylor, B. W., III, L. M. Moore, and R. D. Hammesfahr, "Global Analysis of a Multi-product, Multi-line Production System Using Q-GERT Modeling and Simulation," *AIIE Transactions,* June 1980, pp. 145-155.
37. Wheelwright, S. C., "Japan - Where Operations Really Are Strategic", *Harvard Business Review*, 1981, Vol. 59, No. 4, pp. 67-74.
38. White, J. A., *Production Handbook*, *4th* ed., Wiley, New York, 1987.
39. Wortman, D. B., and J. R. Wilson, "Optimizing a Manufacturing Plant by Computer Simulation," *Computer-aided Engineering*, Vol. 3, 1984, pp. 48-54.

16 PRODUCTION SYSTEMS: SPECIAL TOPICS

16.1 INTRODUCTION

Production planning is a broad field with many special topics. The purpose of this chapter is to explain a few of these topics in terms of network models with the intention of illustrating the flexibility available for both defining and using a network orientation.

The special topics of preemption of a service activity, machine maintenance and failure, queueing situations requiring both workers and machines, and job shop routing are selected for presentation. Models associated with these topics are frequently needed in applications. In addition, new modeling viewpoints are illustrated in the discussions. Only a cursory review of modeling inspection processes and work flow for batches of items is presented since these subjects have been discussed elsewhere in this book although in a different context.

16.2 PREEMPTING A SERVICE ACTIVITY

In planning production, it is sometimes necessary to give priority to one class of jobs, even to the extent of stopping the processing on one job in order to start a different type of job. Consider the situation where jobs of types 1 and 2 are processed by machine 1. The jobs are sequenced in a queue by giving priority to type 1 jobs. Furthermore, when a job of type 1 enters the work area and a job of type 2 is being machined, the machining is stopped, and the type 2 job is returned to the queue so that the higher-priority job can be processed. The network segment in Figure 16-1 models this situation.

Attribute 1 for all job entities arriving at ASSIGN node ASN1 represents its job type. Both types of jobs arrive at ASN1, where a processing time value is

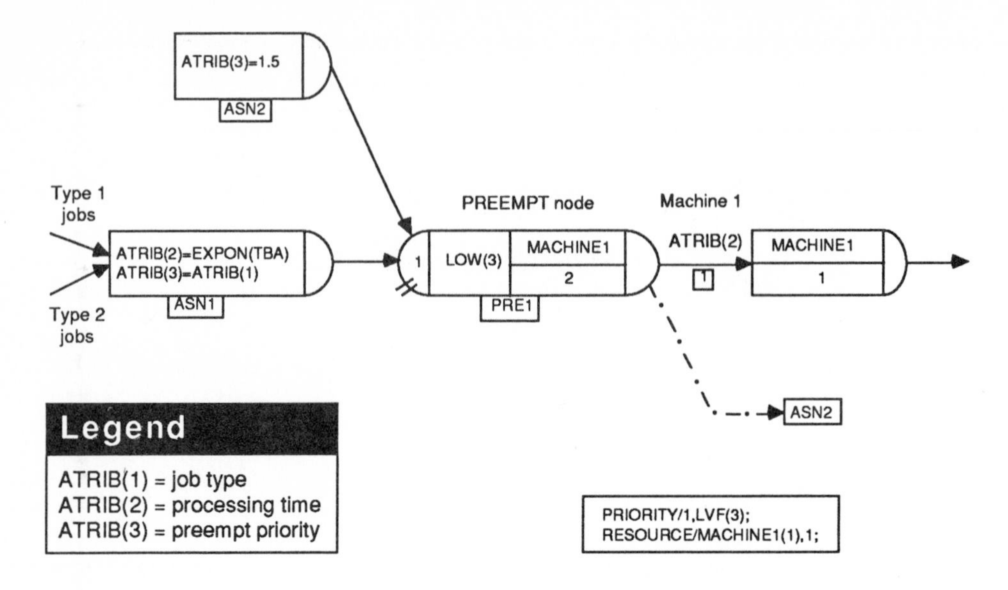

Figure 16-1 SLAM II network segment illustrating preemption.

assigned to attribute 2 for use later as the time specification for machine 1. Attribute 3 establishes a job entity's preemption priority and is initially set equal to attribute 1 (job type). Both job types then enter PREEMPT node PRE1, where they seek access to the resource MACHINE1. When a job is preempted, the entity is routed to ASSIGN node ASN2, with the remaining processing time placed in attribute 2. At ASN2, attribute 3 is set equal to 1.5 and the preempted job is returned to node PRE1. Since priority is given to entities waiting in file 1 at node PRE1 based on low value of attribute 3, LVF(3), preempted jobs with ATRIB(3) = 1.5 are placed ahead of type 2 jobs [ATRIB(3) = 2], but behind any type 1 jobs [ATRIB(3) = 1] that might arrive.

Each job arriving at node PRE1 requires one unit of the resource MACHINE1. This resource is provided through the RESOURCE statement, which establishes the capacity of MACHINE1 as 1. The first job arriving at node PRE1 captures this resource and is placed into service on machine 1 for the time stipulated by its ATRIB(2) value. Jobs arriving with the same or higher value of ATRIB(3) wait in file 1 until MACHINE1 is available. When service is completed for the current job, it exits through the FREE node, where MACHINE1 is freed. A job arriving at node PRE1 with a lower value of ATRIB(3) than the current job in service causes service to stop on the current job, and the new arrival is allocated the resource immediately. This is specified by the LOW(3) at node PRE1. The job removed from service is reentered into the network through ASSIGN node ASN2, as described above.

16.3 MODELING MACHINE MAINTENANCE AND FAILURE

Machine maintenance or failure changes the availability of a machine to process items. To model maintenance operations and machine failures, the machine is modeled as a resource as illustrated in Figure 16-2a. The RESOURCE statement sets the capacity of the resource, MACHINE, to 1. The resource is allocated at the AWAIT node, and activity 1 represents the time required to perform the machine service. When a job has completed service, the machine is made available to the next job by the decision logic of the FREE node and the list of files at the RESOURCE block. Statistics on the time jobs are in the system are collected at the COLCT node.

Now suppose that after every machining operation there is a 10% chance that the machine requires a 5-minute maintenance operation. This situation is modeled by incorporating probabilistic branching in the network as shown in Figure 16-2b. Two GOON nodes are added after the machining operation. The first GOON node GON1 routes the job entity, and clones it to create a machine entity. The upper branch routes the entity to the COLCT node to record the time in the system for the job, as shown in Figure 16-2a. The lower branch sends the machine entity to GON2, where 90% of the time the entity branches to FREE node F1 with a zero time delay, freeing the machine immediately. For the other 10% of the entities, there is a five time unit delay on activity 2, representing a machine maintenance time before the machine is freed for the next job.

This network representation accounts accurately for the added delays encountered due to machine maintenance. However, utilization statistics will not distinguish between maintenance time and processing time as no differentiation between service and maintenance activities has been indicated. To make such a differentiation, the maintenance time should not be included in the busy time for the machine. This is accomplished by decreasing the resource capacity prior to the 5-minute maintenance time through the use of an ALTER node. After maintenance, a second ALTER node is used to indicate that the machine is back in operation again. Figure 16-3 is a network segment with these ALTER nodes added after node GON2. In this network segment, node ALT1 requests that the capacity of the machine be reduced by 1 when maintenance on the machine is to be performed. The FREE node F1 causes this reduction to occur by freeing the resource, which is then used to satisfy the capacity reduction request. This freed machine is not made available because a capacity change request is always satisfied before a resource reallocation is made. The branch from node ALT1 to node ALT2 represents the 5-minute maintenance operation, after which the machine is placed back into service by increasing the capacity of the resource MACHINE by 1 at ALTER node ALT2.

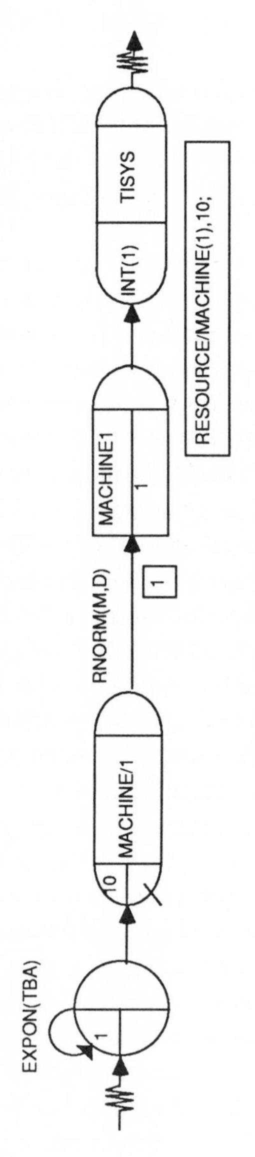

(a) Machine processing without maintenance

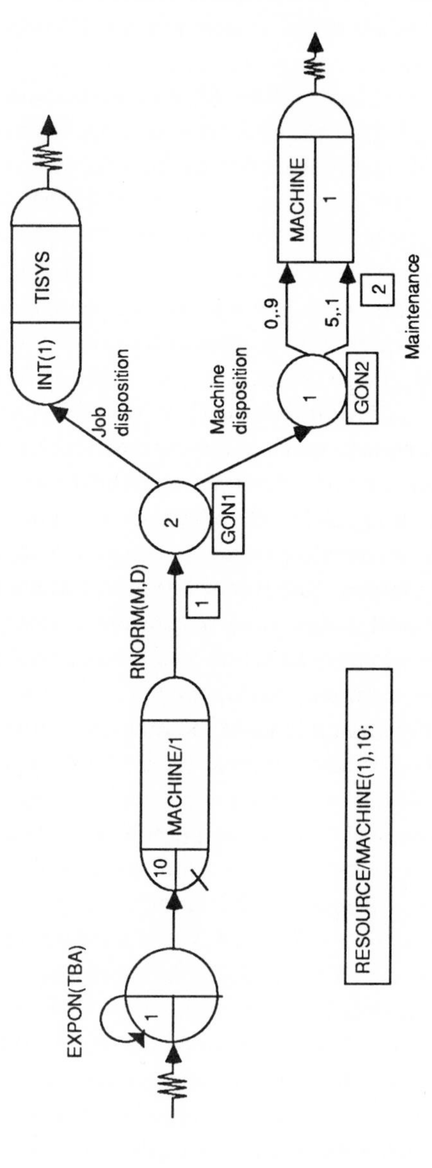

(b) Machine processing with maintenance

Figure 16-2 Machine processing with and without maintenance.

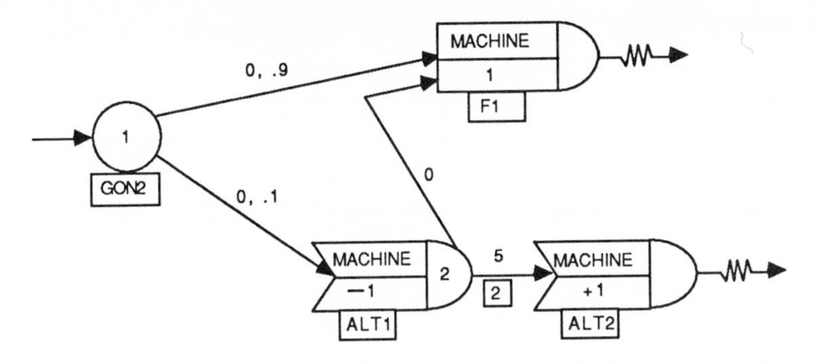

Figure 16-3 Network segment for modeling maintenance activity through resource capacity reduction.

In the examples above, maintenance is modeled as a probabilistic event following a service operation. If it is desired to have maintenance occur periodically, the disjoint network shown in Figure 16-4 used in conjunction with the original network, models a first maintenance operation at time 60, the length of the maintenance activity as 5 time units, and the time until the next maintenance operation as 55 time units. This model causes maintenance to be performed at the first opportunity following the completion of the processing of a job. The time for maintenance is the 5 time units less the processing time remaining after the maintenance request is made. If the remaining processing time is greater than 5 time units, no maintenance time occurs.

If machine failure is to be modeled, processing should be halted and repair action taken immediately at the time of failure. To model failures, a PREEMPT node is used.

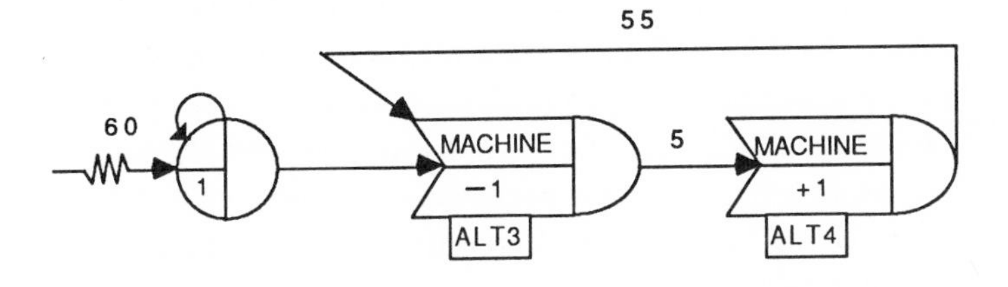

Figure 16-4 Network segment for periodic maintenance requests.

16.4 LABOR-LIMITED QUEUEING SITUATIONS

In the discussion so far, jobs have been modeled as arriving to machines where they wait for one critical resource (machines). Any other resource that may be needed in the service operation, such as a worker to operate the machine, has been assumed to be available. If labor is unavailable but required for a machine operation, then an idle machine remains idle even though a job is waiting for it. This section shows how this situation can be modeled. The concepts presented are applicable to multiconstrained resource situations in general. For illustrative purposes, the discussion is restricted to labor-limited machine operations. Considerable research has been performed in this area [7, 8, 11].

There are two direct ways for modeling a service operation that requires multiple resource types. One involves the use of resource concepts; the second uses the assembly queue selection option, ASM, for SELECT nodes.

16.4.1 Resource Method

Consider a two-work station situation in which each work station has one machine. Both machines require a setup operation for each job and this setup is performed manually. A single operator is assigned the task of setting up the machines. The network shown in Figure 16-5 models this situation assuming that there is an equal likelihood that jobs are routed to either machine; that is, entities are branched from the CREATE node to AWAIT nodes AWT1 or AWT2 with an equal probability. Jobs wait at either node AWT1 or AWT2 for either machine MACH1 or MACH2. Both resources have a capacity of 1. Node AWT1 allocates machine 1 to the first job waiting in file 1 and routes the entity to an AWAIT node, where it waits for the setup operation to be performed by the operator. Machine MACH1 is unavailable until freed at node F1 which represents the completion of processing for that job entity. Therefore, subsequent arriving jobs wait in AWT1 until machine 1 is freed. Likewise, jobs arriving for machine 2 wait for MACH2 and then wait for the operator to perform the setup operation for machine 2.

Since there is only a single machine of each type, one job at a time is removed from an AWAIT node as a machine becomes available. Therefore, the two AWAIT nodes for the OPERATOR will contain one entity at the most. When a machine setup operation is completed, the operator is freed and is allocated to the job entity that is first in file 3. SLAM II processes the job entity along the path of the AWAIT node to which the entity arrived, that is, jobs routed to MACH1 stay in the upper network segment and jobs routed to MACH2 stay in the lower network segment. Machine processing time is specified by PROC1 for machine 1 on activity 1 and PROC2 for machine 2 on activity 2. Upon completion of machine processing, a job is routed from the work station through FREE node F1 or F2, where the file of the associated AWAIT node is checked to see if there is another job waiting for machine setup and processing.

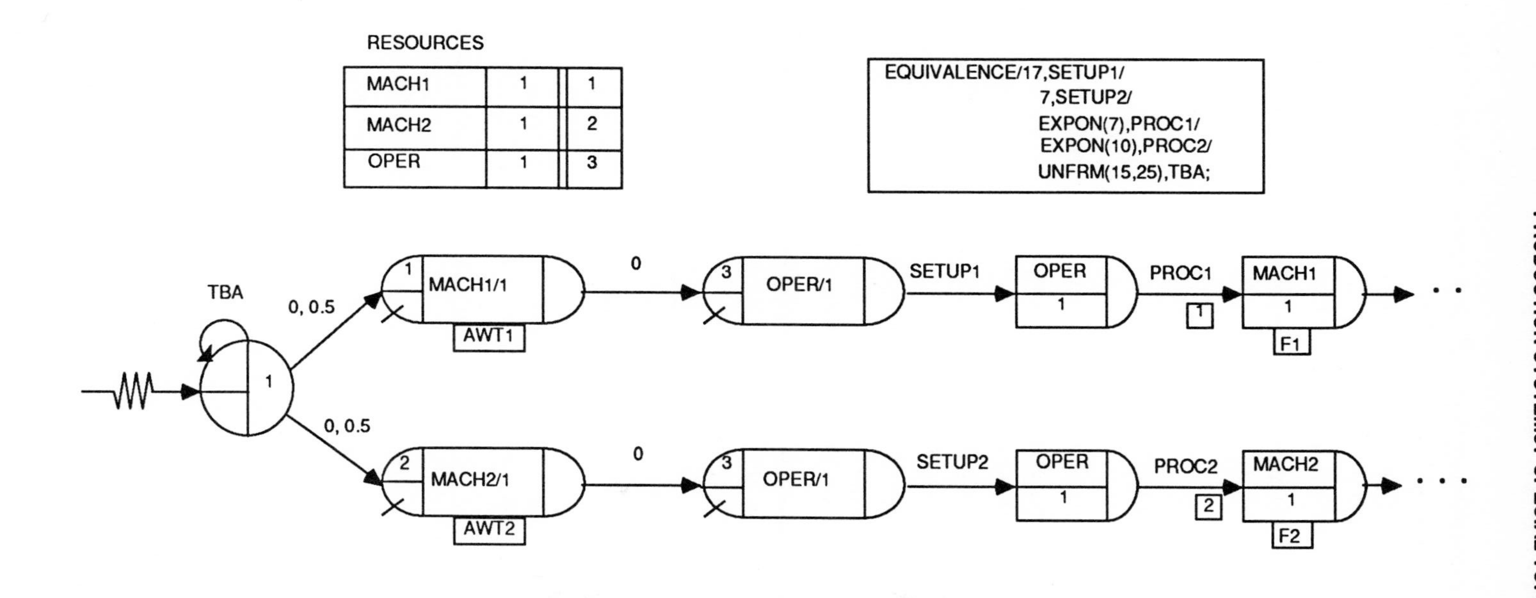

Figure 16-5 SLAM II model of two machines using a single operator for setup.

Through the use of resources, only one setup is ongoing at a time, and each job awaits setup by the operator if necessary. This sequencing constraint will increase the idle time of the machines under heavy job loads and, for this reason, may be an important consideration. This example can be easily extended to situations involving additional machines and operators. Furthermore, additional types of critical resources may also be modeled. The use of the GWAIT and GFREE nodes provides expanded modeling capabilities for this situation.

16.4.2 Modeling Labor Shifts

Another aspect of resource-constrained queueing situations involves a fluctuating supply of one or more resources. A disjoint network comprised of ALTER nodes can be used to model changing machine availability. The network in Figure 16-6 represents shift changes for operators. In this example, the CREATE node generates a first entity at eight time units (hours) into a working day. An additional entity is created every 24 hours. Entities arriving at ALTER node ALT1 increases the resource, OPERATOR, by one. After an 8-hour delay, an entity arrives to ALTER node ALT2 and decreases the resource OPERATOR by 1. The effect of this network segment is to increase the number of operators by 1 during the second shift each working day. Another method that can be used to model work shifts uses gates, as shown in Figure 16-7. This network segment controls the flow of entities in a job entity network by employing AWAIT nodes to model system logic based on the status of the gate SHIFT.

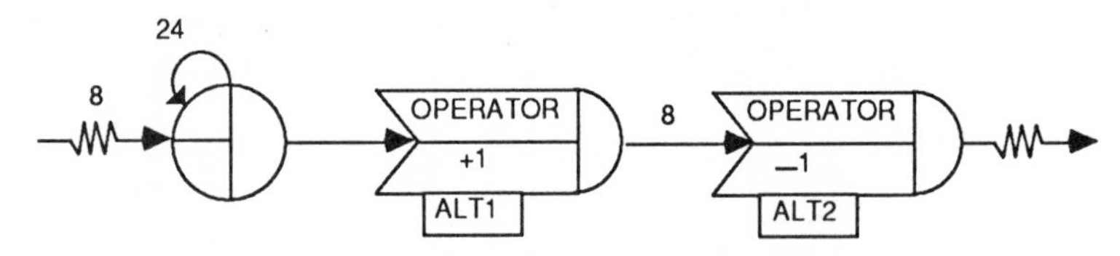

Figure 16-6 SLAM II network to alter operator availability.

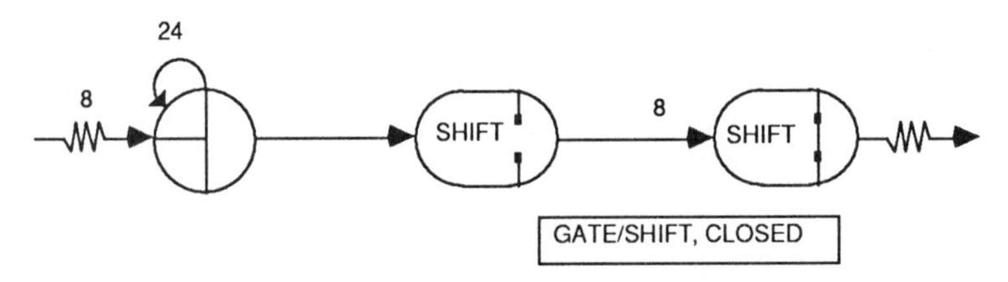

Figure 16-7 SLAM II network to control work shifts.

16.4.3 The SELECT Node Method

The assembly (ASM) queue selection option for SELECT nodes may be used to model multiple resource-constrained queueing situations. This approach recognizes that in a labor-constrained queueing situation a job must be at a station, a machine must be available, *and* an operator must be available. It is the *and* logic that is represented by the SELECT node assembly (ASM) operation. A SELECT node with the ASM rule does not initiate service unless there is an entity in each QUEUE node preceding the SELECT node. The network in Figure 16-8 represents a single-machine, single-operator situation modeled with the SELECT node having the ASM option. An entity in each of the QUEUE nodes on the input side of the SELECT node SASS satisfies the requirement for service to begin. An entity in QUEUE node Q1 represents an incoming job; in node Q2, an available machine; and in node Q3, an available operator. Nodes Q2 and Q3 are pre-loaded with one entity at the start of a simulation run. When a job arrives at node Q1, SELECT node SASS assembles the three entities into one and initiates service activity 1, which represents a machine setup. After setup, the operator becomes available, and GOON node GON1 routes the operator back to node Q3 to represent operator availability. Activity 10 represents either a physical movement or an information signal indicating that the operator is available.

Node GON1 also routes a job entity over activity 2, which represents machine processing. After machining, GOON node GON2 routes the job entity to other work areas (not shown) and also returns an entity to node Q2 to represent the machine's availability.

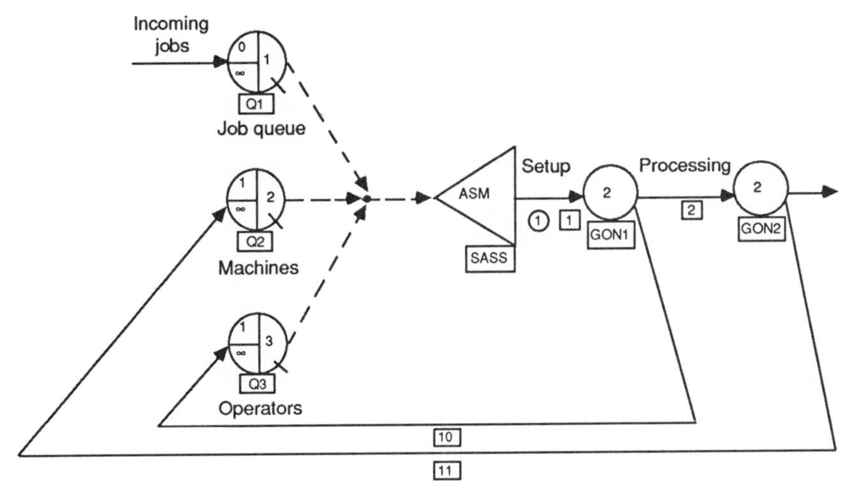

Figure 16-8 SELECT node approach to labor-constrained queueing situation.

In this model, the next setup cannot begin even though there may be jobs waiting in node Q1 until the machine completes its processing on activity 2 and its availability is signaled to node Q2 following activity 11. If setup is allowed during processing, then two SELECT nodes in series are used, assembling the job and operator and then the job and the machine.

Statistics on the number of entities in QUEUE node Q2 provide the fraction of time the machine is not being used for setup or processing. Similarly, the average number in QUEUE node Q3 is the fraction of time the operator is idle. The fraction of time spent in setup is obtained from the average utilization of activity 1, and the fraction of time spent machining is the average utilization of activity 2.

Activities 10 and 11 represent the time to make the operator and machine available after setup and machining, respectively. This time can be zero, or it can be used to model postprocessing delays. Postprocessing delays may be complex, and network segments may be necessary to model them.

One method of modeling shift changes for the operator is to define attribute 1 as the start time of the next shift. By using conditional branching at GOON node GON1, the operator return can be delayed until the condition [TNOW.GE.ATRIB(1)] is satisfied. In addition to delaying the operator when the condition is satisfied, the time of the next shift would need to be updated at the end of the working shift.

The network presented in Figure 16-8 can be enhanced to model multiple machines and multiple operators by making appropriate changes to the initial queue lengths, the queue capacities of QUEUE nodes Q1 and Q2, and the number of parallel servers prescribed for activities 1 and 2. For nonidentical operators, parallel branches emanating from the SELECT node would be employed. If nonidentical machines are to be modeled, a QUEUE node-SELECT node combination would be required to model machine processing.

A direct extension of the model in Figure 16-8 is presented in Figure 16-9, where the operator performs setup operations at two different work stations. In this case, QUEUE node Q3 is associated with two SELECT nodes, and there is a choice of which work station to go to if both are ready for setup. This choice is specified on input by indicating the order in which to poll the SELECT nodes associated with QUEUE node Q3 when an entity arrives at node Q3.

16.5 MODELING INSPECTION PROCESSES AND REWORK ACTIVITIES

An inspection activity may be performed after a machining operation that may result in some jobs being reworked or rejected. The results of inspection, acceptance, rework, and rejection are prescribed by specified probabilities for

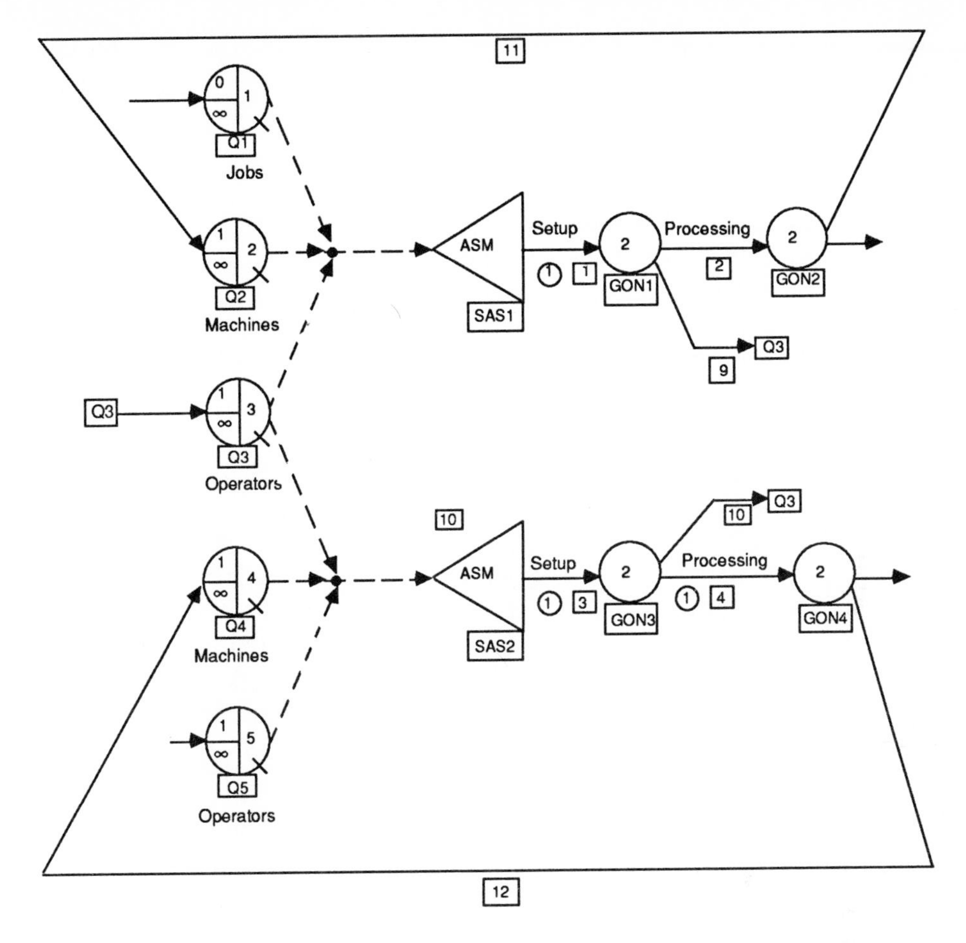

Figure 16-9 Model of a labor-constrained, two-machine queueing situation.

the branches following the inspection operation. To illustrate, consider the case of a single machine operation followed by an inspection process that is performed by either of two inspectors. Past data indicate that 5% of all jobs are rejected, 10% require reworking, and 85% are accepted. This situation is modeled in Figure 16-10a. Work arrives at QUEUE node Q1. Machine processing is represented by server 1. Finished work awaiting inspection waits at QUEUE node Q2. Server branch 2 represents 2 parallel inspectors. Branching from the GOON node GON1 is probabilistic in accordance with the probabilities prescribed for the outcome of an inspection. Accepted work is routed to COLCT node ACPT, rejected work is routed to COLCT node REJT, and jobs that are to be reworked are routed back to QUEUE node Q2 after a time delay for the required processing.

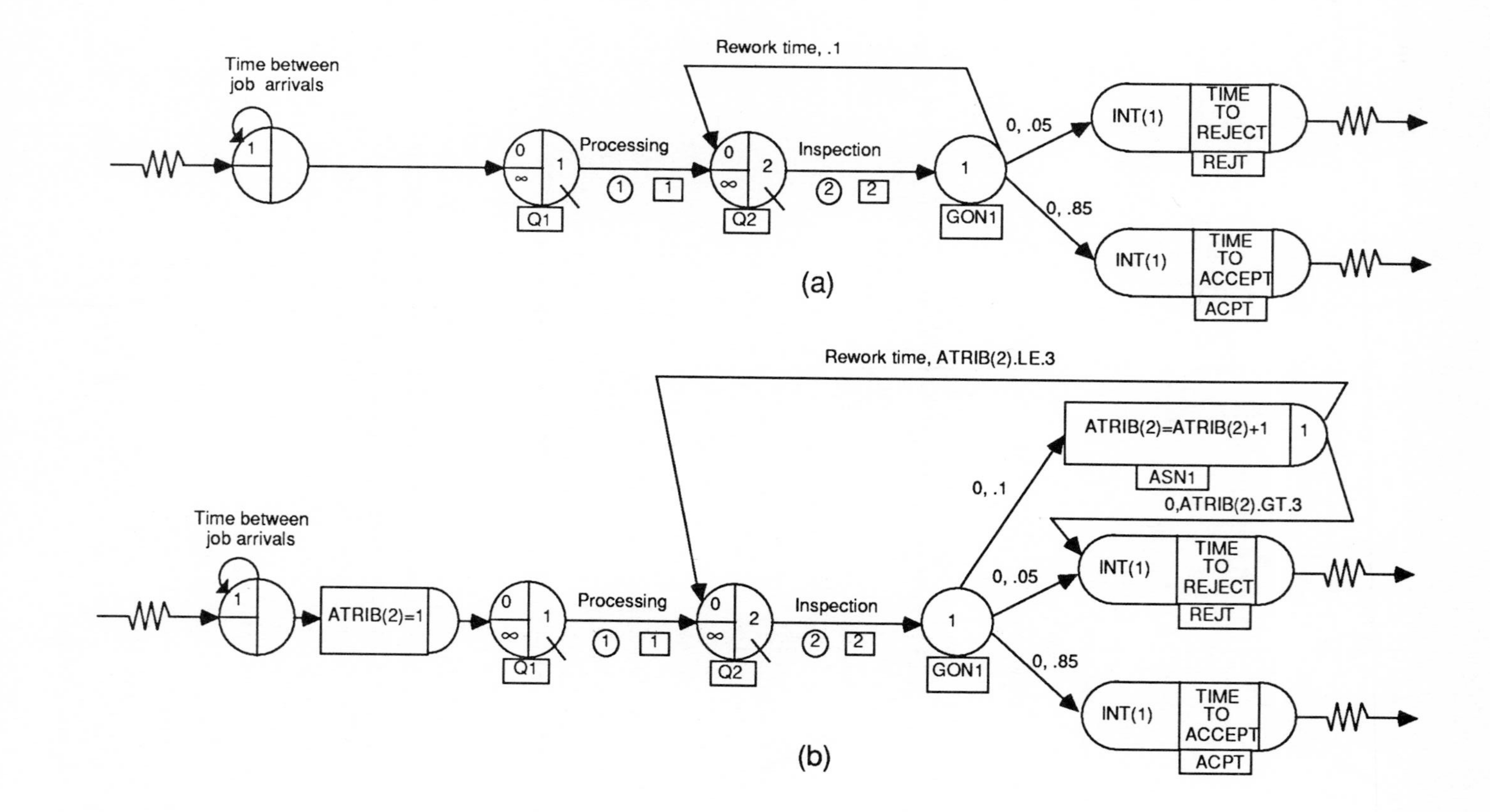

Figure 16-10 Modeling of inspection activities.

Theoretically, an entity can loop indefinitely from node Q2 to node GON1 and then back to node Q3. However, on each loop there is only a 0.10 probability of returning to Q2. To limit the number of loops, attribute assignments can be used as shown in Figure 16-10b. Attribute 2 is used to represent the number of times a job has been machined and returned for rework. The value in attribute 2 is incremented by 1 each time an entity representing rework passes through ASSIGN node ASN1. ASN1 is specified to perform conditional branching to limit the number of rework operations for any one job to three. A job that has failed inspection and been routed for rework for the third time is rejected by routing it from node ASN1 to node REJT.

Note that, by changing the queue ranking procedure of node Q2 to HVF(2), priority can be given to the inspection of jobs that have been reworked. It may also be desirable to have a different inspection time for reworked items. This is accomplished by introducing an ASSIGN node between the CREATE node and node Q1 to assign an initial value to attribute 3 for normal machine processing times. ATRIB(3) is then prescribed as the time duration for service activity 2. ASSIGN node ASN1 would also prescribe a value for attribute 3 when reworking is required.

Another embellishment is to modify the branching probabilities of GOON node GON1 based on the number of times the job has been reworked. This conditional probability may be modeled as shown in Figure 16-11a. As before, an ASSIGN node after the CREATE node sets the initial value of attribute 2 equal to 1. Node GON1 is then used to determine which loop is in progress. GOON nodes G1, G2, and G3, with their associated branching probabilities, specify the chance of each inspection outcome for each possible loop value. An alternative method for representing this situation is to use attribute-base probabilistic branching, as shown in Figure 16-11b. Here, the probabilities associated with each branching condition are specified as data in arrays. Branching from node GON1 is then conditionally based on the value of attribute 2, which stipulates the row of ARRAY to be used and the probability data element in the ARRAY.

16.6 MODELING WORK FLOW IN BATCHES OR LOADS

In some models, it is convenient to use a single entity to represent a batch or a load of individual units. In modeling the flow of checks through a large bank, for example, where hundreds of thousands of checks are processed each day, modeling batches of checks is a reasonable strategy. Since all models are abstractions of reality, there should not be a reluctance to model situations at an aggregate level when necessary. Remember, it is not the one-to-one correspondence between system items and model entities that is of the highest import,

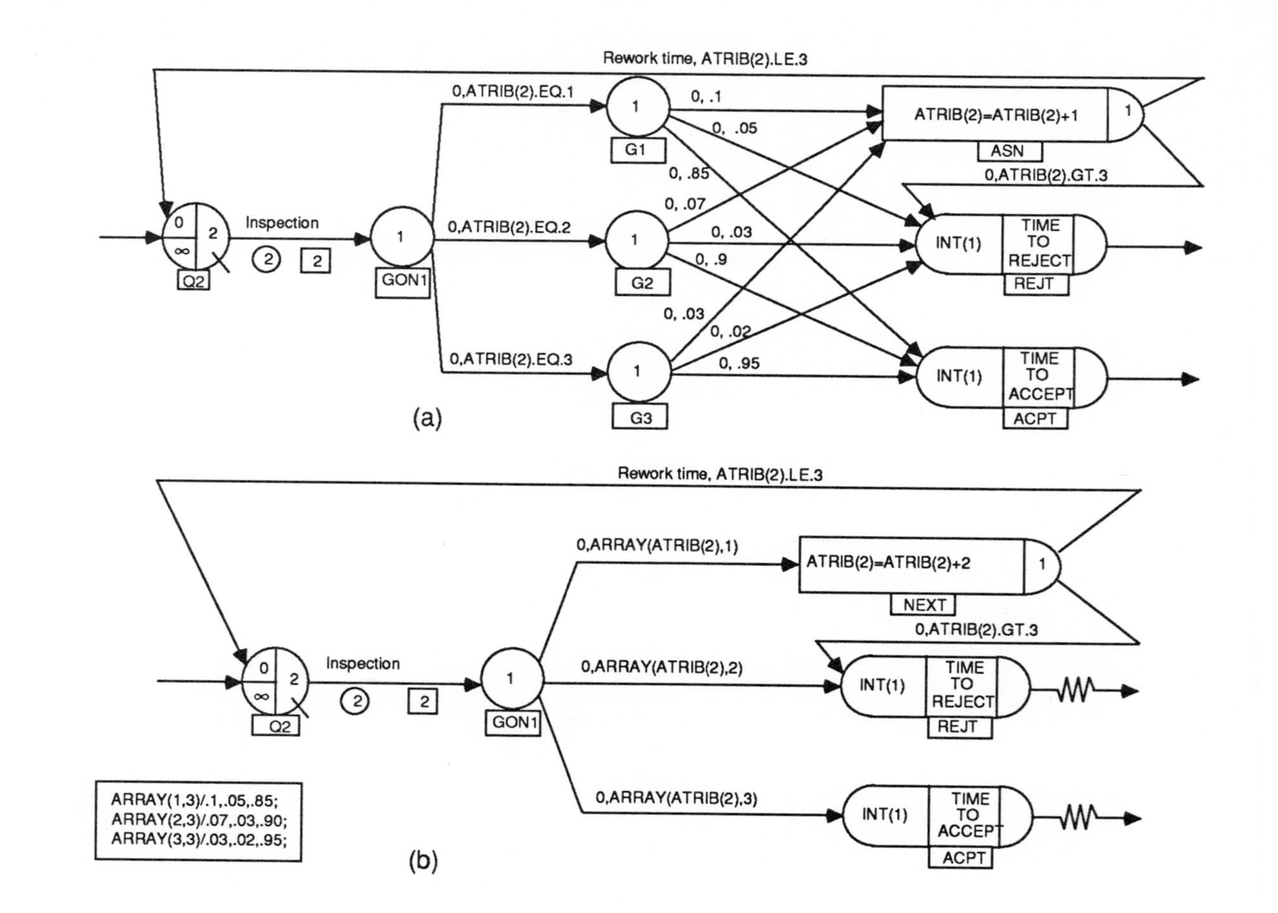

Figure 16-11 Models of inspection activities with variable probabilities of passing inspection.

but the timeliness and usefulness of the model outputs in helping to make decisions.

When a single entity represents a batch of units, an attribute of the entity is set equal to the size of the batch. Since batch size may vary, this attribute identification is important information that is used in defining service times, routing, job loading, and job sequencing. When an entity represents a batch, performance measures must take cognizance of different weights to be attached to the entity based on the batch size. In most cases, this is accomplished by computing statistics on batch size, as was done in the throughput analysis discussed in Section 15.2.1.

It is sometimes necessary to model the flow on individual units through a portion of the network, then accumulate units into loads, route the loads through other portions of the network, and then disassemble loads into units again. This occurs with material handling models, where individual items are assembled into pallet loads prior to transport and then disassembled into individual items later. SLAM II provides procedures for accumulating entities into loads and breaking loads down into units. For example, if ten entities are to be accumulated into one batched entity, an ACCUMULATE node or a BATCH node may be used, as shown in Figure 16-12. Either of these nodes requires ten incoming entities for node release. Regardless of when each incoming unit arrives at the node, a load departs only after each tenth arrival. The attributes of the outgoing load or batched entity are specified by the SAVE criterion. A primary difference between the ACCUMULATE and BATCH nodes is that individual entities included in a batch can be RETAINed by an ALL(NATRR) specification for a BATCH node. The time a batch is created can be marked by inserting an ASSIGN node immediately following the node where the batch is created and setting an attribute value equal to TNOW.

If incoming entities represent loads of varying size and it is desired to accumulate these until a larger load of a specified size is reached, a BATCH node would be employed to perform the accumulation. To illustrate, assume that attribute 1 represents load size and that a resulting load size as close to, but not exceeding, 100 is desired. A BATCH node to meet this specification is shown

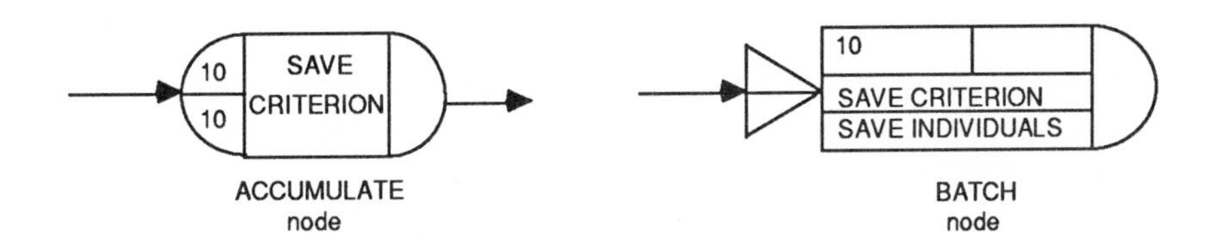

Figure 16-12 ACCUMULATE and BATCH nodes to group 10 entities.

in Figure 16-13. For this node, the threshold is 100 and the attribute specification is 1. Therefore, when the sum of the attribute 1 values for each batch member exceeds 100, a batched entity is released from the node. The assignment of values to attribute 1 is assumed to have been made in the network prior to an entity's arrival at the BATCH node.

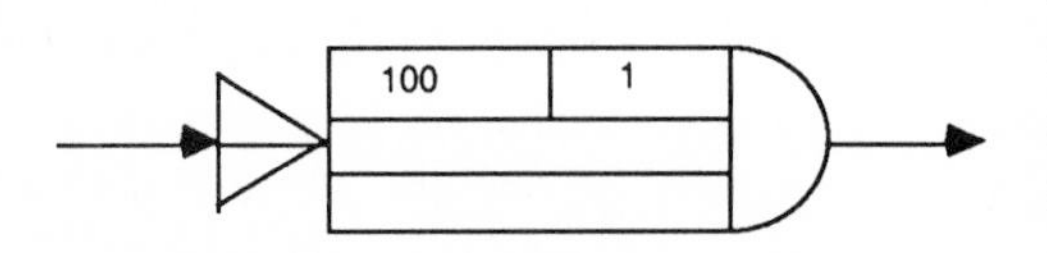

Figure 16-13 BATCH node with a threshold for ATRIB(1).

The ASM option of the SELECT node may also be used to assemble entities into loads. The use of this option for assembling entities in assembly operations was discussed in Section 16.4.3. MATCH nodes also provide a specialized assembly logic, since a MATCH node can be used to halt the flow of entities until a group with the same attribute value is available. The MATCH node can then route all the units to an ACCUMULATE or BATCH node for assembly. The QBATCH node combines features of the BATCH and AWAIT nodes so that a resource is requested when a batch is formed.

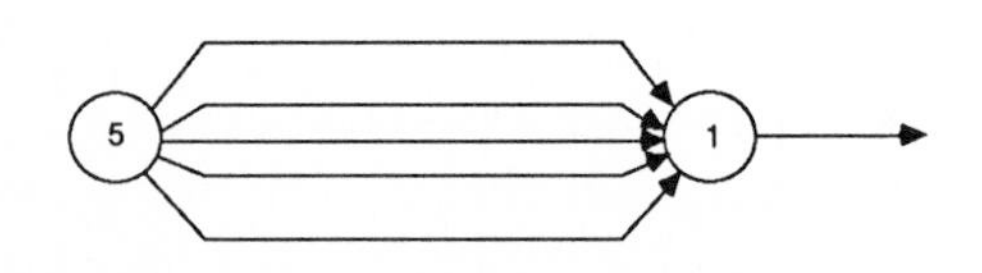

Figure 16-14 Disassembly by selecting multiple branches.

Several procedures are available to model disassembly. A basic method involves the use of multiple routing. To generate five entities from one incoming entity, a GOON node with five emanating branches may be used as in Figure 16-14. An alternative to using multiple branches is to use a user-written subroutine in conjunction with an ENTER node in a network segment. An ENTER node with number NUM is released at the arrival of each entity or whenever subroutine ENTER(NUM) is called from the user-written subroutine. Another alternative is to employ attribute-based routing using an ASSIGN node coupled with an UNBATCH node, as shown in Figure 16-15. Each arriving entity to the UNBATCH node generates ATRIB(2) entities.

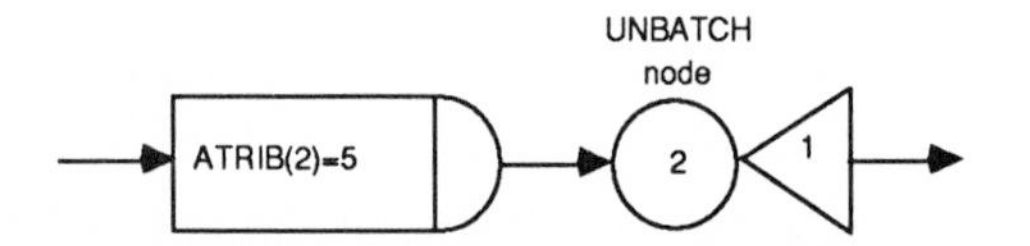

Figure 16-15 UNBATCH node for generating multiple entities.

A similar procedure is used to reinsert unit entities from a batched entity if the RETAIN option is specified at the BATCH node where the batched entity is created. In the example in Figure 16-16, attribute 1 is specified to retain an

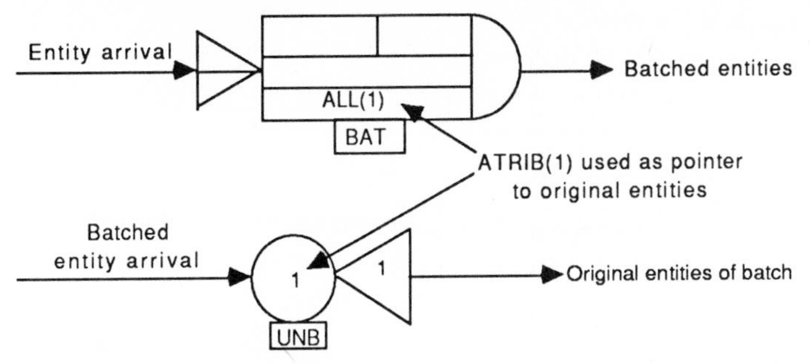

Figure 16-16 UNBATCH node for generating previously batched entities.

internal SLAM II pointer to the entities that are batched at BATCH node BAT. When the batched entity is unbatched at UNBATCH node UNB, the original entities are routed from node UNB.

16.7 JOB SHOP ROUTING USING ROUTE SHEETS

In this section, an illustration is presented to demonstrate how conditional branching may direct jobs through work stations based on a predetermined route sheet that is associated with each job. A route sheet specifies the sequence of work stations that a job is to visit. In Chapter 17, the use of branching as a loading mechanism is discussed along with the use of the queue selection capability of SELECT nodes to extend routing options by taking into account the status of associated QUEUE nodes.

Consider the job shop depicted in Figure 16-17, where a job's route sheet is specified in the SLAM II global variable ARRAY. Job 1 requires processing at station 2 and then at station 1. Job 2 requires processing at station 2 first, then at station 3. Job 3 is to be routed to all three stations in the order of 1, 2, and 3.

Attribute information is combined with the route sheet data to provide the sequencing for job entities. Attribute 1 of an entity is defined as a job's route sheet type. The value of this attribute is used in the network to access the array where the data for the sequencing of the job are stored on input. The value of attribute 2 for the entity represents a job's operation number and is used to access the number of the next work station. These data are stored in ARRAY by route sheet type. In Figure 16-17, three rows of ARRAY define three routes consisting of the work stations in the order: (2, 1); (2, 3); and (1, 2, 3). The initial values of attributes 1 and 2 for an entity with route sheet type 2 are 2 and 1, respectively. The last element of each row of ARRAY is zero. When the route station is zero, the entity exits from the work area.

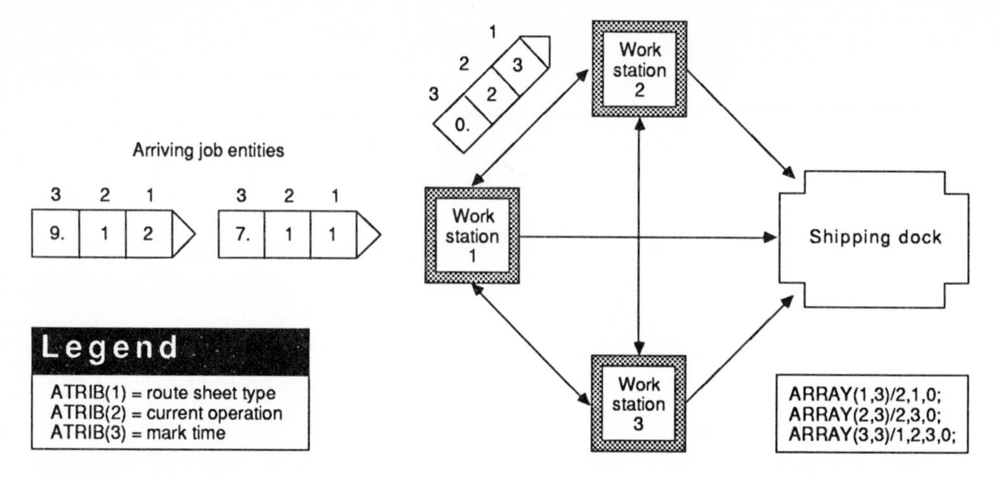

Figure 16-17 Schematic diagram of job route sheets in a work station environment.

The SLAM II network segment depicted in Figure 16-18 illustrates the routing of jobs using ARRAY and attributes as defined above. The initial attribute values for jobs arriving in the network are established prior to the entry of an entity at ASSIGN node ASN1. When a job arrives at node ASN1, the SLAM II global variable II is set equal to the value contained in ARRAY(ATRIB(1), ATRIB(2)). For example, if the first attribute of the entity is equal to 2 and the second attribute is equal to 1, then II = ARRAY(2,1) = 2. The job is then conditionally routed from node ASN1 according to the value of the variable II. Assume as above that II = 2, in which case the job is routed to QUEUE node Q2 to wait

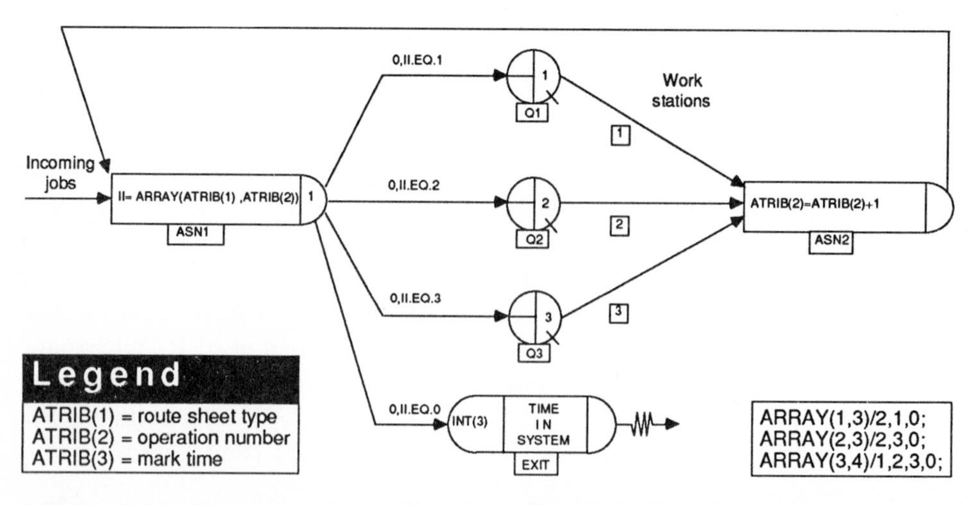

Figure 16-18 Work station routing using route sheets with II as next work station.

for processing by service activity 2. Upon completion of service, each job is routed to node ASN2, where the current value of attribute 2 is incremented by 1. The job is then routed back to node ASN1. The cycle continues for each job until the variable II = 0, indicating that all processing steps have been completed for the job. The job is then routed to the COLCT node EXIT, where statistics are collected, and the entity exits the network segment.

A generalized network segment for work station routing using work sheets is shown in Figure 16-19. This network may be used to model any number of work stations. Attribute 4 represents a work station number, and job sequences are recorded in ARRAY as in the above example. Conditional branching from the ASSIGN node ASNG is based on the value of ATRIB(4). Files of the work station queues are defined by the value of ATRIB(4) which also defines the service activity numbers following the QUEUE nodes. Processing times are established by the variable PROCESSTIME, which may be made equivalent to a row of ARRAY. The network segment in Figure 16-19 illustrates the modeling versatility inherent in SLAM II to model complex routing situations.

Sometimes it is desirable to route entities in a probabilistic fashion. This is similar to classifying entities based on inspection as discussed in Section 16.6. An advantage of this method is that it provides a graphical representation of the networks. An example of the use of a probabilistic-based routing model is shown in Figure 16-20. An entity is generated at the CREATE node in accordance with a time-between-creation specification. Each entity is routed over one of the activities probabilistically to ASSIGN node A1, A2, or A3. These nodes

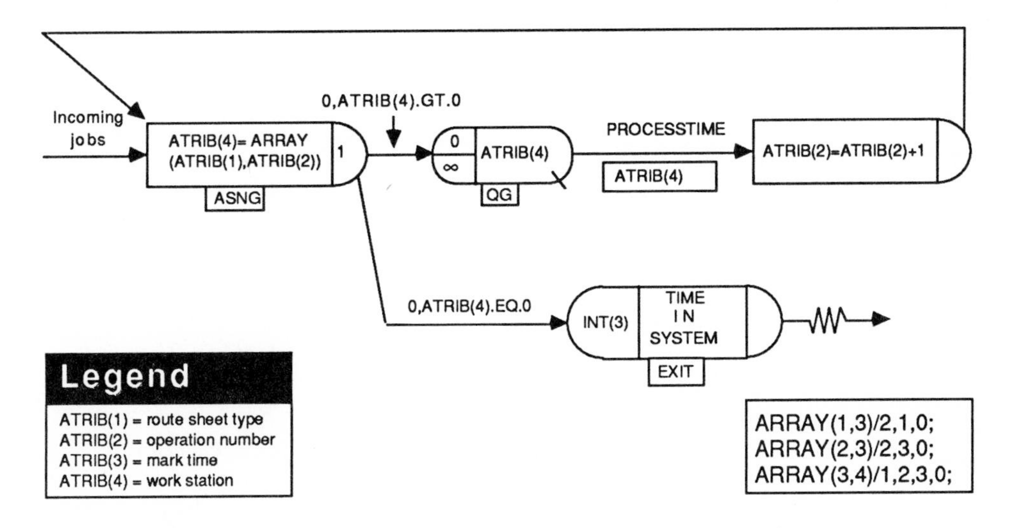

Figure 16-19 Generalized routing model.

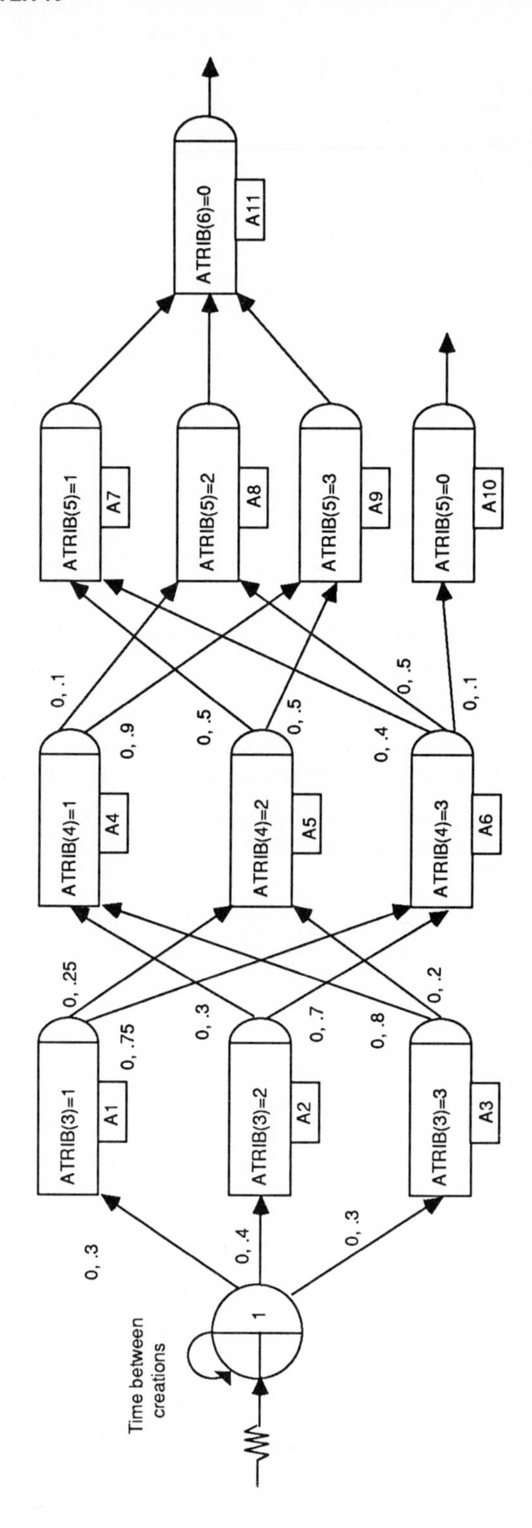

Figure 16-20 Probabilistic determination of work stations for a route sheet.

assign the first station number to attribute 3. From Figure 16-20, there is a 30% chance that station 1 is first, a 40% chance that station 2 is first, and a 30% chance that station 3 is first. Probabilistic routing from these nodes routes entities to ASSIGN nodes, where attribute 4 is assigned a second station number. Similarly, nodes A7, A8, and A9 define attribute 5 as the third station. Note that if station 3 is second on the route sheet there is a 10% chance that a third station will not be visited.

A more compact method for generating probabilistic routing using a network is to define sample route numbers from a probability mass function using DPROBN. The values in the elements of the first row of the array define the possible station numbers. The elements in all other rows of ARRAY contain the cumulative probabilities associated with the order in which a station is to be selected for job processing. The SLAM II function DPROBN(IC,IV) is used to access these probabilities and assign operation sequencing numbers to an entity's attributes. A network segment that can be used to duplicate the logic presented in Figure 16-20 is illustrated in Figure 16-21.

In Figure 16-21, when an entity arrives at ASSIGN node A1, ATRIB(3) is set to the station number for operation 1 using function DPROBN to select an element from ARRAY(1,4) according to the cumulative probabilities in ARRAY(2,4). The global variable II is then set equal to the value in attribute 3 times 10, then incremented by 2; that is, II = 10 * ATRIB(3) + 2. The value of II is 12, 22, or 32 and points to the row in ARRAY where the cumulative probabilities are located for the second operation. The work station number of the second operation is established by drawing a sample value using DPROBN(II,1) and setting ATRIB(4) to the sample. The entity is then conditionally routed from node A1. The entity is then sent to node A2 to determine the value of ATRIB(5), the station number for the third operation, using the same procedure as was

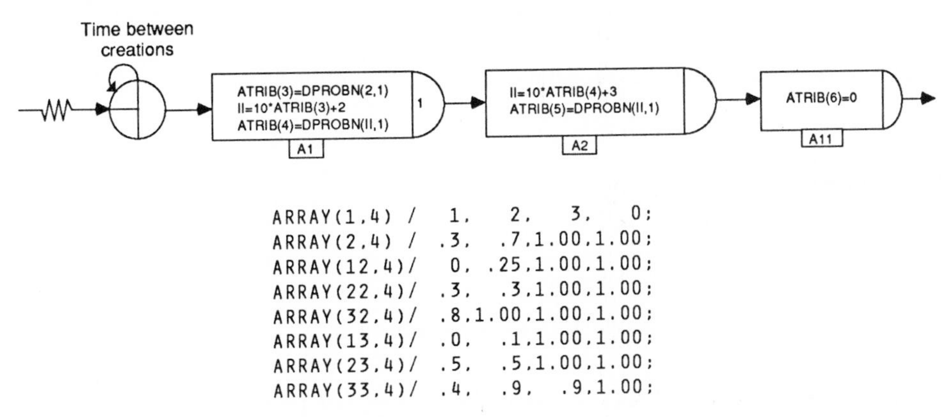

Figure 16-21 Probabilistic determination of work stations for a route sheet.

established for ATRIB(4) where in this case II is set to 13, 23, or 33. If II is 33, then there is a 10% chance that ATRIB(5) is set to 0. At node A11, ATRIB(6) is set to 0 to indicate the end of the route sheet.

The above examples produce routes for entities that included two or three station numbers. The procedure can be generalized as shown in Figure 16-22 for up to nine stations. The variable COUNT is made equivalent to the number of times activity 1 has been traversed. Thus, COUNT starts with the value of 0 and is increased by 1 for each station placed on a route. In ASSIGN node AROUT, II on the first assignment is set equal to 2 as NEXTSTAT [made equivalent to XX(1)] is initially 0. The first station on the route is obtained as a sample from function DPROBN using the probabilities in row 2 of ARRAY and the station number in row 1. This value of NEXTSTAT is placed in ATRIB(3) since COUNT is equal to 0 when the first assignment in AROUT is made. If NEXTSTAT is not 0, then a return to node AROUT is made for the next asssignment of a station to the route. If NEXTSTAT equals 0, the setting of the routing is completed.

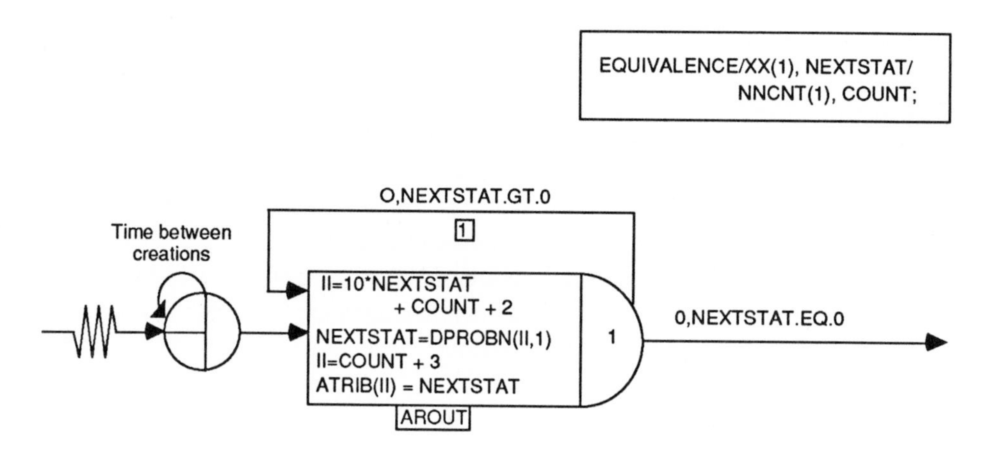

Figure 16-22 Generalized probabilistic work station route specification.

16.8 SUMMARY

The network concepts presented in this chapter may be used in conjunction with standard production planning models for evaluation purposes. The example network segments provide the basis for modeling special production planning topics. The material is consistent with the basic modeling approach proposed in the book: start with basic models and add new features as they are required to meet the purpose of the project.

16.9 EXERCISES

16-1. Consider a system that has two job types, a single work station, and a rule that has type 1 jobs preempting type 2 jobs. Discuss the effects of preemption (a) on the average time for each job type to proceed through a work station, (b) on the utilization of a work station, and (c) on the time spent in a queue by all jobs.

16-2. Specify the changes required in in Figure 16-1 if the job that is preempted is not given priority over other type 2 jobs in the queue, and if the preempted job must be restarted.

16-3. Build a network model for preempting a server when there are three categories of jobs. Jobs that are preempted are to be placed at the front of their job class, and their processing time is to be equal to the remaining processing time plus 0.1. Include in the model the restriction that the new processing time is not permitted to exceed the original processing time.

16-4. Build a model of a single-machine situation where the machine requires maintenance. Assume that maintenance activities are represented as jobs and are given priority over regular jobs. Maintenance jobs, however, do not preempt regular jobs.

16-5. Build a model of a single-machine situation that requires that maintenance be performed after every 25 services are completed on the machine.

16-6. A company produces two types of electronic meters that are installed in residential buildings to measure power consumption. Meters A and B are designed for different voltage and amperage ranges. The meters consist of two subassemblies, labeled C and D, and two parts, labeled E and F. Meter A is assembled out of subassemblies C and D and two part E's. Meter B is assembled out of subassemblies C and D and one part E and one part F. Subassembly D consists of one part E and one part F. The production process involves the building of parts E and F and their inspection. After the inspection, defective parts are removed from the line. Part E has a probability of 0.05 of being rejected, and the probability of part F being rejected is 0.04.

The basic parts E and F are maintained in inventory. They are assembled into subassemblies C and D, which are also inventoried. In addition, the final products, meters A and B, are inventoried. The processing times for all parts have been determined to be normally distributed (truncated). The mean, standard deviation, minimum value, and maximum value of these normal distributions are presented in the following table.

Item	Mean	Standard Deviation	Minimum Value	Maximum Value
A	5.0	0.25	0	10
B	8.0	1.00	0	20
C	6.0	1.00	0	10
D	4.0	1.00	0	10
E	2.0	0.25	0	10
F	1.2	0.50	0	10

Develop a SLAM II model to answer the following questions [9]:

(a) Given 50 units in initial inventory for parts E and F and 50 subassemblies C and D, how long does each process need to operate, on the average, to produce an order for 100 units of meters A and B? (Assume that no ending inventory is desired.)

(b) What is the production lead time required on the average to complete the scheduled orders of meters A and B?

(c) What percent of the time is each of the processes idle do to the lack of preprocess inventories?

(d) Is there sufficient production capacity to meet the order requirements in 1000 time units? Indicate which operations may be potential bottlenecks. How many units of each part are required to satisfy the order?

16-7. Describe how the SELECT node method for modeling labor-limited queues allows for the direct reassignment of operators to different activities.

16-8. Modify the network presented in Figure 16-5 so that there are two type 1 machines. It is desired to use the operator whenever possible to perform a set-up operation.

16-9. Modify the model presented in Figure 16-10b so that after each rework operation the probablity of requiring rework decreases by 0.02 and the probability of passing inspection increases by 0.02.

16-10. (From Reference 12) A production shop is comprised of six different groups of machines. Each group consists of a number of identical machines of a given kind, as indicated below.

Machine Group Number	Types of Machines in Group	Number of Machines in Group
1	Casting units	14
2	Lathes	5
3	Planers	4
4	Drill presses	8
5	Shapers	16
6	Polishing machines	4

Three different types of jobs move through the production shop. These job types are designated as type 1, type 2, and type 3. Each job type requires that operations be performed at specified kinds of machines in a specified sequence. All operation times are exponentially distributed. The visitation sequences and average operation times are shown in the following table. Jobs arrive at the shop with exponential interarrival times with a mean of 9.6 minutes. Twenty-four percent of the jobs in this stream are of type 1., 44% are of type 2, and the rest are of type 3. The type of arriving job is independent of the job type of the preceding arrival. Build a SLAM II model that simulates the operation of the production shop for five separate 40-hour weeks to obtain the distribution of job residence time in the shop, as a function of job type; the utilization of the machines; and queue statistics for each machine group.

Job Type	Total Number of Machines to be Visited	Machine Visitation Sequence	Mean Operation Time (min)
1	4	Casting unit	125
		Planer	35
		Lathe	20
		Polishing machine	60
2	3	Shaper	105
		Drill press	90
		Lathe	65
3	5	Casting unit	235
		Shaper	250
		Drill press	50
		Planer	30
		Polishing machine	25

Embellishments

(a) Employ a shortest-processing-time rule for ordering jobs waiting before each machine group. Compare output values.

(b) Give priority to jobs on the basis of type. Job type 3 is to have the highest priority, then type 2 and then type 1 jobs.

(c) Change the average job interarrival time to 9 minutes and evaluate system performance.

(d) Develop a cost structure for this problem that would enable you to specify how to spend $100,000 for new machines.

16.10 REFERENCES

1. Adam, N., and J. Surkis, "A Comparison of Capacity Planning Techniques in a Job Shop Control System," *Management Science*, Vol. 23, 1977, pp. 1011-1015.
2. Bedworth, D. D., and J. E. Bailey, *Integrated Production Systems*, Wiley, New York, 1982.
3. Bredenbeck, J. E., M. G. Ogdon III, and H. W. Tyler, "Optimum Systems Allocation: Applications of Simulation in an Industrial Environment," *Proceedings, Midwest AIDS Conference*, 1975, pp. 28-32.
4. Felder, R. M., P. M. Kester, and J. M. McConney, "Simulation/Optimization of a Specialties Plant," *Chemical Engineering Progress*, June 1983, pp. 84-89.
5. Gross, J. R., S. M. Hare, and S. Roy, "Simulation Modeling as an Aid to Casting Plant Design for an Aluminum Smelter, *Proceedings, 1982 IMACS Conference*, Vol. 2, 1982, pp. 160-161a.
6. Hancock, W., R. Dissen, and A. Merten, "An Example of Simulation to Improve Plant Productivity," *AIIE Transactions*, Vol. 9, 1977, pp. 2-10.

7. Hogg, G. L.,and others, "GERTS QR: A Model of Multi-resource Constrained Queueing Systems, Part I: Concepts, Notations, and Examples," *AIIE Transactions*, Vol. 7, No. 2, 1975, pp. 89-99.
8. Hogg, G. L., and others, "GERTS QR: A Model of Multi-Analysis of Parallel Channel, Dual Constrained Queueing Systems with Homogeneous Resources," *AIIE Transactions*, Vol. 7, No. 2, 1975, pp. 100-109.
9. Huang, P. Y., E. R. Clayton, and L. J. Moore, "Analysis of Material and Capacity Requirements with Q-GERT," *International Journal of Production Research*, Vol. 20, No. 6, 1982, pp. 701-713.
10. Lilegdon, W. R., C. H. Kimpel, and D. H. Turner, "Application of Simulation and Zero-One Programming for Analysis of Numerically Controlled Machining Operations in the Aerospace Industry," *Proceedings, Winter Simulation Conference,* 1982, pp. 281-290.
11. Nelson, R. T., "A Simulation Study of Labor Efficiency and Centralized Labor Assignments in a Production System Model," *Management Science*, Vol. 17, No. 2, October 1970, pp. B97-B106.
12. Schriber, T., *Simulation Using GPSS*, Wiley, New York, 1974.

17 SCHEDULING AND SEQUENCING

17.1 INTRODUCTION

Scheduling is the process of determining the starting time of jobs and the machines on which jobs are to be performed. The scheduling process involves the fol-lowing stages [2]:

1. Aggregate planning
2. Loading
3. Sequencing
4. Detailed scheduling

Aggregate planning is the activity of determining the overall level of output for a given time period and the level of resources to be deployed during the time period of interest. For example, a production manager's aggregate plan may call for producing 1000 units in July using 10 machines and 20 workers. The aggregate plan does not specify starting times, the order of production, the machines to be used, or the workers to be employed.

The second stage in the scheduling process is called loading. Sometimes, the term shop loading or machine loading is used. ***Loading*** is the action of allocating work to machine areas. Loading also establishes the load each machine center will carry under a given aggregate plan. A typical loading specification is that 50 jobs of type 1 are to be performed in work area B. Loading does not involve a specification of the order to perform jobs.

The ordering of jobs is referred to as ***sequencing***, or ***work dispatching***. Sequencing implies establishing priorities for processing jobs at work centers. The term job sequencing refers to the determination of which job is to be performed next when a machine becomes available. The term machine sequencing is used for the process of determining which machine in a given work center is to be used next when more than one machine is idle.

The establishment of start and finish dates for jobs at a work center is the final stage of scheduling and is called ***detailed scheduling***. Detailed scheduling utilizes the plans developed during aggregate planning, loading, and sequencing and is, therefore, a direct consequence of these activities.

There are many ways to schedule a production system [3, 4]. The success of any scheduling process depends on the ability of the production system to function with respect to the performance measures throughput, meeting due dates, resource utilization, and in-process inventory. In Chapter 16, SLAM II network models illustrated a method for obtaining these system performance measures. In this chapter, the important elements of the scheduling process, including aggregate planning, loading, sequencing, and detailed scheduling, are incorporated into the network modeling process. SLAM II models are used to evaluate the effect of scheduling strategies on a production system's performance measures.

17.2 AGGREGATE PLANNING BASED ON JOB ARRIVALS

The aggregate plan specifies the level of resources to be used and the amount of product to be produced. In SLAM II, the level of resources for a system is specified by the capacity of each resource and the number of servers in the system. The amount of product that can be produced during a given time period is directly related to the quantity of raw material that is available for input to the system, the level of resources available to process the raw material, and the processing procedures used. In this section, SLAM II modeling of raw material inflow is presented.

As an example, assume that the aggregate plan requires that 50 items are to be produced over 1000 time units. In SLAM II, the length of the simulation run would be specified to be 1000 time units on the INITIALIZE statement (TTFIN). To specify that 50 entities are to enter the system at the beginning of the simulation run, the CREATE node in Figure 17-1 may be used. A CREATE node is initially released at the time of first release (TF), which if defaulted is at the start of a simulation. The maximum number of entities created for the node (MC) is set to 50. The time delay between arrivals (TBA) for each entity is specified as zero. Therefore, the 50 entities are placed in the network at the start of the simulation run. If it is desirable to

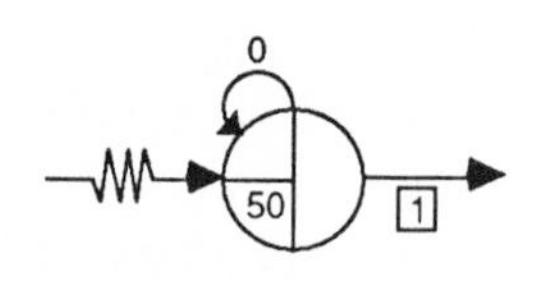

Figure 17-1 CREATE node.

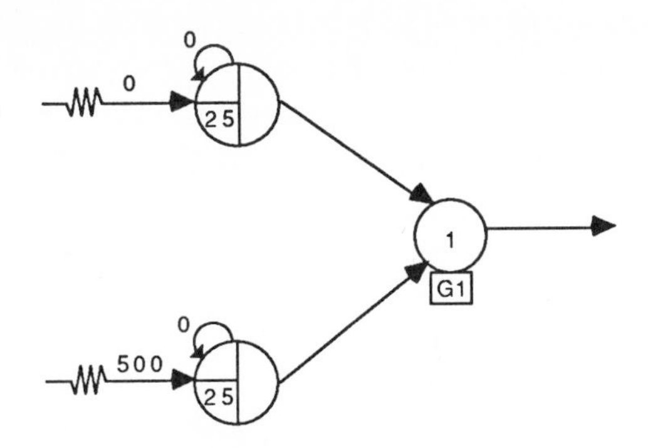

Figure 17-2 Two CREATE nodes

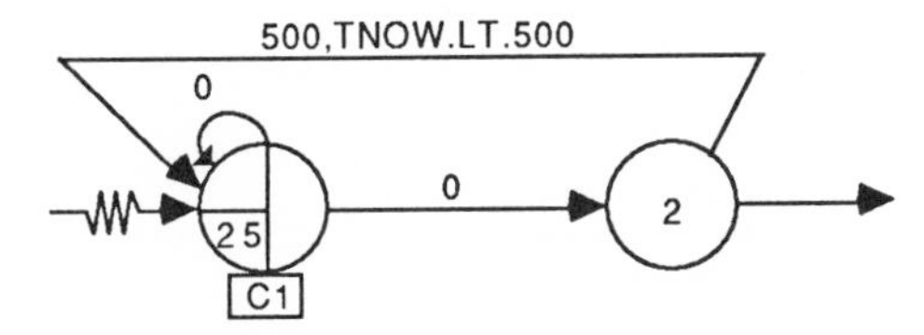

Figure 17-3 CREATE node with feedback

spread the entities equally over the 1000 time units for the simulation run, a TBA specification of 20 would be used.

The flexibility of the CREATE node to enter items into a production system is illustrated by the following two examples. In Figure 17-2, it is assumed that the first 25 entities are to enter the system at time zero, and the last 25 entities are to enter 500 time units later. This arrival pattern may also be modeled with a single create node, as shown in Figure 17-3.

The aggregate planning period, as mentioned above, corresponds to the length of a simulation run. This planning horizon may be directly specified or modeled as an activity on the network. Alternatively, activities representing environmental factors that determine the length of the planning period could be modeled by a network segment to establish a nonconstant planning period. For example, production might be delayed until financing is obtained, or the work is not to be accomplished until the first snowfall. An aggregate planning period could also be defined as the number of units produced by the system. This is modeled by a specification of the number of entities required to arrive at a TERMINATE node to end a simulation run.

17.2.1 Poisson Arrivals

Frequently, the arrival rate of individual items in a system is assumed to follow a Poisson distribution. When this occurs, the time between arrivals for units to the system is exponentially distributed as modeled by a CREATE node as in Figure 17-4 If the mean ar-

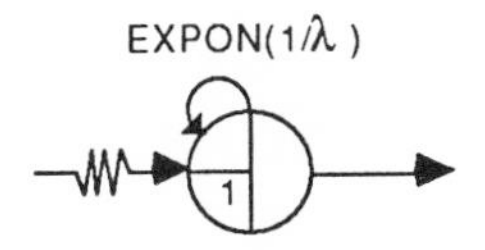

Figure 17-4 Exponentially distributed creations.

rival rate for a Poisson distribution is λ, the mean time between arrivals for the corresponding exponential distribution is 1/λ. Thus, the time between arrivals (TBA) in Figure 17-4 is drawn from the exponential distribution (EXPON) with a mean 1/λ. This discussion presumes that the Poisson arrivals occur one at a time.

17.2.2 Probabilistic Arrival Processes

SLAM II branching capabilities permit the modeling of complex arrival processes. Consider an arrival process in which the next arrival time is a sample from an exponential distribution with mean TBA 30% of the time and a constant TBA 70% of the time. A SLAM II network segment to model this complex mixed arrival distribution is shown in Figure 17-5. The number of creations for the

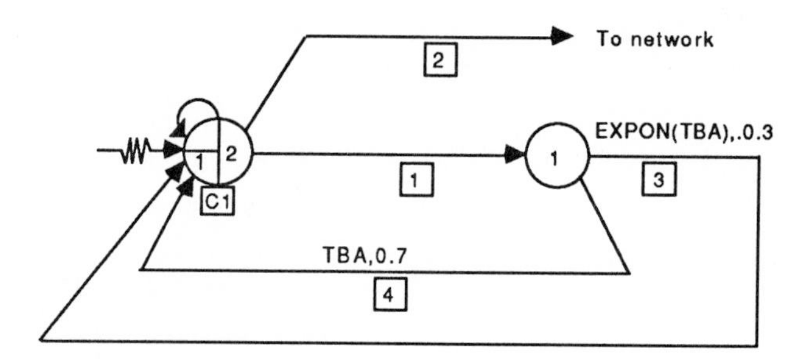

Figure 17-5 Mixed arrival process.

CREATE node is specified as 1 and the time between creations is defaulted to infinity. Two entities emanate from the CREATE node: one representing the current arrival and the other representing the next arrival.

Now consider the situation where one entity is created 60% of the time and two entities 40% of the time. In the network segment shown in Figure 17-6, activity 1 always sends an entity to the GOON node at an arrival time. For 40% of the arrivals, activity 2 is also taken, which causes two entity arrivals. Therefore, one entity arrives 60% of the time and two entities 40% of the time.

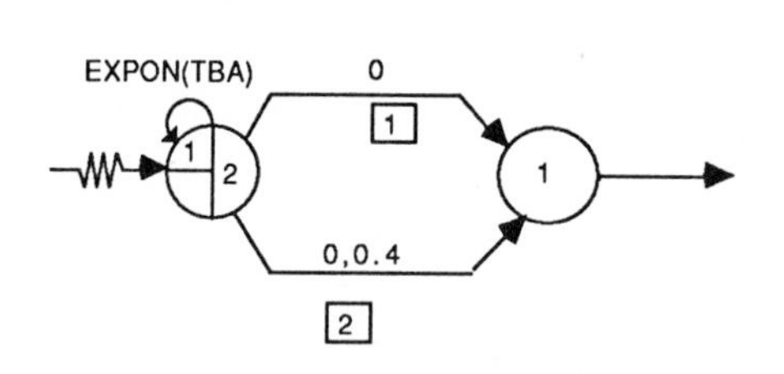

Figure 17-6 Multiple arrivals.

17.2.3. Fixed Arrival Times

In some situations, a list of jobs defines the arrival pattern. There are several techniques available in SLAM II to model this process. For example, a description of job arrival patterns may be stored in a database and accessed in chronological order each time a new interarrival time is required. The access to the arrival list may be accomplished through a call to function USERF, as shown in Figure 17-7.

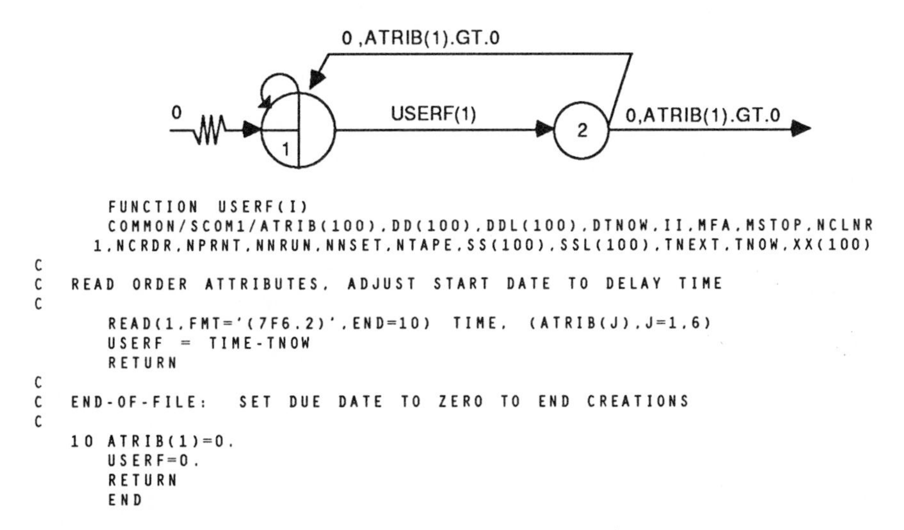


```
      FUNCTION USERF(I)
      COMMON/SCOM1/ATRIB(100),DD(100),DDL(100),DTNOW,II,MFA,MSTOP,NCLNR
     1,NCRDR,NPRNT,NNRUN,NNSET,NTAPE,SS(100),SSL(100),TNEXT,TNOW,XX(100)
C
C  READ ORDER ATTRIBUTES, ADJUST START DATE TO DELAY TIME
C
      READ(1,FMT='(7F6.2)',END=10) TIME, (ATRIB(J),J=1,6)
      USERF = TIME-TNOW
      RETURN
C
C  END-OF-FILE:  SET DUE DATE TO ZERO TO END CREATIONS
C
   10 ATRIB(1)=0.
      USERF=0.
      RETURN
      END
```

Figure 17-7 Historical arrivals.

The arrival times may be established as the outputs from the running of a model by recording the times at which entities reach a node in the network. A model with user-inserted arrival times is sometimes referred to as being trace driven.

17.2.4 Time-dependent Arrival Processes

A commonly employed arrival process involves changing the arrival pattern as a function of time or as a function of activity completions within the network. Such arrival processes can be modeled in SLAM II using a GATE. Consider the situation in which entities arrive at a system during the first 16 hours of operation, do not arrive during the next 8 hours, arrive during the next 16 hours, and so on. A network to model this arrival process is shown in Figure 17-8. The timing sequence for this illustration is modeled by the upper network segment, and entity arrivals to the network are accomplished via the lower segment. At

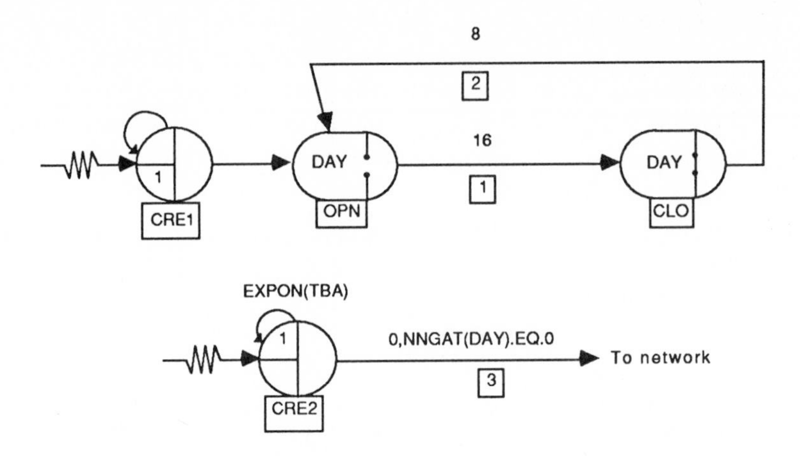

Figure 17-8 Arrivals during the day.

the start of a simulation run, a single entity is entered at CREATE node CRE1. This entity is routed to OPEN node OPN, where gate DAY is opened. The entity is routed over activity 1 with a 16-hour delay to the CLOSE node CLO. After closing the gate DAY, the entity is routed back to node OPN over activity 2, which has an 8-hour duration. This establishes a cycle of 16 hours open and 8 hours closed for gate DAY.

At CREATE node CRE2, arrivals are created. As long as gate DAY is open, that is, NNGAT(DAY).EQ.0, the entity traverses activity 3 into the network. Gate DAY being closed temporarily stops arrivals to the network until it is reopened and another entity arrives to node CRE2.

The structure of the network in Figure 17-8 suggests that arrival processes for different work shifts may be modeled through similar logic. For example, rather than no arrivals during the 8-hour period, a different arrival rate could be modeled. For this case, two separate branches are included in place of activity 3, which are conditioned on NNGAT(DAY) having the value of either 0 or 1. Another embellishment is to specify durations for activities 1 and 2 that are not based on clock time. These activities could employ user functions, attribute values, or resource allocations to model system-related operating conditions that cause a change in the arrival process.

Finally, consider a system where the arrival process requires a fixed number of items in the system. By taking advantage of the SLAM II feature to preload QUEUE nodes, any number of entities may be entered into the system at the start of a simulation run as shown in Figure 17-9. In this illustration, nine items are initially preloaded into the QUEUE node. SLAM II schedules all service activities following QUEUE nodes that have initial items in the queue at the simulation beginning time, which is assumed to be zero in this illustration. The

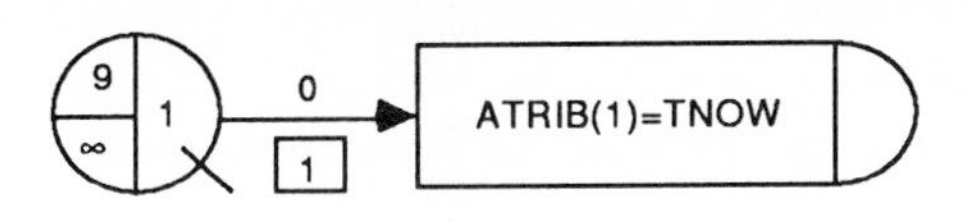

Figure 17-9 Preloaded entities

service time for activity 1 represents a zero time delay; therefore, the first entity arrives at the ASSIGN node at time TNOW = 0.0, and attribute 1 is marked at that time, just as if it had been marked at a CREATE node. Since a service activity immediately starts service on the next entity in its queue upon a completion of service, the nine initial entities in the QUEUE node are entered into the network at time 0.0.

17.3 LOADING

The loading of machines at work centers involves the routing of job entities to queues associated with servers. This routing is modeled through branching operations and queue selection procedures associated with SELECT nodes, and both are described in this section. In loading, the primary interest is in routing a single entity in one of several possible directions.

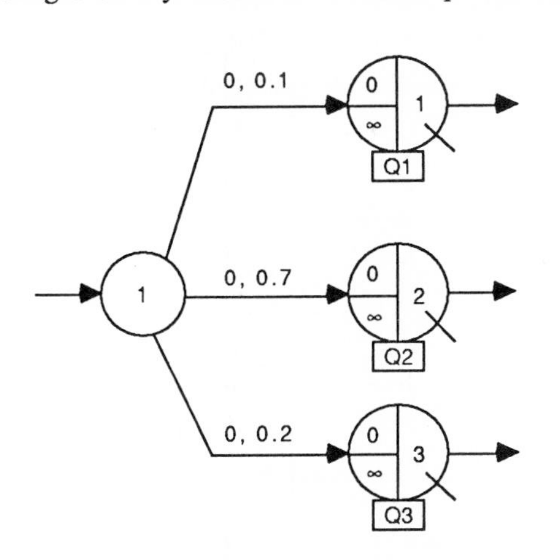

Figure 17-10 Probabilistic work load distribution.

Through probabilistic branching, aggregate machine loading rules may be modeled. For example, assume that 10% of the items that arrive in a system are to go to work center 1, 70% are to be processed in work center 2, and 20% are to be routed to work center 3. This work load distribution pattern may be modeled by the network segment shown in Figure 17-10. This probabilistic branching represents a random routing of an entity and does not take into account the current network status or the values of an entity's attributes. Illustrative network segments for the routing of entities based on current network status and the attributes of an entity are given below. Conditional branching is employed to load the work centers based on the following:

1. Time at which routing is to occur
2. Time at which routing is to occur compared to an attribute value
3. Attribute value as compared to a constant

4. Attribute value as compared with another attribute value
5. Value of SLAM II status variables
6. Value of a SLAM II status variable as specified by an attribute value

Combinations of these illustrations with the probabilistic branching described above provide the modeling capabilities to represent complex loading procedures.

Assume that an entity is to be routed to one of three work stations depending on the current time of a simulation run. In the network segment shown in Figure 17-11, QUEUE node Q1 receives the entity if the current simulation time is less than or equal to 15. If the entity arrives at the GOON node after 25 time units, it is routed to QUEUE node Q3. Otherwise, it is routed to node Q2. Only one branch from the GOON node is selected. The ordering of the branches is important, because when current simulation time, TNOW, is less than 25, two conditions are satisfied. Hence, the first branch, with the condition TNOW.LE.15, will always be selected until simulation time advances beyond 15 time units. If the second branch, TNOW.LE.25, had been listed first, then the branch TNOW.LE.15 would never be selected. This possibility is avoided by redefining the branching conditions as shown in Figure 17-12.

Branching based on attribute values is illustrated next. Let the value of attribute 1 represent the type of part that arrives to the GOON node. That is,

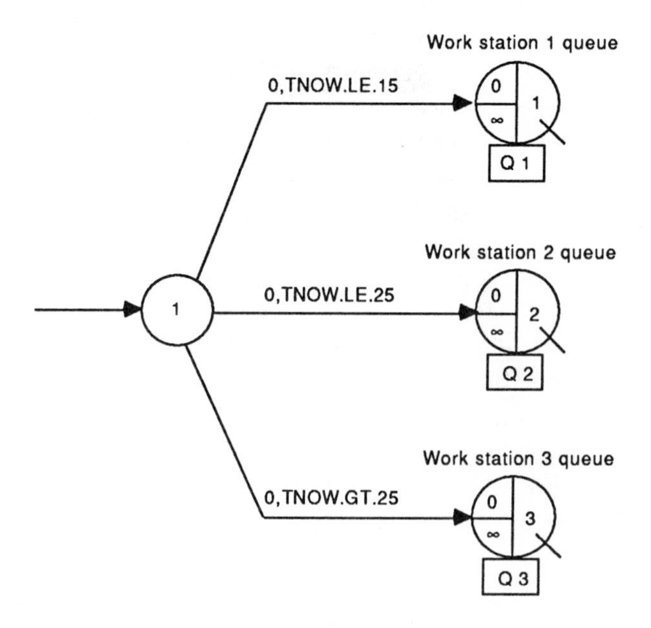

Figure 17-11 Time-based work load distribution.

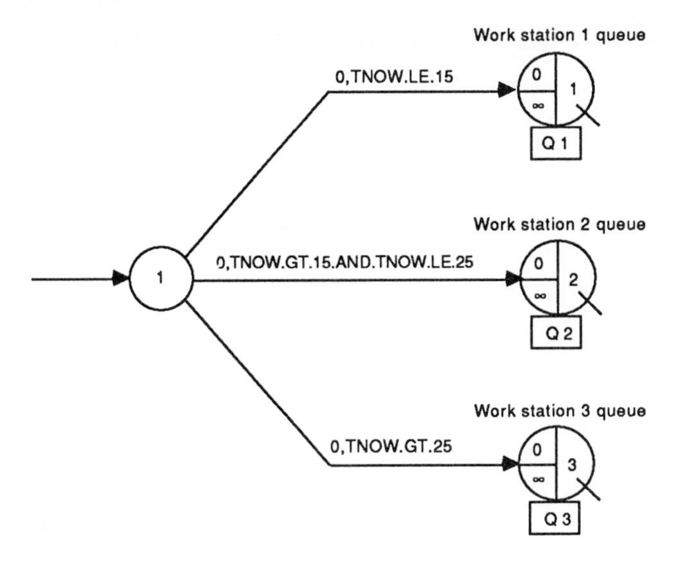

Figure 17-12 Alternate model to Figure 17-11.

if the value of ATRIB(1) equals 1 for an entity, then the part represented by the entity is type 1. Also let the type of part define the work station number on which it is processed. Three types of parts are modeled and are identified earlier in the model to be 1, 2, or 3. A network segment for this loading situation is shown in Figure 17-13. An alternative network segment that makes ATRIB(1) equiva-

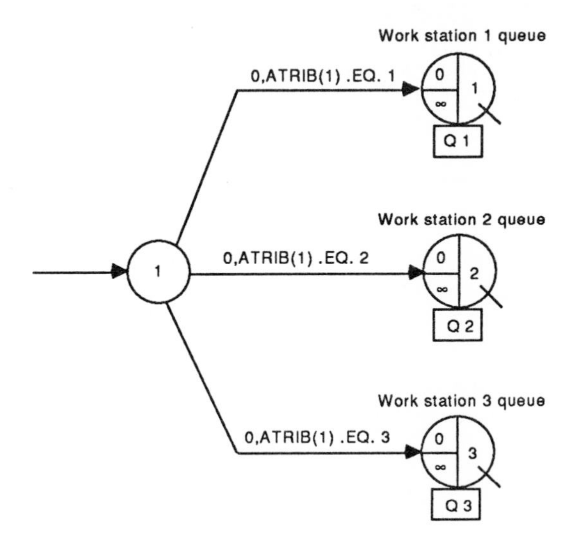

Figure 17-13 Attribute-based work load distribution.

lent to the variable TYPE and that uses TYPE to define the work station number is shown in Figure 17-14.

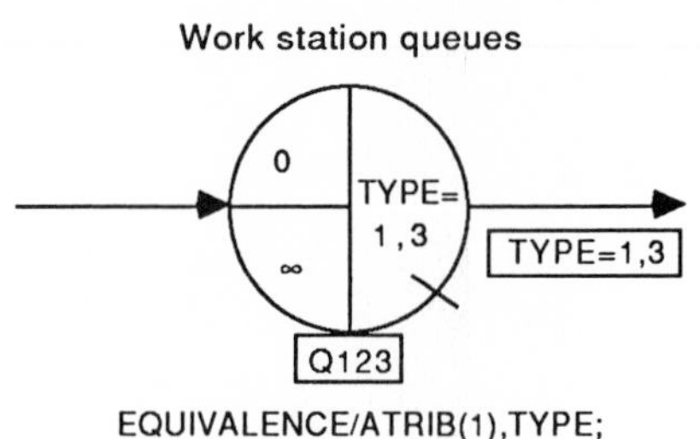

Figure 17-14 Attribute-based work station.

These illustrations can be easily modified to provide for more complex situations. For instance, assume that there are 20 different types of parts in the system. Part types 1 and 2 are to be processed by work station 1, and part types 3 and 4 are to be processed in work station 2. All other parts are to be processed by work station 3. Again letting the value of attribute 1 represent the type of part arriving at the GOON node, this loading requirement may be modeled with the network segment shown in Figure 17-15.

Branching conditions may also be made a function of the values of network status variables as shown in Figure 17-16. In this model, work station 1 processes parts until a certain milestone, represented by activity 6, is completed. Then, after the milestone has been realized, all other parts are to be processed by work center 2. The SLAM II variable NNCNT(NACT) provides a count on the number of entities that have completed activity NACT. Therefore, as long as NNCNT(6) equals 0, entities are routed to work station 1. When NNCNT(6) equals 1, entities are routed to work station 2.

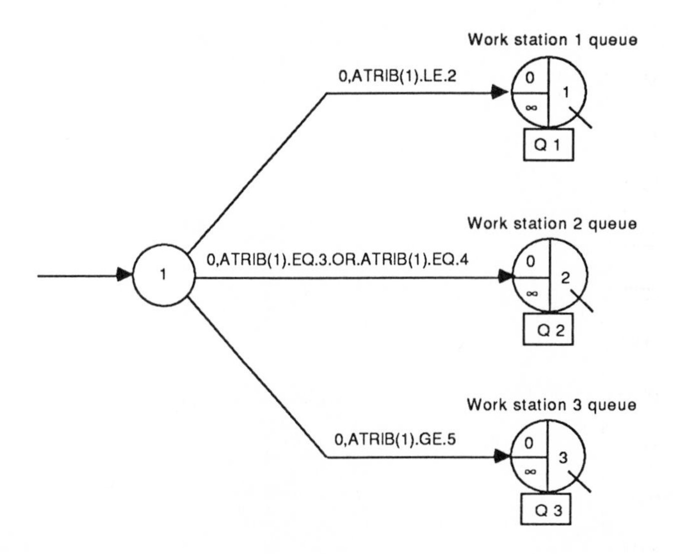

Figure 17-15 Attribute-based loading.

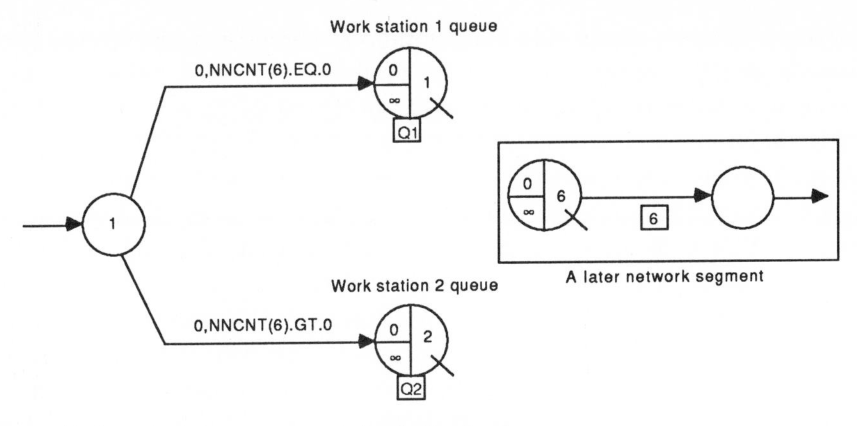

Figure 17-16 Activity status work load distribution.

A wide range of routing and machine loading procedures may be modeled in SLAM II using the conditional branching logic. Additional flexibility is available through the use of SELECT nodes that determine entity routing to QUEUE nodes using predefined or user-coded selection rules. Often, the routing of entities to work stations is a function of the length of a queue (NNQ) at a work station. Routing based on queue length is typically used to balance work loads at various work stations.

In the network segment shown in Figure 17-17, an entity is routed from SELECT node S1 to a following QUEUE node, which has on the average contained the fewest entities. This is accomplished by setting the queue selection rule (QSR) of SELECT node S1 to SAV, the smallest average number. Dashed lines graphically show the direct transfer of an entity from the SELECT node to each QUEUE node. Any one of thirteen queue selection rules could have been chosen to route entities to QUEUE nodes from SELECT nodes (see Chapter 3). The two most commonly used rules for routing entities to QUEUE nodes are based on the smallest number of entities currently waiting, SNQ, and the largest remaining unused capacity, LRC. A SELECT node always selects among available queues; that is, only nodes to which routing is possible are considered. If no queue has available space, the SELECT node causes the arriving entity to balk or be blocked.

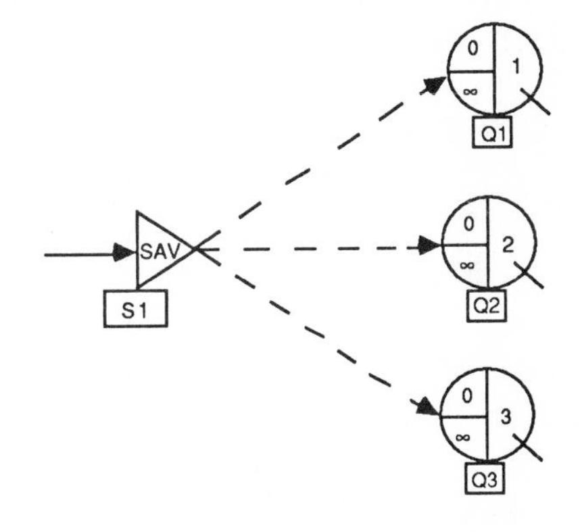

Figure 17-17 SELECT node routing.

Another important queue selection method is based on a specific order of priority for queues to receive entities. The preferred order rule, POR, stipulates that the queue nodes be tested in the order in which QUEUE node labels are placed on the SELECT statement. If POR is specified, the SELECT node always examines the QUEUE nodes in the same sequence until one is found that is not full. A cyclic (CYC) priority causes the SELECT node to choose the next available QUEUE node starting with the last node that was selected. For cyclic selection, the order of the QUEUE nodes is also specified by the placement of their labels on the SELECT statement. A random (RAN) routing causes each available queue to be given an equal chance to be selected. For example, if a selection is to be made from three queues and one is at capacity, then there is a 0.5 probability of selecting one of the other queues.

The queue selection rules for a SELECT node preceding a set of parallel queues are summarized next:

1. Average number of entities in a queue to date (LAV, SAV)
2. Waiting time of a QUEUE node's first entity (LWF, SWF)
3. Number of entities currently in a queue (LNQ, SNQ)
4. Amount of unused capacity remaining at a queue (LRC, SRC)
5. User-defined preferred order (POR)
6. Specific cyclic manner (CYC)
7. Probabilistic routing among available queues (RAN)
8. User-written function to select a queue [NSQ(N)], where N is used to differentiate between different user-written selection procedures

17.4 SEQUENCING

Sequencing decisions relate to both machines and jobs. Job sequencing selects the next job (entity) from a queue or queues when a machine (server) completes an operation (activity). Machine sequencing selects a machine when more than one is available at the time a job arrives. Job and machine sequencing are described in the following subsections.

17.4.1 Job Sequencing

Job sequencing involves the ordering of jobs for processing. Consider the one-queue, single-server network segment shown in Figure 17-18. When server 1 completes service, QUEUE node Q1 is examined. If no items are in the queue, server 1 is made idle. Otherwise, the first item in the queue is removed and server

1 processes it. The ordered arrangement of items in node Q1 determines the job sequence. The order in which items are arranged in a queue is controlled by the file ranking procedure. A PRIORITY statement is used to define the ranking procedure for each file. The options available are as follows:

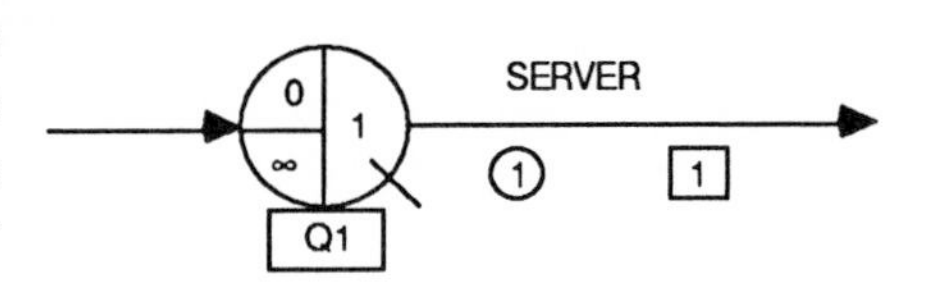

Figure 17-18 Single server.

1. FIFO: first in, first out (default ranking)
2. LIFO: last in, first out
3. LVF(NATR): lowest value first based on the value of attribute NATR
4. HVF(NATR): highest value first based on the value of attribute NATR

Attribute-based queue ranking allows job sequencing to be based on a type of part or the value (worth) of a part. If the desired job sequence is to process type 1 jobs before type 2 jobs, and type 2 jobs before part type 3, a small-value-first rule is specified based on the job type attribute. If attribute 2 is the entity's type, then LVF(2) accomplishes the desired job sequencing. To reverse the job sequencing order, that is, to process job type 3 first, then job type 2, and job type 1 last, the file ranking rule is HVF(2). Should two or more entities have the same value for the ranking attribute, priority is given to the earlier arrival, that is, the tie-breaking rule is based on FIFO.

In SLAM II, entity attribute assignments may be of a general nature. Therefore, attribute-based ranking provides the means for specifying a wide range of job sequencing rules. For example, consider a model of a single machine that receives two different part types, and assume that jobs are to be sequenced according to a shortest-processing-time rule. Let attribute 1 represent the estimated machine processing time for each job and attribute 2 represent the job type. A network that models this situation is shown in Figure 17-19.

Entities arrive in the network via CREATE nodes CRE1 and CRE2. The attribute values for jobs arriving via CRE1 are assigned at ASSIGN node ASN1, where ATRIB(1) = 5 for the processing time and ATRIB(2) = 1 to designate the job type. For type 2 jobs entering ASN2, the attribute values are set to ATRIB(1) = 10 and ATRIB(2) = 2. After the attribute values have been assigned, all arriving entities are routed to the QUEUE node QUE1. A PRIORITY statement for file 1 establishes the shortest-processing-time rule for job service on machine activity 7 by stipulating that entities be ranked in the QUEUE node QUE1 according to the lowest value for attribute 1, LVF(1). Once a job has been processed by service activity 7, it enters COLCT node COL1, where statistics are collected for the time in the system for all jobs, and then the entity is routed

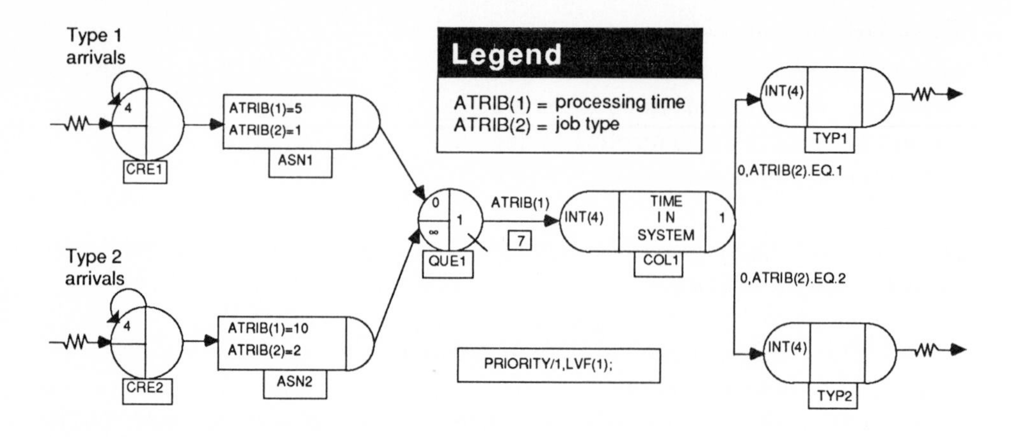

Figure 17-19 Attribute assignments and use.

to the next node according to job type. Entities with ATRIB(2) equal to 1 are routed from COL1 to COLCT node TYP1, and statistics are collected for the time in the system for type 1 jobs. Similarly, entities are routed to COLCT node TYP2, where time in the system values are obtained for type 2 jobs.

Complex ranking rules may be established for entities in queues. For example, if it is desired to rank the entities in QUE1 by part type first and then by job processing time, a special ranking attribute is established that is a function of both job type and processing time. By adding the statement ATRIB(3) = ATRIB(2)*1000.0 + ATRIB(1) at nodes ASN1 and ASN2, the joint priority is established. Low values of attribute 3 rank jobs first by job type and then according to the shortest processing time within each job classification. Hence, LFV(3) as a priority for file 1 achieves the desired job sequencing.

Large job processing times may be given priority by using the statement ATRIB(3) = ATRIB(2)*1000.0 - ATRIB(1) at the ASSIGN nodes. Additional flexibility for job sequencing may be modeled through the introduction of the user-written function, USERF, in the assignment statement should more detailed procedures be required.

Table 17-1 lists ten common job sequencing rules and the corresponding SLAM II queue ranking procedures needed to model them. In this table, it is assumed that five attributes are defined for each job entity:

ATRIB(1) is the estimated machine processing time.
ATRIB(2) is the due date.
ATRIB(3) is the number of remaining operations.
ATRIB(4) is the estimated time for all remaining job operations.
ATRIB(5) is a combination attribute used for priority setting.

Table 17-1 Common Job Sequencing Rules

	Sequencing Rule	Ranking Attribute[a]	Queue Ranking Rule
FIFO	First in, first out	—	FIFO
LIFO	Last in, first out	—	LIFO
SPT	Shortest processing time	A1	LVF(1)
DD	Due date	A2	LVF(2)
SS	Static slack	A2	LVF(2)
SS/PT	Static slack/processing time	A5=(A2-1.)/A1	LVF(5)
SS/RO	Static slack/number of remaining operations	A5=(A2-1.)/A3	LVF(5)
DS	Dynamic slack	A5=A2-A4	LVF(5)
DS/PT	Dynamic slack/processing time	A5=(A2-A4)/A1	LVF(5)
DS/RO	Dynamic slack/number of remaining operations	A5=(A2-A4)/A3	LVF(5)

[a] A1 = estimated processing time for next operation,
A2 = due date,
A3 = number of remaining operations,
A4 = estimated time for all remaining operations,
A5 = ranking attribute if a function of the above attributes is required,

Table 17-1 also includes rules relating to static and dynamic slack for jobs in a system. Static slack is defined as a job's due date minus the current time. Therefore, ranking jobs in a queue based on static slack results in the same priority as ranking jobs on due date. That is, the relative position of an entity is not changed when using a static slack rule instead of a due date rule, because the current time is subtracted from the due date to obtain a job's static slack. However, variants of the static slack rule are used to obtain other rules. For example, to define a rule that gives priority to entities with the smallest static slack relative to the job's processing time (SS/PT), attribute 5 is used to rank jobs and is computed as ATRIB(5) = (ATRIB(2) - TNOW) / ATRIB(1) and using a ranking priority of LVF(5). A similar rule is static slack divided by the number of remaining operations (SS/RO), where the ranking priority again is LVF(5) and ATRIB(5) = (ATRIB(2) - TNOW)/ATRIB(3). When TNOW, the current simulation time, is used in a priority calculation in Table 17-1, an arbitrary constant may be substituted for its value without changing the relative ranking of entities. This procedure is used in Table 17-1, where an arbitrary constant 1.0 is substituted for the value of TNOW. It should also be noted that the value of ATRIB(3), the number of remaining operations, must be updated at the end of each entity processing operation.

Dynamic slack is defined as the due date minus the estimated time required for all remaining operations. To compute dynamic slack, an attribute is updated after each operation to maintain the estimated time for all remaining operations for each job. In this discussion, ATRIB(4) is used for this purpose, and dynamic slack (DS) is computed as ATRIB(5) = ATRIB(2) - ATRIB(4).

Research indicates that the shortest-processing-time (SPT) rule is one of the best sequencing rules [4]. A primary problem with this rule is, however, that lengthy jobs may be delayed in the system for long periods of time. To compensate for this drawback, the SPT rule is often modified to form the truncated SPT rule. Under this rule, the SPT sequencing procedure is applied within two job categories. The job categories are established according to a value of slack where slack is equal to the time remaining until the due date minus the processing time for all remaining operations. Should this slack value be less than or equal to zero, the job is given a high priority. Within each priority category, the jobs are then ranked by the smallest value of estimated processing time [6].

To model this situation, two QUEUE nodes are placed before a SELECT node, as shown in Figure 17-20. Late jobs arriving to GOON node GON1 are routed directly to QUEUE node Q1. Jobs are late if their DUEDATE [ATRIB(3)] is prior to the current time. Jobs that are not late are routed to node Q2. Jobs in both queues are ranked on the low value of the estimated processing time, which is assumed to be the value assigned to ATRIB(1).

The SELECT node S1 employs the user queue selection rule, NQS. The user written function NQS is called by the SLAM II processor at the completion of each job on activity 1 to select a file number of a QUEUE node where the next entity to be processed resides. If there are jobs waiting in Q2, these jobs are first checked to see if any of them have become late while they have been in the queue. Each late job in Q2 is taken from Q2 and placed in Q1 where it is ranked according to the estimated processing time. NQS is set to the file number of Q1 if it has an entity. If not, NQS is set to 2 if Q2 has an entity. Otherwise, it is set to zero to indicate no QUEUE node could be selected.

The FORTRAN code for function NQS(N) to model the truncated SPT rule is included in Figure 17-20. The search operation for late jobs in Q2 locates all late jobs by searching file 2 starting at the first entity, which is referenced by MMFE(2). Pointers to successor entities in file 2 are obtained using the SLAM II function NSUCR. Subroutine COPY is used to obtain the attribute values, and the negative of the pointer NEXT is used to indicate that NEXT is a pointer and not a rank. Subroutines ULINK and LINK are employed to take late jobs from file 2 and insert them into file 1. The last section of code in Figure 17-20 sets the value of function NQS to be the file number of the QUEUE node selected or sets NQS to 0 if no entities reside in files 1 or 2. The code presented in function NQS serves as a basis for more advanced decision logic regarding the disposition of different classes of jobs.

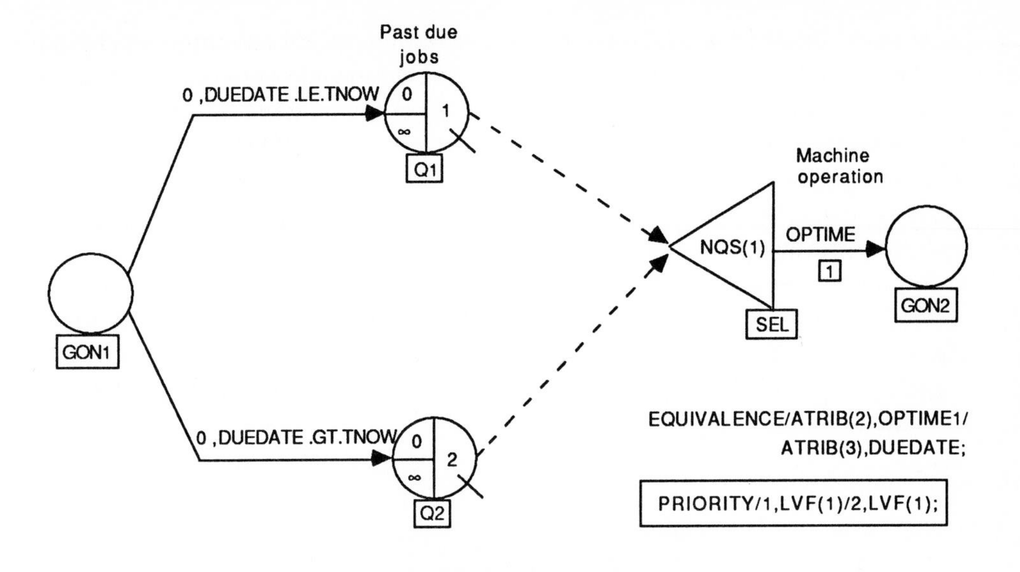

Legend
ATRIB(1) = estimated job processing time
ATRIB(2) = actual job processing time
ATRIB(3) = job due date

```
      FUNCTION NQS(N)
      COMMON/SCOM1/ATRIB(100),DD(100),DDL(100),DTNOW,II,MFA,MSTOP,NCLNR
     1,NCRDR,NPRNT,NNRUN,NNSET,NTAPE,SS(100),SSL(100),TNEXT,TNOW,XX(100)
      DIMENSION A(10)
      NEXT = MMFE(2)
5     IF (NEXT .EQ.0) GO TO 10
      CALL COPY (-NEXT,2,A)
      NEXT = NSUCR(NEXT)
      IF (A(3) .LE.TNOW) THEN
      CALL ULINK (-NEXT,2)
      CALL LINK (1)
      ENDIF
      GO TO 5
10    IF (NNQ(1).GT.0) THEN
          NQS = 1
      ELSE
          IF (NNQ(2).GT.0) THEN
          NQS=2
          ELSE
          NQS=0
          ENDIF
      ENDIF
      RETURN
      END
```

Figure 17-20 Network and function NQS for modeling the truncated SPT rule.

When multiple waiting areas precede a server (or a group of servers), job sequencing is determined by both the queue selection and the selection of a job from the queue. (In the previous model, two QUEUE nodes were used to partition jobs not to represent two waiting areas.) To select a queue, a SELECT node is used in a look-backward mode. A queue selection rule (QSR) is used to identify the file number of the QUEUE node on the input side of the SELECT node from which the next entity is to be processed. A QUEUE node is selected according to one of the available rules, and the first entity in the file of the QUEUE node is selected. If the queue selection rule is specified by NQS(N), then NQS is set equal to a file number or zero, where zero indicates that no file could be selected. In function NQS, subroutines ULINK and LINK can be used to make an entity the first in a file.

17.4.2 Machine Sequencing

Machine sequencing involves the determination of the order in which machines, work stations, resources, or operators are used when a choice exists. When each machine has a separate queue, an implied machine sequence is created. In the network shown in Figure 17-21, conditions on branches are used at a GOON node to route entities to service activities 1 or 2. The conditions placed on the branches from the GOON node to node Q9 or Q10 route entities to machines. The conditions on the branches explicitly specify a machine selection procedure. A similar procedure to sequence resources involves replacing the QUEUE nodes in Figure 17-21 with AWAIT nodes.

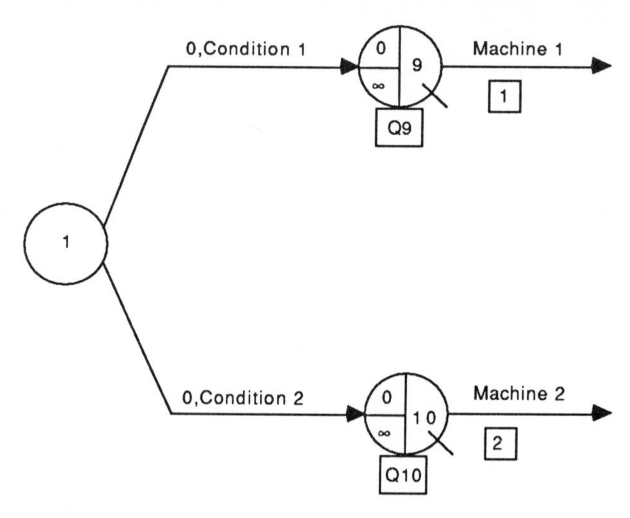

Figure 17-21 Conditional routing.

The conditional specification may be replaced by a SELECT node with a queue selection procedure. A network segment representing this situation is shown in Figure 17-22. The queue selection rule at the SELECT node S11 specifies the procedure for machine sequencing by selecting a QUEUE node-service activity combination.

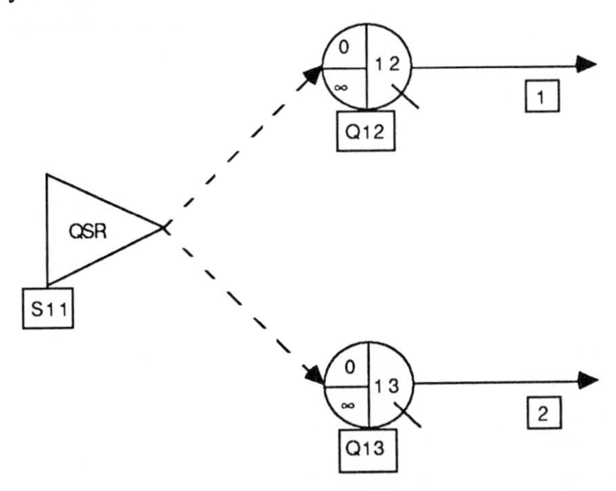

Figure 17-22 Work station sequencing.

A more direct machine sequencing procedure is to use a SELECT node to select from among service activities that emanate from it. This is shown in Figure 17-23, where node S15 uses a server selection rule (SSR) to select among the service activities 1, 2, or 3. This network segment provides a general scheme for machine sequencing as the user-written server selection function, NSS(N), may be employed.

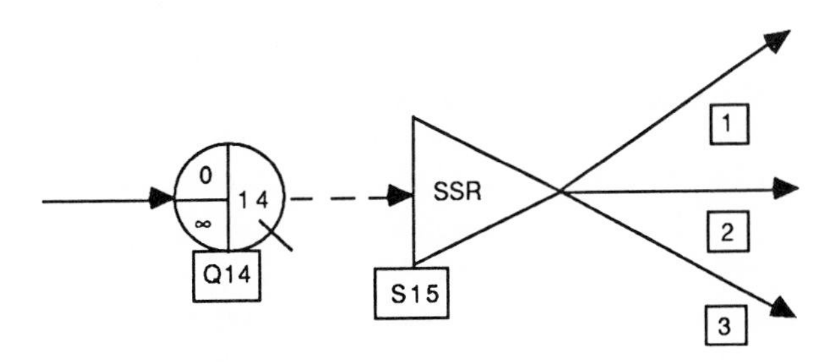

Figure 17-23 Machine sequencing.

A two-level machine selection procedure is illustrated in the network segment in Figure 17-24 with a QUEUE node followed by two SELECT nodes that can each select from among parallel servers. In this example, SSR_1 at node S15

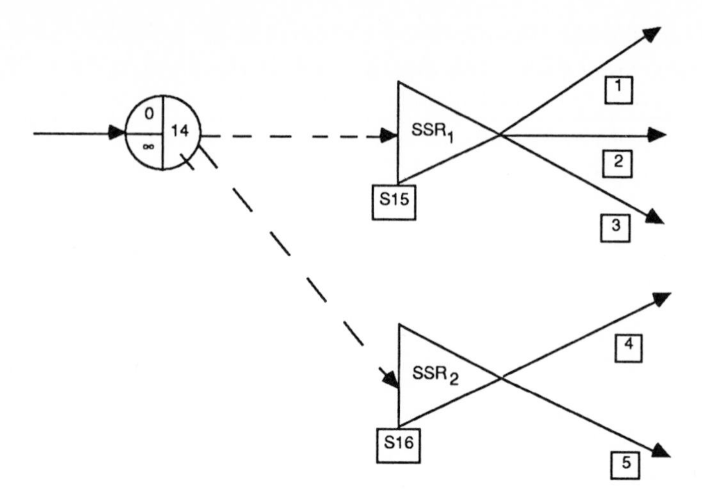

Figure 17-24 Multiple machine sequencing.

specifies the sequence for machines 1, 2, and 3. SSR_2 at node S16 specifies the sequence for machines 4 and 5. The polling sequence of the SELECT nodes is the order they are listed on the QUEUE node's input statement.

Modeling machine sequencing using AWAIT nodes and subroutine ALLOC is also available. In this case, the selection of a resource or set of resources to allocate to an entity arriving to the AWAIT node is determined directly through the coding of subroutine ALLOC and using subroutine SEIZE to acquire the resources selected.

17.5 DETAILED SCHEDULING

The aggregate planning, loading, and sequencing aspects of scheduling provide information about jobs and work centers on a relative basis. An analysis of this information provides information such as: Work center A should perform jobs 1, 3, and 7, while work center B should do jobs 2, 4, 5, and 6; job 1 should be performed before job 3 in work center A. ***Detailed scheduling***, however, involves the specification of the start or completion dates for all jobs at each work center in the shop. In the past, schemes for detailed scheduling have been notoriously ineffectual. A primary reason for this failure has been the lack of flexibility and adaptability when scheduling completion times for an interrelated set of jobs. This resulted in a detailed schedule becoming obsolete as soon as new jobs arrived, unexpected machine delays occurred, or some other disruption in the production process was experienced. Inaccurate data and the absence of up-to-date job-in-progress information also contributes to the dilemma.

Currently, computers are employed to perform detailed scheduling on an adaptive basis. The FACTOR system [7] employs simulation to develop and evaluate detailed schedules on line. Thus, instead of preparing a detailed schedule a week in advance, a daily or even a job-by-job completion date specification may be accomplished. Such systems perform well as long as accurate records are kept and the integrity of the database supporting detailed scheduling is maintained. FACTOR includes modules to maintain accurate data on current operations, which is the only way a detailed scheduling system can succeed.

To use SLAM II within a detailed scheduling system, a user-written function is included at each node in the network where a job entity leaves a work center. In the user function, a statement is inserted to record the job identifier, the work station, and the time the job entity passes through the node. The time that is recorded should be in actual time units. Therefore, simulation time, TNOW, must be converted to a date and specific hour. A subroutine to make this conversion has been written. The subroutine requires the starting date, the number of shifts to be worked per day, the starting times and lengths of each shift, the number and length of worker breaks, and the number of working hours per week as input data.

A detailed schedule consists of the completion dates for each job at each work center computed through the above procedure. This detailed schedule uses as its basis, the aggregate planning, loading, and sequencing rules incorporated in the SLAM II network. Changing any one of these planning procedures results in a new detailed schedule. The detailed schedule may be organized chronologically, by work center, and/or by job numbers by using a sorting routine. In many situations, schedules according to all three organizations are made available. Gantt charts for presenting the schedule provide a good visual display of the time sequence of individual machine operations.

If machine times are random variables, any completion date observed in an actual system operation or obtained through simulation techniques represents just one sample observation of the random variable describing the job completion time. Detailed scheduling procedures have not been developed for this situation. The development of techniques to provide a detailed schedule to the shop floor when job completion dates, and therefore job start dates and slack values, are random variables is a good research topic. An estimate of these random variables can be obtained with SLAM II by performing multiple runs of the system's network (an example of this procedure for a project scheduling environment was presented Chapter 9). A rule to translate estimated dates obtained through network simulation into job completion dates for a detailed schedule must be developed and tested. Some possible rules might be the following:

1. Use the expected date.
2. Use a date such that 95% of the job completion times are less than the date selected.
3. Use the expected date plus two standard deviations.

In any event, the application of SLAM II for detailed scheduling in the face of random machine times is a fertile research area.

17.6 EVALUATING THE EFFECTS OF RANDOM VARIATION ON SOLUTIONS OBTAINED BY ANALYTIC METHODS

Analytic methods have been developed to determine optimal solutions for special problem situations within production planning. Typically, the use of an analytic procedure involves linear and deterministic assumptions. The use of analytic procedures can be considered as a first stage of the problem solution process. For example, an analytic solution is obtained to develop a first-pass schedule or sequence. This schedule or sequence is then used for evaluation of the system under more realistic conditions, where the assumptions of linearity and constant times are removed. Simulation provides an easy-to-use methodology to evaluate system performance when analytical assumptions are relaxed. This is the approach used in this section to evaluate proposed job sequences in a flow shop.

17.6.1 Johnson's Sequencing Rule

In a shop consisting of two machines, jobs are processed first on machine 1 and then on machine 2. The processing time for each job on both machines is known with certainty, although the times may be different for different jobs. The problem of sequencing jobs to minimize the total time to complete all jobs (makespan) is known as ***Johnson's problem***, since Johnson developed an optimal sequencing rule [3]. The rule can be stated as follows: Let t_{j1} represent the processing time on machine 1 of job j, and let t_{j2} represent the processing time on machine 2 for job j. Let U be the set of jobs such that $t_{j1} < t_{j2}$, that is, $U = \{j \mid t_{j1} < t_{j2}\}$, and let $V = \{j \mid t_{j1} \geq t_{j2}\}$. If the elements of U are ordered with small values of t_{j1} first, and the elements of V are ordered with large values of t_{j2} first, then an optimal sequence for performing the jobs on both machines is the ordered set U followed by the ordered set V.

A SLAM II network model for the implementation of Johnson's sequencing rule is given in Figure 17-25. Procedures are included to evaluate random

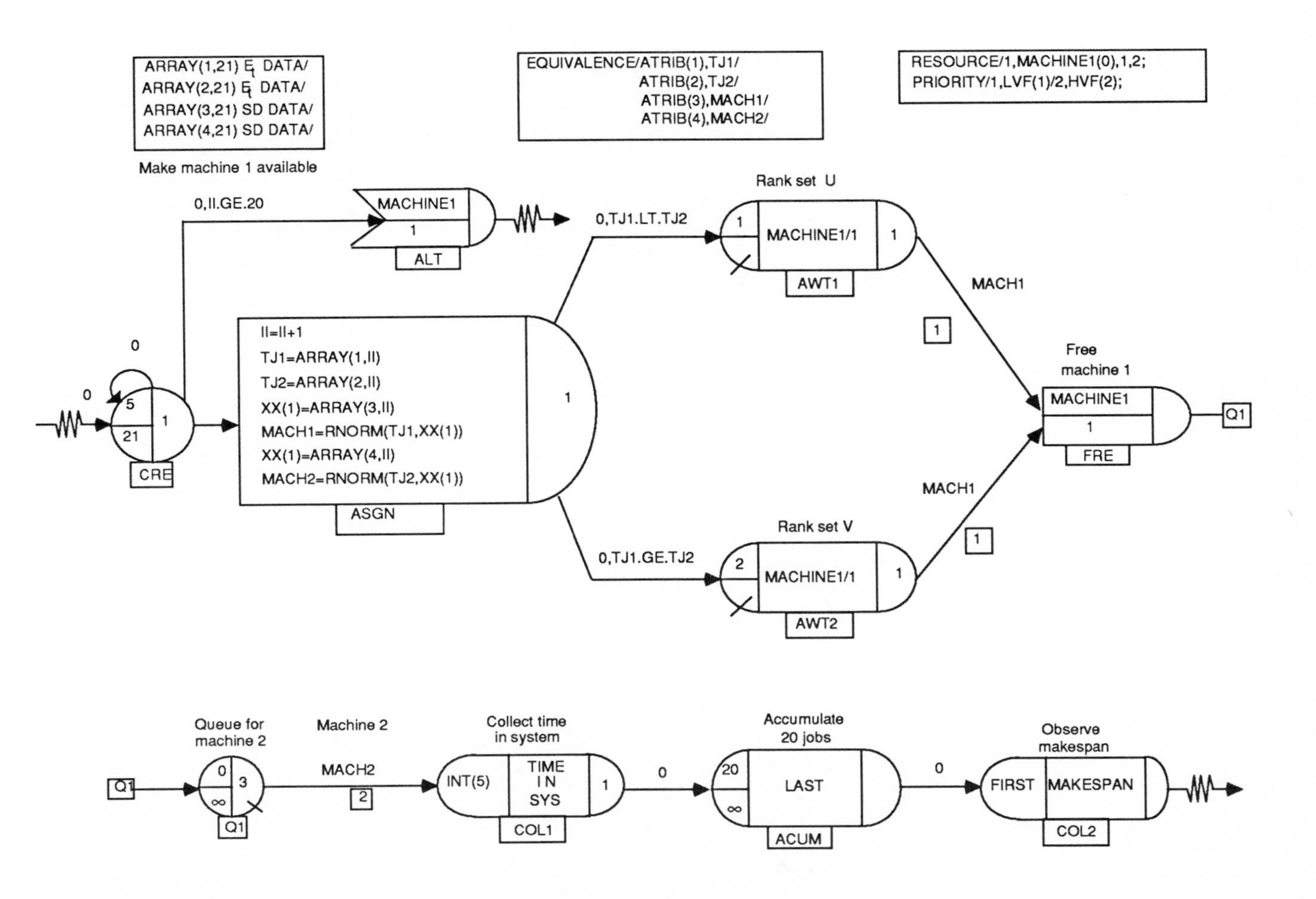

Figure 17-25 SLAM II network for the implementation of Johnson's sequencing rule.

processing times. ARRAY is used to store the expected processing time and the standard deviation of the processing time on the *ith* machine for 20 jobs. ARRAY(1, 20) and ARRAY(2, 20) contain the data for the expected processing times of each job on machines 1 and 2, respectively. The standard deviations on machine 1 are stored in ARRAY(3, 20) and in ARRAY(4, 20) for machine 2. The variable names TJ1 and TJ2 are given to the expected processing times on machines 1 and 2, respectively, and MACH1 and MACH2 are the names for the actual (sampled) processing times. These four variables are made equivalent to the first four attributes of a job entity. The actual machine processing times for the simulation are randomly selected from a normal distribution with the mean time and standard deviation assigned to each job for each machine according to the parameters stored in ARRAY.

Twenty entities (jobs) and one control entity are entered into the system at CREATE node CRE at time zero. Each job causes index II to be increased by 1 at ASSIGN node ASGN. The actual processing time of job II is set on machines 1 and 2. After 20 jobs enter the system, a twenty-first entity is routed to an ALTER node to make machine 1 available.

After attribute assignments, jobs are routed to either AWAIT node AWT1 or AWT2 depending on the relative value of TJ1 to TJ2. Jobs in AWT1 have TJ1 less than TJ2 and are members of set U. They are ordered lowest value first in file 1 based on the value of TJ1 (as specified on the PRIORITY statement). Jobs in AWT2 have TJ1 greater than or equal to TJ2 and are members of set *V*. They are ordered highest value first in file 2 based on the value of TJ2. The ordering of jobs in the AWAIT nodes takes place at time zero, as machine 1 is not made available until all 20 jobs are waiting in the network.

By the RESOURCE statement

```
RESOURCE/1, MACHINE1(0), 1, 2;
```

the capacity of machine 1 is set at 0. The last entity to enter the system branches to ALTER node ALT when the number of entities traversing activity 3 is greater than or equal to 20. Node ALT makes one unit of MACHINE1 available. Machine 1 is then allocated to jobs waiting in AWT1 and then AWT2. This order of allocation is based on the order of the file numbers listed on the RESOURCE statement. Since file number 1 is listed first, priority is given to jobs waiting in node AWT1. The jobs are processed in order based on their mean processing times due to the priority established for file 1.

The processing of a job on machine 1 is represented by the activities following the AWAIT nodes. The processing time is equal to MACH1 set at the ASSIGN node. At the completion of job service by machine 1, FREE node FRE releases the resource, MACHINE1, and the next job is scheduled. The processed job then proceeds to machine 2, which has its waiting area modeled by QUEUE node Q1. The time for the job on machine 2 is specified by MACH2. The first COLCT node COL1 collects statistics on the job's flow time. ACCUMULATE node ACUM requires that all 20 jobs be completed before a single entity is routed to the second COLCT node COL2, which collects the value of the makespan for a given simulation run. This time is collected by specifying statistics type FIRST at COLCT node COL2.

The SLAM II statement model for evaluating Johnson's sequencing rule is given in Figure 17-26. Expected processing times have been set to be increasing

```
GEN,PRITSKER,JOHNSON RULE,03/02/88,400,,NO,,NO,YES/400;
LIMITS,3,5,25;
EQUIVALENCE/ATRIB(1),TJ1/
            ATRIB(2),TJ2/
            ATRIB(3),MACH1/
            ATRIB(4),MACH2;
ARRAY(1,20)/10,10.5,11,11.5,12,12.5,13,13.5,14,14.5,15,15.5,16,16.5,
            17,17.5,18,18.5,19,19.5;
ARRAY(2,20)/19.5,19,18.5,18,17.5,17,16.5,16,15.5,15,14.5,14,13.5,13,
            12.5,12,11.5,11,10.5,10;
ARRAY(3,20)/20*1/(4,20)/20*1;
PRIORITY/1,LVF(1)/2,HVF(2);
NETWORK;
       RESOURCE/1,MACHINE1(0),1,2;
CRE    CREATE,0,,5,21,1;
       ACT,0,II.GE.20,ALT;
       ACT;
ASGN   ASSIGN,II=II+1,
            TJ1=ARRAY(1,II),TJ2=ARRAY(2,II),
            XX(1)=ARRAY(3,II),MACH1=RNORM(TJ1,XX(1)),
            XX(1)=ARRAY(4,II),MACH2=RNORM(TJ2,XX(1));
       ACT,0,TJ1.LT.TJ2,AWT1;
       ACT,0,TJ1.GE.TJ2,AWT2;
AWT1   AWAIT(1),MACHINE1/1,1;
       ACT/1,MACH1,,FRE;
AWT2   AWAIT(2),MACHINE1/1,1;
       ACT/1,MACH1,,FRE;
FRE    FREE,MACHINE1/1;
Q1     QUEUE(3);
       ACT/2,MACH2;
COL1   COLCT,INT(5),TIME IN SYS;
ACUM   ACCUMULATE,20,20,LAST;
COL2   COLCT,FIRST,MAKESPAN,20/295/1;
       TERM;
;
ALT    ALTER,MACHINE1,1;
       ASSIGN,II=0;
       TERM;
       END;
INIT,,,YES/2;
FIN;
```

Figure 17-26 SLAM II statement model for evaluating Johnson's sequencing rule.

for machine 1 and decreasing for machine 2 by an increment of 0.5. For this illustration, the makespan is 305 time units if the processing times are exactly the expected times, since the time for the last job on machine 2 will begin immediately upon its completion on machine 1. By setting the standard deviations to zero, the SLAM II model duplicates this result. A small standard deviation of 1 is set for all jobs. The output from one run of the model yielded a makespan of 308.1 time units.

Four hundred runs of the model were made to investigate the statistical variation due to the sampling of the processing times and with the sequencing rule based on the expected value. To make the 400 runs, the GEN control statement was modified to set the number of runs at 400 and the report indicators to request only a summary report after the four hundredth run. The revised GEN statement is

```
GEN,PRITSKER,JOHNSON RULE,03/02/88,400,,NO,,NO,YES/400;
```

A histogram is specified for COLCT node COL2 to obtain an estimate of the distribution of the makespan times. To collect makespan times over the multiple runs, the INIT statement

```
INIT,,,YES/2;
```

is used to specify that statistics for the first COLCT node is to be cleared following each run. Thus, for all other COLCT nodes, statistics are to be based on observations over all runs.

The average makespan of the 400 runs is 307.3 with a standard deviation of 4.11. The minimum makespan was 296.2 and the maximum makespan was 320.1. The minimum value is smaller than the optimum value of 305 due to the smaller processing times obtained from sampling from the normal distribution. It is hypothesized that the closeness between the average makespan value and the optimal makespan value is due to the wide variation in the average processing times for the jobs, that is, 10 to 19.5. With only a one time unit standard deviation, it is unlikely to have the jobs processed in an order too different from the optimal order. The histogram and statistics of the makespan values are shown in Figure 17-27.

If it is desired to order the jobs on the sample processing times, the PRIORITY statement is changed to rank jobs for machine 1 on low value of attribute 3 and for machine 2 on high value of attribute 4. The revised PRIORITY statement is

```
PRIORITY/1,LVF(3)/2,HVF(4);
```

```
                          **HISTOGRAM NUMBER 2**

                                 MAKESPAN

OBSV   RELA    CUML        UPPER
FREQ   FREQ    FREQ    CELL LIMIT   0         20        40        60        80        100
                                    +    +    +    +    +    +    +    +    +    +    +
   0   0.000   0.000   0.2950E+03   +                                                 +
   0   0.000   0.000   0.2960E+03   +                                                 +
   1   0.002   0.002   0.2970E+03   +                                                 +
   3   0.007   0.010   0.2980E+03   +C                                                +
   5   0.013   0.023   0.2990E+03   +*                                                +
   3   0.007   0.030   0.3000E+03   + C                                               +
  10   0.025   0.055   0.3010E+03   +* C                                              +
  17   0.043   0.098   0.3020E+03   +**  C                                            +
  25   0.063   0.160   0.3030E+03   +***    C                                         +
  20   0.050   0.210   0.3040E+03   +***      C                                       +
  30   0.075   0.285   0.3050E+03   +****         C                                   +
  38   0.095   0.380   0.3060E+03   +*****            C                               +
  43   0.108   0.488   0.3070E+03   +*****                 C                          +
  31   0.078   0.565   0.3080E+03   +****                      C                      +
  46   0.115   0.680   0.3090E+03   +*******                         C                +
  33   0.083   0.763   0.3100E+03   +****                                C            +
  19   0.047   0.810   0.3110E+03   +**                                     C         +
  28   0.070   0.880   0.3120E+03   +****                                      C      +
  11   0.027   0.908   0.3130E+03   +*                                          C     +
  13   0.032   0.940   0.3140E+03   +**                                           C   +
   9   0.023   0.963   0.3150E+03   +*                                             C  +
  15   0.038   1.000          INF   +**                                               C
   -                                +    +    +    +    +    +    +    +    +    +    +
 400                                0         20        40        60        80        100
```

STATISTICS FOR VARIABLES BASED ON OBSERVATION

	MEAN VALUE	STANDARD DEVIATION	MINIMUM VALUE	MAXIMUM VALUE	NUMBER OF OBSERVATIONS
MAKESPAN	0.3073E+03	0.4111E+01	0.2962E+03	0.3201E+03	400

Figure 17-27 Makespan results for 400 runs.

The makespan obtained with this new priority statement was 307.6. This is a slight improvement over the makespan of 308.1 obtained for the same sample values and indicates only a minor change in the optimal ordering of jobs due to sequencing on the actual rather than the expected processing times.

17.6.2 Jackson's Sequencing Rule

Jackson [9] extended Johnson's problem by considering the two-machine situation where jobs may require the machines in either order and, in addition, where a job may require only one of the two machines. For this discussion, jobs that must start on machine 1 and then go to machine 2 are referred to as AB jobs. Jobs that must start on machine 2 and are then routed to machine 1 are referred to as BA jobs. Jobs that are to be processed only on machine 1 are called A jobs, and those processed only on machine 2 are called B jobs. Given this situation Jackson has prescribed a rule to sequence the jobs optimally. The rule consist of two steps. The first step specifies that the following sequencing be performed.

1. Sequence AB jobs by Johnson's rule

2. Sequence BA jobs by Johnson's rule
3. Sequence A and B jobs in any order

The second step of the rule states that machine 1 is to process all AB jobs first, all A jobs next, and BA jobs last. Machine 2 is to process all BA jobs first, all B jobs next, and the AB jobs last.

The SLAM II model for Jackson's rule is an extension of the previous network for Johnson's rule and is shown in Figure 17-28. The CREATE node generates 41 entities. Type A and AB jobs are represented by entities 1 to 20, type B and BA jobs by entities 21 to 40, and entity 41 is used to make machines 1 and 2 available for processing jobs after all jobs have been entered into the network. The ASSIGN node functions in a similar manner to the one described for Johnson's rule. If a job is a type A job, its processing time for machine 2 [ATRIB(2)]is specified as zero. Likewise, type B jobs are indicated by a zero value for processing time on machine 1. As before, the values for attributes 3 and 4 are sampled from the normal distribution to obtain actual machine processing times. The arrays used to enter the data for each job are dimensioned to accommodate the 40 jobs [ARRAY(I,40)].

Following the ASSIGN node, ASGN, jobs are branched conditionally to the GOON nodes, GON1 and GON2, according to type. The first 20, representing type A, are routed to GON1. At GON1, the jobs are routed according to expected machine processing time to set U and set V and are placed in their respective AWAIT nodes, where they wait for machine 1 to become available for processing. Similarly, entities 21 to 40, representing type B jobs, are routed through GON2 to AWAIT nodes to wait for machine 2 to become available. The forty-first entity is routed to ALTER node ALT1 and then ALT2 to make machines 1 and 2 available for processing after all jobs have been entered.

AWAIT nodes, AWT2 and AWT3, order jobs according to Johnson's rule as stipulated on the PRIORITY statement. Node AWT1 contains jobs that require processing by machine 1 and are identified by a zero value for attribute 2. On input for the network statements, the activity from node GON1 to node AWT1 is placed first so that this routing condition is the first one tested. Similarly, nodes AWT6 and AWT7 order BA jobs for machine 2, and the input statement describing the branch to AWT5 is placed first to route B jobs to this node.

Resource 1 represents machine 1 and resource 2 represents machine 2. Both of the machines are initially unavailable. The sequencing of job types according to Jackson's rule is incorporated into the model through the order in which the AWAIT node file numbers are listed on the RESOURCE statements. For machine 1, AB jobs are processed first (in Johnson's order), A jobs next, and BA jobs last. For machine 2, BA jobs are processed first (in Johnson's order), B jobs next, and AB jobs last.

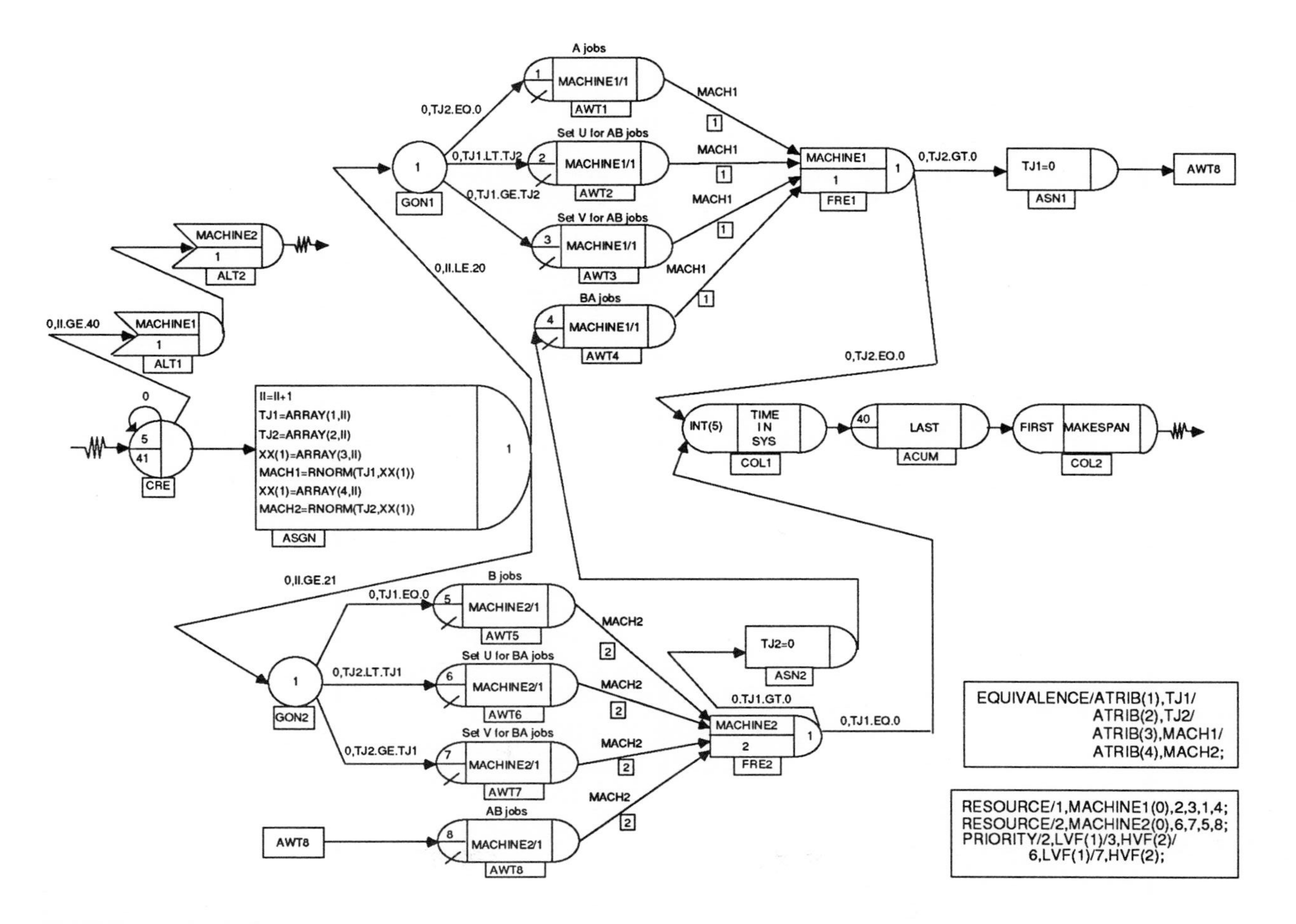

Figure 17-28 SLAM II model for evaluating Jackson's rule.

When a job finishes at machine 1, FREE node FRE1 makes machine 1 available for the next entity, and routes the job out of the system if no further processing is required (TJ2 = 0). If processing on machine 2 is required (TJ2 0), the job is routed to ASSIGN node ASN1, where attribute 1 is set equal to zero to indicate that processing has been completed by machine 1 (TJ1 = 0). The job is then placed into node AWT8, where AB jobs wait for the availability of machine 2.

Node FRE2 performs routing based on whether TJ1 is zero or not. Thus, B and AB jobs leave the system immediately after being processed at machine 2. However, BA jobs are routed to node ASN2 and then to node AWT4. Node ASN2 sets TJ2 to zero for BA jobs, which leave the system after machine 1 processing. COLCT node COL1 collects statistics on flowtime, and node COL2 records the makespan for the 40 jobs. Resource utilization statistics provide information on machine usage. The statement model for Jackson's rule is given in Figure 17-29. The makespan from one run of the model was 795.7 time units.

```
GEN,PRITSKER,JACKSON RULE,03/3/88,1;
LIMITS,8,5,45;
EQUIVALENCE/ATRIB(1),TJ1/ATRIB(2),TJ2;
EQUIVALENCE/ATRIB(3),MACH1/ATRIB(4),MACH2;
;
;      ASSIGN TIMES FOR MACHINE 1
ARRAY(1,40)/10,10.5,11,11.5,12,12.5,13,13.5,14,14.5,15,15.5,16,16.5,
            17,17.5,18,18.5,19,19.5,20,20.5,21,21.5,22,22.5,23,23.5,
            24,24.5,25,25.5,26,26.5,27,27.5,28,28.5,29,29.5;
;      ASSIGN TIMES FOR MACHINE 2
;
ARRAY(2,40)/29.5,29,28.5,28,27.5,27,26.5,26,25.5,25,24.5,24,23.5,23,
            22.5,22,21.5,21,20.5,20,19.5,19,18.5,18,17.5,17,16.5,16,
            15.5,15,14.5,14,13.5,13,12.5,12,11.5,11,10.5,10;
ARRAY(3,40)/40*1/(4,40)/40*1;
;       ESTABLISH RANKING FOR SETS U AND V
PRIORITY/2,LVF(1)/3,HVF(2)/6,LVF(1)/7,HVF(2);
NETWORK;
;      SET ORDER FOR ALLOCATING TO FILES
       RESOURCE/1,MACHINE1(0),2,3,1,4;
       RESOURCE/2,MACHINE2(0),6,7,5,8;
CRE    CREATE,0,,5,41,1;
       ACT,0,II.GE.40,ALT1;
       ACT,,,ASGN;
ASGN   ASSIGN,II=II+1,TJ1=ARRAY(1,II),TJ2=ARRAY(2,II),
             XX(1)=ARRAY(3,II),MACH1=RNORM(TJ1,XX(1)),
             XX(1)=ARRAY(4,II),MACH2=RNORM(TJ2,XX(1)),1;
;      FIRST 20 JOBS START AT MACHINE 1
       ACT,0,II.LE.20,GON1;
       ACT,0,II.GE.21,GON2;
```

Figure 17-29 SLAM II statement model for evaluating Jackson's sequencing rule.

```
;       MACHINE 1 ARRIVALS
GON1    GOON,1;
        ACT,0,TJ2.EQ.0,AWT1;
        ACT,0,TJ1.LT.TJ2,AWT2;
        ACT,0,TJ1.GE.TJ2,AWT3;
;       MACHINE 2 ARRIVALS
GON2    GOON,1;
        ACT,0,TJ1.EQ.0,AWT5;
        ACT,0,TJ2.LT.TJ1,AWT6;
        ACT,0,TJ2.GE.TJ1,AWT7;
AWT1    AWAIT(1),MACHINE1/1,,1;
        ACT/1,MACH1,,FRE1;
AWT2    AWAIT(2),MACHINE1/1,,1;
        ACT/1,MACH1,,FRE1;
AWT3    AWAIT(3),MACHINE1/1,,1;
        ACT/1,MACH1,,FRE1;
AWT4    AWAIT(4),MACHINE1/1,,1;
        ACT/1,MACH1,,FRE1;
AWT5    AWAIT(5),MACHINE2/1,,1;
        ACT/2,MACH2,,FRE2;
AWT6    AWAIT(6),MACHINE2/1,,1;
        ACT/2,MACH2,,FRE2;
AWT7    AWAIT(7),MACHINE2/1,,1;;
        ACT/2,MACH2,,FRE2;
AWT8    AWAIT(8),MACHINE2/1,,1;
        ACT/2,MACH2,,FRE2;
;       REALLOCATE MACHINE 1
FRE1    FREE,MACHINE1/1,1;
        ACT,0,TJ2.GT.0,ASN1;
        ACT,0,TJ2.EQ.0,COL1;
;       REALLOCATE MACHINE 2
FRE2    FREE,MACHINE2/1,1;
        ACT,0,TJ1.GT.0,ASN2;
        ACT,0,TJ1.EQ.0,COL1;
ASN1    ASSIGN,TJ1=0;
        ACT,,,AWT8;
ASN2    ASSIGN,TJ2=0;
        ACT,,,AWT4;
COL1    COLCT,INT(5),TIME IN SYS;
        ACCUMULATE,40,,LAST;
;       WHEN ALL JOBS DONE, RECORD TIME
        COLCT,FIRST,MAKESPAN;
        TERM;
;       MAKE MACHINES AVAILABLE TO PROCESS JOBS
ALT1    ALTER,MACHINE1,1;
        ALTER,MACHINE2,1;
        TERM;
        END;
FIN;
```

Figure 17-29 SLAM II statement model for evaluating Jackson's sequencing rule (concluded).

17.7 ASSEMBLY LINE BALANCING

An assembly line is a sequence of work stations where operators assemble units into subassemblies and finished products. Subassemblies are passed from work station to work station, usually on a conveyor. Typically, a production rate is specified for the entire line. For example, the line may have to produce 480 units in an 8-hour shift; that is, one unit must be produced each minute. Thus, the time interval from the completion of one finished part to the completion of the next is 1 minute. This time interval between the completion of finished units is called the ***cycle time***.

In this initial discussion, it is assumed that there is a single operator at each work station. Each operator performs a set of tasks on each unit that enters the work area. The completion of all tasks at all work stations results in a final assembled unit. Although some tasks must be performed prior to other tasks, a great deal of flexibility exists in assigning tasks to operators at work stations. The objective in line balancing is to find the minimum number of groupings of tasks into stations so that (1) each station consumes no more time than the cycle time in completing assigned tasks and (2) precedence requirements among tasks are maintained.

The time it takes to perform all assigned tasks at a station for a given unit is called the station time. The cycle time minus the station time for station i is called the delay time for station *i*. The sum of all station delays is called the ***balance delay***. It can be shown that the optimal grouping of tasks within stations results in a minimum balance delay. Thus, a restatement of the line balancing problem is: Given a certain number of stations, task precedence relations, and the condition that the total station time not exceed the cycle time, assign tasks to stations to minimize the cycle time or, equivalently, the balance delay.

Many approaches exist to line balancing [11, 13]. These approaches assume constant task times. For a prescribed balancing scheme, SLAM II can be used to simulate the assembly line to estimate delays and/or cycle times. In this way, balancing schemes may be evaluated even when random fluctuations in task times exists. Furthermore, once a SLAM II network for an assembly line is constructed, it provides a basis to investigate alternative balancing strategies.

The use of SLAM II in evaluating line balancing strategies is illustrated by the following example. Consider an assembly operation requiring nine tasks with constant task times as given in Figure 17-30, where both an activity network and a precedence diagram are given. The latter is used extensively in assembly line balancing studies and uses a node to represent a task. Branches are used to indicate precedence relations among tasks. Hence, in Figure 17-30b, the branches indicate that task 4 must be preceded by tasks 2 and 3. Tasks 2 and 3 may be completed in any order, but both must follow task 1. Similarly, task

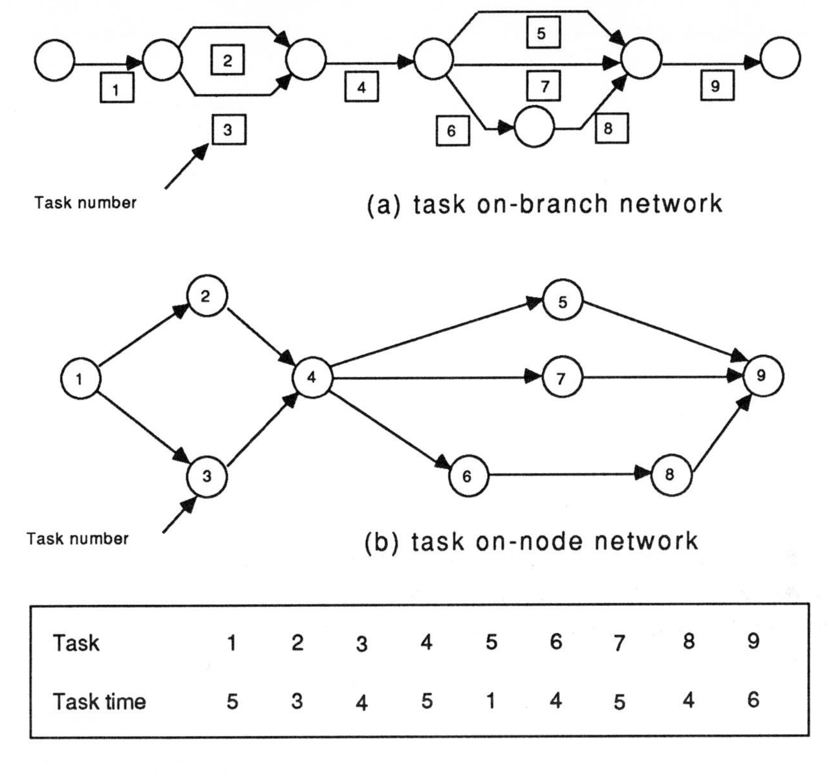

Task	1	2	3	4	5	6	7	8	9
Task time	5	3	4	5	1	4	5	4	6

Figure 17-30 Activity and precedence networks and task times for assembly line balancing example.

8 must follow task 6. Tasks 5 and 7 start times are not dependent on tasks 6 and 8. However, task 4 must precede tasks 5, 6, and 7. Finally, task 9 cannot start until tasks 5, 7, and 8 are completed.

Table 17-2 presents an allocation of tasks to stations for this line balancing problem under the assumption of a cycle time of 14 time units. Tasks 1, 3, and 2 are assigned to station 1; tasks 4, 7, and 6 to station 2; and tasks 5, 8, and 9 to station 3. The station times are computed as the sum of station task times. The station delay is the cycle time minus the station time. A SLAM II network for this example is shown in Figure 17-31.

The CREATE node represents the entry of units onto the assembly line. The time between arrivals of this node is set to the cycle time of 14 time units. The nodes AWT1 through COL2 represent the first work station. The activity numbers correspond to the task numbers. Activity times are shown as constants and are equal to the task times prescribed in Table 17-2. Node AWT1 holds units waiting for the station 1 operator. An entity arrives from the CREATE node at

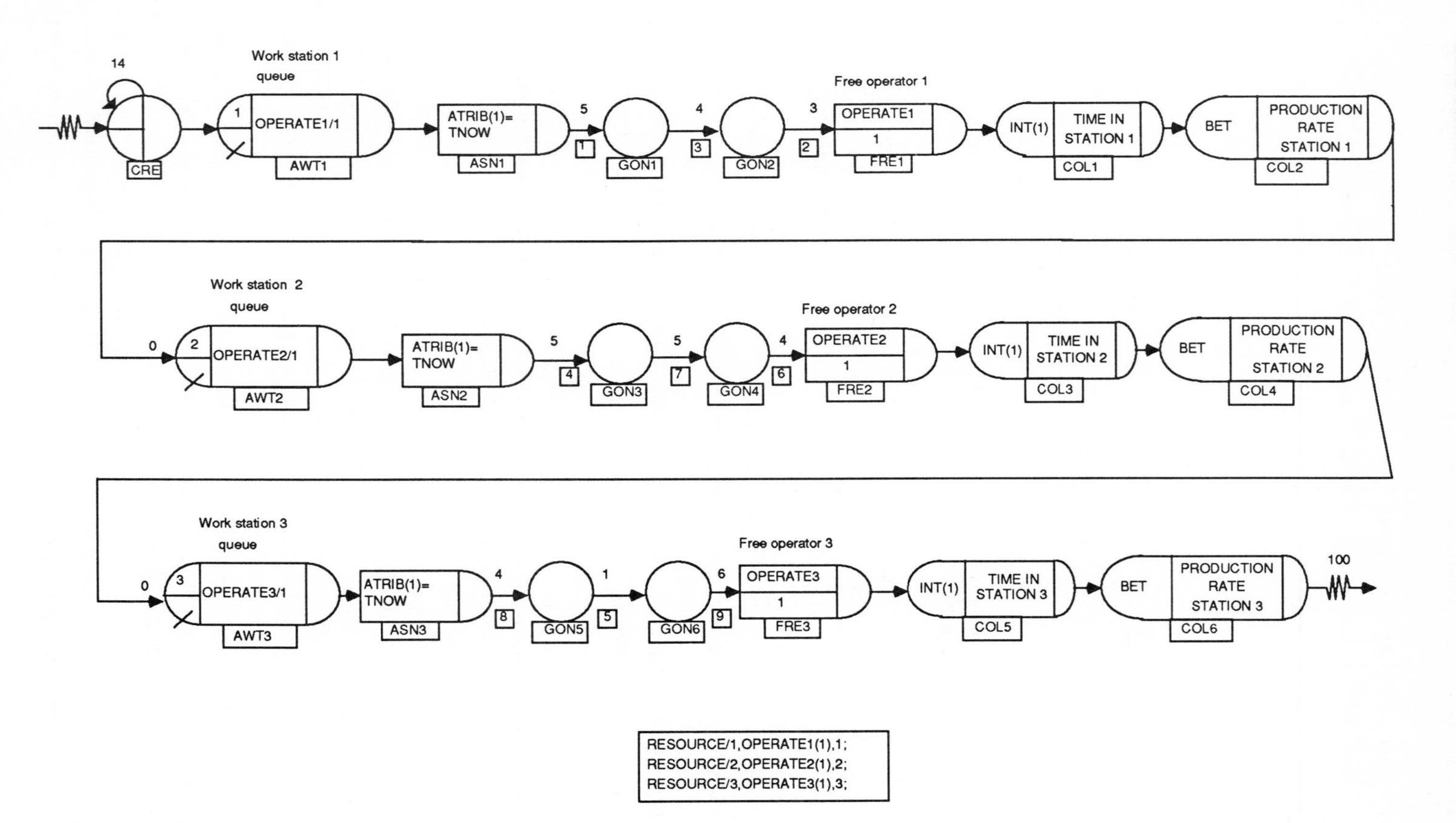

Figure 17-31 SLAM II network model of an assembly line balancing situation.

Table 17-2 A Solution to the Assembly Line Balancing Problem

Station	Task	Task Time	Station Time	Station Delay[a]
1	1	5		
	3	4		
	2	3	12	2
2	4	5		
	7	5		
	6	4	14	0
3	8	4		
	5	1		
	9	6	11	3

[a] Cycle time is assumed to be 14.

time zero to be processed. The operator is modeled as the resource OPERATE1 which is required at node AWT1. If a unit arrives at node AWT1 and OPERATE1 is available, the entity captures one unit of OPERATE1 and passes through ASSIGN node ASN1 to activity 1. The sequence 1, 3, and 2 represents the tasks at station 1. Upon completion of task 2, FREE node FRE1 frees OPERATE1 to begin work on the next unit waiting in AWT1. If there are no entities waiting at node AWT1, OPERATE1 becomes idle.

Attribute 1 for each entity is set equal to TNOW at node ASN1, and interval statistics are collected at node COL1, which represents the station time for station 1. For the constant times used in this illustration, the station time will always be 12 for station 1. However, when the task times are random variables, the station time becomes a random variable, and node COL1 provides statistics on this quantity. The closer the station time is to the cycle time, the better the utilization of the work station. However, for random times, a higher utilization results in a higher probability of not completing all tasks in a cycle. The production rate at which the station is actually operating is revealed by the between statistics that are collected at node COL2. The average number of entities in node AWT1 represents the backlog for the station, and resource utilization statistics for resource type 1 represent the usage and idle statistics for the operator.

The SLAM II statement model corresponding to the network of Figure 17-31 is shown in Figure 17-32. The assembly line is run for 100 cycles as indicated on the TERMINATE statement. The ENTRY/2 statement places a job in file 2 for station 2, and the ENTRY/3 statement places a job in file 3 for station 3. At the beginning of the simulation, the CREATE node provides a job for station 1 and the ENTRY statements provide jobs for stations 2 and 3. Station 1 finishes

```
GEN,PRITSKER,ASSEMBLY LINE,3/3/88,1;
LIMITS,3,1,100;
;
NETWORK;
        RESOURCE/1,OPERATE1(1),1;
        RESOURCE/2,OPERATE2(1),2;
        RESOURCE/3,OPERATE3(1),3;
;
CRE     CREATE,14;
;
;       WORK STATION 1
AWT1    AWAIT(1),OPERATE1/1;
ASN1    ASSIGN,ATRIB(1)=TNOW;
        ACT/1,5;
GON1    GOON;
        ACT/3,4;
GON2    GOON;
        ACT/2,3;
FRE1    FREE,OPERATE1/1,1;
COL1    COLCT,INT(1),TIME IN STN 1;
COL2    COLCT,BET,PROD TIME STN 1;
;
;       WORK STATION 2
AWT2    AWAIT(2),OPERATE2/1;
ASN2    ASSIGN,ATRIB(1)=TNOW;
        ACT/4,5;
GON3    GOON;
        ACT/7,5;
GON4    GOON;
        ACT/6,4;
FRE2    FREE,OPERATE2/1,1;
COL3    COLCT,INT(1),TIME IN STN 2;
COL4    COLCT,BET,PROD TIME STN 2;
;
;       WORK STATION 3
AWT3    AWAIT(3),OPERATE3/1;
ASN3    ASSIGN,ATRIB(1)=TNOW;
        ACT/8,4;
GON5    GOON;
        ACT/5,1;
GON6    GOON;
        ACT/9,6;
FRE3    FREE,OPERATE3/1,1;
COL5    COLCT,INT(1),TIME IN STN 3;
COL6    COLCT,BET,PROD TIME STN 3;
        TERM,100;
        END;
;       PUT ENTITIES INTO FILES 2 AND 3
ENTRY/2;
ENTRY/3;
FIN;
```

Figure 17-32 SLAM II statement model of an assembly line balancing situation.

its first job at time 12, station 2 at time 14, and station 3 at time 11. All stations start their second job at time 14. To get 100 jobs completed through station 3, there will be 99 cycles of 14 time units and the last job will take 11 time units. This sums to 1397 time units. At time 1397, stations 1 and 2 will still be working on their 100*th* job. The SLAM II summary report for this assembly line balancing situation is shown in Figure 17-33, and agrees with the above analysis.

The current formulation of the network permits the comparison of the station time with the cycle time and the monitoring of any violation of the desired condition that the station time be less than the cycle time. The mean value for the interval statistics can be subtracted from the cycle time for an estimate of station delay. From histograms for nodes COL1, COL3, and COL5 (not presented), an estimate of the probability that a station time exceeds the cycle time can be obtained. If statistics on station delay are desired, the delay for each entity could be computed as the cycle time minus TNOW - ATRIB(1). This value could be collected directly at an inserted COLCT node. Furthermore, statistics could be segregated according to whether this delay is positive or negative. A negative delay indicates a violation of the cycle time constraint.

To study the impact of a change in the number of operators at a station, resource capacities are modified. To investigate different task assignments to stations, network branches and activity durations are modified.

17.8 SUMMARY

This chapter has illustrated the complexity of scheduling and sequencing concepts. It has provided definitions of the different types of scheduling that are performed in a production context. The process of scheduling has been decomposed down to the level of SLAM II elements, which were then integrated in network models for studying scheduling procedures. Building on these modeling concepts, the management scientist can explore the impact on the performance measures described in Chapter 16 for new procedures for aggregate planning, scheduling, sequencing, assembly line balancing, and just-in-time inventory systems for the production process.

17.9 EXERCISES

17-1. Prepare a set of forms that can be used to portray the results (outputs) obtained from aggregate planning, loading, sequencing, and detailed scheduling. Define each output operationally and specify the inputs required to generate the values that would be inserted on your forms.

```
                    S L A M   I I   S U M M A R Y   R E P O R T

              SIMULATION PROJECT ASSEMBLY LINE              BY PRITSKER

               DATE  3/ 3/1988                              RUN NUMBER    1 OF    1

              CURRENT TIME    0.1397E+04
              STATISTICAL ARRAYS CLEARED AT TIME   0.0000E+00
```

STATISTICS FOR VARIABLES BASED ON OBSERVATION

	MEAN VALUE	STANDARD DEVIATION	COEFF. OF VARIATION	MINIMUM VALUE	MAXIMUM VALUE	NUMBER OF OBSERVATIONS
TIME IN STN 1	0.1200E+02	0.0000E+00	0.0000E+00	0.1200E+02	0.1200E+02	99
PROD TIME STN 1	0.1400E+02	0.0000E+00	0.0000E+00	0.1400E+02	0.1400E+02	98
TIME IN STN 2	0.1400E+02	0.0000E+00	0.0000E+00	0.1400E+02	0.1400E+02	99
PROD TIME STN 2	0.1400E+02	0.0000E+00	0.0000E+00	0.1400E+02	0.1400E+02	98
TIME IN STN 3	0.1100E+02	0.0000E+00	0.0000E+00	0.1100E+02	0.1100E+02	100
PROD TIME STN 3	0.1400E+02	0.0000E+00	0.0000E+00	0.1400E+02	0.1400E+02	99

FILE STATISTICS

FILE NUMBER	LABEL/TYPE	AVERAGE LENGTH	STANDARD DEVIATION	MAXIMUM LENGTH	CURRENT LENGTH	AVERAGE WAITING TIME
1	AWAIT	0.0000	0.0000	1	0	0.0000
2	AWAIT	0.1417	0.3488	1	0	1.9800
3	AWAIT	0.0000	0.0000	1	0	0.0000
4	CALENDAR	3.6457	0.7157	4	3	3.1811

REGULAR ACTIVITY STATISTICS

ACTIVITY INDEX/LABEL	AVERAGE UTILIZATION	STANDARD DEVIATION	MAXIMUM UTIL	CURRENT UTIL	ENTITY COUNT
1	0.3579	0.4794	1	0	100
2	0.2140	0.4101	1	1	99
3	0.2863	0.4520	1	0	100
4	0.3579	0.4794	1	0	100
5	0.0716	0.2578	1	0	100
6	0.2842	0.4510	1	1	99
7	0.3579	0.4794	1	0	100
8	0.2863	0.4520	1	0	100
9	0.4295	0.4950	1	0	100

RESOURCE STATISTICS

RESOURCE NUMBER	RESOURCE LABEL	CURRENT CAPACITY	AVERAGE UTILIZATION	STANDARD DEVIATION	MAXIMUM UTILIZATION	CURRENT UTILIZATION
1	OPERATE1	1	0.8583	0.3488	1	1
2	OPERATE2	1	1.0000	0.0000	1	1
3	OPERATE3	1	0.7874	0.4091	1	0

RESOURCE NUMBER	RESOURCE LABEL	CURRENT AVAILABLE	AVERAGE AVAILABLE	MINIMUM AVAILABLE	MAXIMUM AVAILABLE
1	OPERATE1	0	0.1417	0	1
2	OPERATE2	0	0.0000	0	1
3	OPERATE3	1	0.2126	0	1

Figure 17-33 SLAM II summary report for assembly line balancing situation.

17-2. Prepare an aggregate plan for a computer center with which you are familiar.

17-3. Discuss how rules for job sequencing may influence the procedures used for machine sequencing.

17-4. Build a SLAM II model to represent the arrival of two types of jobs. There are 1000 units of job type 1 and 500 units of job type 2 to be processed. Three operations are to be performed on type 1 jobs, with the machining times being exponentially distributed with a mean of 10. A due date for type 1 jobs is prescribed as 1.5 times the sum of the processing times for its three operations. A type 2 job requires one operation, which may have to be repeated with probability 0.15. The time to perform the operation is normally distributed with a mean of 40 and a standard deviation of 5. For a type 2 job, a due date of 70 time units after its arrival is prescribed. In addition to building the SLAM II model for the arrival of jobs, estimate (guess) the fraction of jobs of each type that will be late. Also estimate the makespan for this situation.

17-5. Describe the advantages and disadvantages associated with the following rules for loading jobs to queues:

(a) Load the job to the queue that has the smallest average number of transactions passing through it to date (SAV).

(b) Load jobs to queues in a preferred order (POR).

(c) Load jobs in a cyclic manner (CYC).

(d) Load jobs to the area that has had the smallest number of balkers to date (SNB).

17-6. Develop a loading rule using function NQS that attempts to equalize the waiting time for jobs at work centers. Code FUNCTION NQS for the loading rule that you developed.

17-7. Test the job sequencing rules first in, first out (FIFO), last in, first out (LIFO); shortest processing time (SPT); smallest due date (DD); and smallest slack for the single-queue, single-server system under the following conditions:

(a) Exponential arrivals with mean 20; exponential service times with mean 8; due dates assigned as 20 units from arrival time.

(b) Same as part (a) with due dates equal to arrival time plus a sample from an exponential distribution with mean 20.

(c) Same as part(a) except for a constant interarrival time of 10.

(d) Same as part (a) except for a constant service time of 9.

17-8. Given the following seven jobs, apply Johnson's rule to determine the sequencing of the jobs.

Job	Tooling	Production
A	6	9
B	8	6
C	8	7
D	1	3
E	5	4
F	16	9
G	11	16

Compare machine idle times if the jobs are done in alphabetical order against the sequence obtained using Johnson's rule. Assuming that the times given in the table are normally distributed with a standard deviation equal to one-tenth of the mean, use SLAM II to evaluate sequencing by Johnson's rule. Make statements concerning the probability that this makespan is less than the makespan obtained for constant job times.

17-9. Using the procedures described in Section 17.2.4, prepare for the arrival of the 20 jobs shown in the accompanying table, where each job is processed through three operations (A, B, and C) in the sequence given. Evaluate each of the rules specified in Table 17-1 for this particular job input stream.

Job Number	Arrival Time	Operation Sequence	Operation Times	Due Date
1	0	A, B, C	2, 3, 4	12
2	0	B, C, A	3, 4, 2	12
3	0	B, C, B	3, 3, 3	12
4	3	A, C, B	6, 1, 4	20
5	7	A, B, A	1, 3, 5	25
6	12	B, C, A	1, 2, 2	30
7	13	B, C, A	4, 1, 3	30
8	20	A, B, C	2, 3, 4	35
9	21	B, C, B	4, 3, 2	40
10	22	A, B, A	3, 1, 2	42
11	24	C, A, B	4, 1, 1	45
12	30	B, C, B	6, 3, 2	50
13	36	C, A, C	3, 4, 2	55
14	38	C, A, B	4, 1, 1	58
15	41	B, C, A	4, 1, 3	60
16	49	A, B, A	1, 3, 5	65
17	52	B, C, B	3, 3, 3	70
18	59	A, C, B	6, 1, 4	70
19	60	B, C, A	3, 4, 2	72
20	62	B, C, A	1, 2, 2	72

17-10. Run the SLAM II model shown in Figure 17-28 to obtain statistics that evaluate the proposed assignment of tasks to stations. Develop a new task assignment and evaluate it.

17-11. Change the duration specification for the activities in Figure 17-28 so that they are normally distributed with a mean equal to the constant value and a standard deviation equal to one-tenth of the mean values. The model is to be changed so that a job not completed within the cycle time is routed to an off-line station. Evaluate the proposed assignement of tasks to operators presented in Figure 17-28 by assessing the number of jobs that must be routed off-line.

17-12. Build a SLAM II model for the following situation. Items flow through a paced assembly line. There are three stations on the assembly line, and the service time at each station is exponentially distributed with a mean of 10. Items flow down the assembly line from server 12 to server 2 to server 3. A new unit is provided to server 1 every 15 time units. If any of the servers have not completed processing their current unit within 15 minutes, the unit is diverted to one of two off-line servers, who complete the remaining operations on the job diverted from the assembly line. One time unit is added to the remaining time of the operation that was not completed. Any following operations not performed are done by the off-line servers in an exponentially distributed time with a mean of 16. Draw the SLAM II network to obtain statistics on the utilization of all servers, and the fraction of items diverted from each operation.

Embellishments

(a) Assume that the assembly line is paced and that the movement of units can occur only at multiples of 15 minutes.
(b) Allow one unit to be stored between each assembly line server.
(c) If a server is available, route units back to the assembly line from the off-line servers.

17-13. Develop a SLAM II network model for a 60-job, three-machine problem (assume all jobs require processing by the three machines) based on comparing the times on machines 1 and 3 first; that is, create two sets. Next divide each set into two sets by comparing times on machines 1 and 2 and machines 2 and 3. Evaluate different 60-job examples to determine the effectiveness of the algorithm.

Embellishments

(a) Modify the network model to allow one-machine jobs.
(b) Modify the network to allow one and two-machine jobs.

17.10 REFERENCES

1. Abbot, R. A. and T. J. Greene, "Determination of Appropriate Dynamic Slack Sequencing Rules for an Industrial Flow Shop via Discrete Simulation," *Proceedings, Winter Simulation Conference*, 1982, pp. 223-230.
2. Adam, E. E., Jr. and R. J. Ebert, *Production and Operations Management: Concepts, Models and Behavior*, Prentice-Hall, Englewood Cliffs, NJ, 1978.
3. Baker, K. R., *Introduction to Sequencing and Scheduling*, Wiley, New York, 1974.
4. Conway, R. W., W. L. Maxwell, and L. W. Miller, *Theory of Scheduling*, Addison-Wesley, Reading, MA, 1967.
5. Conway, R. W., and others, *User's Guide to XCELL+ Factory Modeling System*, 2*nd* ed., The Scientific Press, Redwood City, CA, 1987.
6. Eilon, S., I. G. Chowdhury, and S. S. Serghiou, "Experiments wtih the SI^x Rule in Job Shop Scheduling," *Simulation*, Vol. 24, 1975, pp. 45-48.
7. *FACTOR Implementation Guide*, FACTROL, P.O. Box 2569, West Lafayette, IN 47906, 1987.
8. Huang, P. Y., L. P. Rees and B. W. Taylor III, "A Simulation Analysis of the Japanese Just-in-Time Technique (with Kanbans) for a Multiline, Multistage Production System," *Decision Sciences*, Vol. 14, 1983, pp. 326-344.
9. Jackson, J. R., "An Extension of Johnson's Results on Job-Lot Scheduling," *Naval Research Logistics Quarterly*, Vol. 3, September 1956, pp. 201-203.

9. Johnson, R. V., "Optimally Balancing Large Assembly Lines With 'FABLE'," *Management Science*, Vol. 34, 1988, pp. 240-253.
11. Kao, E. P. C., "Computational Experience with a Stochastic Assembly Line Balancing Algorithm," *Computers and Operations Research*, Vol. 6, 1979, pp. 79-86.
12. Mansoor, E. M., "Assembly Line Balancing - An Improvement on the Ranked Positional Weight Technique," *Journal of Industrial Engineering*, Vol. 15, 1964.
13. Mastov, A. A., "An Experimental Investigation and Comparative Evaluation of Production Line Balancing Techniques," *Management Science*, Vol. 16, July 1970, pp. 728-742.
14. Schrage, L., and K. R. Baker "Dynamic Programming Solutions of Sequencing Problems with the Precedence Constraints", *Operations Research*, Vol. 26, 1978, pp. 444-459.
15. Standridge, C. R., and A. A. B. Pritsker, *TESS: The Extended Simulation Support System*, Wiley, New York, 1986.

18 MODELING MATERIAL HANDLING SYSTEMS

18.1 INTRODUCTION

A major cause for delay in manufacturing operations is the transportation of jobs and material from one work center to another [1, 2]. This transportation delay has created a need for improved facility design, as well as detailed analyses of material handling (MH) equipment. Estimates are that over 85% of manufacturing time is spent in a waiting or transportation state and less than 10% in processing. Caution should be exercised in basing decisions on these values [19] as greater emphasis is now being placed on machine layout and the design of materials handling equipment during the system design process. For example, new facility layouts are based on group technology concepts where machines are grouped in work cells by job function, rather than by machine type, to reduce or eliminate the waiting and materials handling delays. Including material handling processes in a design adds complexity to the design process, which increases the need for models to evaluate alternative designs. The benefits are a reduction in inventory and higher throughputs.

Detailed models of material handling equipment use the geometry of a facility layout to analyze the transport operations of material handling equipment [3, 8, 14]. Thus, models specific to individual systems are typically developed. Based on experience, general modeling constructs have been derived from the system-specific models.† In SLAM II, these general constructs are included in the material handling extension to SLAM II discussed in Chapter 19.

In this chapter, basic concepts are presented for modeling material handling equipment. Basic models are developed for the movement of material by conveyors,

† Steven Duket of Pritsker & Associates performed much of the original modeling of material handling systems described in this text.

carts, cranes, and pipelines. As will be seen, the detailed analysis of a manufacturing system including material handling equipment requires modeling constructs for both the facility layout and the control logic associated with MH operations. The simulation concepts and capabilities presented in this chapter represent developments that have proved useful in practice. [5, 6, 9, 10]

18.2 BASIC MATERIAL HANDLING CONCEPTS

A material handling resource moves a job or material between two points. If infinite resources are available to perform the movement, then no waiting occurs, and job movement can be represented by a single activity. This could also be assumed if the job movement time is large compared to the time required to acquire a material handling resource, such as may occur in truck, train, or airplane movement. In normal manufacturing situations, the number of material handling resources is limited, and the delay due to their acquisition is not insignificant compared to the movement time. For this reason, models involving job movement require that jobs must wait for a material handling resource at AWAIT nodes. The move activity is then performed, and the material handling resource is then returned or rerouted before it is free to perform another move. The job entity that was moved is ready to begin its next processing step after the move activity is completed. A SLAM II network model for this situation is shown in Figure 18-1 where the material handling resource is given the name MHUNIT. In this model, MHUNIT could be a forklift, a human carrier, or any conveyance related to a discrete move from one location to another.

Since material handling resources are normally limited, jobs are, in many instances, grouped together so that the movements may perform for a group of jobs. The modeling of this situation is accomplished using the BATCH and UNBATCH nodes of SLAM II, as shown in Figure 18-2. For this network segment, a BATCH node is placed before the AWAIT node, which assigns jobs to one or more batches according to the parameters of the BATCH node. By specifying RETAIN(3), each entity in the batch is retained for later insertion into the network.

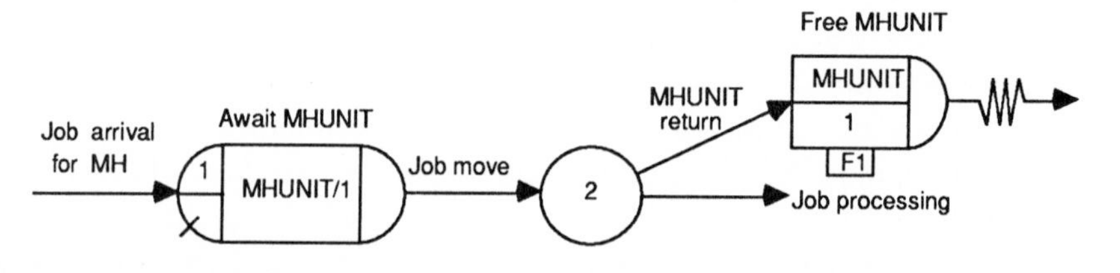

Figure 18-1 Basic material handling model segment.

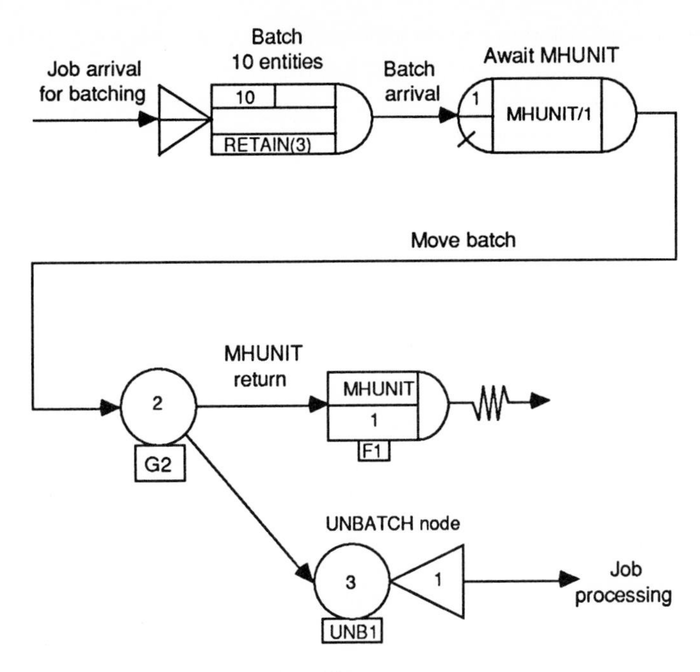

Figure 18-2 Batching for material handling.

The batched entity is routed to the AWAIT node where MHUNIT is acquired. An activity models the movement of the batch. At GOON node G2, MHUNIT moves unloaded before being freed at node F1 to move another batch. The batched entity is routed to UNBATCH node UNB1 where the entities making up the batch are reinserted into the network.

18.3 MODELING CONVEYOR SYSTEMS

Conveyors are important methods of transporting materials within a production system. A general modeling capability for conveyors is difficult to develop because of their diversity and multiple functionality. Although transportation is the basic function of conveyors, they are also used to accummulate, sort, and provide in-process storage. In addition, conveyors are designed to handle individual items, loose parts, packages, containers, or unitized loads. For example, packaging unit load conveyors may be further identified by the load-carrying device and include roller, belt, skate wheel, slat, and carrier. These conveyor types are normally mounted to the floor, with merge and diverge points for routing and sorting included in the conveyor system. Other conveyors are installed overhead, and may be of the power and free type.

Conveyors for parts handling are used in packaging lines for boxes, cans, and bottles and require a different modeling approach. These conveyors can process thousands of parts per minute and contain sensors to permit the monitoring of the part buildup on the conveyor. Control systems for controlling conveyor speed and machine processing speed are now commonplace in industry.

The above discussion indicates the complexity of conveyor systems. The approach to presenting conveyor models is to use standard SLAM II modeling elements to build simple models to which complexity can be added. Consider first a one-way conveyor moving between two points, a loading point and an unloading point. In loading a conveyor, there is a maximum amount of material that may be placed on the conveyor within a given time period. That is, once a unit of material has been loaded on a section of the conveyor, the loader must wait until that section of the conveyor moves forward before another unit can be loaded. The length of time required until another load can be put on the conveyor is a function of the conveyor speed. If a unit load occupies 50 centimeters of conveyor and the conveyor moves at a speed of 100 centimeters/minute, the wait time must be at least 0.5 minute. This time is referred to as the ***one-unit move time***.

Another aspect of one-way conveyor transport is that the time interval between departing unit loads from the conveyor is identical to the time interval between the placement of loads on the conveyor. This assumes that when the loading process terminates the conveyor transports all material on the conveyor to the exit point. Given this background, a series of models is presented, with each model introducing a new feature. Consider the network segment shown in Figure 18-3. AWAIT node A1 models the buildup of material waiting to be loaded on to the conveyor. Activity 1 is the conveyor, and its time duration is the time to traverse from an entry point to an exit point on the conveyor. This representation is inadequate, however, since it allows only a single unit load to occupy the conveyor. One might consider changing the specification for the number of parallel servers from 1 to N, where N represents the number of unit loads the conveyor can hold. This would also be inadequate, as it allows the simultaneous loading and transporting of units on the conveyor. That is, it does not model the one-unit move time referred to earlier.† The network segment in Figure 18-4 models the one-unit move time. In this network segment, unit loads arrive at node A1, but before entering the conveyor, a delay may occur

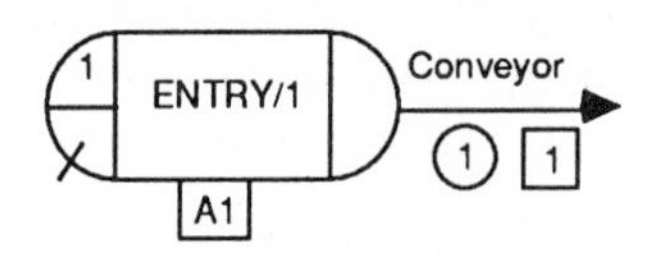

Figure 18-3 Conveyor entry.

† If this delay is not significant, a single AWAIT node with N parallel servers approximates conveyor flow.

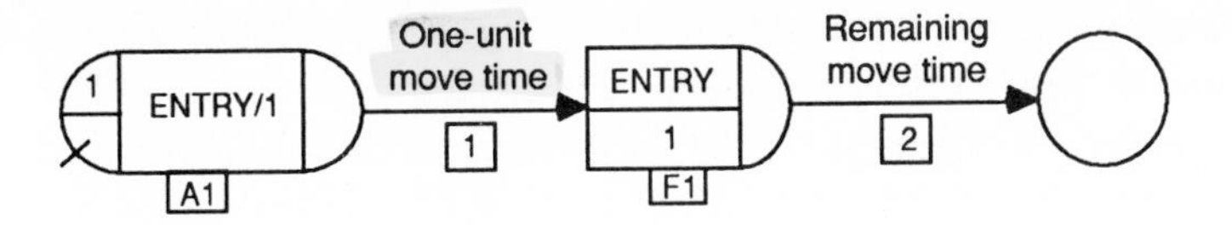

Figure 18-4 Conveyor model.

depending on the availability of the ENTRY to the conveyor. This time represents the time for the conveyor to move one unit away from the loading area, that is, the one-unit move time. When a unit reaches node F1, the conveyor entry is freed and another unit may be loaded on the conveyor. The unit arriving to node F1 continues on the conveyor, which is modeled by activity 2.

As an extension of this model, consider a situation in which excessive buildup may occur in the area where units exit the conveyor. If the conveyor cannot be unloaded, all units on the conveyor become blocked as each load on the conveyor reaches a nonmoving load in front of it. A method for modeling this situation is to divide the conveyor into sections, each having a resource associated with it. The model shown in Figure 18-5 divides the conveyor into ENTRY, MIDDLE, and EXIT sections. The number of resources assigned to the middle section of the conveyor prescribes the number of entities that can be carried by the conveyor that are not included in the ENTRY and EXIT portions of the conveyor. In the network shown, a total of 12 parts can be on the conveyor. By specifying a FIFO priority for file 3 at the exit AWAIT node, the entities are processed in the order in which they traverse the conveyor. In the illustration, an EVENT node is used to indicate that complex end-of-conveyor logic can be performed before the EXIT portion of the conveyor is freed.

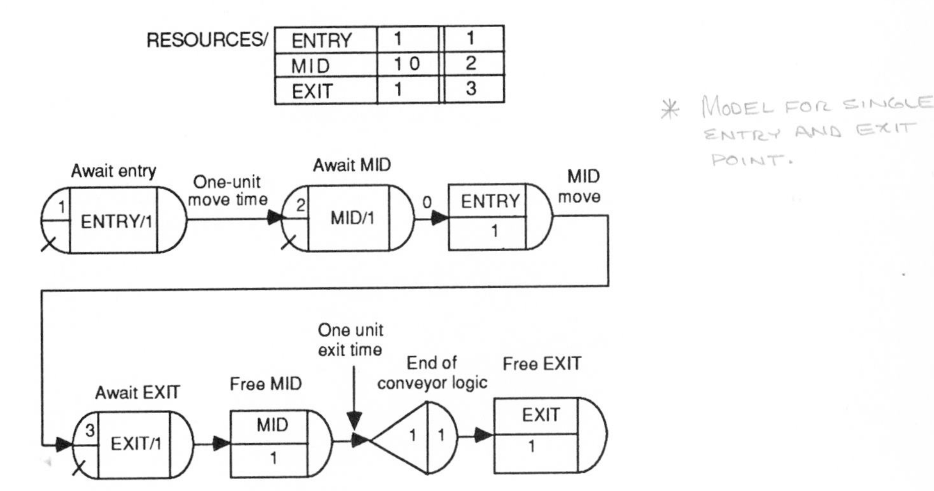

Figure 18-5 Conveyor model with blocking.

Up to this point, conveyors with only a single entry and exit position have been discussed. Some conveyors operate in a circular fashion and have a number of exit positions. To illustrate, consider the case of five servers stationed along a circular conveyor belt [11]. Assume that items to be processed by the servers arrive at the conveyor belt with a interarrival time that is exponentially distributed with a mean of 1 minute. After being placed on the conveyor belt, it takes 2 minutes for a new arrival to reach the first server. Service time for each server averages 3 minutes and is also assumed to be exponentially distributed. No storage space for items is provided before any of the servers. If the first server is idle, the item is processed by that server. If the first server is busy when the item arrives, the item continues down the conveyor belt until it arrives at the second server. The delay time between servers is 1 minute. If an item encounters a situation in which all servers are busy, it is recycled to the first server with a time delay of 5 minutes. At the completion of service for an item, the item is removed from the system. A diagram of the conveyor system is shown in Figure 18-6. Described next is a SLAM II model designed to collect statistics on the system residence time for an item and the utilization of each server based on the processing of 200 items.

In this example, an item arrives and is transferred to the first server with a 2-minute time delay. The size of an item is small and the one-unit move time is considered insignificant. If the first server is free, the item is taken off the

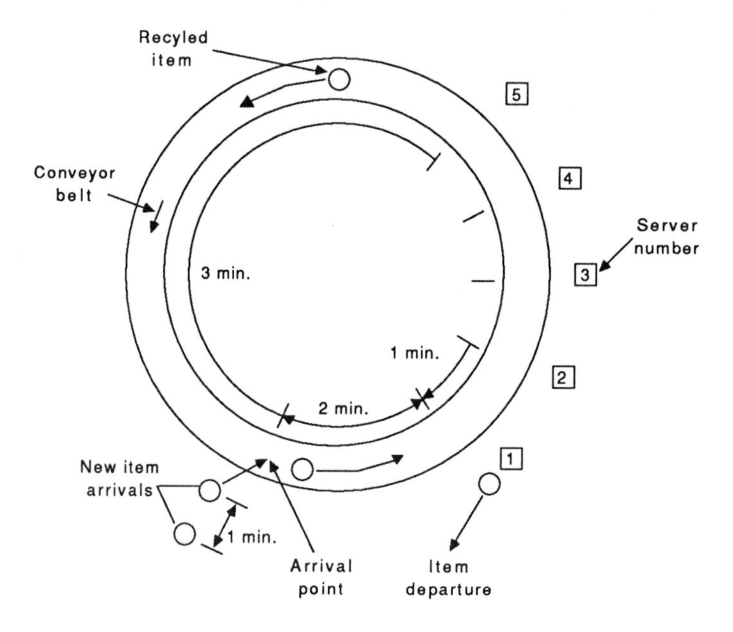

Figure 18-6 Schematic diagram of a circular conveyor.

conveyor belt and service is initiated on it. If the first server is busy, the item must bypass the first server and continue along the conveyor belt to the second server. The decision as to whether the item is processed by the first server may be modeled in SLAM II using a QUEUE node with a zero queue capacity. If the server is idle, the item flows through the QUEUE node directly to the server. However, if the server is busy, the item cannot stay at the QUEUE node because it has a zero capacity; hence, it balks. In this situation, balking from the server involves continuing on the conveyor toward the next server. The remainder of the SLAM II model follows this pattern of attempting to gain access to a server through a QUEUE node and then balking from it if the server is busy. The SLAM II network model for the conveyor system is shown in Figure 18-7.

Arrivals of items are modeled using a CREATE node. Since it takes 2 minutes for the item to reach the first server, the branch from the CREATE node to QUEUE node Q1 is given a 2-minute duration. Node Q1, preceding server number 1, has no initial entities and a capacity of zero. When an entity arrives at node Q1 and server 1 is busy, the entity balks to node G1. The activity from node G1 to node Q2 represents the transit time on the conveyor belt required to move to the second server. In this example, it is assigned a 1-minute time delay. When

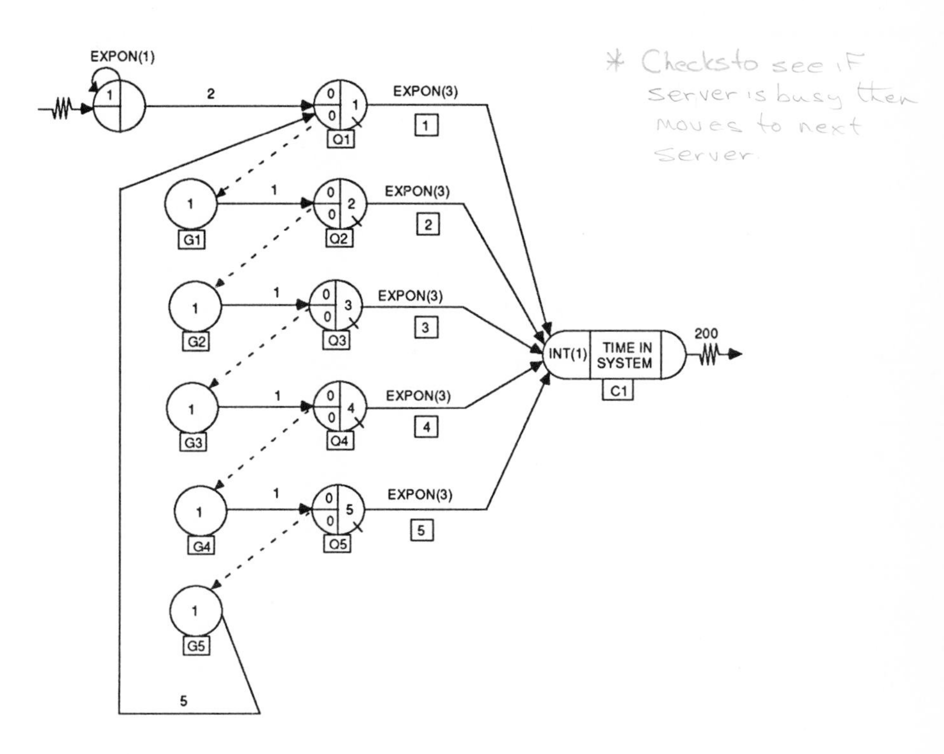

Figure 18-7 SLAM II network model of circular conveyor.

an entity arrives at COLCT node C1, service has been completed and the time spent on the conveyor system is collected. The entity is then terminated and a count of the number processed by the conveyor system is increased by 1. Two hundred exiting entities are required to complete a run.

If an item balks from all five QUEUE nodes, it reaches node G5. The activity from node G5 to node Q1 represents the recycling of items and requires a 5-minute time delay. By returning to node Q1, another attempt to process the item through the five-server conveyor system is initiated.

For circular conveyors of this type, service activity 1 will be busier than service activity 2, which in turn will be busier than service activity 3, and so on. This occurs because arriving items attempt to gain access to servers in a prescribed order [15]. One method for balancing the work load of the servers is to incorporate in-process storage areas prior to the servers. If larger storage areas are provided for servers with higher activity numbers, a balancing of the busy times of the servers may be obtained. Control systems could be used to assign items upon their arrival to a particular server (centralized control) or to make a decision at each service station as to whether the item should be taken off the belt and served or remain on the conveyor belt (decentralized control). In-process storage and control procedures are easily incorporated into the network model presented (see Exercises 18-2 and 18-3).

18.4 CONTINUOUS PRODUCT FLOW THROUGH A PIPELINE

Continuous product flow, such as crude oil through a pipeline, may be represented in network form by considering the continuous flow as made up of blocks or batches that sequentially flow through the pipeline. With this transformation, pipeline flow seems analogous to conveyor transport. However, there is an important characteristic of pipeline flow that distinguishes it from conveyor flow. When input to a conveyor ceases, the output continues until the conveyor is empty. Furthermore, input to a conveyor does not cause an output from the conveyor. Thus, output from a conveyor can be obtained with or without input even if the conveyor is only half full. With pipeline flow, however, not only must the pipeline be full, but there must also be an input to the pipeline before output is obtained. Units of material entering a pipeline push units ahead of them, and output is realized only when the pipeline is full. Once input ceases, no output flows. The key to modeling pipeline flow is to recognize that this full requirement of pipeline flow is an *and* condition. When the pipeline is at capacity *and* there is an incoming unit of flow, the pipeline produces one unit of flow as output.

One method of modeling this *and* condition is to have a resource with zero capacity and an AWAIT node that has a capacity equal to the number of units required to fill the pipeline. When the file of the AWAIT node is full and another arrival occurs, the arriving unit is routed to an ALTER node where the capacity of the resource is increased by 1. The network segment that models this condition for a pipeline that can carry ten units is shown in Figure 18-8.

File 3 contains the entities that model the blocks of material to be moved. Activity 1 is the time to pump one unit into the pipeline. At the GOON node G1, an entity is routed to ALTER node AL1 if the pipeline is full. The ALTER node allows one unit of the resource PIPE to be allocated at AWAIT node AW1, permitting one unit to flow out of the pipeline, which is modeled by activity 2. Activity 2 has a zero duration, as the pipeline exit time is assumed equal to the time to pump 1 block into the pipeline, which has transpired due to activity 1. Following activity 2, the block of material is put into output storage represented by QUEUE node Q5. The new block of material is then sent to AWAIT node AW1 to refill the pipeline by the second activity leaving GOON node G1. If the pipeline is not full, this block of material waits in file 3. Note that the model moves the blocks of material in a different order than is done in the actual system. A block of material is taken from the pipeline by altering the resource PIPE and then the new block is placed in the pipeline.

18.5 CAPACITY PLANNING FOR AN OIL DISTRIBUTION SYSTEM

In recent years the production of crude oil in the United States has declined and the importation of foreign crude has increased. The processing of foreign crude significantly affects crude distribution and storage facilities in two respects: (1) large volumes (tanker loads) arrive at irregular intervals, and (2) the quality

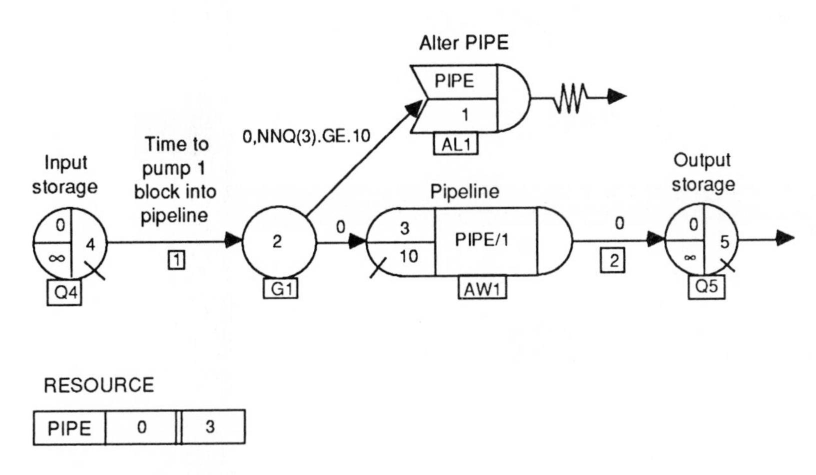

Figure 18-8 Pipeline model.

of crude requires additional segregation at pipeline stations and refineries.

The increased levels of crude from foreign imports have strained the facilities for handling the crude oil, especially the crude tankage at refineries and pipeline stations. It is commonly thought that additional crude tankage is highly desirable, but it is not obvious where the additional tankage should be located and how much additional tankage could be justified on an economic basis.

This section describes a network simulation study conducted for a large oil company that assessed the capacity requirements of a crude oil distribution system consisting of pipelines, wells, tankers, and pumping stations [5].

18.5.1 The System and Management Concerns

Figure 18-9 contains a schematic drawing of the crude oil distribution system. The circles represent storage areas. Storage areas exist at pumping stations, docks,

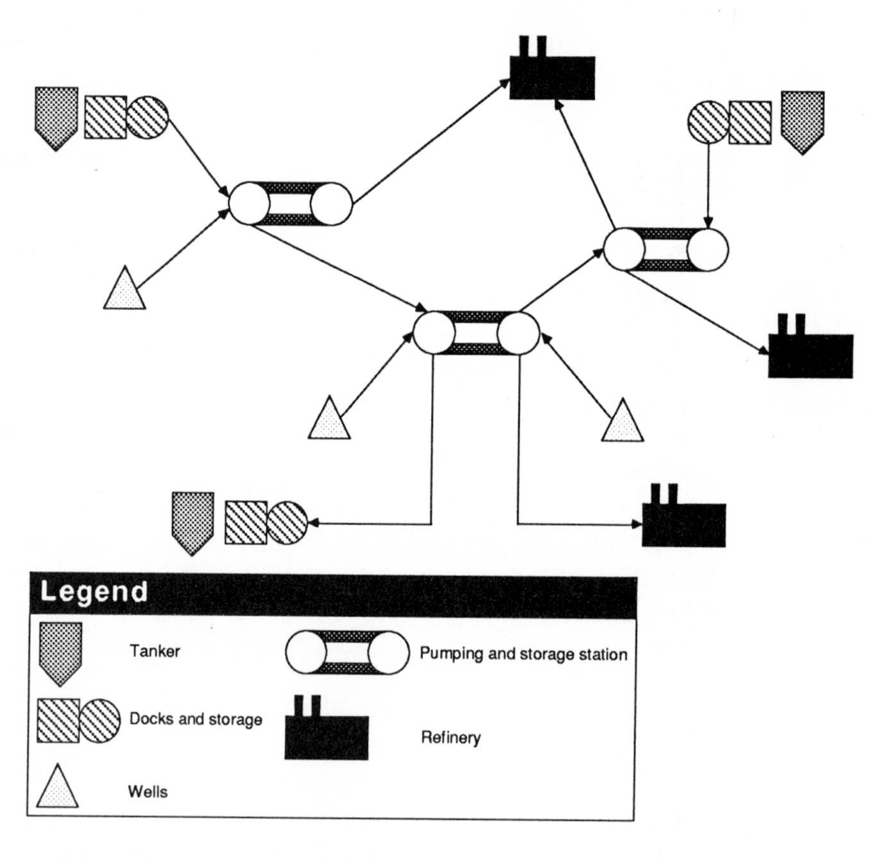

Figure 18-9 Schematic diagram of a crude oil distribution system.

and refineries. Pipelines connect the storage areas and provide the link between the various system inputs and outputs. There are two forms of input to the pipeline. First, tankers arrive at docks and deliver crude to the pipeline. This form of input is not continuous, but rather is turned on and off as a function of the arrival of tankers. Another form of input is the delivery of crude oil to the system from the wells. This is a continuous input into the system. Once crude enters the pipeline from any of the input sources, it flows continuously through the system. As shown, crude may flow to a pumping station and reside in a storage facility before being pumped to the next destination. Each pumping station redirects flow to other stations or to refineries. The goal of the distribution system is to deliver crude oil to the refineries, and therefore the amount of crude delivered represents a system output. A complication to be considered is that the distribution system is designed to transport crude oil of different types, and it is necessary to model the flow of each type of crude oil.

The objective of the project is to determine whether the proposed storage capacity is adequate. If it is not, the amount and location of the needed storage are to be determined. Management desires that the system be analyzed under a variety of input conditions, as specified by projected ranges of values for the tanker and well inputs. New storage capacity requirements for combinations of input values are to be determined.

18.5.2 Modeling Oil Distribution

Crude oil flow is transferred between locations in batch units where one unit represents 5000 barrels of crude. The tanker and well input are modeled as discrete arrivals. Continuous pipeline flow is modeled as discussed in Section 18.3. Identification of crude oil type is maintained by an attribute value, and the storage areas are represented by QUEUE nodes. The maximum capacity of each QUEUE node is set to the capacity of the proposed storage tank. Balking in the model corresponds to an overflow. By measuring the overflows, a determination of new storage capacity requirements can be made. In this section the important model features of tanker input, well input, pipeline flow, and crude oil overflow are discussed.

Figure 18-10 shows the SLAM II representation of the arrival of tankers loaded with crude oil. The CREATE node controls the arrival of tanker loads at a given dock. Two attribute assignments are made at the ASSIGN node to describe each arriving tanker load. Attribute 1 is the crude type and is set equal to 1. Attribute 2 represents the load size in units of 5000 barrels. In this model, tanker load size is assumed to be a normally distributed random variable.

Loads enter the system at the CREATE node and are routed via the ASSIGN

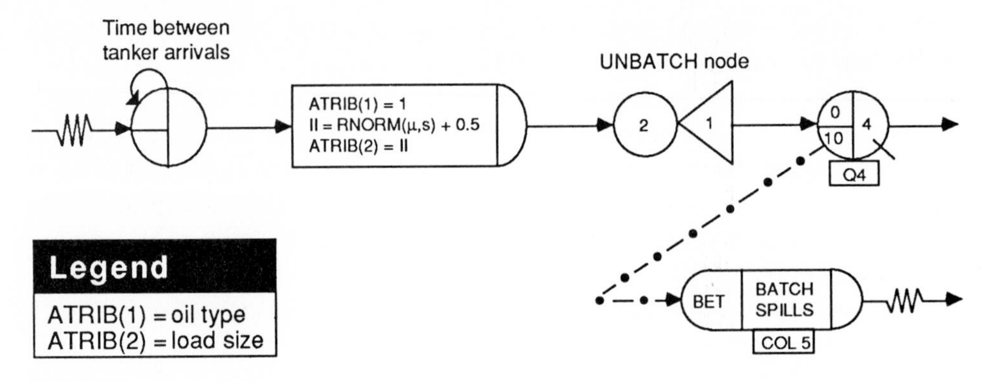

Figure 18-10 SLAM II representation of tanker arrivals.

node to the UNBATCH node. The purpose of this node is to divide the tanker load into units of 5000 barrels and place them into storage at QUEUE node Q4. Since the number of units is represented by attribute 2, the UNBATCH node splits the arriving entity into several identical entities corresponding to the value of attribute 2 and routes them to node Q4. The time duration on the activity between the ASSIGN node and the UNBATCH node is set equal to the time required to unload the tanker. The capacity of node Q4 is set to 10 units, which limits the storage at this dock to 50,000 barrels. If storage of more than 50,000 barrels is required at this location, then balking from node Q4 will occur. The number of units that overflow (balk) per time unit is reported by the SLAM II processor. Units that balk are routed to COLCT node COL5 where between statistics are maintained to provide information on the time between spillage of batches. Hence, balking statistics and between statistics are used to measure overflow.

The above representation of the unloading dock assumes that resources are always available for unloading tankers at the time of their arrival. Examples of dock resources are docking berths, unloading equipment, and personnel. When resources cannot be assumed to be available, the representation in Figure 18-11 is used. In this example, a constraint that limits the dock to only five berthing spaces is introduced. As before, the CREATE node controls tanker arrivals, the ASSIGN node sets the attribute values of arriving entities, the UNBATCH node splits the tanker load into discrete unit loads, node Q4 represents storage, and node COL5 computes overflow statistics. Before unloading begins at the UNBATCH node, however, a berth resource, represented by resource type 1, must be available. Tankers wait in the AWAIT node until a unit of the resource is available and is allocated to the entity. The maximum number of tankers that can be unloaded simultaneously is equal to the maximum number of resources (berths) that are available at the dock, which is specified as five at the RESOURCE

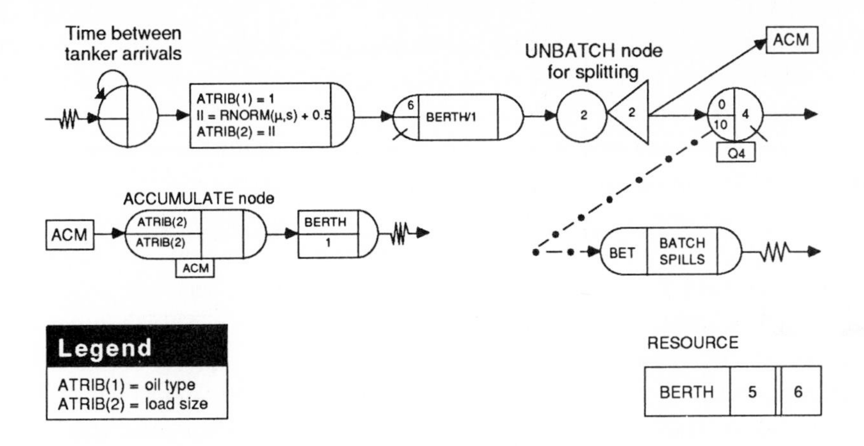

Figure 18-11 Model of tanker arrivals when unloading resources are limited.

block. As unloading is completed for each unit of 5000 barrels of oil in the tanker, a branch to the ACCUMULATE node is taken, which specifies that the UNBATCHed units are to be recombined into one entity before the berthing resource is freed for the next tanker at the FREE node. Therefore, unloading delays due to resource requirements are modeled. In the actual project, multiple resource requirements were modeled by including multiple resource types and additional AWAIT and FREE nodes.

The representation of the unloading dock given in Figure 18-11 can be embellished to account for dock shutdowns due to shifts, tides, or failure by incorporating ALTER nodes in disjoint subnetworks.

Consider now the modeling of the continuous flow of crude oil into the system from a well. Modeling this flow involves the creation of an entity representing 5000 barrels based on the time to pump this amount out of the well. This is shown in Figure 18-12, where the CREATE node creates entities representing one unit,

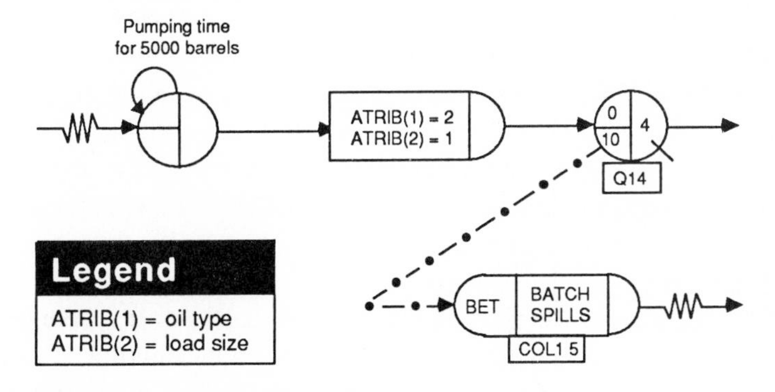

Figure 18-12 Continuous flow from wells.

and the time between creations is the pumping time for 5000 barrels. The storage tank is represented by QUEUE node Q14, which provides oil to the pipeline. COLCT node COL15 is used to compute overflow statistics. As before, attribute 1 represents the crude oil type and attribute 2 represents the load size, which in this case is 1 since unit loads come from the well.

The model for continuous pipeline flow is presented in Figure 18-13 and is based on the discussion given in Section 18.4. A GOON node is used to route crude of different types to different storage areas. Tanker loads for crude of type 1 are stored at node Q1, which represents dock storage. AWAIT node AWT2 represents the pipeline segment, with a capacity of 100 units; QUEUE node Q3 represents a pumping station's storage tanks. If there are any units in dock storage and space is available in the pipeline, service activity 1 pumps a unit into the pipeline. The disjoint network, consisting of a DETECT node and an ALTER node, serves as a switch to control the flow of oil in the pipeline. Whenever the number of units in the pipeline reaches 100 units [NNQ(2) = 100], the DETECT node routes an entity to the ALTER node, where the resource FILLUP, initially set at zero, is increased by 1. When NNQ(2) reaches 100, the first unit waiting in node AWT2 is allocated this resource and is pumped through the pipeline on activity 2 to the pipeline storage area at node Q3. Consequently, when the pipeline

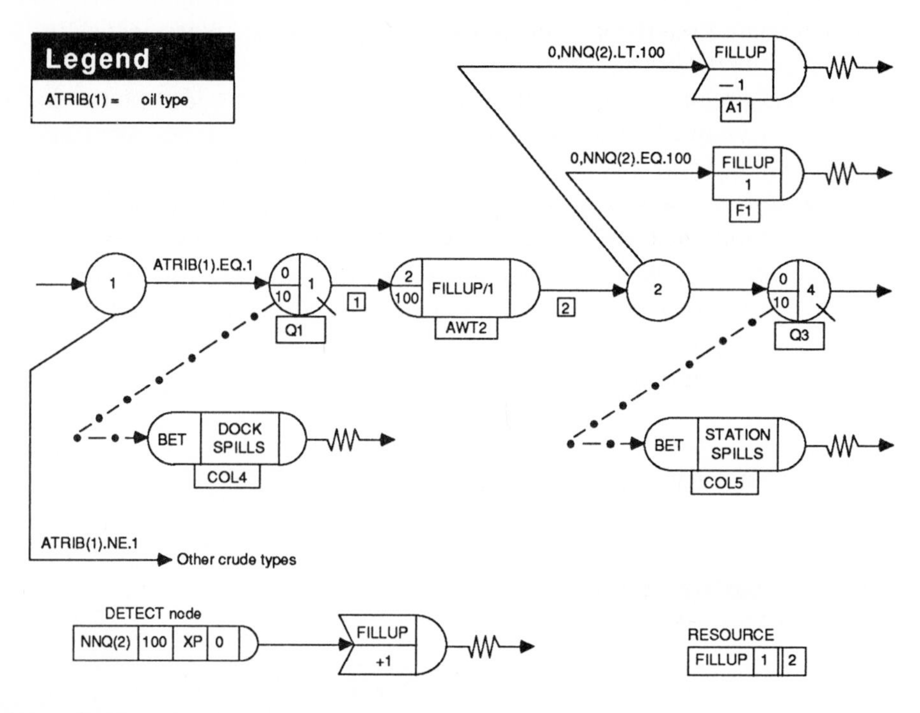

Figure 18-13 Continuous pipeline flow.

is full, any unit arriving at node AWT2 causes the next unit in the pipeline to be deposited into the storage tank at node Q3. When the pipeline is not full [NNQ(2) < 100], the FILLUP resource capacity is altered to zero at node A1. After a batch is processed over the pipeline, the next batch is allowed to proceed by freeing FILLUP at node F1. COLCT nodes COL4 and COL5 collect statistics on overflow at each of the storage tanks.

The network segments given in Figures 18-10 through 18-13 form the basic elements of the distribution system. These segments can be connected to represent the entire distribution system depicted in Figure 18-9. The actual model of the proposed system consisted of over 100 nodes. The modeling concepts presented above capture the significant aspects of the network model. Extensive user function code was employed in the original model to portray actual refinery demand and crude oil routing and mixing. The refinery demand established the type of crude oil required, which, in turn, was an input to the decision as to which tank should serve as an input to a pipeline. A decision of this type is required for each pipeline, and it is necessary to consider such decisions simultaneously.

18.5.3 Model Output and Use

The important outputs for this study are the balking statistics, average values in storage tanks (QUEUE nodes), and the between statistics that characterize the overflow at storage areas. The model was run under various input conditions concerning estimates of tanker arrivals and well production. Essentially, the model converted these inputs into storage usage and requirements. The ideal situation is to have no overflow, which corresponds to no balking. For most runs, however, overflow was obtained at several storage areas. In some runs, new routing rules could alleviate the overflows. In other situations it was necessary to add new storage tanks. In this way, capacity requirements were designed and evaluated for various input conditions. Discussions of model validation and output results for a sample problem are reported by Bandy and Duket [5].

18.6 MODELING OVERHEAD CRANES

Many material handling systems involve vehicles that move on tracks. Examples of such systems are overhead cranes, transporter carts, and automated guided vehicles. In this section, cranes are modeled using the basic SLAM II network elements. This example provides an introduction to the need for special network capabilities to model material handling systems.

Consider the schematic drawing of an overhead crane system shown in Figure

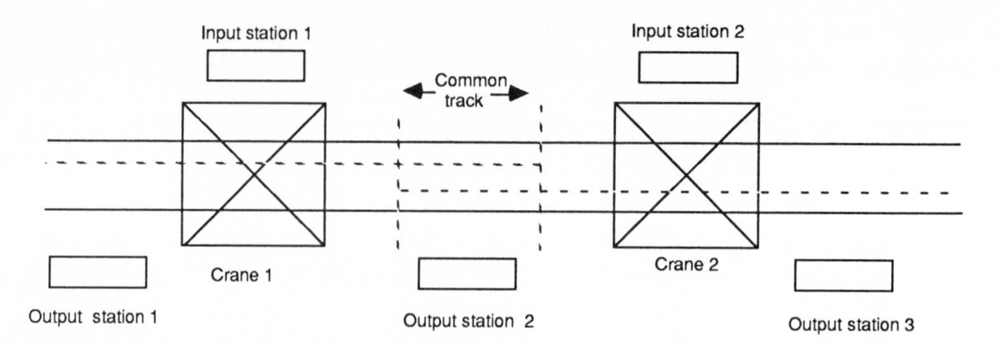

Figure 18-14 Schematic drawing of an overhead-crane system.

18-14. There are two cranes that operate on a single runway. The dashed line shows the movement pattern for crane 1, which only serves input station 1 and is used to transport material to either output station 1 or 2. Crane 2, whose movement pattern is indicated by a dotted line, transports material from input station 2 to either output station 2 or 3. There are two complexities involved with modeling this type of system. First, crane interference in reaching output station 2 needs to be modeled. Second, travel times differ between input and output stations and between loaded and unloaded crane movements.

The SLAM II network model for the overhead crane system illustrated in Figure 18-14 is shown in Figure 18-15. In this model, the crane number and the input station number are the same. They are both prescribed by attribute 1, made equivalent to CRANE, of the entity and specify the location of the job for which material handling is desired. OUTST is equivalenced to attribute 2 and defines the output station number. Rows 1 and 2 of ARRAY define crane travel times either from input stations 1 and 2 to output stations 1 and 3 or to the beginning of the common track location. Rows 3 and 4 of ARRAY define crane return times from output stations 1 and 3 or from the beginning of the common track to an input station. The EQUIVALENCE statement in Figure 18-15 specifies ARRAY(1,1) as the travel time from input station 1 (CRANE1) to output station 1. The crane number plus 2 defines a row of ARRAY for the return times. For example, ARRAY(4,2) specifies the return time for CRANE2 from the common track to input station 2.

The resources for the model are as follows:

1. Crane 1, for movement from input station 1 to output station 1 or 2
2. Crane 2, for movement from input station 2 to output station 2 or 3
3. COMTR, for movement along the common track to output station 2

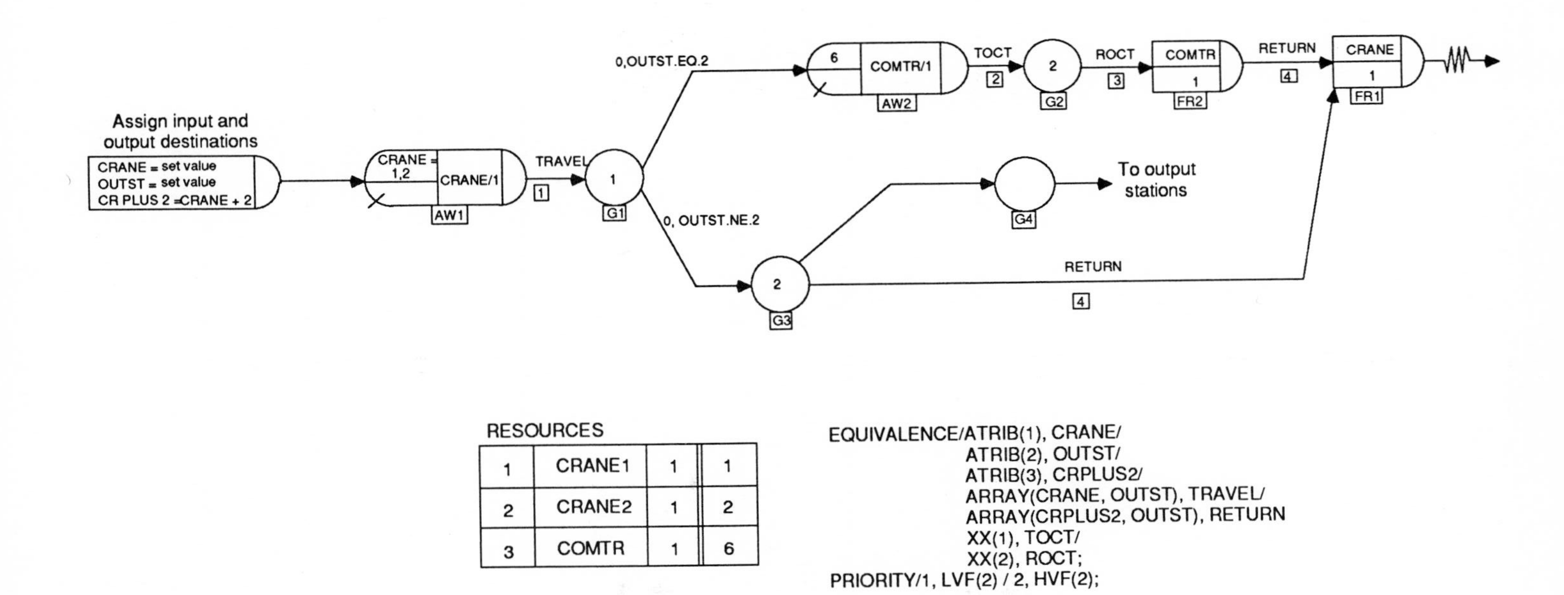

RESOURCES

1	CRANE1	1	1
2	CRANE2	1	2
3	COMTR	1	6

```
EQUIVALENCE/ATRIB(1), CRANE/
            ATRIB(2), OUTST/
            ATRIB(3), CRPLUS2/
            ARRAY(CRANE, OUTST), TRAVEL/
            ARRAY(CRPLUS2, OUTST), RETURN
            XX(1), TOCT/
            XX(2), ROCT;
PRIORITY/1, LVF(2) / 2, HVF(2);
```

Figure 18-15 Network model for overhead crane system.

An ASSIGN node is shown at the beginning of the network segment to illustrate that attributes defining the crane (input station) and output station are established prior to entering the crane-modeling subnetwork. At AWAIT node AW1, an entity arrives at one of the input stations and awaits the availability of a crane. Attribute 1, CRANE, has a value of either 1 or 2 to specify which crane is desired. Resource numbers are shown in the resource block of the model to provide a correspondence between the numeric index and the resource type. With a crane allocated to an entity, it is moved over activity 1 with a delay time that is taken from either row 1 or row 2 of ARRAY with the column number, OUTST, defining its desired output station. At GOON node G1, entities requiring the common track are routed to AWAIT node AW2 to wait for the availability of the common track resource COMTR. If the entity is assigned to output stations 1 or 3, it has arrived at its destination and is made available for output processing at GOON node G4. In addition, the crane is sent back to its input station over activity 4, with a duration of RETURN obtained from either row 3 or 4 of ARRAY.

If the entity is routed to station 2, then the travel time for activity 1 is the travel time to the common track. The entity waits in file 6 specified at AWAIT node AW2. When the common track is allocated to the entity, the travel time on the common track, TOCT, is modeled as activity 2, which routes the entity to its output station. The entity is then made available at the output station at node G4. The crane returns on the common track with a duration of ROCT given to activity 3. The common track is freed at node FR2, and the crane returns to its input station over activity 4. The crane is freed at node FR1 following the RETURN time.

The network model shown in Figure 18-15 is a highly condensed representation of the system, which takes advantage of SLAM II attribute-based resource allocations, ARRAY statements, and EQUIVALENCE statements. There are many statistical quantities included in the model that characterize crane performance. The time spent in files 1 and 2 estimates the waiting time for a material handling movement. The time for the movements is estimated from the statistics derived for activities 1 and 2. These values represent loaded crane movement times. Empty crane movement times are associated with activities 3 and 4. Interference time occurs when cranes are waiting in file 6 for the common track. The utilization of the cranes is obtained from the utilization of resources 1 and 2 and includes the movement times both loaded and unloaded. If desirable, the time for an entity to flow through this material handling system could be obtained by assigning the time of arrival at node AW1 as an attribute and collecting interval statistics when the entity reaches the output stations.

Crane modeling becomes more difficult when a crane can serve multiple input and output stations. When this occurs, it is necessary to move cranes out of the way so that a crane may have access to the input and output stations. Furthermore, cranes tend to be large units making short moves, and therefore the time for the movement depends on the acceleration and deceleration characteristics of the crane. For this reason, it has been found necessary to provide special capabilities to model multiple cranes serving multiple input and output stations. This situation is described in Chapter 19. However, at an aggregate level, the model presented in this section provides the capability to approximate the delays associated with cranes moving entities in a multiple station and common track environment.

18.7 EVALUATING CAPITAL EXPENDITURE IN THE STEEL INDUSTRY

In the steel industry, as in most industries, capital expenditures are made in new facilities and equipment for the purpose of increasing the rate of production or throughput of a given facility. It is rare that a planned investment is known to meet with certainty new output requirements. Modeling has played a significant role in the evaluation of proposed capital expenditures by predicting their impact on productive capacity [6].

This section describes a SLAM II study of a capital expenditure to be made by a major steel company.† The expenditure is for a new facility that would upgrade the quality of current production as well as increase throughput. From among several alternative production designs, the company desired to choose the one with the lowest cost that would be able to meet the throughput and quality requirements. Management specified a desired level of throughput, and a SLAM II model was constructed to estimate the probability that this level of production could be met.

18.7.1 The System and Concerns of Management

A schematic drawing of the proposed system is shown in Figure 18-16. Heated steel parts enter the system in the upper left corner of the figure. The first operation to be performed is rolling. Through a conveyor line transfer, the material is moved to a cooling area. Cranes then transfer the steel parts to another roller line where a processing operation is performed. After processing, the individual steel parts are accumulated into loads and enter an overhead-crane material handling and storage system. Each load goes through a preliminary inspection process at station B. If the load fails the inspection, it waits for crane 1 to move it to station A. Rejects are removed from this station. If a load passes the initial inspection, it waits for a crane to move it to station C. Either crane 1 or crane 2 can perform

† In the interest of confidentiality, the model has been simplified and hypothetical values employed.

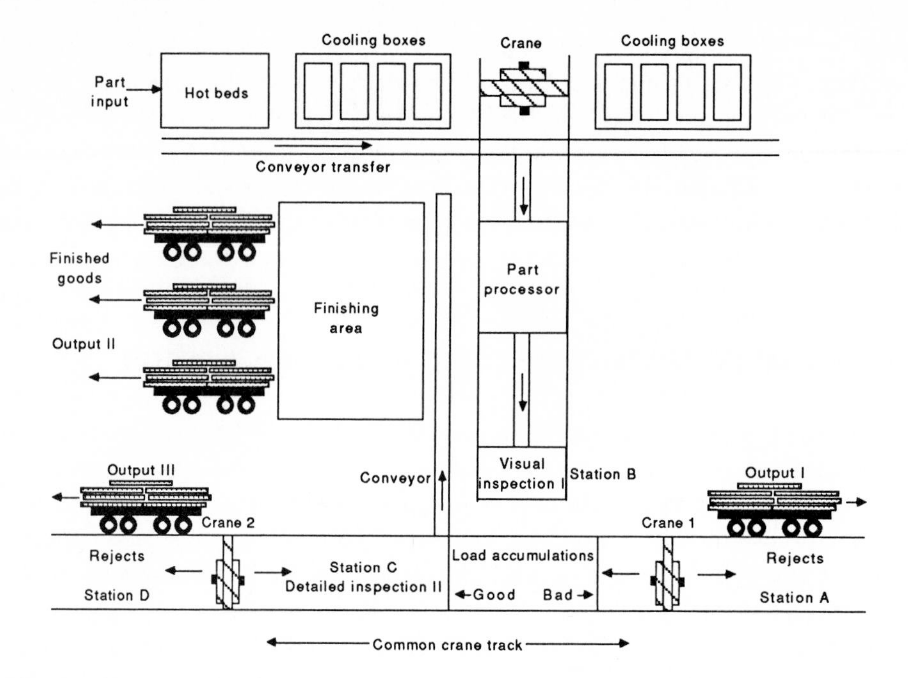

Figure 18-16 Schematic diagram for steel industry example.

this move. At station C, the load is stored as it awaits a second inspection process. If a load passes this inspection, it is automatically moved along a roller line to the finishing area. Rejects are moved by crane 2 to station D, where a roller line removes them.

Management is concerned that delays due to crane interference along the track between stations B and C might prevent the system from achieving the required throughput. For this reason, an alternative crane system was designed and only the overhead-crane subsystem modeled. A delivery rate was specified for steel entering the load accumulation center from the processing operation. The question posed was: What throughput can be expected from station C for the specified input rate and proposed design of the crane system?

18.7.2 Model of Steel Movements

A SLAM II model of the system is shown in Figure 18-17. In the network, resources 1 and 2 are used to model cranes 1 and 2, respectively. Resource 3 is used to represent the portion of track between stations B and C that is common to both cranes. Inspection decisions are modeled by probabilistic routing at ACCUMULATE node ACM1 and GOON node GON2.

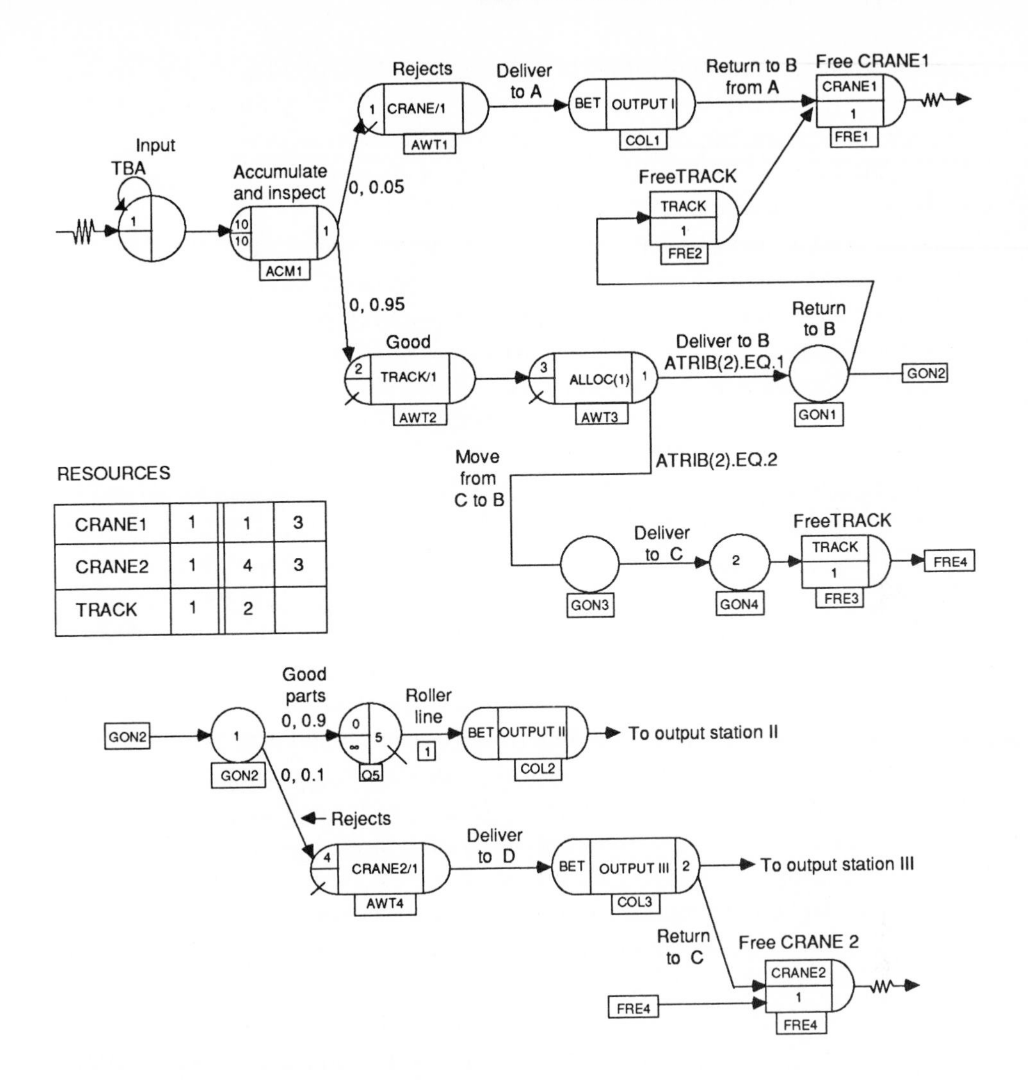

Figure 18-17 SLAM II model of overhead crane system.

The CREATE node represents the output of the part processor operation and thus provides the inputs to the model. Node ACM1 accumulates loads of ten parts by requiring ten incoming entities for release. Following the release of node ACM1, probabilistic routing of a load to AWAIT node AWT1 or AWT2 occurs. This represents the first inspection process. It is estimated that 5% of the loads are rejected at this inspection station and that 95% are accepted. Rejected loads wait at node AWT1 for crane 1 to become available. When it does, the branch to COLCT node COL1 is taken, representing the time for crane 1 to deliver the load to station A. Node COL1 collects statistics on the time between load arrivals to give an estimate of the throughput rate at station A. In Figure 18-17, this variable is labeled as OUTPUT I. The activity from node COL1 to FREE node FRE1 represents the return trip of crane 1 to station B from station A. At the end of this trip, the crane is freed, and files 1 and 3 at AWAIT nodes AWT1 and AWT3 are checked to determine the next task for crane 1.

Loads that pass inspection at station B, node ACM1, are routed to node AWT2. Since these loads are destined for station C, the portion of track common to both cranes must be available prior to transport. A resource is used to represent this common track as in Section 18.6 (the Material Handling Extension described in Chapter 19 could also be used). AWAIT node AWT2 assigns this resource to arriving loads on a FIFO basis. Loads arriving while the track is busy wait at node AWT2 until it is available. When the track is free, a load is removed from node AWT2 and routed to node AWT3. At this point the load can be moved to station C by either CRANE1 or CRANE2. Consider CRANE1 first. Subroutine ALLOC(1) is invoked at node AWT3 to test for the availability of this resource. If CRANE1 is available, attribute 2 is set equal to 1, CRANE1 is made busy, and the load conditionally branches to node GON1 in the network. At node GON1, two activities are initiated. The first represents the crane traveling back to station B. On arrival at station B, the common portion of the track is freed at node FRE2 and the crane is made available at node FRE1. The second activity emanating from node GON1 carries the entity representing the newly delivered load to station C where it is inspected. The outcome of the inspection is determined at node GON2. As mentioned above, this load could also have been carried to station C by crane 2. However, crane 2 must first be moved from station C to pick up the load and then moved back to station C. This process is modeled by nodes GON3 and GON4. When crane 2 is used, node FRE3 frees the common track and node FRE4 frees the crane. Should both cranes be in use when a load arrives at node AWT3, the load waits until one of the cranes becomes available.

Rejects from the second inspection are routed to node AWT4,where they await crane 2 for transfer to station D. COLCT node COL3 collects statistics on the time between rejects at station D, which is labeled output III in Figure 18-17. After a return to station C, node FRE4 frees CRANE2. Loads that pass the second inspection enter QUEUE node Q5 and are removed from station C using a roller line. The number of loads reaching node COL2 is an important output from the model since it represents the throughput of finished product. Thus, the number of entities completing activity 1 leading into node COL2 represents the throughput on a run. The number of entities passing through activity 1 is maintained as the variable NNCNT(1). The length of each run is specified on input as 1000 time units. At the end of each network run, the variable NNCNT(1) is used to calculate the throughput of finished product for the 1000 time units in subroutine OTPUT. A STAT statement is used to define the variable TPUT for the collection of these observations and to generate output statistics, including a histogram, for throughput over multiple runs. To obtain output statistics and histograms over multiple runs of the network, the user specifies that the statistical arrays should only be cleared up to the third collect variable at the end of each run on the INITIALIZE statement (see Section 4.3.6). For this model, these statements are as follows:

```
STAT,4,THROUGHPUT,10/10000/200;
INIT,0,1000,YES/4;
```

The following code is included in subroutine OTPUT and executed after each network run to collect the observations on throughput.

```
      SUBROUTINE OTPUT
      COMMON/SCOM1/ATRIB(100),DD(100),DDL(100),DTNOW,II,MFA,MSTOP,NCLNR
     1,NCRDR,NPRNT,NNRUN,NNSET,NTAPE,SS(100),SSL(100),TNEXT,TNOW,XX(100)
C     OBTAIN NUMBER OF OBSERVATIONS AT NODE COL2
      X = NNCNT(1)
C     CALCULATE THROUGHPUT FOR RUN
      TPUT = X/TNOW
C     COLLECT OBSERVATIONS ON THROUGHPUT FOR EACH RUN
      CALL COLCT(TPUT,4)
      RETURN
      END
```

The sequencing of crane operation is important and requires careful and precise specification. In Figure 18-17, the sequencing is specified through AWAIT nodes and their associated resources. For example, the resource CRANE1 is associated with AWAIT nodes AWT1 and AWT3. The file sequence specification listing the order of the files on the RESOURCE block requires that node AWT1 be checked first when CRANE1 returns to station B at FREE node FRE1. This ensures that CRANE1 is kept busy with loads between stations A and B as long as rejected loads are waiting at node AWT1. Similarly, when CRANE2 is freed at node FRE4, its disposition is controlled by the RESOURCE block specification that node AWT4 be checked before node AWT3. Therefore, CRANE2 travels between stations C and D unless there are no rejected loads at station C.

Incoming loads that pass the initial inspection are routed to AWAIT node AWT2, where they wait for the resource TRACK. Therefore, AWAIT node AWT3 will contain only one load at a time, since moving a load from station B to station C requires the use of that part of the track that is common to both cranes. When a load arrives at node AWT3, subroutine ALLOC(1), shown in Figure 18-18, is called to determine the availability of cranes. The load is assigned to CRANE1 if it is available. ATRIB(2) of the load entity is set to the crane number for this purpose. If it is not available and CRANE2 is idle, then this crane is moved to station B from station C to transport the load. If no crane is available, the

```
      SUBROUTINE ALLOC(I,IFLAG)
      COMMON/SCOM1/ATRIB(100),DD(100),DDL(100),DTNOW,II,MFA,MSTOP,NCLNR
     1,NCRDR,NPRNT,NNRUN,NNSET,NTAPE,SS(100),SSL(100),TNEXT,TNOW,XX(100)
C**** SET IFLAG  TO INSERT LOAD INTO NODE AWT3
      IFLAG = 0
C**** CHECK FOR AVAILABILITY OF CRANE 1
      IF (NNRSC(1).LE.0) GO TO 10
C**** CRANE 1 IS AVAILABLE.  SET ATTRIBUTE 2 TO BRANCH FROM NODE AWT3 TO
C**** NODE GON1 AND ALLOCATE CRANE 1 RESOURCE TO LOAD.
      ATRIB(2) = 1.0
      IFLAG = 1
      CALL SEIZE(1,1)
      RETURN
C**** CRANE 1 NOT AVAILABLE.  CHECK CRANE 2 AVAILABILITY
   10 IF (NNRSC(2).LE.0) GO TO 20
C**** CRANE 2 IS AVAILABLE.  SET ATTRIBUTE 2 TO BRANCH FROM NODE AWT3 TO NODE
C**** GON3, MOVING CRANE FROM STATION C TO STATION B
C**** AND ALLOCATE CRANE 2 TO LOAD
      ATRIB(2) = 2.0
      IFLAG = 1
      CALL SEIZE(2,1)
C**** NO CRANE AVAILABLE.  PLACE LOAD IN NODE AWT3 (IFLAG = 0)
   20 RETURN
      END
```

Figure 18-18 Subroutine ALLOC for overhead-crane model.

load waits at node AWT3 until a crane is freed. After the load is moved from station B to station C, the resource TRACK is freed (at node FRE2 or node FRE3, depending on which crane was used), and the next load is moved from node AWT2 to node AWT3. The crane used is then made available for the next load as described above at FREE nodes FRE1 or FRE4.

18.7.3 Model Output and Use

Table 18-1 provides a summary of the output obtained from the model presented in Figure 18-17. Throughput statistics for finished product are obtained through user statistics. Between statistics give estimates of the reciprocal of the rate at which rejects are delivered to stations A and D. This information is useful in determining the rate at which the interfacing roller lines must be able to remove units from the system. In-process inventory statistics are provided by the file statistics for AWAIT nodes AWT1, AWT2, and AWT4 and QUEUE node Q5. Resource utilization statistics provide estimates for the fraction of time each crane and the common track portion are in use.

An illustrative histogram representing throughput after 1000 hours is shown in Figure 18-19. This output is for 100 runs. The histogram shows that 5% of the time, the throughput for the period will be in the range of 9800 to 10,000 loads. Management requires a throughput of 10,000 loads in 1000 hours, and,

Table 18-1 Summary of Performance Variables for the Model in Figure 18-17

Performance Variable	SLAM II Collection Method
Time between deliveries	
output station II	COLCT node COL2
rejected loads to station A	COLCT node COL1
rejected loads to station D	COLCT node COL3
In-process inventory	
rejects at station B	Statistics for file 1
accepted loads at station B	Statistics for file 2
rejects at station C	Statistics for file 4
accepted loads at station C	Statistics for file 5
Utilization	
crane 1	Statistics for resource 1
crane 2	Statistics for resource 2
common track	Statistics for resource 3
Throughput of finished product	Call to subroutine COLCT in subroutine OTPUT

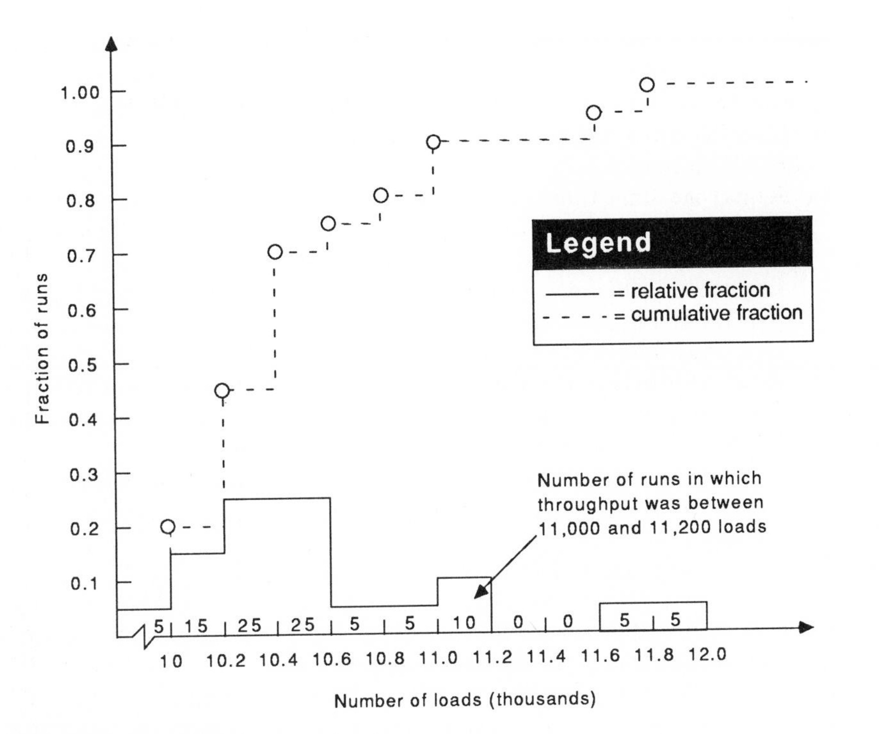

Figure 18-19 Example of throughput histogram for 100 observations.

from the simulation, it is estimated that this requirement will be met 95% of the time. The use of the outputs from this model enabled decision makers to evaluate proposed capital expenditures for this production system.

18.8 SUMMARY

This chapter illustrates how to build models of material handling systems using SLAM II. A representation of material handling equipment depends on the level of detail to be included in the model and the purpose for modeling. Using the basic SLAM II symbols, models may be built at any level to represent material handling aspects of a system.

18.9 EXERCISES

18-1. A conveyor system involves five servers stationed along a conveyor belt. Items to be processed by the servers arrive at the first server at a constant rate of four per minute. The service time by each server is 1 minute on the average and is exponentially distributed. Space for one item is provided before each server. At an end-of-service time, the item is removed from the system. The delay time to travel between servers on the conveyor is 1 minute. If an item cannot gain access to any of the servers, it is recycled to the first server with a time delay of 5 minutes. Model this conveyor system to determine statistics on the time spent in the system by an item and the utilization of each server. Perform a simulation to obtain statistical estimates over a 500-minute period.

Embellishments

(a) Repeat the simulation with a time delay of 2 minutes between servers. Is there an effect on the utilization of the servers because of a change in the time delay between servers, that is, the speed of the conveyor?

(b) Evaluate the situation in which the last server has sufficient space for storage so that all items passing servers 1, 2, 3, and 4 are processed by server 5. Simulate this situation.

(c) Assess the increased performance obtained by allowing a two-item buffer before each server. Based on the results of this embellishment, specify how you would allocate ten buffer spaces to the five servers.

(d) Discuss how you would evaluate the trade-offs involved between reducing the number of servers in the conveyor system versus increasing the buffer size associated with each server.

18-2. For the conveyor system described in Exercise 18-1, a server number is assigned to an item at its arrival time. Build the SLAM II models and obtain statistics using the SLAM II processor for each of the following assignment procedures.

(a) Random assignment of server numbers to items.

(b) Assign server who has the smallest number in its queue.

(c) Assign the server who has the lowest average utilization up to the time of arrival of the item.

(d) Assign servers in a cyclic manner.
(e) Assign to the first server who is idle when the item arrives.

18-3. For the conveyor system presented in Exercise 18-1, include within the SLAM II model, the logic that allows for decentralized decision making by deciding if a server should process an item when it arrives at a server station. Process the item at the station if:
(a) the server is idle.
(b) there is no server idle without recycling and space is available at the current service station.

18-4. Simulate the crane model given in Figure 18-15 to collect statistics on the throughput from each input station to each output station for the following data. The arrival rate to input station 1 is four units per hour, with 50% of the units to be sent to output station 1 and 50% to output station 2. The arrival rate to input station 2 is two units per hour, with every other unit to be sent to output station 2. The travel and return times (hours) from input stations to outputs stations are given in the accompanying table.

			Output Stations	
		1	Common Track[a]	3
Input Stations 1	Travel to	0.12	0.05	—
	Return from	0.10	0.04	—
Input Stations 2	Travel to	—	0.08	0.24
	Return from	—	0.06	0.22

[a] Each crane movement holds the common track an additional 0.1 hour.

Run the model assuming that all the times in the travel and return time matrix are constants. Repeat the analysis assuming that the times are normally distributed with a mean equal to the value given in the table and a standard deviation equal to one-tenth of the mean.

Embellishments

(a) Evaluate the impact of giving priority to jobs that do not require the common track, that is, to jobs destined to output stations 1 and 3.
(b) Develop a procedure for investigating the impact of having two separate crane systems that do not require a common track and hence involve no interference. In this procedure, establish a cost structure to evalute the potential savings due to reduction of interference time. Develop a SLAM II model from which an assessment of this new system operation can be evaluated.

18.10 REFERENCES

1. Adam, E. E., Jr., and R. J. Ebert, *Production and Operations Management: Concepts, Models and Behavior*, Prentice-Hall, Englewood Cliffs, NJ, 1978.
2. Apple, J. M., *Material Handling Systems Design*, Wiley, New York, 1972.
3. Auterio, V. J., "A Q-GERT Simulation of Air Terminal Cargo Facilities," *Proceedings, Pittsburgh Modeling and Simulation Conference*, Vol. 5, 1974, pp. 1181-1186.
4. Auterio, V. J., and S. D. Draper, "Aerial Refueling Military Airlift Forces: An Economic Analysis Based on Q-GERT Simulation," Material Airlift Command, presented at ORSA/TIMS Conference, 1974.
5. Bandy, D. B., and S. D. Duket, "Q-GERT Model of a Midwest Crude Supply System," *ORSA/TIMS Joint National Meeting*, October 1979.
6. DeJohn, F. A., and others, "The Use of Computer Simulation Programs to Determine Equipment Requirements and Material Flow in the Billet Yard," *Proceedings, 1980 AIIE Spring Annual Conference*, 1980, pp. 402-408.
7. Duket, S., and D. B. Wortman, "Q-GERT Model of the Dover Air Force Base Port Cargo Facilities," MACRO Task Force, Material Airlift Command, Scott Air Force Base, IL, 1976.
8. Francis, R. L., and J. A. White, *Facility Layout and Location: An Analytical Approach*, Prentice-Hall, Englewood Cliffs, NJ, 1974.
9. Jarvis, G. L., and R. M. Waugh, "A GASP IV Simulation of an Automated Warehouse," *Proceedings, Winter Simulation Conference*, 1976, pp. 541-547.
10. Nagy, E. A., "Intermodal Transhipment Facility Simulation: A Case Study," *Proceedings, 1975 Winter Simulation Conference*, 1975, pp. 217-223.
11. Pritsker, A.A.B., "Application of Multichannel Queueing Results to the Analysis of Conveyor Systems," *Industrial Engineering*, Vol. 17, 1966, pp. 14-21.
12. Pritsker, A. A. B., *Modeling and Analysis Using Q-GERT Networks*, 2nd ed., Wiley, New York, 1979.
13. Pritsker, A. A. B., *Introduction to Simulation and SLAM II*, 3rd ed., Wiley and Systems Publishing , New York and West Lafayette, IN, 1986.
14. Sims, E. R. Jr., "Materials Handling Systems," in *Handbook of Industrial Engineering*, G. Salvendy, ed., Chapter 10.3, Wiley, New York, 1982.
15. Tompkins, J. A., *Facilities Design*, N. Carolina State Univ., Raleigh, NC, 1975.
16. Tompkins, J. A., and J. A. White, *Facilities Planning*, Wiley, New York, 1984.
17. Waugh, R. M., and R. A. Ankener, "Simulation of an Automated Stacker Storage System," *Proceedings, Winter Simulation Conference*, 1977, pp. 769-776.
18. White, J. A., and H. D. Kinney, "Storage and Warehousing," *Handbook of Industrial Engineering*, ed. by G. Salvendy, Chapter 10.4, Wiley, New York, 1982.
19. White, J. A., "The 85% Rule of Thumb," *Modern Materials Handling*, July 1980.
20. White, J. A., ed., *Production Handbook*, 4th ed., Wiley, New York, 1987.

19 MATERIAL HANDLING EXTENSION TO SLAM II

19.1 INTRODUCTION

The modeling orientation of the Material Handling (MH) Extension to SLAM II provides for nodes and resource types tailored for material handling systems. Specialized resource definitions are used to represent the MH equipment, the area for pickup and delivery of material, and the paths in which the MH equipment can move. The modeling of detailed control features required for directing the operations of MH equipment is performed in user-written subprograms.

Three types of MH capabilities are presented in this chapter. The first type is the QBATCH node, which combines the capabilities of the BATCH node and the AWAIT node and can be used to model transporters like forklift trucks.

The second capability models cranes and cranelike MH devices, which move material from a pickup pile to a drop-off pile. A ***pile*** is a part of an area resource that has a physical location and a capacity. Piles are used to segregate material in an area. A ***crane*** is the generic name given to a resource that picks up, moves, and drops off material. Crane movements are controlled by the generalized wait and free nodes, GWAIT and GFREE. The SLAM II MH Extension has built-in control logic to decide on crane movements to avoid crane interference (collisions) that results from multiple cranes operating on the same runway.

A third set of nodes provides a modeling capability for ***automatic guided vehicle systems (AGVS)***. The performance of an AGVS is highly dependent on the physical layout and control policies used to direct a fleet of vehicles. Resources in an AGVS include vehicles, vehicle control points, and track segments connecting the control points. In an AGVS, the movement of vehicles from control point to control point over segments may be managed or controlled by a central computer, computers on board vehicles, or a combination of the two.

The logic associated with the movement of vehicles is modeled through vehicle wait, VWAIT, and vehicle free, VFREE nodes. At a VWAIT node, an entity requests a vehicle for transportion to its next station or inventory location. A VFREE node models a completed move by a vehicle. The movement of a vehicle occurs in a VMOVE activity in accordance with the characteristics of the vehicle fleet and the guidepath system, which are defined through resource blocks.

In summary, the modeling world views provided by the MH Extension provide representative approaches to the design and evaluation of MH systems. The details of each approach are provided in subsequent sections of this chapter.

19.2 QBATCH NODE

The QBATCH node is used to accumulate entities until a threshold, THRESH, is attained, and then to request one unit of resource, RES. Like the BATCH node, the values contained in attribute NATRS of arriving entities are summed to compute the amount accumulated for the batch. When this sum exceeds THRESH, a batch is ready. If resource RES is available, the batch is made, one unit of RES is seized, and the entity is routed from the node. Otherwise, the batch continues to grow until the resource is available. If the sum of the ATRIB(NATRS) values exceeds a MAXSIZE specification before the resource becomes available, the entities are still maintained as a single set. However, at the time the resource becomes available, a batched entity is formed consisting of those entities for which the sum of the ATRIB(NATRS) values does not exceed MAXSIZE. A batched entity requires resource RES for it to leave the QBATCH node. Only one unit of RES may be requested.

The entities included in the batched entity are selected based on their order of arrival. The symbol and statement for the QBATCH node are shown in Figure 19-1. Examples of the use of the QBATCH node are the filling of a tote box, the preparation of a pallet load, and the grouping of people for a tour.

For a QBATCH node, the attributes of the original entities are always

QBATCH(IFL), RES, NBATCH/NATRB, THRESH, MAXSIZE, NATRS,
SAVE/ I, J, ..., RETAIN(NATRR), M;

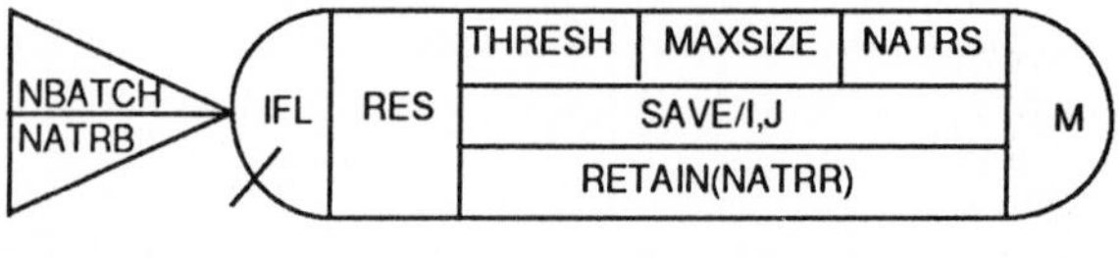

Figure 19-1 QBATCH node.

maintained and are pointed to by the attribute value assigned to ATRIB(NATRR) of the batched entity. The RETAIN attribute index, NATRR, must be specified for QBATCH nodes (in contrast to BATCH nodes where the field is optional). The batched entity released from a QBATCH node has the attribute values of the FIRST arriving entity for the batch unless a list of attribute indexes is specified as indicated by I, J, ... on the QBATCH statement. For each index specified, the corresponding attribute of the batched entity is the sum of the attribute values of each entity included in the batch. Statistics on the number of entities waiting at the QBATCH node are maintained.

A QBATCH node may be used to sort and accumulate entities based on an entity type, that is, it may accumulate multiple batches. A maximum of NBATCH batches may accumulate simultaneously. Entities placed in a batch must have the same ATRIB(NATRB) value. A maximum of M activities is initiated at each release of the QBATCH node.

19.3 CRANE MODELING WITH MATERIAL HANDLING EXTENSION

A crane system consists of a runway and a crane. The crane is a MH device that spans across and moves up and down a runway. The cab of the crane moves along the crane rails perpendicular to the runway. Movement along the runway is referred to as movement in the bridge direction and is defined as the x-coordinate direction in the MH Extension to SLAM II. The movement of the cab is referred to as movement in the trolley direction and is defined as the y-coordinate direction. To initiate a pick or a drop, the cab must move to the x- and y-coordinates of the entity to be transported. Cranes are assumed to have a width of ten units in the x-direction, and a maximum of ten cranes can be placed on a single runway. In the x-direction, a crane position is defined by the center point of the crane, which requires that five units of x are on either side of the crane.

Crane movements in the bridge direction are characterized by an acceleration to a maximum velocity, a period of no acceleration, and a period of deceleration. The duration of each of these phases depends on the distance between pickup and dropoff points and the locations of other cranes on a runway. The MH Extension maintains the location in the bridge direction of all cranes and automatically adjusts a crane's velocity to accommodate acceleration and deceleration characteristics specified for the crane resource. If a zero acceleration or deceleration is prescribed, instantaneous start-up and stopping for the crane resource is employed. Trolley movements in the y-direction are assumed to have instantaneous acceleration and instantaneous deceleration. Bridge and trolley movements are accomplished concurrently.

19.3.1 Area, Pile, and Crane Resources

Definitions of three resource types are used to model crane movements. These are the AREA, PILE, and CRANE. An AREA resource block describes an area in which entities wait to be picked up by a crane or where entities are to be dropped off by a crane. The statement for an area resource block is

AREA, ARLBL/ARNUM, NPILES, NATRA, IFLs;

An area is identified by a resource label, ARLBL, and an optional area number, ARNUM. NPILES specifies the number of piles contained within the area. Each pile is subsequently described by a PILE resource block, which gives the location and size of the area within the crane coordinate structure. The amount of space required by an entity is defined by the NATRA attribute of the entity. A list of files, IFLs, is defined in the AREA resource block that specifies the file numbers at GWAIT nodes where areas can be requested. When units of an AREA resource are freed, the files listed in its resource block are interrogated in the order listed to determine if entities are waiting for the allocation of the AREA resource.

A PILE resource block is used to describe the piles located within a previously defined area. A pile is given a unique identification number, PNUM, and must be identified as belonging to an area by specifying an area label, ARLBL. A label for a pile, PLBL, is optional. The PILE resource block statement is

PILE, PNUM/PLBL, ARLBL, CAP, XCOORD, YCOORD;

An entity requests an area and is placed in a pile of the area depending on the available space in the piles of the area and piling logic rules. The amount of space in a pile is defined on the PILE resource block by CAP. The location of the pile is defined by prescribing an x-coordinate value, XCOORD, and a y-coordinate value, YCOORD. Only cranes that can reach these coordinate values are allowed to pickup or dropoff entities to a pile.

The CRANE resource block statement is

CRANE, CRLBL/CRNUM, RWAY, XCOORD, YCOORD,
MXVEL, ACC, DEC, TVEL, IFL;

where a crane is referenced by a label CRLBL or optionally by a crane number CRNUM. A runway number, RWAY, is defined. Runway numbers must be assigned starting with 1. CRANE resource blocks must be listed in the statement model with consecutively numbered runways and crane numbers.

The initial coordinates for a crane are prescribed by the values XCOORD and YCOORD. As previously discussed, the movement in the bridging direction is accomplished in accordance with a maximum velocity, MXVEL, acceleration, ACC, and deceleration, DEC. Cranes move at their maximum velocity until they are required to decelerate due to crane interference or to arrive at their destination with a zero velocity. Both acceleration and velocity are controlled to avoid collision with other cranes. Instantaneous acceleration and deceleration are accomplished by specifying zero values for ACC and DEC. Trolley speed, TVEL, is used to determine if the trolley travel time in the y-direction is longer than the travel in the bridge direction. Whenever possible, trolley travel time is performed concurrently with bridge travel time. It is assumed that there is instantaneous acceleration and deceleration in the y-direction. All entities waiting for a crane are maintained in a single file specified by IFL.

19.3.2 GWAIT Node

The GWAIT node is a generalized AWAIT node where an entity waits for an amount of space in an area, ARLBL, one crane from a list of crane resources, CRLBL, and/or a set of regular resources, RES. An arriving entity waits in file IFL, and no resource is seized until enough units of all requested resource types are available. When all resources are available and allocated to an entity, the crane begins its move to the pickup location of the entity. The symbol and statement of the GWAIT node are shown in Figure 19-2. When an area is requested, the entity specifies the amount of pile space required. The amount of space is defined by attribute NATRA of the arriving entity, where NATRA is an input on the AREA resource block. When an area consists of more than one pile, a pile is selected by a piling logic rule specified by PLOGIC(NATRM) on the GWAIT statement [4, 5].

```
GWAIT(IFL),AREA/PLOGIC(NATRM),
CRANE/repeats,RES/UR,repeats, MODE,M;
```

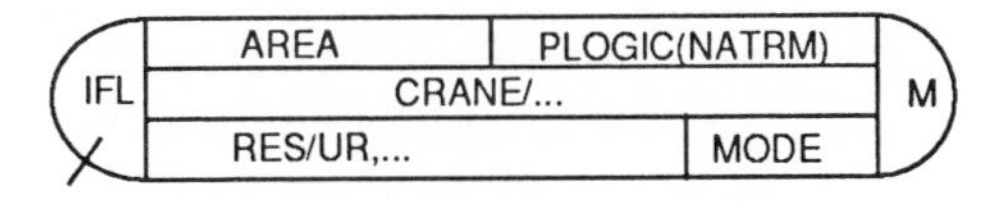

Figure 19-2 GWAIT node.

If an entity requires a crane, cranes capable of performing the MH operation are listed. Only one crane is required for the move, and alternative crane labels are separated by slashes. Regular resources requested at the GWAIT node are listed along with the amount of each resource type required, UR. A MODE specification, AND or OR, indicates whether all or any one of the listed regular resources may be used to satisfy the request. For the OR specification, the list is searched starting from the left, and the first resource that has the available number of units is allocated to the entity.† An entity released from the GWAIT node may be routed over M activities at most.

19.3.3 GFREE Node

The GFREE node frees an area resource, one of a set of crane resources, regular resources, or a combination of these resource types. When an area is specified, the GFREE node releases the amount of area capacity specified by the value of the arriving entity's ATRIB(NATRA). The AREA field indicates that the area held by the entity is to be released. Either FRAREA, YES, or the area label may be input for this purpose. NOFRA or NO indicates that an area is not to be freed. Only one crane may be released at a GFREE node. However, more than one crane may be listed. The crane released is the one allocated to the arriving entity. Multiple regular resources may be listed to be freed along with the number of resources of each type to be released. A MODE specification of ALL allows all such regular resources to be freed, and OR indicates that only the resource units allocated to the arriving entity are to be released. The symbol and statement for the GFREE node are shown in Figure 19-3.

If an OR mode is used on a GFREE node, the entity arriving at that GFREE node must have seized one of the regular resource types listed at a GWAIT node with an OR mode. Because of this, a GWAIT node may specify an OR mode even if the selection list consists of only one resource. The OR mode specification ensures that the specific resource allocated to the entity is released at a GFREE node. The number of units of the resource freed is specified by the value of UF.

The optional MOVE field on the GFREE node indicates that a crane movement is to begin. Prior to arriving at the GFREE node, the entity must have seized a crane resource and its destination area must be specified. Upon entering a

GFREE/MOVE, AREA, CRANE/repeats, RES/UF, repeats, MODE,M;

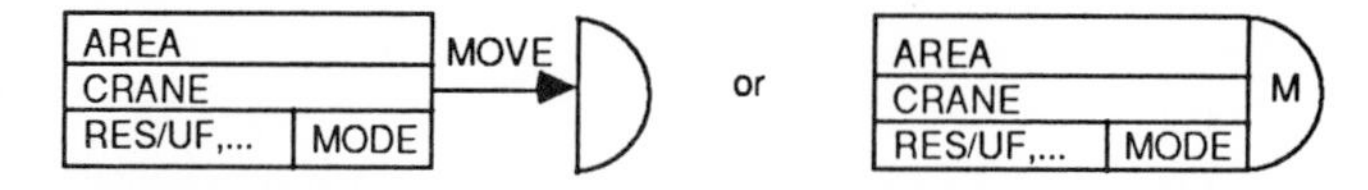

Figure 19-3 GFREE node.

† Several restrictions pertain to the regular resources requested at GWAIT nodes. A maximum of five regular resource types may be allocated at a GWAIT node. At any time only one resource type may be held by an entity when the MODE is specified as OR.

GFREE node with the MOVE specification, the arriving entity is removed from the network and placed in control of the MH Extension. The MH Extension continually determines the crane's movement toward the entity's destination as defined by the area the entity has most recently seized. Upon completion of the crane movement, the entity branches from the GFREE node in accordance with its M value.

As each resource is released, an attempt is made to reallocate the resource. If a regular resource is freed, the files listed in its RESOURCE definition are polled in the order listed. If space within an area is freed, the files listed in the AREA block definition are polled in the order listed. A freed crane examines the job list in the file specified in the CRANE block definition.

To summarize the use of the GWAIT and GFREE nodes in the modeling of a crane movement, an entity must seize an area to obtain an initial location. It then requests both a crane resource and an amount of a desired destination area prior to beginning the crane transport. When all resources requested at the GWAIT node are available, they are seized by the waiting entity. The GWAIT node initiating the crane transport process is followed by an activity representing a load or pick operation. The activity time is the time required to connect the material to the crane and lift it off the ground. This operation may consist of one activity or a sequence of activities that will end at a GFREE node, which releases the area resource the entity held and initiates the crane transport with the MOVE specification at the GFREE node. After completing the movement by the crane, the crane resource and other resources used in the movement are freed at a second GFREE node.

19.3.4 Crane Modeling Illustration

The crane system modeled in Section 18.6 is modeled here to illustrate the MH Extension capabilities. A physical diagram showing the XY coordinates of the input and output station and the location of piles for material at these stations is given in Figure 19-4. The definition of the area, pile, and crane blocks for characterizing this physical layout and the material handling equipment is given next.

The network segment to model the material flow from the input stations to the output stations is shown in Figure 19-5. An equivalence is made between ATRIB(1) and CRANE and between ATRIB(2) and OUTST. Note that CRANE can be either 1 or 2, which is the input station number and the AREA number it serves. For differentiation of parameters on the network, ATRIB(1) is used as the variable to identify the area of the input station and CRANE for specifying a crane number. Before the MH segment is entered by an entity, the definitions

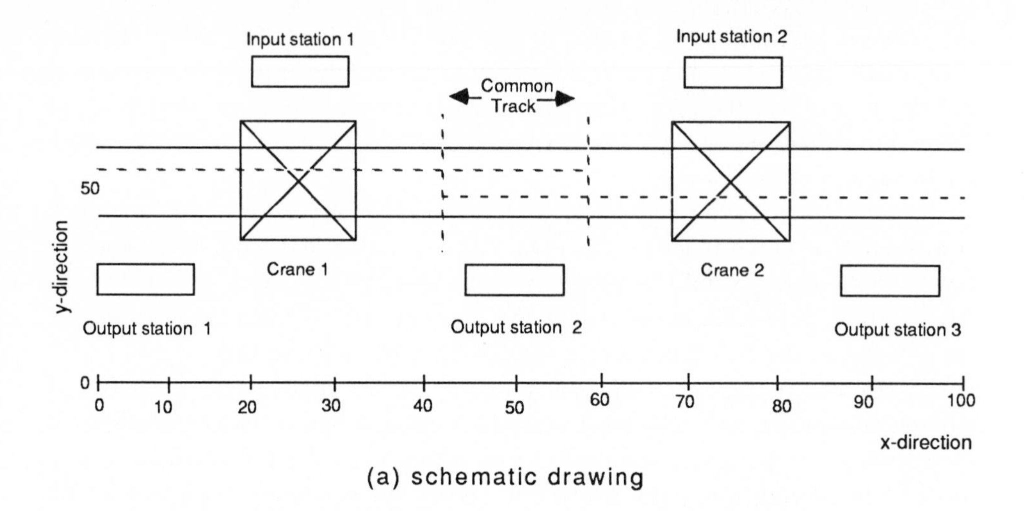

(a) schematic drawing

```
C**** ATRIB(1) IS BOTH THE STATION AREA AND CRANE NUMBER

      EQUIVALENCE/ATRIB(1),CRANE/
                  ATRIB(2),OUTST/
                  ATRIB(3),QUANTITY;

      CRANE,CRANE1/1,1,30,50,4,12,15,,8;
      CRANE,CRANE2/2,1,70,50,4,12,15,,9;

C             AREA        AREA     NUMBER      QUANTITY      ASSOCIATED
C             LABEL       NUMBER  OF PILES     ATTRIBUTE       FILES

      AREA,   STATION1  ,   1   ,   1    ,         3      ,      3     ;
      AREA,   STATION2  ,   2   ,   1    ,         3      ,      4     ;
      AREA,   STATION3  ,   3   ,   1    ,         3      ,      4     ;
      AREA,   STATION4  ,   4   ,   3    ,         3      ,      6     ;
      AREA,   STATION5  ,   5   ,   1    ,         3      ,      7     ;

C     PILES   PILE        AREA        PILE         X-COOR        Y-COOR
              NUMBER      LABEL     CAPACITY      POSITION      POSITION

      PILE,   1     ,  STATION1 ,     10    ,       30     ,       50     ;
      PILE,   2     ,  STATION2 ,     10    ,       70     ,       50     ;
      PILE,   3     ,  STATION3 ,      6    ,       10     ,       50     ;
      PILE,   4     ,  STATION4 ,      2    ,       40     ,       50     ;
      PILE,   5     ,  STATION5 ,      2    ,       50     ,       50     ;
      PILE,   6     ,  STATION6 ,      2    ,       60     ,       50     ;
      PILE,   7     ,  STATION7 ,      6    ,       90     ,       50     ;
```

(b) Definitions and x-y coordinates

Figure 19-4 Schematic drawing and x-y coordinates of an overhead-crane system.

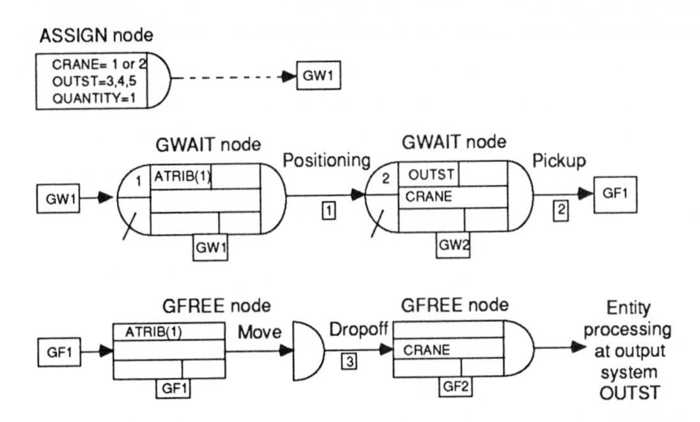

Figure 19-5 Network segment to model overhead crane system using MHE contructs.

for its input station (crane) and its output station are established at an ASSIGN node. At GWAIT node GW1, an area, defined by the value of ATRIB(1), is requested. This GWAIT node defines a beginning position for the entity that requests a move. Activity 1 following node GW1 is for positioning the entity in the area. (In the model of Chapter 18, positioning time is assumed to be zero. If desired, the duration of activity 1 could be set to zero.) After positioning, a request is made for space in area OUTST, which is the destination for the material. When space at the output station becomes available, a crane is requested to move to the location of the material so that it can be moved from its input station to its output station. When the crane arrives at the input station, activity 2, representing a pickup time, is included in the network. GFREE node GF1 is then used to free the space taken by the material in the area of the input station and to initiate a move by the crane to the desired output station. This move is made in accordance with the crane characteristics defined on the CRANE resource block for the crane that is assigned to make the move. When the crane and material arrive at the output station, an activity is shown on the network segment to represent the time required to dropoff the material at the output station. Following this dropoff activity, the crane is freed at GFREE node GF2. The entity representing the material is then ready for processing at its output station.

The network segment does not explicitly show the logic that is required for cranes to be returned to their input stations to pick up material. This is accomplished by the crane control logic included in the MH Extension. The time required to accomplish the return is determined by the distances between area piles and the maximum velocity, acceleration, and deceleration characteristics of the crane. The interference that occurs when deliveries are made to output station 4 is automatically computed by the interference algorithm included in the MH Extension. In this model, no restrictions have been placed on the output stations that can receive material from the input stations. If restrictions are to be modeled, then the values for the output station would be restricted to 3 or 4 for input station 1 and to 4 and 5 for input station 2. If crane 1 is to deliver material to output station 5, then the MH capabilities included in the MH Extension must provide for the movement of crane 2 below output station 5 to allow crane 1 to make deliveries to output station 5.

In addition to the standard SLAM II outputs, the performance measures obtained for this model of crane movement are the statistics on area/pile utilization, the number of entities arriving at and leaving from a pile, and a crane movement and use report. The values reported for a crane include the number of pickups and dropoffs and the fraction of time a crane spends in each of the following states:

1. Moving to pickup an entity
2. At pickup location
3. Waiting due to interference while making a pickup
4. Moving to dropoff an entity
5. At dropoff location
6. Waiting due to interference while making a dropoff

19.4 AUTOMATED GUIDED VEHICLE SYSTEM (AGVS) WITH MATERIAL HANDLING EXTENSION

The use of automated guided vehicles (AGVs) is increasing at a rapid rate [2]. An AGV provides a means to tightly govern material control policies and is a cost-effective alternative to labor-intensive and floor space-consuming methods of material handling. The implementation of an AGVS requires special consideration by system designers and managers. The impact of an AGVS on the entire production system is critical, especially when the production system has been designed to maintain little in-process inventory. The expense of the vehicles, the movement restrictions on a guidepath, and the complexity of the computer control system prohibit any tendency toward excessive overdesign.

Several industries use automated guided vehicles and the automotive industry is one of the prime users. In this industry, automobiles are assembled directly on AGVs, which move them as they are being assembled from one station to another. Physically, an AGVS is comprised of vehicles and guidepaths. Guidepaths may be wire installed under the surface of the floor through which radio signals are sent, tape on the surface of the floor, or other visual or chemical means by which a vehicle's path may be defined. Vehicles travel along these paths to reach destinations at rates depending on the acceleration, maximum speed, and deceleration of the vehicle unit. A vehicle may encounter portions of the guidepath that are congested by slower moving AGV traffic, traffic at intersections, vehicles that are stopped for loading or unloading, and interference from other MH devices.

Vehicles are controlled by either a central computer, an on-board computer, or a combination of the two. The control system instructs a vehicle to travel to a particular position and describes the route the vehicle is to take to get to that position. The control system also determines system logic rules, such as how to dispatch a vehicle for a job when more than one vehicle is available, which vehicle has priority when there is contention at an intersection, and other decisions related to traffic control.

The AGVS is an integral part of a larger system and can directly affect total system performance. For example, the number of vehicles selected to operate on a guidepath can cause blocking at a loading station in one of two ways: too few vehicles may result in the slow removal of material from a loading station, or too many vehicles may increase congestion and prevent vehicles from efficiently reaching the loading station. Control policies can have a similar effect on total system throughput, utilization of equipment and personnel, and the level of in-process inventory.

Three resource types are defined for modeling an AGVS with the SLAM II MH Extension: control points, segments, and vehicle fleets. Vehicle control points are locations on the guidepath network where an AGV begins traveling on a segment, stops to load or unload, or waits for instructions from a controlling computer. Control points are also called communication or broadcast points. All intersections as well as load or unload points, are inherently control points. Segments begin and end at control points.

Vehicles are part of a fleet and are used to pickup and dropoff loads at control points. The movement of a vehicle is in accordance with its velocity, acceleration, and deceleration characteristics. Its routing is determined by specifying rules for selecting segments that can be taken to reach a destination control point. In the MH Extension, the names of three special resource types are VCPOINT, VSGMENT, and VFLEET, and they define control points, segments, and vehicle fleets, respectively.

19.4.1 Control Point Resource: VCPOINT

Control points are described by the VCPOINT resource, whose statement is

VCPOINT, CPNUM/CPLBL, RCNTN, RROUT, CHARGE;

The control point is given a reference number, CPNUM, for use in other portions of the model, in user-written programming code, and on output reports. An optional field, CPLBL, may be specified to provide a label for the control point on the output reports.

There are two logic rule types for the control point resource: RCNTN, the rule for contention resolution, and RROUT, the rule for routing decisions. Contention for a vehicle control point resource occurs when more than one vehicle is waiting to enter the control point. A list of the waiting vehicles is maintained in their order of arrival at the control point. The contention rule specifies which vehicle is to be given the right-of-way when the control point becomes available, that is, does not have a vehicle at the control point. The rule for vehicle routing, RROUT, is used to decide which path segment a vehicle is to traverse from a control point.

The CHARGE field indicates that a control point has in-line battery-charging capabilities. A YES input for CHARGE causes a vehicle that is stationary at the control point to have its charging flag set to ON. The vehicle accumulates battery-charging time until it exits the control point. The charging flag is then set to OFF when the vehicle departs the control point.

19.4.2 Segment Resource: VSGMENT

The guidepath track between control points is defined as a resource of type VSGMENT. The AGV must travel on a segment to move from one control point to another. One VSGMENT resource is required for each segment connecting two control points. The statement for a VSGMENT resource is

VSGMENT, SGNUM/SGLBL, IDCP1, IDCP2, DIST, DIR, CAP;

The segment is given an identifying number SGNUM for use in reports and user-written subprograms. An optional label SGLBL is used for output reports. The control points of the ends of a segment are specified as CPNUM1 and CPNUM2. The length of the segment is defined as DIST, that is, the distance from CPNUM1 to CPNUM2. The minimum value for DIST is twice the check zone radius plus the vehicle length. The check zone radius is a vehicle characteristic that specifies the size of a control point. Check zone radius is described further in the next section.

The DIR field specifies the directional flow on the segment. The options for this field are UNIDIRECTIONAL or BIDIRECTIONAL. If UNIDIRECTIONAL is specified, vehicles may only travel from CPNUM1 to CPNUM2. The final field, CAP, is used to specify the maximum number of vehicles that may travel on the segment concurrently. A vehicle is on a segment until it enters the ending control point CPNUM2.

19.4.3 Vehicle Fleet Resource: VFLEET

The VFLEET resource is used to model MH devices such as driverless tractors, wire-guided pallet trucks, unit-load transporters, and platform carriers. Vehicles like forklift trucks that are not automatically guided may also be modeled with the VFLEET resource.

Each VFLEET resource block defines a fleet of AGV units. The definition includes a vehicle fleet's physical attributes, next job selection rule, rule for idle disposition logic, and the initial location of the individual vehicle units. The statement syntax is

VFLEET, VFLBL/VFNUM, NVEH, ESPD, LSPD, ACC, DEC, LEN, DBUF,
CHKZ, IFL/RJREQ, RIDL, IIDCP(NOV,IDSG)/ repeats, REPIND;

The label of the vehicle fleet, VFLBL, is used to reference the vehicle fleet on the network and on output reports. The optional field VFNUM specifies a vehicle fleet identification number.

The number of vehicles in the vehicle fleet is NVEH. Each vehicle can travel at a maximum speed of ESPD when it is empty and a maximum speed of LSPD when it is carrying a load. The acceleration of the vehicles is ACC. The deceleration is DEC. The length of the vehicle is specified as LEN. DBUF defines the distance buffer that must be maintained between any two vehicles. The parameter CHKZ defines the radius of a check zone area for vehicles and is used to define the size of the control point resource.

Job requests waiting to be performed by the vehicle fleet are maintained in file IFL. Job requests are not loads but are entities representing signals that a vehicle is desired. When a vehicle has completed a previous job, it interrogates its job file to determine if there are outstanding requests.

When an AGV completes delivery of material to a dropoff location, the vehicle is released to perform other tasks. The vehicle becomes idle if there are no transport jobs waiting to be performed. The parameter RIDL specifies a logic rule (STOP or CRUISE) to be applied to idle vehicles. Generally, an idle vehicle either stops at a specified position on the guidepath network or cruises on the guidepath network in a predefined pattern until requested.

Each vehicle is assigned an initial location by the ICPNUM(NOV,SGNUM) field. One vehicle is located at control point ICPNUM, and (NOV-1) vehicles are placed on segment number SGNUM waiting to gain access to control point ICPNUM.

The final field REPIND for the VFLEET statement requests a utilization report for the individual vehicles. If YES is specified for REPIND, observations on each vehicle are collected and reported.

19.4.4 AGVS Network Elements

Three network elements are included in the MH Extension to model an AGVS. The VWAIT node assigns a control point to an arriving entity, requests an AGV from a vehicle fleet, and allocates a vehicle to an entity. An entity requiring a vehicle waits in a file at a VWAIT node in a manner similar to an entity waiting for a resource at an AWAIT node. When an entity arrives at a VWAIT node, a request for a vehicle is issued and an idle vehicle is dispatched to the control point specified for the VWAIT node. When a vehicle arrives at the control point of the VWAIT node, it is allocated to an entity waiting at that VWAIT node (not necessarily the requesting entity). The entity allocated a vehicle leaves the VWAIT node. To move the entity over the guidepath, a VMOVE activity is required. The VMOVE activity assigns a control point destination to the entity and models the traveling of the entity from one control point to another. The duration of the move is calculated based on (1) the characteristics of the vehicle allocated to the entity, (2) the control point location of the VWAIT node where the entity starts its move, (3) the control point destination assigned at the VMOVE activity, and (4) the congestion and status of control points and segments over which the entity/vehicle travels.

A VFREE node is used to free a vehicle that has been allocated to an entity. After being freed, the vehicle is dispatched from the control point destination of its most recent VMOVE activity. The disposition of the vehicle is determined in accordance with its current list of job requests. If there are no job requests for the vehicle, then the prescribed rule for idle vehicle movement is invoked. The entity arriving at the VFREE node is routed from it without the vehicle resource. The following sections describe the statements and symbols for VWAIT and VFREE nodes and the VMOVE activity.

19.4.5 VWAIT Node

The VWAIT node models a request for transport by a vehicle at a control point. When the entity arrives at the VWAIT node, a signal is placed in the vehicle fleet's request file to notify the AGV controller that there is a demand for transportation services. This is comparable to an operator entering a request for a vehicle or an automatic signal being sent to the AGV controller from a photoelectric device. This load request is satisfied by dispatching an idle vehicle to the load's control point location. The travel time to the control point is based on the location from which the AGV is dispatched, its vehicle dynamics, and the congestion it encounters. A load entity is released from the VWAIT node when the AGV arrives at the control point from which the request was made. The VWAIT node symbol and statement are shown in Figure 19-6.

Entities wait in file IFL until an idle vehicle from vehicle fleet VFLBL moves to control point CPNUM. When more than one vehicle is idle at the time an entity arrives at the VWAIT node, a rule for vehicle request, RVREQ, is used to select from among idle vehicles. Examples of RVREQ are the closest vehicle or the vehicle that has been idle the longest. When a vehicle arrives at CPNUM, an entity is removed from the file IFL in accordance with the release rule, REREL. For example, if REREL is MATCH, the entity issuing the request for this particular vehicle is removed from the file. REREL can also be specified as TOP, in which case the entity ranked first in the file is removed.

VWAIT(IFL), VFLBL, CPNUM, RVREQ, REREL, M;

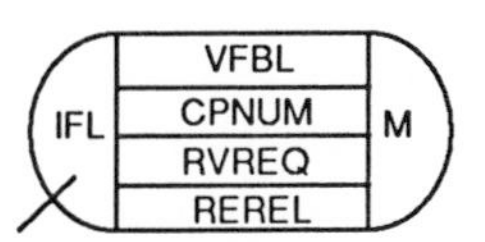

Figure 19-6 VWAIT node.

The states of a vehicle are traveling to load, loading, traveling to unload, unloading, traveling empty blocked, traveling full blocked, traveling idle, and stopped idle. Utilization statistics are reported on each of these states and the number of loads and unloads for each vehicle fleet.

19.4.6 VFREE Node

When an AGV completes its transport and unloading operations, it is released to be dispatched to another job request. The VFREE node releases a vehicle from fleet VFLBL. The symbol and statement for the VFREE node are shown in Figure 19-7.

The vehicle held by an entity is freed and reassigned in accordance with the rule for job requests, RJREQ, or the rule for idle vehicle disposition, RIDL. The default values for these rules are taken from the rules given on the VFLEET resource block for VFLBL. Entities arriving at the VFREE node must have seized a vehicle at a VWAIT node. Unloading and other activities that require the vehicle should be performed before the entity is routed to the VFREE node.

VFREE, VFLBL, RJREQ, RIDL, M;

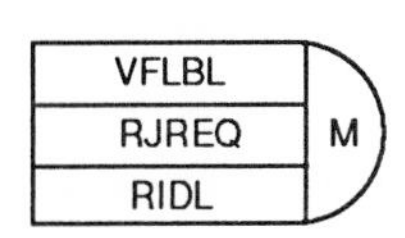

Figure 19-7 VFREE node.

19.4.7 VMOVE Activity

To model the movement of an entity with a vehicle over the guidepath, it is necessary to specify a destination control point. This is done at a VMOVE activity. The movement of the entity is then controlled by the AGVS characteristics as specified through the definition of the VFLEET resource, the guidepath network, and the movement rules. The movement time is the duration of the VMOVE activity. Since the typical mode of operation involves a loading activity before the move activity and an unloading activity after the move activity, a grouping of symbols to accomplish the vehicle move is used. That is, a GOON node would normally be required before and after the VMOVE activity. To avoid these extra nodes, a node-activity-node symbol is used for the VMOVE activity, which is a split GOON node with a branch connecting the two halves of the GOON node, as shown in Figure 19-8. The parameter CPNUM is a destination control point number and may be specified as a constant or SLAM II variable.

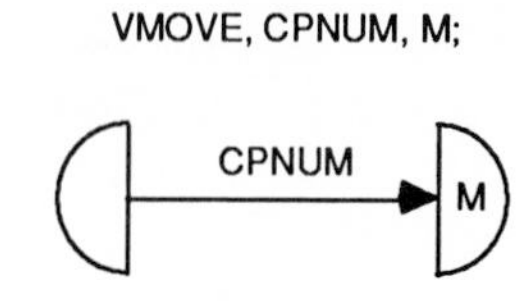

Figure 19-8 VMOVE node.

19.5 AGVS MODEL OF A FLEXIBLE MANUFACTURING SYSTEM [3-6]

As an illustration of AGVS modeling constructs, consider the system where a manufacturer of castings desires to evaluate alternative milling machine center configurations to achieve a production goal of 3520 finished castings per two-shift week (80 hours). The *flexible manufacturing system (FMS)* in Figure 19-9 is designed to perform machining operations on the castings. Castings are initially loaded onto pallets that carry 16 castings each and then sent by conveyor to one of two lathes. The castings are turned and then conveyor-transported in pallet loads to a wash/load area. After washing, the castings are sent to the machining center on a wire-guided vehicle.

The machining center design consists of ten identical horizontal milling machines that can perform any one of three operations: OP1, OP2, or OP3. For a particular casting, the milling machines may be dedicated to one operation or provided with sufficient tooling so any of the three operations could be performed. The latter mills are referred to as flexible mills. Two fixture types, A and B, are used in this system. Fixture A is used for OP1, and Fixture B is used for OP2 and OP3.

Before a pallet is routed for OP1 machining, each casting on the pallet is attached to fixture A and sent through the wash station. When a machine capable

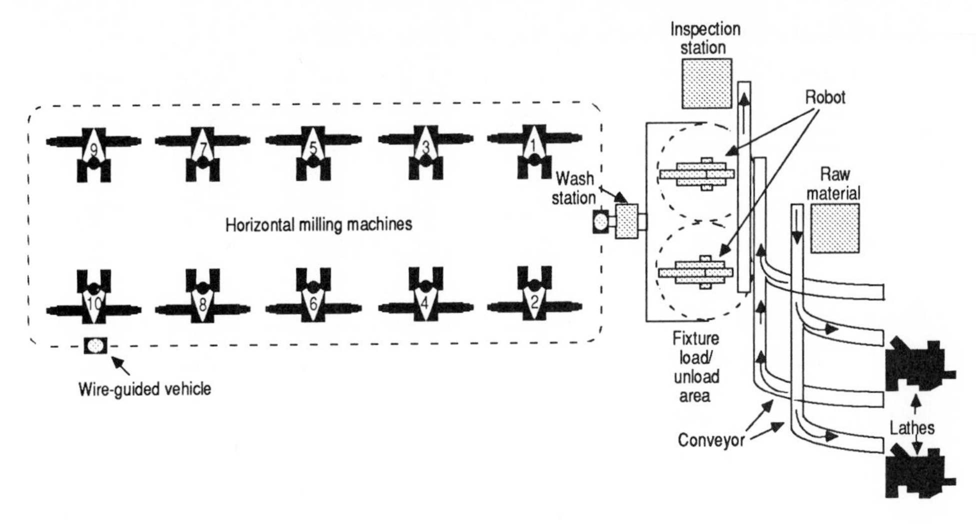

Figure 19-9 Diagram of a flexible milling-machine system.

of performing OP1 becomes available, the pallet is transported to it. After the castings are machined, they are returned to the wash/inspection area and attached to fixture B to await OP2. The same procedure is used for OP3, except the castings are rotated 180 degrees on the same fixture. After all three machining operations have been completed, the pallets are sent to final inspection and then depart the system.

The modeling objectives are to (1) evaluate system balance and productivity, (2) determine additional equipment needs, (3) determine which resources, if any, could be eliminated, and (4) establish the number of dedicated and flexible mills required. Because tooling costs are much higher for flexible mills, a decrease in the number of required flexible mills reduces manufacturing costs.

19.5.1 Preliminary Analysis

The FMS designer identified the greatest concerns within the system as the number of horizontal milling machines to purchase, the number of these machines to dedicate to each of the three operations, and the number of machines that should be tooled for flexibility. A review of operation and travel times confirmed that operating the milling machines involved the longest times and that the mills could indeed create a bottleneck. Based on these and other data observations, the following initial assumptions were developed:

1. It is necessary to concentrate the modeling on the mills.

2. The mills' setup times could be included in the processing times.
3. A dedicated or a flexible mill could be assigned when a pallet arrives. Therefore, no reassignment algorithm would be required.
4. Fixturing need not be included in the analysis.

19.5.2 Modeling the FMS with the Wire-guided Vehicle

The wire-guided vehicles for the system could be modeled in detail using AGV constructs. This would involve:

1. Defining each mill as a resource
2. Providing a description of the guidepath that has control points before each mill and segments connecting the control points
3. Describing the guided vehicles fleet

The SLAM II statements to model the FMS system are shown in Figure 19-10. The corresponding network model is given in Figure 19-11.

Two ARRAY rows are used with the SLAM II model to define mill processing times and machine type. The first, ARRAY row 4, provides the mill processing times for the three operations. ARRAY row 5 defines the type of machine at each location. Machine types 1, 2, and 3 perform only OP1, OP2, and OP3, while a type 4 machine is defined as a flexible mill. The ARRAY(5,10) statement

ARRAY(5,10)/1,1,1,1,1,2,3,3,4,4;

establishes that mills 1 through 5 are dedicated to operation 1, mill 6 is dedicated to operation 2, mills 7 and 8 are dedicated to operation 3, and mills 9 and 10 are flexible.

The transport times are automatically included in this model and based on vehicle speed characteristics. After each operation, a wash/inspection time of ten time units is included in the model. The wash/inspection station is assigned the resource name INSP and employs two workers.

During the simulation of the network, entities are created every 22 time units and sent to an ASSIGN node where OPERATION is set to 1. The next set of statements from node GON1 to AWAIT node AWA1 identifies an available mill. If a dedicated mill is available, the entity is assigned that mill. If not, the flexible mill is assigned if available. If neither is available, the entity is placed in activity 1, 2, or 3, or in QUEUE node Q123 to wait for a mill that can perform its operation. The durations of activities 1, 2, and 3 are indeterminate and specified by STOPA.

```
GEN,PRITSKER,FMS & AGV,7/24/87,1;
LIMITS,8,7,200;
INIT,0,4800;
VCONTROL,0.01,0.1;
CONTINUOUS,0,1,,,,N;
ARRAY(4,3)/120.0,40.0,56.0;
ARRAY(5,10)/1,1,1,1,1,2,3,3,4,4;
EQUIVALENCE/ATRIB(1),OPERATION/
            ATRIB(2),MILL/
            ATRIB(4),TYPE/
            ARRAY(4,OPERATION),PROCESS_TIME/
            ARRAY(5,MILL),MILL_TYPE;
NETWORK;
;  DEFINE RESOURCES
      RESOURCE/1,MILL1(1),4;
      RESOURCE/2,MILL2(1),4;
      RESOURCE/3,MILL3(1),4;
      RESOURCE/4,MILL4(1),4;
      RESOURCE/5,MILL5(1),4;
      RESOURCE/6,MILL6(1),4;
      RESOURCE/7,MILL7(1),4;
      RESOURCE/8,MILL8(1),4;
      RESOURCE/9,MILL9(1),4;
      RESOURCE/10,MILL10(1),4;
      RESOURCE/11,INSP(2),7;
;
;  DEFINE VEHICLE CONTROL POINTS
;           CONTROL
;             POINT      LABEL
      VCPOINT,    1/      MILL#1;
      VCPOINT,    2/      MILL#2;
      VCPOINT,    3/      MILL#3;
      VCPOINT,    4/      MILL#4;
      VCPOINT,    5/      MILL#5:
      VCPOINT,    6/      MILL#6;
      VCPOINT,    7/      MILL#7;
      VCPOINT,    8/      MILL#8;
      VCPOINT,    9/      MILL#9;
      VCPOINT,   10/      MILL#10;
      VCPOINT,   11/      INSPECT;
;
;  DEFINE VEHICLE GUIDEPATH SEGMENTS
;                      CONTROL  CONTROL
;           NUMBER    POINT 1  POINT 2    DISTANCE    DIRECTION
      VSGMENT, 1,        1,       2,         20,         UNI;
      VSGMENT, 2,        3,       3,         20,         UNI;
      VSGMENT, 3,        3,       4,         20,         UNI;
      VSGMENT, 4,        4,       5,         20,         UNI;
      VSGMENT, 5,        5,       6,         60,         UNI;
      VSGMENT, 6,        6,       7,         20,         UNI;
      VSGMENT, 7,        7,       8,         20,         UNI;
      VSGMENT, 8,        8,       9,         20,         UNI;
      VSGMENT, 9,        9,      10,         20,         UNI;
      VSGMENT, 10,      10,      11,         30,         UNI;
      VSGMENT, 11,      11,       1,         30,         UNI;
```

Figure 19-10 SLAM II statement model of a flexible milling machine system.

```
;  DEFINE AUTOMATED GUIDED VEHICLE FLEET
;
      VFLEET,AGV,1,100,100,0,0,4,,,8/CLOSEST,CRUISE;
;
;  DEFINE NETWORK
;
      CREATE,22,,3;
      ASSIGN,OPERATION=1,MILL=1,TYPE=1;
GON1  ASSIGN,II=MILL,1;
      ACTIVITY,,TYPE.EQ.MILL_TYPE.AND.NNRSC(II).GT.0,AWA1;
      ACTIVITY,,MILL_TYPE.EQ.4.AND.NNRSC(II).GT.0,AWA1;
      ACTIVITY,,MILL.EQ.10,Q123;
      ACTIVITY,,,NXT1;
NXT1  ASSIGN,MILL=MILL+1;
      ACTIVITY,,,GON1;
Q123  QUEUE(TYPE=1,3);
      ACTIVITY(1)/TYPE=1,3,STOPA(TYPE);
      ASSIGN,MILL=XX(1);
AWA1  AWAIT(4),MILL;
VW1   VWAIT(5),AGV,11,CLOSEST;
      VMOVE,MILL;
VF1   VFREE,AGV;
      ACTIVITY/4,PROCESS_TIME;
VW2   VWAIT(6),AGV,MILL,CLOSEST;
FRE1  FREE,MILL,2;
      ACTIVITY,,,VM1;
      ACTIVITY,,,ASN3;
VM1   VMOVE,11;
      VFREE,AGV;
      AWAIT(7),INSP;
      ACTIVITY/5,10;
      FREE,INSP,1;
      ACTIVITY,,OPERATION.EQ.3,TRM1;
      ACTIVITY,,,ASN2;
ASN2  ASSIGN,OPERATION=OPERATION+1,MILL=1,TYPE=OPERATION;
      ACTIVITY,,,GON1;
TRM1  COLCT,INT(3),TIME IN SYSTEM;
      TERMINATE;
ASN3  ASSIGN,XX(1)=MILL,1;
      ACTIVITY,,MILL_TYPE.LE.3,ASN4;
      ACTIVITY,,MILL_TYPE.EQ.4,GON2;
ASN4  ASSIGN,STOPA=MILL_TYPE;
      TERMINATE;
GON2  GOON,1;
      ACTIVITY,,NNACT(2).GT.0,ASN5;
      ACTIVITY,,NNACT(3).GT.0,ASN6;
      ACTIVITY,,NNACT(1).GT.0,ASN7;
ASN5  ASSIGN,STOPA=2;
      TERMINATE;
ASN6  ASSIGN,STOPA=3;
      TERMINATE;
ASN7  ASSIGN,STOPA=1;
      TERMINATE;
      ENDNETWORK;
FIN;
```

Figure 19-10 Statement model of a flexible manufacturing system (concluded).

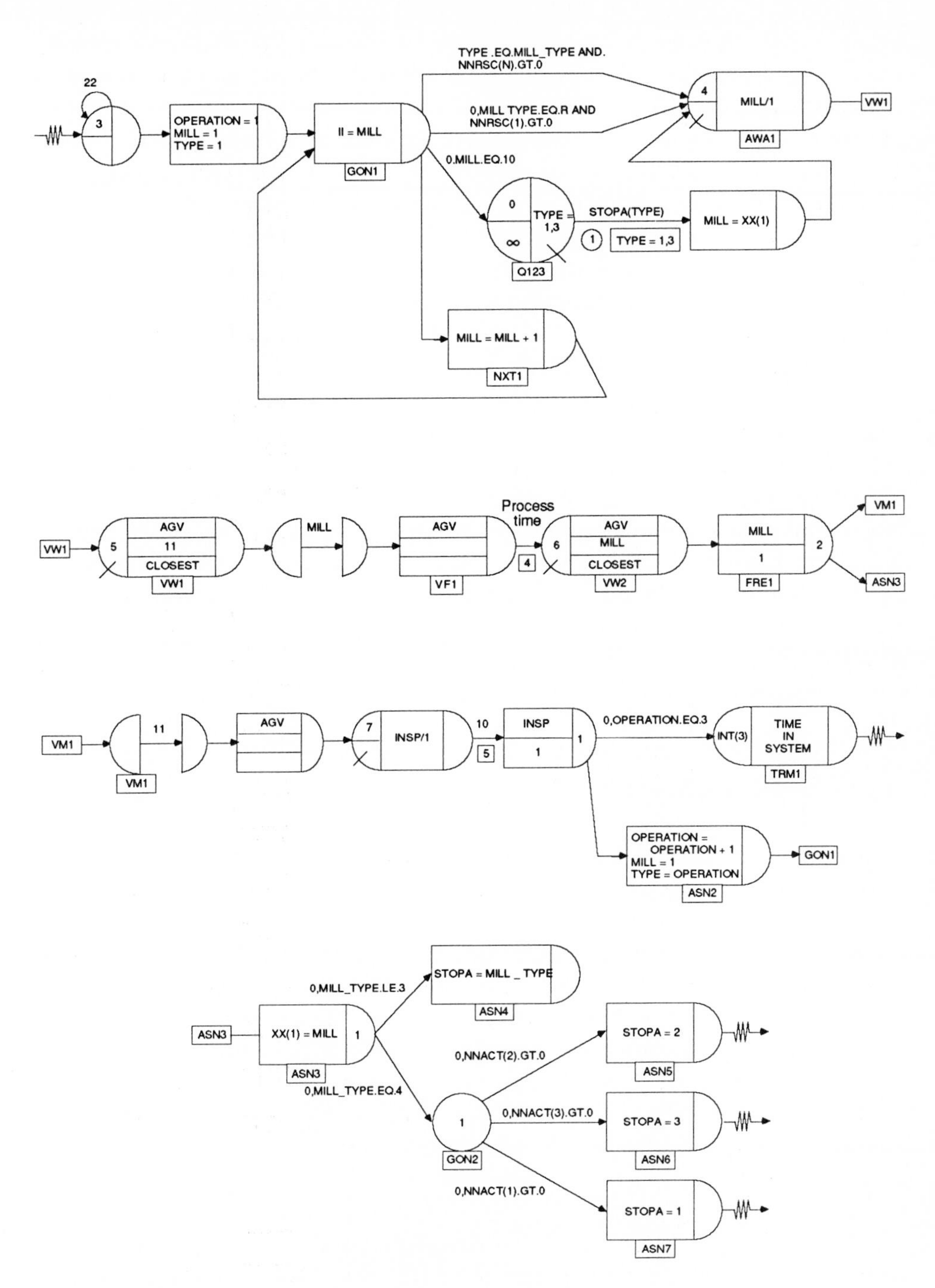

Figure 19-11 SLAM II network model of a flexible milling machine system.

The mill is allocated at AWAIT node AWA1, and a request for the closest available AGV is made at VWAIT node VW1 to come to the inspection station, which is located at control point 11. When the AGV arrives, the entity is moved to the allocated mill by the VMOVE activity. Following the VMOVE activity, the entity is at the mill. The entity frees the AGV by releasing VFREE node VF1. The entity continues on the network; the AGV is reassigned according to the VFLEET resource statement. If no entities are waiting for the AGV, the AGV cruises the guidepath as specified on the VFLEET statement.

The pallet entity is processed by its allocated mill in activity 4. After processing, a request for an AGV is made at the VWAIT node VW2. The location of the pallet is the control point associated with MILL, which is attribute 2 of the pallet entity. When the AGV arrives, it picks up the pallet entity. The MILL is made available by routing the entity to FREE node FRE1. The network segment starting at node ASN3 reallocates the MILL to a pallet waiting for the type of mill that completed the operation. The AGV moves the pallet to the inspection station (control point 11) and is then freed. The pallet load awaits inspection, parts are inspected and washed, and then the pallet load either leaves the system because all three operations have been performed or cycles back for its next operation. Outputs for this model include the standard SLAM II statistics for activities and files. The outputs from the MH Extension provide detailed information on the use of vehicles and the number and type of movements over each segment of the guidepath.

19.5.3 Model Embellishments

Model embellishments for this system include increasing the available number of vehicles and modifying their allowed direction of movement. To increase the number of guided vehicles in the system to 2, the number of vehicles prescribed on the VFLEET statement is changed to 2. To change the guided vehicles from unidirectional to bidirectional movement, the VSGMENT statement is modified to allow each segment to have bidirectional flow. To accomplish this embellishment, the UNI specification is replaced with a BID input. In the current design, mill types are located sequentially along the guide path. This is established by the definition of ARRAY(5,MILL). By changing the values in this ARRAY row, the mills may be located anywhere along the guide path. For example,

```
ARRAY(5,10)/1,2,1,3,1,3,1,4,1,4;
```

intersperses operation 1 mills with mills performing other operations and the flexible mills.

19.6 SUMMARY

The MH Extension includes predefined resource types and new network elements to model the diverse special features and problems relating to material handling systems. Subprograms are available to interface with these new network constructs [4]. The MH Extension concepts are used extensively in the modeling and analysis of manufacturing systems.

19.7 EXERCISES

19-1. A transfer car moves steel pipe that needs to be straightened from a manufacturing area to an off-line straightening area. The transfer car is not allowed to carry more than 30 tons of pipe. Policy is not to make a trip unless there is 20 tons of pipe to be moved. There is only one transfer car. The weight of a load of steel pipe is between 5 and 8 tons, uniformly distributed. Pipe requiring straightening arrive in an exponential fashion with a mean of 2.7 time units. The time for making a trip from the manufacturing area to the straightening area is normally distributed with a mean of 3.5 time units and a standard deviation of 0.3 time units. The time for the transfer car to return to the manufacturing area is triangularly distributed with a mode of 1.9 time units, a minimum of 1 time unit and a maximum of 2.3 time units. The time to straighten the pipe is a function of the weight of the pipe. On the average it takes between 0.4 and 0.8 time units per ton to straighten the pipe. Model this situation to estimate the time between pipe units leaving the straightening area, the time required for pipe to go from the manufacturing area to the time it leaves the straightening area, and the utilization of the straightening area and the transfer car.

19-2. Define the components of an automatic storage and retrieval system and develop concepts that facilitate the modeling of such systems.

19-3. A logging operation involves cutting a tree which takes 25 minutes on the average. This time is normally distributed with a standard deviation of 10 minutes. The weight of a tree varies uniformly from 1000 to 1500 pounds based on the density of Southern pine. The logging company has five trucks with a capacity of 30 tons each. The truck will not move the trees to the sawmill until at least 15 tons of trees are loaded. The trip to the sawmill requires 2 1/2 hours and the return trip requires 2 hours. The time to unload at the sawmill is 20 minutes. Develop a SLAM II model of this situation to determine if the logging operation has sufficient capability to deliver all the trees cut in a one month period.

Embellishments

(a) Determine the effect of breakdowns of the unloading unit every 40 hours plus or minus 10 hours. It takes 2 hours to repair the unloading unit.

(b) Include in the model the logging of oak trees in addition to pine trees. The oak trees are also cut and segregated at the cutting site. The oak trees have weights which vary between 1500 to 2000 pounds, uniformly distributed.

(c) Vary the minimum requirement of the trees that are loaded on a truck from

15 to 20 to 25 tons to determine the effect on throughput of this decision variable.

19-4. Discuss how the use of *group technology* reduces the requirements for material handling equipment when compared to the grouping of machines by machine function. Group technology organizes machines to produce products by locating them in a manner that resembles product flow. This reduces the potential use of a machine as not all products that require a machine type will flow past machines performing the desired function. Build a model that portrays the differences in material handling between a group technology based manufacturing design and a manufacturing plant which is organized by machine function.

19-5. Develop a set of specifications for a SLAM II extension to model a specific manufacturing or processing plant with which you are familiar.

19-6. Develop performance measures for the flexible manufacturing system described in Section 19-5. Add nodes to the network model to indicate how performance measures can be collected. Specify how decisions would be made based on the performance measures obtained.

19.8 REFERENCES

1. DeJohn, F. A., and others, "The Use of Computer Simulation Programs to Determine Equipment Requirements and Material Flow in the Billet Yard," *Proceedings, AIIE Spring Annual Conference*, 1980, pp. 402-408.
2. Muller, T., *Automated Guided Vehicles*, Springer-Verlag, New York, 1983.
3. Musselman, K., "Computer Simulation: A Design Tool for FMS," *Manufacturing Engineering*, September 1984, pp. 117-120.
4. O'Reilly, J., "Simulation Constructs for Material Handling," *Proceedings, 2nd European Simulation Congress,* Antwerp, Belgium, 1986, pp. 641-645.
5. Pritsker, A. A. B., *Introduction to Simulation and SLAM II*, 3rd ed., Wiley and Systems Publishing, New York and West Lafayette, IN, 1986.
6. Pritsker, A. A. B., "Model Evolution II: An FMS Design Problem," *Proceedings, Winter Simulation Conference*, 1987, pp. 567-574.
7. Ranky, P. G., *The Design and Operation of FMS*, Elsevier North-Holland, New York, 1983.
8. Wortman, D. B., and J. R. Wilson, "Optimizing a Manufacturing Plant by Computer Simulation", *Computer-aided Engineering*, September 1984, pp. 48-54.

20 SERVICE SYSTEMS

20.1 INTRODUCTION

Service systems are an important sector of the U.S. economy. Service systems differ from manufacturing systems in that they typically do not produce a tangible product or involve construction. The output of a service system is often consumed at the time it is produced and represents value because of its information content, comfort or health, amusement, or convenience [17]. The service sector of the U.S. economy is growing rapidly and, by one estimate, now accounts for 71% of the GNP and approximately 75% of its employment [22]. This trend of increasing services exists in all the major industrialized countries. There is no generally accepted definition of the service sector of the economy. However, it is common to include all the major industrial groupings of transportation, communications, public utilities, wholesale and retail trade, finance, insurance, real estate, health care, and other business, personal, and government services [11]. For example, the Bureau of Labor Statistics uses this broad definition to ensure inclusion of as many industries as possible. Some researchers use more restricted definitions of the service industries, excluding, for example, government activities.

This chapter discusses measures for evaluating service systems and the difficulties associated with defining good service. Service sector companies are organized to manage a large labor force and to respond quickly to customer demands. This environment complicates the modeling process. Three projects in the health care, banking, and insurance industries are presented. More detailed studies of models built to analyze network communication systems are presented in Chapter 21. Based on current trends, there will be increased use of modeling and simulation techniques to improve the delivery of services to people, corporations, and government organizations.

20.2 PERFORMANCE MEASURES FOR SERVICE SYSTEMS

At first glance, performance measures for service systems are similar to those for manufacturing systems. Throughput, order service timeliness, and resource utilization all have relevance in both settings as performance measures. Service systems, however, have a critical emphasis on customer timeliness. In many cases, that emphasis overshadows all other measures of evaluating performance. In a steel mill, for example, having a coil wait for four hours to be annealed is not a critical problem. In a bank, a four-minute wait by a customer for a teller is not acceptable.

Along with service delays and timeliness, variability is also a critical element in evaluating service systems. In our bank example, for instance, waiting an average of forty five seconds for a teller may be acceptable. It would not be acceptable if twenty percent of the bank's customers, however, had to wait longer than two minutes. Many of these customers would become disgruntled and find another bank. Another bank with the same average delay time, but a lower percentage of customers waiting longer than the two minutes undoubtedly would be viewed as having better service.

The service system environment has the following underlying differences from the manufacturing environment that managers should recognize.

1. The service industry has been primarily labor intensive, with few capital expenditures and little technology employed in the delivery of services.
2. Service systems exhibit a high variability in the demand for service both in quantity and quality, and service is typically provided in a highly variable manner.
3. Customer complaints vary over time; priorities in satisfying these complaints are often unclear and based primarily on the loudest complainer.

This environment is changing as the cost for providing services increases. These increased costs require greater management attention. Consequently, technological advances and innovations are coming to the forefront of the service industry as management is confronted with larger capital expenditures. In addition, governmental constraints on who can provide service and how it is provided are also causing managers to evaluate the consequences of their decisions.

Many problems exist in performing system analysis projects for service systems. For example, research and development funds frequently are limited within service organizations. Even though funds may exist for data collection, they are not typically budgeted for system analysis and improvement. Furthermore, the identification of key individuals who can make decisions regarding change in service delivery is not an easy task. The characteristics of the service system environment for performing modeling and simulation projects is reminiscent of the manufacturing environment of one or two decades ago. Although the need exists and great benefits are possible, the infrastructure is not in place for investing in such projects. However, the fact that some projects are now beginning to be funded and solutions implemented indicates that change is on the way [13].

20.3 EVALUATING SURGICAL BLOCK SCHEDULES FOR A HOSPITAL [19]

With the advent of the prospective payment system and the growth of alternative delivery systems, hospitals are searching vigorously for ways to reduce costs and improve productivity. The surgical suite is a major target for cost containment [12] because, typically, surgical suites have a high capital investment and large operating costs. In addition, surgical patients generate a significant portion of the demand for other hospital services, such as laboratory, radiology, and nursing.

Hospital managers have recognized that new types of information, techniques, and skills are needed to improve productivity and efficiency in the operating room. Operating room demand cannot be completely controlled; hence the use of expected demand values does not provide a true representation of the high and low use patterns that occur [3, 4]. Similarly, procedures are never of constant length, and this uncertainty makes scheduling operating rooms difficult.

Computer simulation is one technique that has been used by these managers to evaluate changes in scheduling practices, the number of operating rooms, the number of recovery beds, and changes in demand or case mix. Simulation can improve operating room management's ability to anticipate the impact of changes, as well as evaluate the effectiveness of current practices. Another advantage of computer simulation is that it can provide the capability to analyze the effect of random fluctuations in case load and procedure length. It provides the manager the ability to consider service and cost tradeoffs in the operating suite.

20.3.1 Inputs, Outputs and "What If" Questions

An overview for a surgical scheduling model is shown in Figure 20-1. Surgical demand for cases is either planned (elective) or emergent. The elective surgical demand is typically generated by the surgeon's office as a request for time in the operating suite. The scheduler must allocate an estimated amount of time on the schedule for each case. The actual length of the operation is a random variable and depends on the type of operation, patient, and often, the specific surgeon that is to perform the surgery. All patients arrive at the

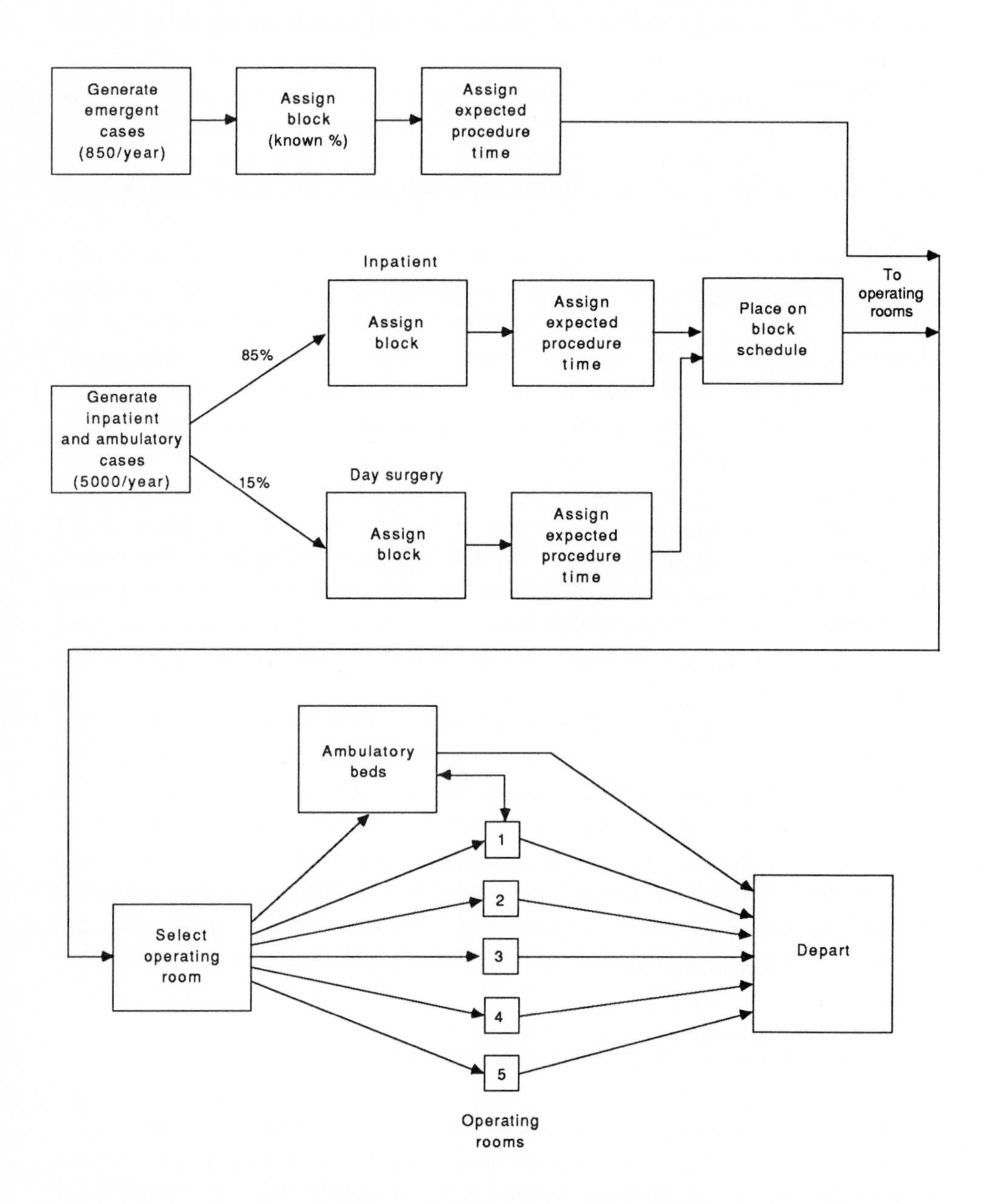

Figure 20-1 Overview of an operating room scheduling model

operating area and await an available operating room. When an operation is completed, the patient is moved to a recovery room. A certain amount of turnover (cleanup and setup) time is required before the next procedure can be performed in the operating room. The patient's stay in the recovery room is also a random variable, after which the patient is transported to a surgical ward in the hospital.

Since operating procedure times depend on the type of case (gynecologic, general surgery, and so on), it is important to differentiate case types in the model. The type of case also plays a role in which room is to be used, since some operating rooms are reserved for certain types of cases because of specialized equipment. Typical inputs to the model include:

1. Categorization of procedures by specialty
2. Case arrival pattern: a description of the interarrival time by specialty
3. Procedure length estimates: a description of how long each operation takes
4. Room selection by case type: the use of rooms for specific procedures
5. Turnover time: the time to clean up after and set up for the next operation
6. Scheduling logic and policies
7. Number of operating rooms
8. Number of recovery beds
9. Recovery time estimates

For surgical suites, performance measures fall into two basic categories: utilization and service. The following list serves as a guideline of outputs desired from simulation models:

1. Operating room utilization
2. Recovery room utilization
3. Number of cases performed per shift
4. Booking delays: the delay encountered by the surgeon in scheduling a case
5. Service delays: the number of times and length of time a scheduled case is delayed in starting
6. Overtime: time added at the end of the day due to emergencies and procedures taking longer than expected

Some examples of the types of questions that need to be posed by management regarding operating room scheduling are:

1. What if the case load and/or the mix of cases change?
2. What if the turnover time is decreased?

3. What if new operating rooms or new recovery room beds are added?
4. What if improved scheduling or sequencing methods are implemented; that is, what if a block schedule is used or the existing one is changed?
5. What if outpatient surgery patients use the same facility?

Using a simulation model, the effect of changing case load, altering schedule practices, reducing personnel, varying case sequencing, reducing setup times, or building or closing operating rooms can be evaluated. The simulation process supports the difficult decision making required to balance service and cost objectives. For example, higher operating room utilization must be balanced against larger booking delays and overtime justified in terms of allowable patient processing delays. Understanding these tradeoffs is the key to successful decision making in the hospital environment.

20.3.2 A SLAM II Model for Evaluating Operating Room Block Scheduling

This section describes a SLAM II model used to determine how to improve surgical scheduling in a 175-bed hospital. This hospital has five operating rooms where 7000 surgeries are performed per year. The surgeons of the hospital felt that delays in scheduling elective surgery had become unacceptable. The hospital administration wanted to determine if additional time should be added to the surgical schedule or whether a new scheduling procedure should be implemented. They were concerned that changes would lower room utilization and increase staffing expenses. A simulation study was initiated to evaluate the impact of proposed scheduling changes on these performance measures. The model developed on this project contained four major components.

1. *Block schedule input*: Block scheduling is a frequently used method for scheduling cases in an operating room. In a block scheduling system, a segment of operating room time (a block) is reserved for an individual surgeon or a designated group of surgeons. Prior to execution of the simulation, data associated with the block schedule are input to the model. These data include start and end times for surgical blocks, the identification of the surgeons assigned these blocks, and group scheduling preference information.

   Schedule data are input on a day of week basis by surgical groups using the same block start and end times. This structure allows the user to account for any anomalies in the schedule. This would occur, for example, if on Wednesday morning only two of the facility's five available operating rooms are scheduled to be open.

All relevant schedule data are presented to the user to verify the specifics being evaluated by the model. An example of such a report is shown in Figure 20-2. This schedule shows four different periods of time, called block periods, when rooms are operational. During block period 1 (8:00 to 12:00), four rooms are available on Monday, Tuesday, Thursday, and Friday. On Wednesday, only three rooms are available during this time. Looking at block period 4, it can be seen that one room is allocated time between 13:30 and 17:00. This is a continuation of service by the room available in block period 2 between 12:00 and 13:30. The remaining four rooms in this schedule are available between 12:00 and 15:30 according to the block period 3 schedule.

2. *Room request generation*: This component of the model issues the requests to the operating room scheduler for time to perform a surgical procedure. In the model, requests for regular and day surgeries are generated on a daily basis by drawing a sample from a Poisson distribution. The mean of this distribution was determined from the hospital's surgical records. Once generated, the room requests are scheduled to reach the scheduler randomly throughout the operating room workday of 8:00 to 17:00.

   Emergent patient arrivals are generated using an exponential interarrival time distribution based on the hospital's historical data. Emergent

BLOCK SCHEDULE USED TO ASSIGN PROCEDURES

			MONDAY	TUESDAY	WEDNESDAY	THURDSDAY	FRIDAY
PERIOD NO.	1		ORTHO-1	ORTHO-1	ORTHO-1	ORTHO-1	ORTHO-1
PERIOD START TIME	800	HRS.	GENERAL-1	GENERAL-1	GENERAL-1	GENERAL-1	GENERAL-1
PERIOD END TIME	1200	HRS.	ORTHO-2	UROLOGY	GENERAL	OB/GYN	ORTHO-2
PERIOD LENGTH	240	MINS.	NEURO	ORTHO-2		ORTHO-2	OTHER
PERIOD NO.	2		ORTHO-1	GEN/PRAC	GEN/PRAC	ORTHO-1	GEN/PRAC
PERIOD START TIME	1200	HRS.					
PERIOD END TIME	1330	HRS.					
PERIOD LENGTH	90	MINS.					
PERIOD NO.	3		GENERAL-1	GENERAL-1	GENERAL-1	GENERAL-1	GENERAL-1
PERIOD START TIME	1230	HRS.	OB/GYN	GENERAL-1	UROLOGY	ORTHO-2	OB/GYN
PERIOD END TIME	1530	HRS.	GENERAL-2	OB/GYN	NEURO	ORAL-PODIATRY	UROLOGY
PERIOD LENGTH	180	MINS.	ORTHO-2	ENT	ENT	PLASTIC	ORTHO-2
PERIOD NO.	4		ORTHO-1	ORTHO-1	ORTHO-1	ORTHO-1	ORTHO-1
PERIOD START TIME	1330	HRS.					
PERIOD END TIME	1700	HRS.					
PERIOD LENGTH	210	MINS.					

Figure 20-2 Block schedule produced by model.

surgeries are scheduled to begin as soon as possible without regard to the block schedule.

For each scheduling request generated, its characteristics, such as the type of surgery (for example, orthopedic or OB/GYN), and estimated procedure time are determined. In the model, these patient characteristic assignments are accomplished using a SLAM II network fashioned in a treelike structure. At the top of the tree, a patient is assigned a patient type, either inpatient or outpatient. Then, based on their type, the patient is probabilistically assigned to a surgeon or group. Finally, based on that group and the patient type, a procedure time is computed.

3. *Patient scheduling*: Once a request is generated and its assigned characteristics are determined by the network, it is processed through an EVENT node representing the surgical scheduling function. At the scheduler, the request type is matched with the groups in the schedule. When a match is found, the request's estimated procedure time is compared with the available time remaining in the block. If sufficient time is available (including room turnover), the case is booked on the schedule. If time is not available in this block, the search process continues.

   Besides matching group ID and searching for available time, the scheduler must also consider other factors. These include using released time (time available to any group) on the schedule if possible, restricting certain groups to access in their block(s), and not allowing groups to access blocks set aside for day surgeries.

4. *Operations*: When a scheduled surgery enters the network portion of the model, it is placed in a queue to await an available operating room. Emergent patients have priority in this queue. When an operating room becomes available, these scheduled patients are assigned the room based on their scheduled start time.

A network segment to illustrate a model that can schedule operating rooms is shown in Figure 20-3. At ASSIGN node A0, the value of ATRIB(7) is set to TNOW to mark the operating room request time by a patient entity. If a special room is required, the patient entity is routed to AWAIT node SPRM. The attribute CASE defines the file number in which the patient entity will wait for a special room. In this way, patients are segregated by case type in files 2 through 6. At AWAIT node ROOM, patients wait in file 1 if they do not require a special room. The allocation of an operating room is done in subroutine ALLOC. Each operating room is treated as a separate resource and the file numbers included in a room's resource block identifies the case types that can be treated in that room. For the operating rooms that are not for special cases, only file 1 is listed in their resource block. The allocation procedure is written in subroutine ALLOC and will be described later. When an operating room is allocated, the

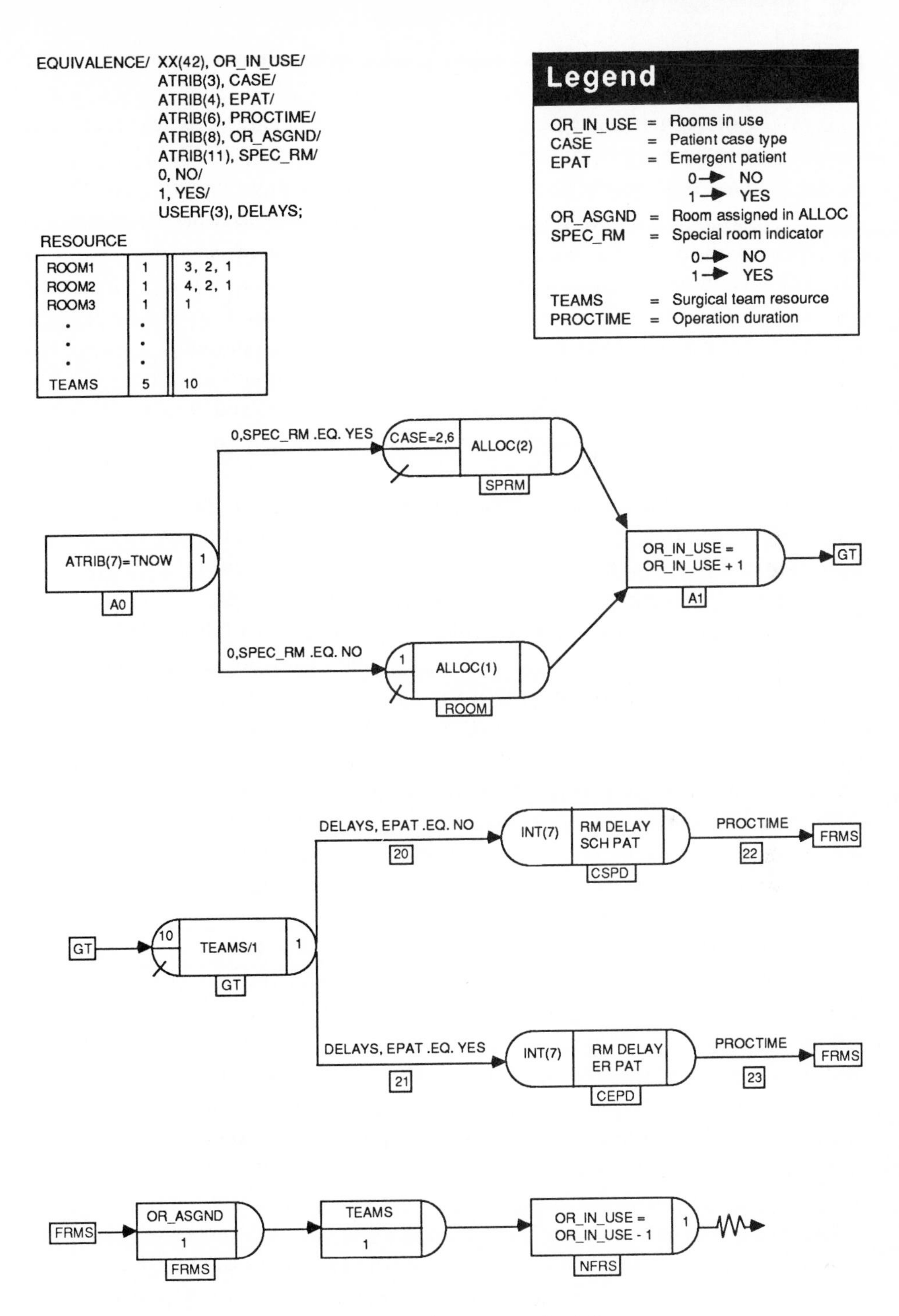

Figure 20-3 Network segment for operating room scheduling.

number of rooms in use, OR_IN_USE, is increased by 1 at ASSIGN node A1.

After an operating room is allocated, a surgical team is requested at AWAIT node GT. Activities 20 and 21 model the delays encountered before the operation can be started for scheduled patients and emergent patients, respectively. Statistics on the waiting time and delay time are collected at nodes CSPD and CEPD. Activities 22 and 23 model the duration of an operation. Following the operation, the operating room assigned the patient entity is freed. The value of OR_ASGND, the operating room assigned, is set in subroutine ALLOC and is attribute 8 of the patient entity. When this room is freed, the files in which patients are waiting for the room are searched and, if a patient is waiting, the room is allocated at either AWAIT node SPRM or ROOM. A surgical team is then freed and the number of OR rooms in use is decreased by 1 at node NFRS. If the operating room is reallocated, the number of rooms in use is increased by 1 at node A1.

Subroutine ALLOC for allocating operating rooms to patients is shown in Figure 20-4. The argument IFLG is set to 0 to indicate that a resource has not been allocated as yet in the subroutine. IROM is the number of the room resource allocated and is initially set to 0. UTRM is a utilization value and is used to test room utilizations until the lowest value is obtained. It is initially set to a high value.

If subroutine ALLOC is called with IK equal to 1 then a special room is not required. A search is made to find the free room with the lowest utilization. If the room is available [NNRSC(IRX) .GT.0] then its average utilization is compared to the lowest utilization observed for other available rooms. If room IRX has a lower utilization, it becomes the candidate for allocation to the patient. The average utilization of a room is obtained from the SLAM II function RRAVG.

If subroutine ALLOC is called for a special room, then IK equals 2. In this case, function ISPCRM is called to determine if a room is available for special case ICASE. Function ISPCRM either returns a room number of an available special room for case ICASE, or zero.

In the last section of subroutine ALLOC, if IROM is not equal to zero then one unit of resource IROM is seized by a call to subroutine SEIZE. ATRIB(8) of the patient entity is assigned the room number IROM. IFLG is set to 1 to communicate to SLAM II that a room has been allocated, and that the first patient entity should be removed from the appropriate file. A positive IFLG value indicates that the attributes of the entity were changed in subroutine ALLOC.

The model was validated against a schedule and case mix situation for a prior year. Data for the model were obtained from the hospital's operating room log. These data permitted the estimation of the number of case requests per day by specialty. In addition, procedure times by specialty were characterized from this

```
      SUBROUTINE ALLOC(IK,IFLG)
C
C     THIS SUBROUTINE TRIES TO ALLOCATE AN AVAILABLE ROOM
C     TO A PATIENT. IF THE PATIENT REQUIRES A SPECIAL ROOM,
C     THE ALLOCATION IS MADE BASED ON THE PATIENT'S CASE TYPE
C     IF THE PATIENT DOES NOT REQUIRE A SPECIAL ROOM, THE ROOM WITH THE
C     LOWEST UTILIZATION WHICH IS AVAILABLE IS ALLOCATED.
C
      COMMON/SCOM1/ATRIB(100),DD(100),DDL(100),DTNOW,II,MFA,MSTOP,NCLNR
     1,NCRDR,NPRNT,NNRUN,NNSET,NTAPE,SS(100),SSL(100),TNEXT,TNOW,XX(100)
C     NRMS IS NUMBER OF OPERATING ROOMS
      COMMON/VCOM1/NRMS
      IFLG = 0
      IROM = 0
      UTRM = 200
C
C     IK = 1, NO SPECIAL ROOM IS REQUIRED, FIND AN AVAILABLE ROOM
C     IF RESOURCE IS AVAILABLE, AND LOWEST UTILIZED,
C        MAKE IT CURRENT ROOM FOUND.
C
      IF (IK.EQ.1) THEN
         DO 200 IRX = 1 NRMS
         IF (NNRSC(IRX).LE.0) GO TO 200
         IF (RRAVG(IRX).GE.UTRM)GO TO 200
         IROM = IRX
         UTRM = RRAVG(IRX)
  200 CONTINUE
C
C     IK = 2, SPECIAL ROOM REQUIRED,
C
      ELSE IF (IK .EQ. 2) THEN
C
C     FUNCTION ISPCRM RETURNS AVAILABLE SPECIAL ROOM FOR ICASE,
C     OR ZERO
C
        ICASE = ATRIB(3)
        IROM = ISPCRM(ICASE,I)

      ELSE
      ENDIF
C
C     IF ROOM FOUND, SEIZE IT
C
      IF(IROM.NE.0) THEN
         CALL SEIZE(IROM,1)
         ATRIB(8) = IROM
         IFLG = 1
      ELSE
      ENDIF
      RETURN
      END
```

Figure 20-4 Subroutine ALLOC for allocating operating rooms to patients.

database into statistical distributions. The results from the validation runs agreed closely with the utilization and delay situation experienced by the hospital. These runs pinpointed one practice group whose access delays to the schedule were becoming intolerable due to the group's preference to utilize only its own block time. The model also indicated that several blocks in the surgical schedule were inefficiently sized for the groups using them. Adjustments of the block times for each group were made and tested with the model. A new block schedule was developed that balanced booking delays with satisfactory room utilization.

20.3.3 Examples of Model Output and Use

Using the modeling approach previously discussed, scheduling procedures have been evaluated at other hospitals. One study involved a 350-bed hospital performing 9000 surgeries per year in an eleven-room suite. Inputs were the case mix data from the previous year. The model was used to evaluate several possibilities, including

1. the use of blocked and semi-blocked scheduling techniques
2. the cost and service impacts of increasing and decreasing scheduled room hours
3. changes in case mix to a significantly higher percentage of outpatient surgeries
4. changes to room allocations for specific patient types and procedures

Figure 20-5 shows the simulation results of the utilization for the eleven rooms in the surgical suite. As can be seen, large variations exist between the rooms. The variations are due to procedural rules that restrict the type of patient that can be placed in a room.

In a study of a 5-room, 135-bed hospital with over 5000 procedures per year, the percentage of time that 0 to 5 rooms are still operating at 15:30 P.M. was estimated as shown in Figure 20-6. Since the shift ends at this time, the histogram represents a measure of overtime and was useful in estimating the effects of scheduling changes on overtime staffing levels. The graph reveals that 56.6% of the time at least one room is still operating at 15:30 and that 20% of the time four or five rooms are active at this time. Based on these results, management explored alternate ways of decreasing the amount of overtime required.

The graph in Figure 20-7 shows the percentage of time an orthopedics group at the hospital waited between one and ten days to get a case on the schedule. The model was used to explore changes in the elective schedule procedure to

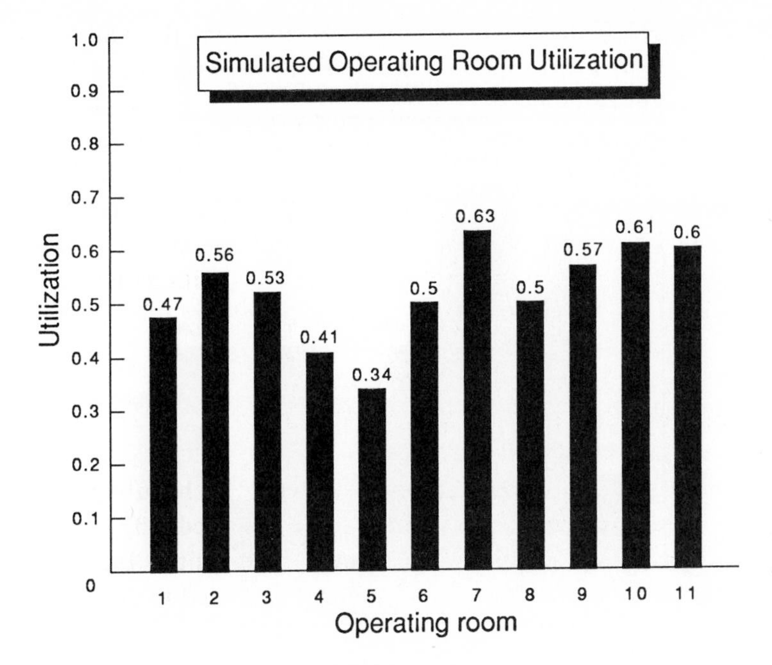

Figure 20-5 Model output of operating room utilization.

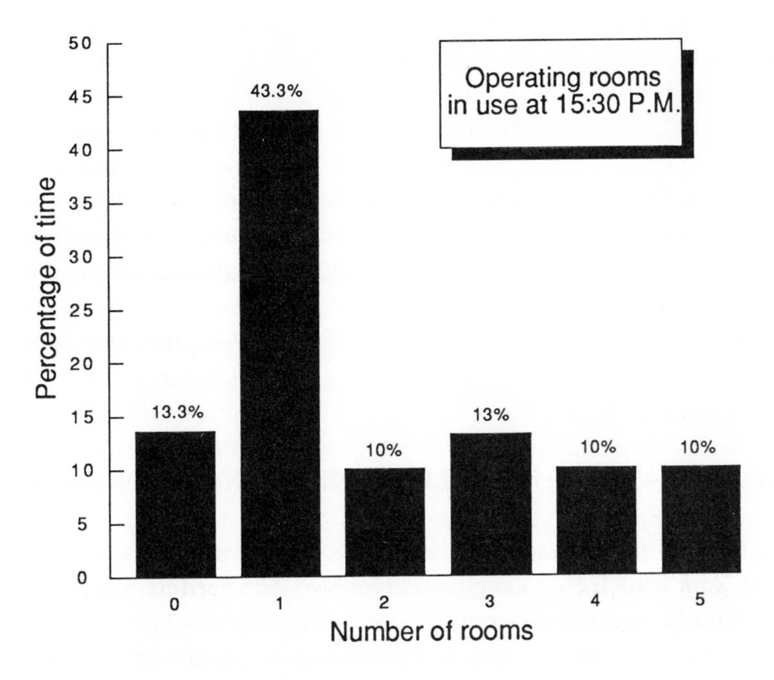

Figure 20-6 Model output of operating rooms in use at 15:30 P.M.

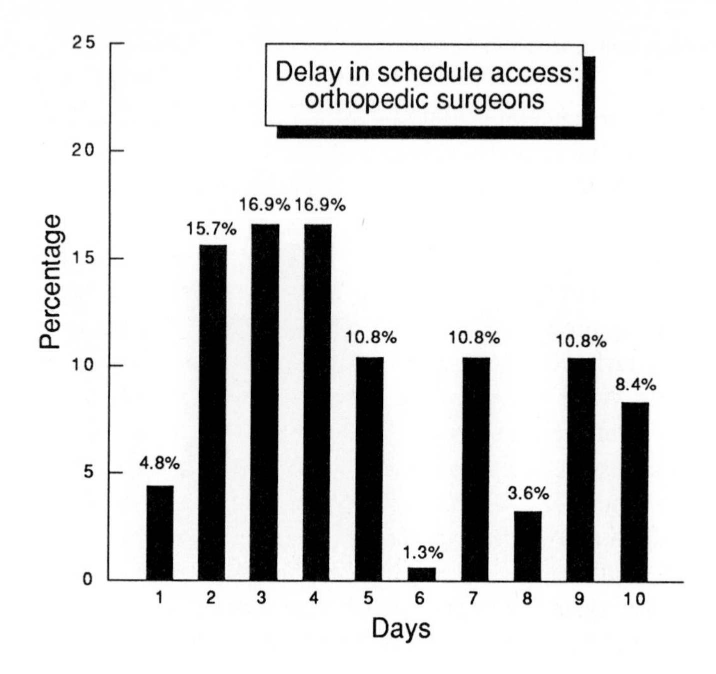

Figure 20-7 Model output of delay in orthopedic surgeons accessing schedule.

reduce this delay in booking. The results were compared with the impact on operating room utilization and overtime.

In another study using case mix projections for 1990, analysis was performed to examine the effects of dedicated outpatient surgical facilities and the scheduling of outpatients. These runs suggested workable room and scheduling alternatives based on a projected change in day surgeries from 15% to 40% of the total procedures that occurred. The analysis projected that a more efficient utilization of the facilities is achieved if outpatient blocks are added to the schedule on certain mornings during the week. A recommendation was made that there be no special distinction between inpatient and outpatient operating rooms.

20.4 ANALYZING A CHECK-PROCESSING FACILITY OF A MAJOR BANK

Over a million checks and other account transactions arrive each day at a central New York bank from its regional branches. The need to process this large volume of items has turned large banks into huge data-processing facilities. This section describes one such facility and a simulation study that was performed to evaluate check-processing procedures.

20.4.1 The System and Management Concerns

Figure 20-8 is a schematic diagram of a check-processing facility. Checks and other items arrive in batches from regional branches and from other departments within the bank. Although deliveries to the bank occur at regular hours, there are delivery delays within the bank in getting the checks to different departments. The information from each batch of checks that arrive is entered on a computer tape. These tapes are sent to a tape-processing department. While the tapes are being processed, the checks are audited. The auditing function is performed only during daytime shifts, whereas the other departments operate on a 24-hour basis. After auditing, the checks are matched against the computer tapes released from the tape-processing department. This matching and verification process results in a verification delay. After the delay, the checks are moved to a machine room, where operators feed them into one of four machines for automatic debit and credit transfers. This computerized machining operation represents the final processing stage for the checks at the facility.

A study of the system was initiated by bank management for two reasons. First, new regional banks were opening and the work load of the facility was expected to increase by 20%. Second, it was desirable for personnel reasons to do all auditing during a single 8-hour shift (current auditing was performed on two shifts). Management posed questions relating to the effects due to the proposed changes on the following performance measures:

1. Time a batch spends in the system
2. Rate at which batches are processed

Both of these measures provide a quantitative assessment of customer-bank relations. Of significant import is that, if a check is processed before a specific

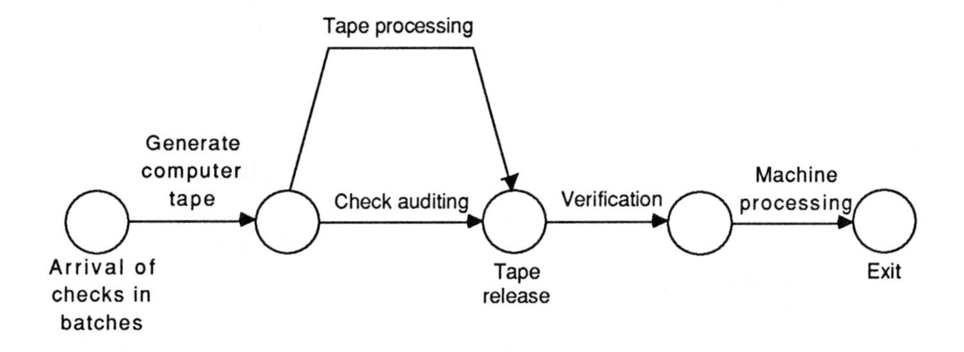

Figure 20-8 Schematic diagram of check processing facility.

time, it becomes part of a bank's float, which, determines its lending capacity.

Preliminary analysis and opinion indicated that the machine-processing operation was the current bottleneck of the system and that ways to increase the efficiency of this last stage of the process needed to be investigated to accommodate the expected increase in work load. However, sufficient uncertainty regarding this opinion and the need for a quantitative analysis led to the development of the simulation model to assess management's concerns. The model is presented in the next section in simplified form in the interest of confidentiality and ease of presentation.

20.4.2 The SLAM II Model

Figure 20-9 contains a network model of the check-processing facility. ASSIGN nodes ASN1, ASN2, and ASN3 represent the arrival of batches of checks to the system. Two attributes are assigned to each batch for identification purposes. Attribute 1 represents the type of item arriving (the actual study involved over 30 classifications of bank items). Attribute 2 represents the number of items in the batch, that is, the volume of the batch. Data analysis revealed that the volume is a random variable with a uniform density function whose parameters depend on item type. Thus, the size of arriving batches is set by taking a sample from a uniform distribution. Arrivals of each type of check occur at regular intervals during the day as follows: type 1 checks arrive at 8 A.M., type 2 checks arrive at 2 P.M., and type 3 checks arrive at 8 P.M. These regular intervals are modeled by the activities emanating from ASSIGN node ASN2. A batch priority value is assigned to attribute 4 through the user function USERF. A user function is employed so that different batch priority algorithms can be coded and evaluated.

For each type of check, there is a delay in delivery to the facility. These delays are represented by activity 1 using ARRAY(3,II) as the activity duration. At node ASN3, a unique identifying number is assigned to attribute 3. This assignment permits a match to be made of a check batch entity with a computer tape entity. To accomplish this assignment, attribute 3 is simply set to the number of entities completing activity 1.

A check batch entity and a computer tape entity are established by routing over the two activities emanating from node ASN3. Activity 10 represents the time for tape processing. After processing, released tapes are routed to node Q11. The corresponding check batch entity passes through an auditing network segment before it is routed to node Q10. The two entities are combined after MATCH node M12 which requires an entity in node Q10 and an entity in node Q11 with the same value for attribute 3.

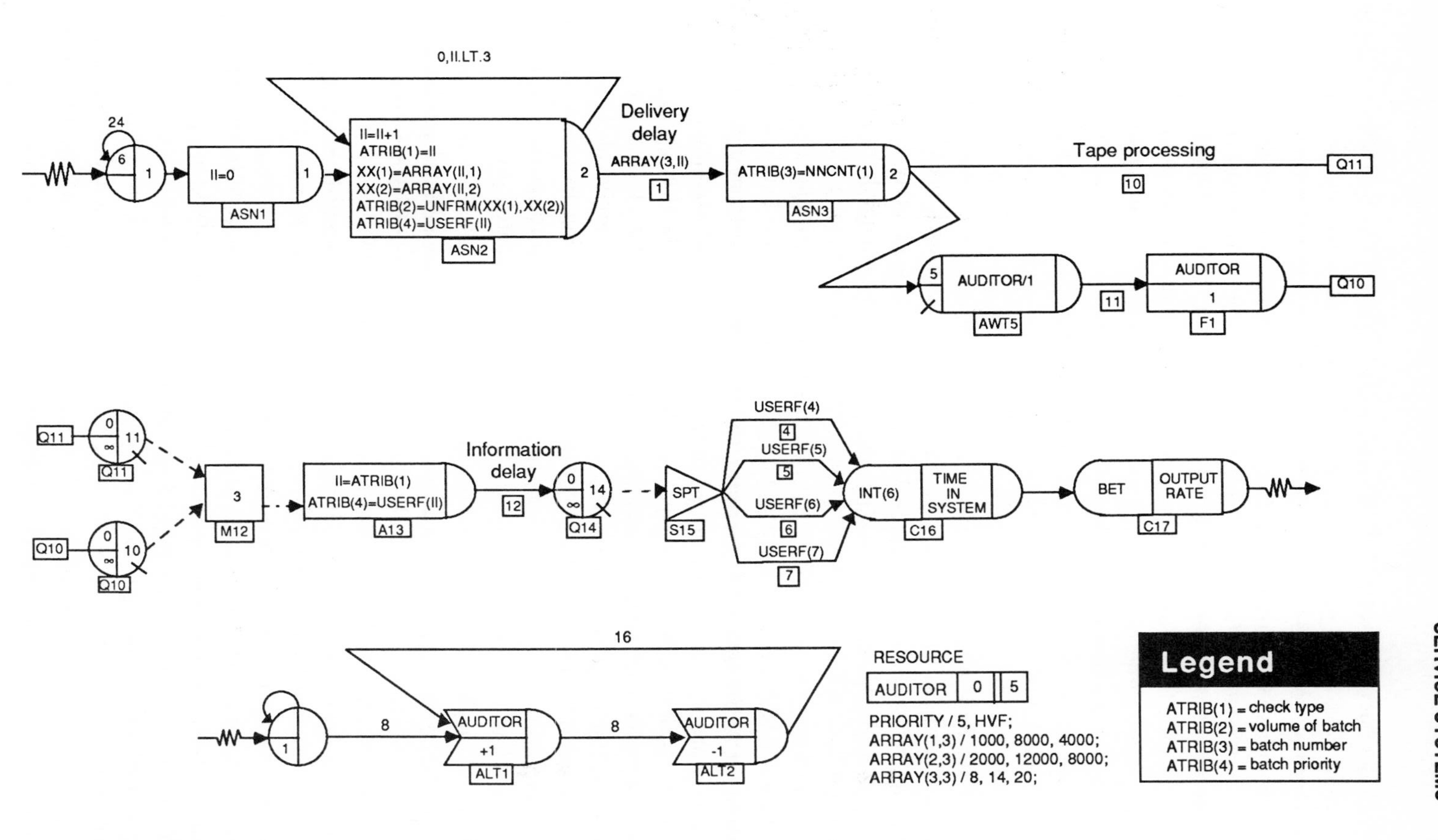

Figure 20-9 SLAM II model of a check processing facility.

Through the use of a PRIORITY statement, entities in node AWT5 are ranked high value first based on attribute 4, that is, HVF(4). The value of attribute 4 is assigned at node ASN2 in USERF(II). Therefore, attribute 4 can be thought of as an index, with higher values of the index receiving greater priority. By setting the index as a function of a batch's size, type, and arrival time, different sequencing rules can be incorporated into the model. The availability of auditors during specific working shifts is scheduled through a disjoint network using ALTER nodes as described in Chapter 3. The number of auditors available is controlled by the number of resources that are added or subtracted at the ALTER nodes, and the working shift is controlled by the activity times between nodes.

The processing of checks starts at node AWT5, where an auditor resource is required to perform the auditing function modeled by activity 11. An auditor is freed at FREE node F1, which permits the start of service on the next batch of checks if the auditor is still working; that is, ALTER node ALT2 has not been released. After the current batch is audited, it is routed to node Q10.

When both the tape-processing and auditing functions are finished for a specific batch, a match occurs at MATCH node M12. A single entity is then routed to ASSIGN node A13. At node A13, II is reset to the batch type and attribute 4 is redefined in USERF(II) to be the batch priority index for ranking work in QUEUE node Q14. Activity 12 is a verification task that ensures that there is a proper association between tape and batch information.

Server activities 4, 5, 6, and 7 represent four processing machines. The service time is calculated in function USERF so that it can be determined as a function of the batch type and size (attributes 1 and 2). SELECT node S15 selects among available machines using the smallest-busy-time (SBT) server selection rule. By varying this selection rule and the priority ranking in file 14, different machine sequencing procedures can be modeled and evaluated.

COLCT nodes C16 and C17 are used to collect statistics for each batch leaving the system. Interval statistics at node C16 provide information on the time in the system. Between statistics from node C17 yield information on the frequency of batches leaving the system.

The control of the auditor shift is depicted in the disjoint network in Figure 20-7. The start of the first work shift is controlled by the time delay on the activity from the CREATE node to ALTER node ALT1, and the length of each workshift is controlled by the activity from node ALT1 to node ALT2. The time to start subsequent work shifts is controlled by the time delay loop from node ALT2 back to ALT1. The number of auditors available for a work shift is represented by the increase in the resource AUDITOR at node ALT1. This resource is decreased at the end of the shift to stop processing check batches at node AWT5 in the main network. The length of a shift and the number of auditors available can be modified by changing the parameters of this disjoint network. Additional shifts can be included by increasing the number of ALTER nodes and activities.

20.4.3 Model Output and Use

The main performance measures of interest in this application are (1) the total time it takes to process a batch of work, (2) the rate at which batches depart the system, and (3) the utilization of the four machines. Estimates of these quantities are obtained from the SLAM II model from the interval statistics collected at node C16, the between statistics at node C17, and the utilization of servers 4, 5, 6, and 7.

This model was used to investigate sequencing methods to improve the utilization of the four machines by varying the batch priority rule defined in USERF(II) at node A13. The effects of the different rules were measured in terms of the effect on the time a batch spends in the system. The effect of projected additional work load was observed by increasing the arrival rates of batches, and the impact of the batch priority rules was further measured in light of this increase in work load.

The model was also used to evaluate personnel shift changes in the auditing department. Management was interested in the effect of proposed shift changes on the time to process work. The model was run under the case of the new shift change policy. An example histogram for the time to process batches under this situation is shown in Figure 20-10. From this histogram, it is observed that under the new shift policy that 268 batches or 82% of the work was processed within 8 hours. Fifty-three percent of the work was processed within 3 hours. Management was encouraged by these estimates and instituted the proposed one-shift operation.

20.5 WORK FLOW ANALYSIS IN A REGIONAL SERVICE OFFICE OF A PROPERTY AND CASUALTY INSURANCE COMPANY

Network simulation has been used to evaluate operational procedures within a regional office of a large property and casualty insurance company that processes various types of claims, endorsements, and new business items [14]. The office can be viewed in terms of paper flow through 14 distinct operational units or departments, including a centralized, computer-based information system. Over 150 personnel are involved in the processing of hundreds of transactions each day. The office is a complex queueing situation, which has been modeled to identify the bottlenecks in the work flow and to assist in investigating the effects of selected managerial decisions.

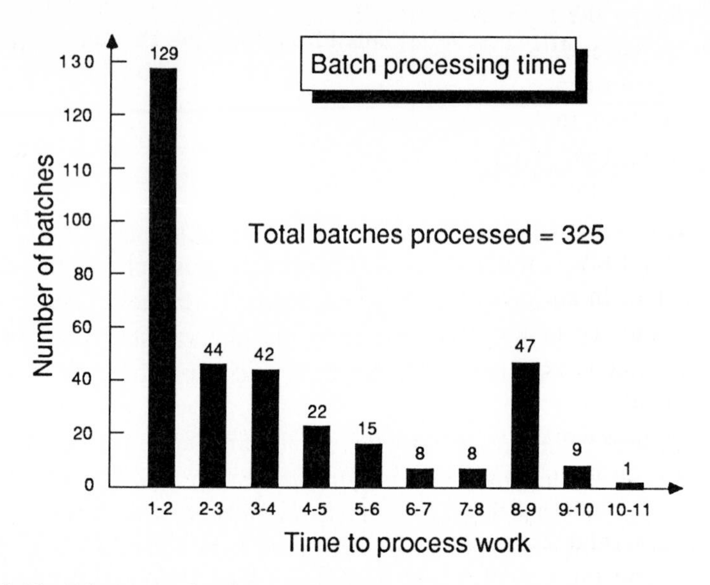

Figure 20-10 Histogram for time to process work.

Management was concerned about the effect of projected business increases on the quality of service provided the claimants. New operating procedures had been designed and an evaluation of the new design was needed. The following questions were posed by management:

1. What will personnel requirements be at the regional office in the future?
2. How should proposed service centers in other areas be designed or operated to maximize service and effectiveness?

20.5.1 The System and Management Concerns

The regional service center involves a paper work flow system that contains many interface points with a computer-based information system. The service center handles six basic work flows, each of which involves many decision-making points and record-keeping activities. A preliminary analysis indicated that the computerized information system appeared to be a critical component in the system and that only four of the six work flows interfaced with the information system. These four work flows are endorsements, new business, internal changes, and claims. A brief description of each of the work flows is given next.

The endorsement work flow consists of activities that are concerned with making adjustments to existing insurance policies. For example, the paper work required to change the amount of liability coverage on an insured automobile is classified as an endorsement. The endorsement work flow begins with a policyholder or an agent contacting the regional service office by telephone or mail and requesting that a change be made to a policy. After the initial processing, the request for a change in existing coverage undergoes a thorough review to determine the feasibility of the requested policy change. If acceptable, the endorsement is keyed into the computer system and the change is made on the customer's policy. This activity involves communication with a central computing system in another city. When computer processing is completed, a document is produced and mailed out to the policyholder. If the endorsement is rejected, a letter is sent to the customer citing the reason the endorsement could not be made.

The new business work flow involves processing requests of potential customers for automobile or home insurance coverage. The new business work flow contains a comprehensive underwriting review procedure.

The work flow related to internal changes includes updates in record-keeping classifications and the correction of administrative errors. These changes are initiated in the policy services unit, the underwriting unit, or the verification and assembly unit. The flow from these departments is to the computer system for processing. Eventually, a document is produced that is added to the policyholder's file.

The claims work flow begins with the policyholder initiating a claim action by telephone or through the mail. After initial records are filled out, a claims reserve is established by a claims representative. Both manual and computerized files are created for the claim to store incoming information. These files are periodically updated as more information is obtained. When the claim is settled, payment is sent out and the file is closed.

An overview of the system is presented in Figure 20-11. Although the exact work flows are not drawn, the overview illustrates how the office accommodates all four work flows and the number of personnel involved in processing the paper flows. A network model was developed to estimate the consequences of specific actions contemplated or anticipated by management. These actions involved the following:

1. Changes in the volumes of the type of work processed by the regional service office
2. Changes in priority rules for the processing of work in an operating unit
3. Changes in work flow paths
4. Reallocation of personnel among the different departments
5. Changes in processing times for specific items due to the introduction of training programs

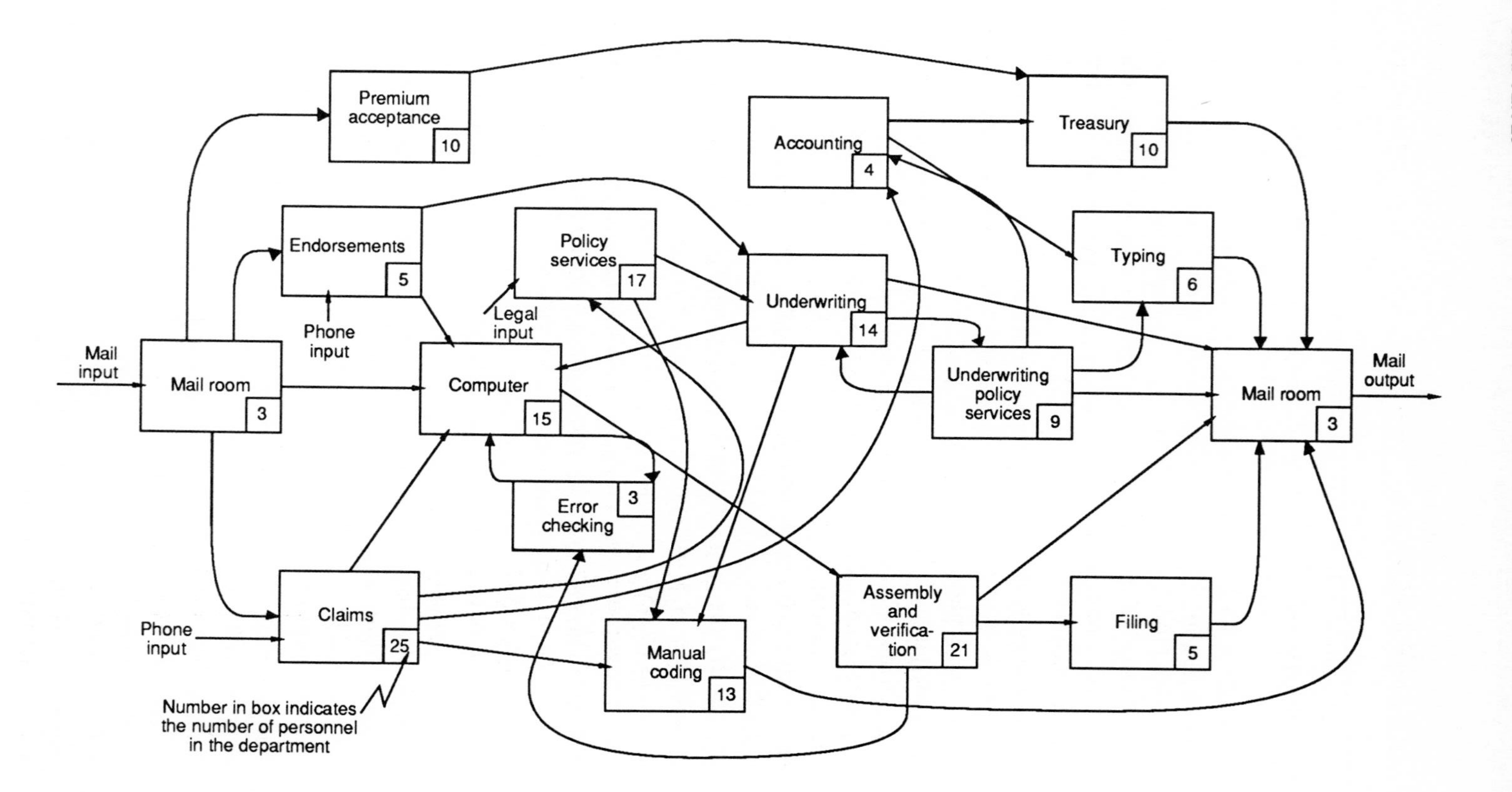

Figure 20-11 Diagram of work flow through the regional service office.

20.5.2 The SLAM II Model

The network model developed on this project consists of over 250 nodes and 500 branches; hence, only the basic model components can be presented in this book. Knowledge of this basic model provides sufficient insight into the development of the entire network model.

Because of the large volume of paper work flow, batches of work, rather than single items, are modeled. The entities flowing through the network are batches of 100 items of work classified as endorsements, new business, internal changes,or claims. CREATE nodes generate arrivals of each type of work to the system, and values are assigned to attribute 1 to establish the work type and to attribute 5 to provide a unique batch number at ASSIGN nodes. Each department has a queue of work to be processed and a set of servers corresponding to the personnel in the department. The flow paths through the network depend on work type (attribute 1). Routing of work is controlled by branching based on a route tag (attribute 2) that identifies the type of work and where in the network it was last processed. This allows different entities to enter a department and be properly differentiated when they exit a department. Service times and work priorities are also a function of the work type.

Figure 20-12 shows a SLAM II representation of the underwriting department. To simplify the graphic presentation, only the processing of new business and endorsements is shown in the figure. When a batch representing new business arrives at ASSIGN node A156 from the policy services department, it is assigned a new route tag (attribute 2). Upon emerging from the department at GOON node G14, entities are routed based on attribute 2. Node A8 similarly tags an entity from the endorsement unit before it enters the underwriting department. Nodes A156 and A8 also assign to attribute 3, the time required for the underwriters to process this particular work. This value is used by the service branches emanating from SELECT nodes S17 and S18. Node A156 assigns the service time according to a beta distribution. Node A8 assigns attribute 3 according to a normal distribution.

QUEUE node Q12 represents the stack of unprocessed new business and node Q13 represents unprocessed endorsements. All work is ranked on attribute 4, which is assigned to all incoming entities by user function USERF(1) at node A11. This function uses the time the entity entered the system and the work type (attribute 1) to assign a priority value for each entity. Small values are used to indicate a high priority. Different procedures for ranking work were modeled by modifying the user function.

Service activities 1 and 2 represent the underwriters, which totaled 14. Underwriters are arranged into two groups. Four highly trained underwriters, represented by activity 1, primarily handle new business, but they can work in

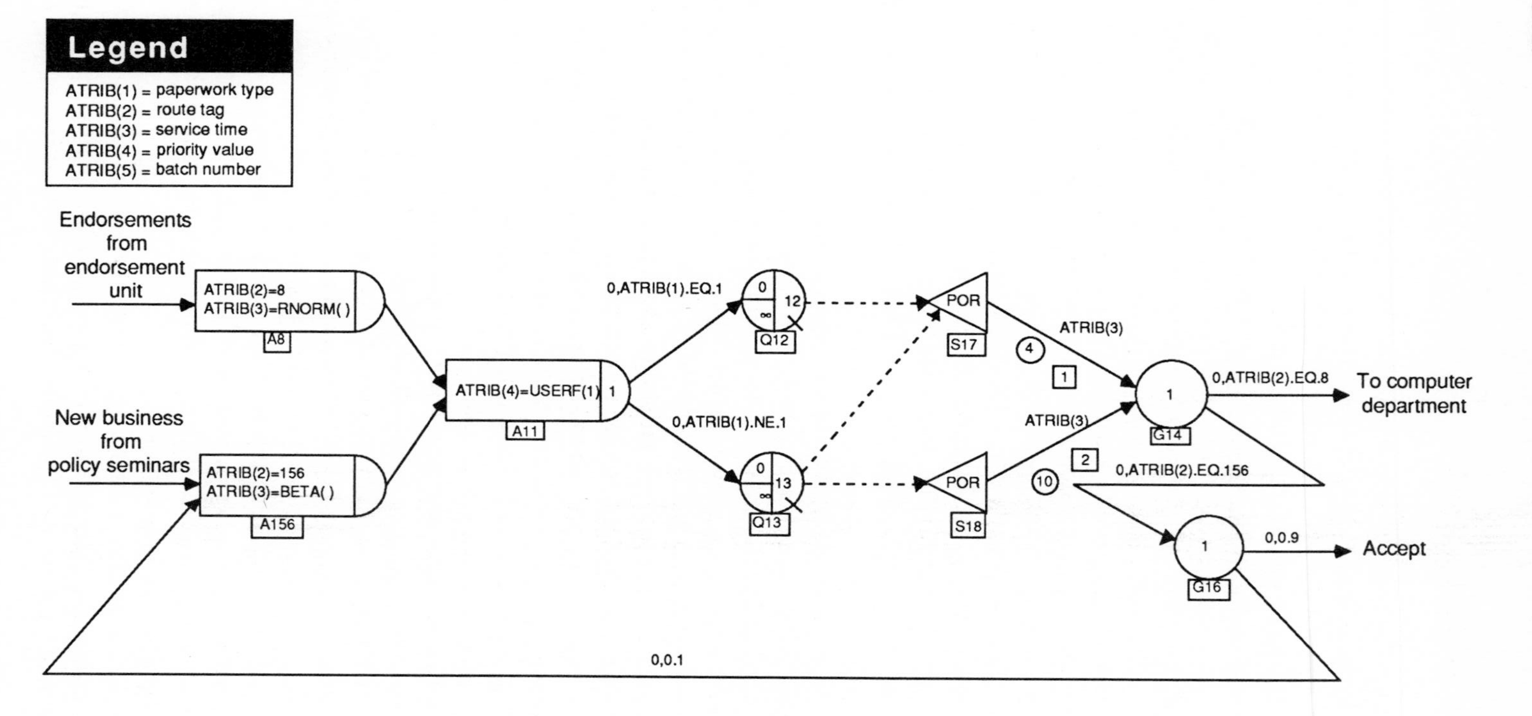

Figure 20-12 SLAM II model representation of underwriting department.

endorsements. Preference on the part of these underwriters is given to new business, as indicated by the POR queue selection rule for node S17. The ten other underwriters are less experienced and are used exclusively for endorsement work.

Each type of work performed is established by a separate entry point (node) to the department. The time to perform the service is established by the third attribute value, which is set at the entry node to be a function of both the type of work and the routing of the work.

To summarize, endorsements that come to underwriting wait in accordance with the prescribed priority in node Q13 until they can be processed by one of the 14 underwriters in the department. After processing, endorsements are routed to the computer department by node G14. New business entities arrive at node A156 and wait in node Q12 to be processed by one of the four highly qualified underwriters. After processing by the underwriters, new business is routed to node G16, where probabilistic branching is used. Ninety percent of the time a new business transaction is sent to the next department. Ten percent of the time it is rerouted to node A156 for additional processing by the underwriters. This is similar to the inspection process described in Section 16.5. As discussed there, this situation could be embellished to have service time and reprocessing probabilities that reflect the number of reworkings. In addition, excessive reworking of a new business application could be avoided by keeping track of the number of times an entity is reprocessed.

The other departments and work flow on the regional service center were modeled in a fashion similar to the underwriting department. All the departments are connected through entry and exit nodes as illustrated, with the routing to and from a department based on the work flow specifications.

One additional network modeling feature deserves to be mentioned. In the computer department, claim information is recorded in a database, while the actual claim is processed concurrently in the policy service unit. The claim is not allowed to proceed to other departments until both operations are performed. To model this, a duplicate entity is created and sent to policy services prior to entry into the computer department. A MATCH node is used to control the flow of the two entities. After a match is made, they are merged back into a single entity. The matching attribute is attribute 5, the unique batch number. The network concepts for modeling this matching are analogous to those presented in the bank check-processing model (see Figure 20-9).

20.5.3 Model Output and Use

The network representing the operation of the regional service office was simulated for a one-month period using data compiled from an on-site study. The simulation output provided the following types of information.

1. Statistics on utilization for personnel in each department (server statistics)
2. Statistics on work congestion: the average number of items waiting to be processed in each department (QUEUE node statistics)
3. Statistics and histograms on work flow time (interval statistics from COLCT nodes); this information was further categorized by:

   - Time to process accepted endorsements
   - Time to process rejected endorsements
   - Time to process accepted new business
   - Time to process rejected new business
   - Time to process internal changes
   - Time for internal changes to reach the premium acceptance department

Using the model, an analysis was performed that provided the basis for the following statements:

1. The projected new work load caused a significant degradation in system throughput.
2. Different work priorities decreased the time required to serve new customers.
3. Personnel utilization was improved by reallocation of personnel based on in-process work loads.
4. The number and type of new positions that were needed to provide a given level of service under a projected increased work load were determined.

Quantitative estimates of each improvement were provided to management.

Another use of the model was to measure the impact of training programs on the efficiency of personnel. The effect of training programs is readily quantifiable, as it is modeled by a decrease in service times. Hence, runs of the model with these new service times evaluated the training programs. In fact, service time can be incorporated in the model as a function of learning, that is, a learning curve. If, however, training is given by personnel on the job, then a decrease in the number of individuals processing the work must also be modeled. Another embellishment involves the use of work load processing guides. This and other information obtained from the model enabled management to better plan the operation of the proposed new service centers, as well as to redesign the existing centers. In summary, the network approach provided the basis for the systems analysis required to probe, formulate, convey, and study this work flow problem.

20.6 SUMMARY

The service sector of the economy of industrialized nations is growing at a rapid pace. A requirement exists for measuring the quality of services provided. The use of modeling and simulation in the service industries will expand dramatically in the coming years.

20.7 EXERCISES

20-1. Discuss the similarities of the case studies on check processing and work flow analysis in an insurance company. Specify the requirements of a network language for modeling systems of paper work flow.

20-2. A bank uses a distributed data-processing network to support its nationwide on-line banking operations. This highly complex, interactive network is controlled by two distributive computing facility centers. These two centers are the focal points of the bank's data communications network. Information is transmitted between these two centers and the bank's branch offices by means of communications channels. Teller access to these channels is regulated by programmable control units located in each branch office. At the branch offices, teller entity terminals are used to service customers through the communications network. The bank has a concern about the functional and physical expansion of the communications network. Long delays in response to teller requests could seriously jeopardize customer services. The bank, having an acute interest in customer satisfaction, needs to determine the inherent limitations and performance characteristics of this system before making any managerial decision regarding its expansion. Develop a modeling strategy that will help bank management analyze this complex, on-line, computer communications system. Specify performance measures for evaluating strategies.

20-3. Evaluate the similarities and differences in the modeling and simulation process when applied to service systems and manufacturing systems.

20-4. Design a network modeling language for describing the operations and procedures of a pyschiatric hospital.

20-5. Discuss quality of service issues and specify quality performance measures for the three models presented in this chapter.

20.8 REFERENCES

1. Barnoon, S., and H. Wolfe, "Scheduling a Multiple Operating Room System: A Simulation Approach," *Health Services Research*, Vol. 3, No. 4, 1968, pp. 272-288.
2. Clark, T. D., "A Systems Analysis and Model of Driver Licensing in the State of Florida," *Proceedings, Winter Simulation Conference*, 1986, pp. 842-849.

3. Esogbue, A. O., "Simulation in Surgery, Anesthesia and Medical Interviewing," *Simuletter*, Vol. 10, No. 4, 1979, pp. 19-24.
4. Fetter, R. B., and J. D. Thompson, "The Simulation of Hospital Systems," *Operations Research*, Vol. 13, No. 5, 1965, pp. 689-711.
5. Gibson, D. F., and E. L. Mooney, "Analyzing Environmental Effects of Improving System Productivity," *Proceedings, AIIE Systems Engineering Conference*, 1975, pp. 76-82.
6. Goldman, J., H. A. Knappenberger, and E. W. Moore, "An Evaluation of Operating Room Scheduling Policies," *Hospital Management*, 1969, pp. 40-51.
7. Hancock, W. M., J. Martin, and R. H. Storer, "Simulation-based Occupancy Recommendations for Adult Medical/Surgical Units Using Admissions Scheduling Systems," *Inquiry*, Vol. 15, March 1978, pp. 25-32.
8. Jones, A. W., K. V. Sahney, and A. Kurtoglu, "A Discrete Event Simulation for the Management of Surgical Suite Scheduling," *Annual Simulation Symposium*, 1984, pp. 147-152.
9. Kendrick, J. W., "Output, Inputs, and Productivity in the Service Industries," in *Statistics about Service Industries: Report of a Conference*, National Academy Press, Washington DC, 1986, pp. 60-89.
10. Kendrick, J. W., "Service Sector Productivity," *Business Economics*, April 1987, pp. 18-24.
11. Kutscher, R., and J. Mark, "The Services Sector: Some Common Perceptions Reviewed," *Monthly Labor Review*, April 1983, pp. 21-24.
12. Kwak, N. ., P. J. Kuzdrall, and H.H. Schmitz, "Simulating the Use of Space in a Hospital Surgical Suite," *Simulation*, Vol. 25, No. 5, 1975, pp. 147-152.
13. Larson, R. C., "Operation Research and the Service Industries," in *Technology in Service Industries: The Next Economy*, National Academy of Engineering, Washington, DC, 1988.
14. Lawrence, K. D., and C. E. Sigal, "A Work Flow Simulation of a Regional Service Office of a Property and Casualty Insurance Company with Q-GERT," *Proceedings, Pittsburgh Modeling and Simulation Conference*, Vol. 5, 1974, pp. 1187-1192.
15. Lu, K. H., and L. Van Winkle, "A Critical Evaluation of Some Problems Associated wtih Clinical Caries Trials by Computer Simulation," *Journal of Dental Research*, May 1984, pp. 796-804.
16. Magerlein, J. M., and J. B. Martin, "Surgical Demand Scheduling: A Review," *Health Services Review*, Vol. 13, No. 4, Winter 1978, pp. 418-433.
17. Mark, J., "Measuring Productivity in the Services Sector," *Monthly Labor Review*, June 1982, pp. 3-8.
18. Meier, L., Sigal, C. E., and F. Vitale, "The Use of a Simulation Model for Planning Ambulatory Surgery," *Proceedings, Winter Simulation Conference*, 1985, pp. 558-563.
19. Murphy, D. R., S. D. Duket, and C. E. Sigal, "Evaluating Surgical Block Schedules Using Computer Simulation," *Proceedings, Winter Simulation Conference*, 1985, pp. 551-557.
20. Pritsker, A. A. B. and E. Sigal, *Management Decision-making: A Network Simulation Approach*, Prentice-Hall, Englewood Cliffs, NJ, 1983.

21. Pritsker, A. A. B., "Applications of Simulation," in *Operational Research '84: Proceedings, IFORS*, J. P. Brans, ed., Elsevier, New York, 1984, pp. 908-920.
22. Quinn, J. B., "The Impacts of Technology in the Services Sector," *Technology and Global Industry*, in B. Guile and H. Brooks, eds., National Academy Press,Washington, 1987, pp. 119-159.
23. Quinn, J. B., "Technology and the Services Sector: Past Myths and Future Challenges," in *Technology in Service Industries: The Next Economy*, National Academy of Engineering, Washington, 1988.
24. Reitman, J., *Computer Simulation Applications*, John Wiley, New York, 1971.
25. Sigal, C. E., "Designing a Production System with Environmental Considerations," *Proceedings, AIIE Systems Engineering Conference*, 1973, pp. 31-39.
26. Swanberg, G., and B. Fahey, "More Operating Rooms or Better Use of Resources?" *Nursing Management,* Vol. 14, No. 5, May 1983, p. 16.
27. Zilm, F., L. Calderaro, and M. D. Gande, "Computer Simulation Model Provides Design Framework," *Hospitals*, Vol. 50, No. 16, 1976, pp. 79-85.

21 MODELING COMMUNICATIONS NETWORKS

21.1 INTRODUCTION

The communications industry is a service industry that directly supports manufacturing industries as well as other service industries. Telecommunications is considered a key element in the world's shift to an information economy. For example, the drive to automate manufacturing requires the linking of subsystems through data communication networks. The banking, insurance, and other service industries already rely heavily on communication networks. Figure 21-1 illustrates the extent to which communications systems will affect commercial activities of the future [14].

Communications systems, like many systems, generally evolve to meet current requirements. Often, in the design of these systems, insufficient thought is devoted to the problem of future requirements. An analysis of a communications system helps to understand current system user needs and to provide for future requirements.

Considerable research has been performed on electronic communications and data transfer. The evolution of communication systems spans nearly a century and the literature relating to the development of communication technology covers volumes. Today's communication systems technology takes full advantage of the advances in electronic switching and detection, fiber optics, microwave, laser transmission, and the availability of satellites for the transfer of data. A communication system consists of a network of nodes for the receipt and transmission of data and links with limited capacity that require time to process and route messages. SLAM II networks can be used to model communications systems to obtain estimates of their performance under changing operating conditions. The SLAM II outputs provide the communication system designer with estimates of system performance under existing conditions, and provide a

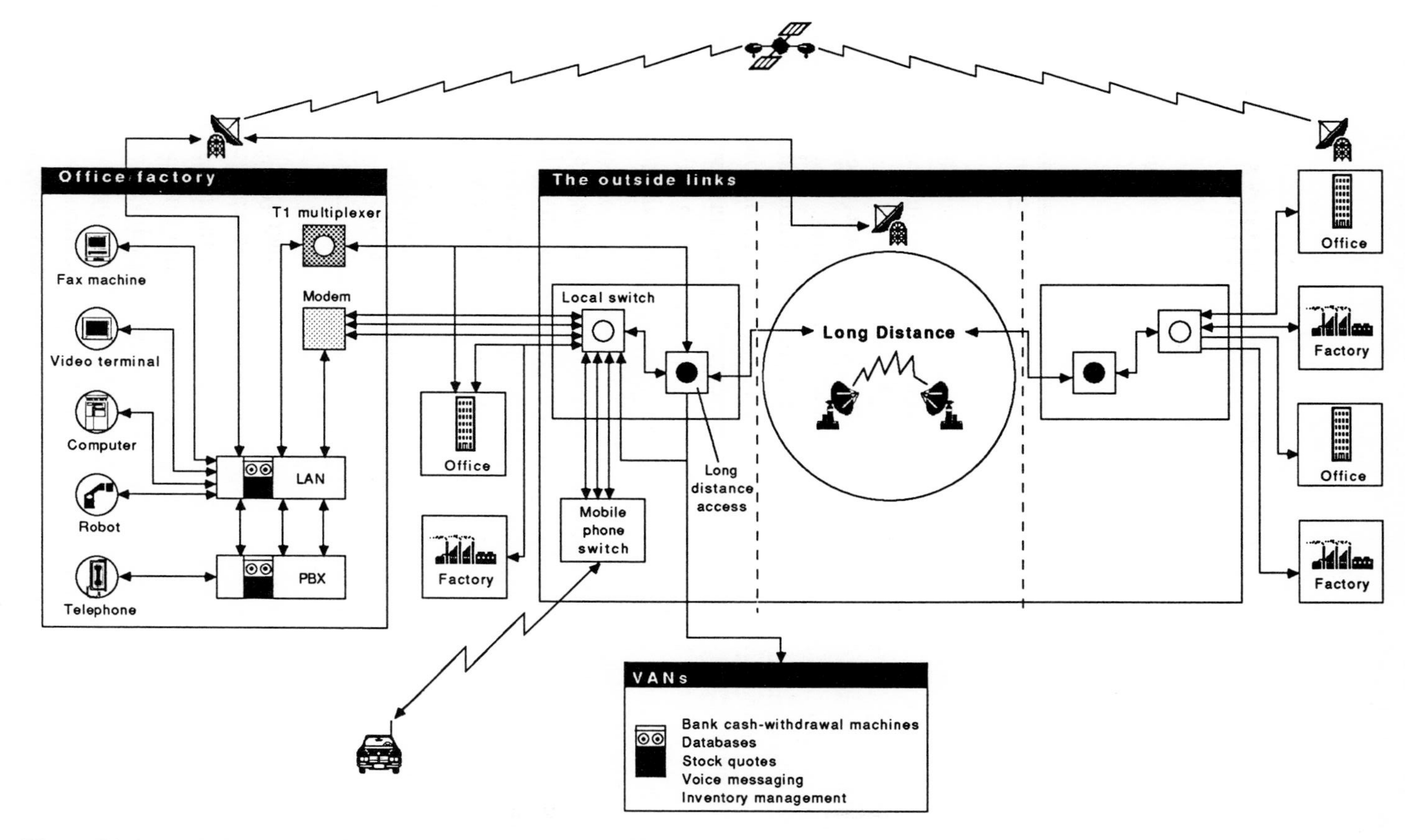

Figure 21-1 Telecommunications network of the future. Source [14] Reprinted with permission from *The Economist,* ©1985. Distributed by Special Features.

means to forecast system performance for increasing demand levels.

The presentation in this chapter is intended to introduce the subject of the analysis of communication systems through the application of computer simulation techniques using the SLAM II processor. A brief discussion is given of the basic characteristics of communication networks and four performance measures. Two examples of SLAM II modeling are described. The first example is a model of a computer communication network. The second example is an analysis of a local area network (LAN) used to link workstations and computers using an Ethernet system.

21.2 CHARACTERISTICS OF COMPUTER COMMUNICATION NETWORKS

A primary concern in the design of any communication system is the decision relating to the routing of messages on the network. Numerous data communication network routing algorithms are available, and each routing network algorithm has advantages [7]. The system designer is faced with trade-off decisions relating to cost, speed of message transfer, message capacity of the system, and system reliability. These decisions are strongly influenced by a network's topology, the transmission medium employed to link the stations on the network, and the access method used to enable the transfer of messages on the network.

A network's topology defines its physical layout and is the most distinguishing characteristic of the network. The linking transmission medium for sending and receiving messages between working stations on a network usually consists of twisted-pair wire, fiber optics, or coaxial cable for distances ranging from a few feet up to several miles. Two major methods have been developed to provide working station access to the network [9]. These are discussed in the next section.

21.2.1 Common Computer Communication System Topology

A local area network (LAN) is commonly employed in offices, manufacturing environments, and other business establishments to allow multiple access to computers and to connect working stations on a computer communication system. A LAN consists of hardware and software that enable communications between connecting stations. LANs are becoming more prevalent as microcomputer technology and applications expand. Three widely used LAN topologies are the **star**, **ring** and **bus** [9].

In a star network, all messages from the connecting stations are routed through a central station. This network is typical of many mainframe computer installations. The central station processes each message and controls the communications process between the central station and all outlying stations. One major advantage of the star topology is its simplicity. Each outlying station is served by a distinct link, and a routing table is used to match each message to its destination. A disadvantage of the star topology is that it is the most difficult to modify, since links are established between the central and outlying stations. Therefore, this topology is usually employed in communication systems that are not subject to frequent configuration changes.

In a ring topology, all the stations are physically connected through a looping arrangement that provides a communication link between any of the stations on the loop. Message control is distributed in a ring network because they do not have to pass through a central station. However, messages transmitted over the loop may have to pass many other stations before reaching their destination. A ring topology has the advantage of flexibility since stations can be added to or deleted from the network with little difficulty once it is in place. The disadvantages include an increased probability of collisions between messages on the loop and a limited message capacity.

A bus topology consists of a single line to which individual stations are connected in parallel. Each station has independent access to the bus. This topology provides the greatest flexibility under changing network installation configurations such as might be encountered in manufacturing environments. However, as in the ring topology, stations compete for access to the main line, and special access methodology must be employed. Care is required to avoid a rapid deterioration of system performance as additional demands or stations are placed on the network. Ethernet is typical of a LAN that employs this topology. A model to evaluate the performance of an Ethernet in a specific communication network is presented in Section 21.4.

21.2.2 Common Computer Communication Systems Transmission Links

The three primary data transmission links are twisted-pair wire, optic fiber cable, and coaxial cable. The selection of any of these linking methods depends on the hardware and software selected for the communications network. A brief discussion of the relative merits of these transmission media is given below.

Twisted-pair wire consists of two copper wires twisted together. It is used extensively by telephone communications systems and is, therefore, installed over wide areas. Twisted-pair wire, the least expensive data-transmission alternative, has limited message-handling capacity and is subject to electrical noise interference. Consequently, it is not recommended for use in networks with a high message capacity requirement or in environments where electrical interference may be encountered.

Fiber-optic cable offers the most advanced technology for data transmission, however, it is also the most expensive. The primary advantage of fiber-optic cable is that a single cable has the signal-carrying capacity of hundreds of twisted-pair wires. In addition, since the data are transmitted in the cable as pulses of light, there is virtually no electrical interference. Fiber-optic cable data transmission, like twisted-pair wire, can be used to send data over long distances. One major disadvantage to fiber-optic cable, in addition to its cost, is that it is not as easy to tap into or to repair as twisted-pair wire or coaxial cable. Therefore, adding work stations or performing maintenance on an existing fiber-optic network could be difficult.

Baseband and broadband coaxial cable data transmission links provide reasonable alternatives for communications systems where the limitations of twisted-pair wire are unacceptable. Baseband coaxial cable is normally used when only a single channel is needed and the length of the network is less than 3 miles. Broadband coaxial cable, the standard cable for cable television (CATV), is usually employed when multiple channels are required. Therefore, multiple communication networks can utilize the same broadband coaxial cable where each network is assigned its own frequency range. Coaxial cable shielding eliminates most types of electrical interference and, although it costs more than twisted-pair wire, it is less expensive than fiber-optic cable.

21.2.3 Work Station Access To Computer Communication Systems

Work station access to star topology communication networks is direct since all messages are routed through a central servicing center. In this topology, messages arriving at the central servicing center are stored in a buffer and directed to their destinations according to a predefined priority system. However, in the ring and bus network topologies, several work stations can be competing for the single main line at the same time. Therefore, a network accessing method is essential to the operation of the system. CSMA/CD and token passing are the two primary LAN access methods; they are defined in the following paragraphs.

In the CSMA/CD network access method, a transmitting station first listens to the line. If the network is free, the station transmits its data. Otherwise, the station waits until the line is available. The primary limiting characteristic of this access method becomes evident when two or more stations attempt to access the line simultaneously. In this event, the accessing stations transmit their data and a collision occurs. When a collision occurs, the transmitting station detects excess activity on the line and ceases transmission, waits for the line to become free, and then attempts to retransmit the entire data string. The process continues until each message has been successfully transmitted. Thus, the CSMA/CD network access method works well in communications systems where message traffic is dispersed and access delays are not critical. It is not efficient in high-density message traffic systems.

The token-passing network access method is more complex than the CSMA/CD method. Collisions are avoided on the network by passing a token to each work station and allowing it to transmit only when it has the token. The token is actually a code that circulates the network in a predefined logical sequence. A station with nothing to transmit simply passes the token to a next station. Since the order in which stations can transmit is sequential and predefined, as opposed to the random network access of the CSMA/CD method, the maximum time for any station to transmit can be calculated for any given network load. Therefore, the token-passing network access method enables the design of priority levels for the network. By using a token, this network access method requires more overhead and, in some cases, may lead to single-station dominance.

21.3 COMPUTER COMMUNICATION SYSTEM PERFORMANCE MEASURES

The design of computer communication systems involves many decisions related to network topology, linking capacity, message access and routing procedures, and message flow control. Performance is normally evaluated in terms of message delay time and system capacity for a given system cost. The delay time and capacity both depend on the number of messages or load imposed on the system. As the number of messages increases per unit of time, the delay time per message increases. When the number of messages approaches system capacity, the delay time increases rapidly and at some point the system becomes totally saturated. In networks that employ the CSMA/CD access method, the probability of a collision on the network and the collision rate are also important performance measures. These performance measures are a function of the message rate and also the length of the message that is transmitted. Definitions of these four performance measures are:

1. Network wait time: the average time that a message waits to access the network after it is ready to be transmitted
2. Network response time: the average time required to complete the transmission of a message since the network was first checked for access
3. Probability of a collision: the probability that a transmitted message collides at least one time

4. Collision rate: the total number of messages that are involved in collisions per unit of time

21.4 EVALUATION OF COMPUTER COMMUNICATION NETWORKS

Many models of computer communication systems are reported in the literature. Green and Fox used the GASP IV simulation language to model a circuit-switching system [5], Musselman and Hannan simulated a distributed data-processing network of the Bank of America's on-line banking operation with Q-GERT [10], and Schwetman demonstrated the use of CSIM for the simulation of computer systems [12]. Models developed by Garcia [4] and Erdbruegger and Bonin [2] demonstrate the versatility of the SLAM II language through the simulation of complex computer communication networks to obtain estimates of the performance measures described in Section 21.3. Their work is presented in the next two sections.

21.5 MODELING COMPUTER COMMUNICATION NETWORKS

Garcia [4] developed a generalized simulation model to evaluate the performance of computer communication networks. The model portrayed working stations as nodes connected by data communication links. The nodes on the network received, routed, acknowledged, detected, and corrected messages for transmission. If outgoing links from a node were busy, messages waited in a queue until the link became free. The primary performance measure for the model is network wait time, although throughput, cost, and reliability are also considered.

The SLAM II model for Garcia's computer communication network consists of four modules. The modules are presented in Figure 21-2 and are the message generation module, node-to-node transfer module, statistics collection module, and the external effects module. In general, the external effects module is used to generate a response message for some, but not all, of the original messages that are transmitted over the network. In this regard, each response message places an additional demand on the network. In this model, no response to a response message is considered. An entity in the network represents a message to be transmitted. It is uniquely identified by values assigned to seven attributes. The attributes of an entity are (1) mark time, (2) destination, (3) message type, (4) precedence or priority, (5) origin, (6) identification number, and (7) message length. Entity attributes and XX variables are defined in Table 21-1. Eighty-seven XX(I) variables are used to store data and computational values.

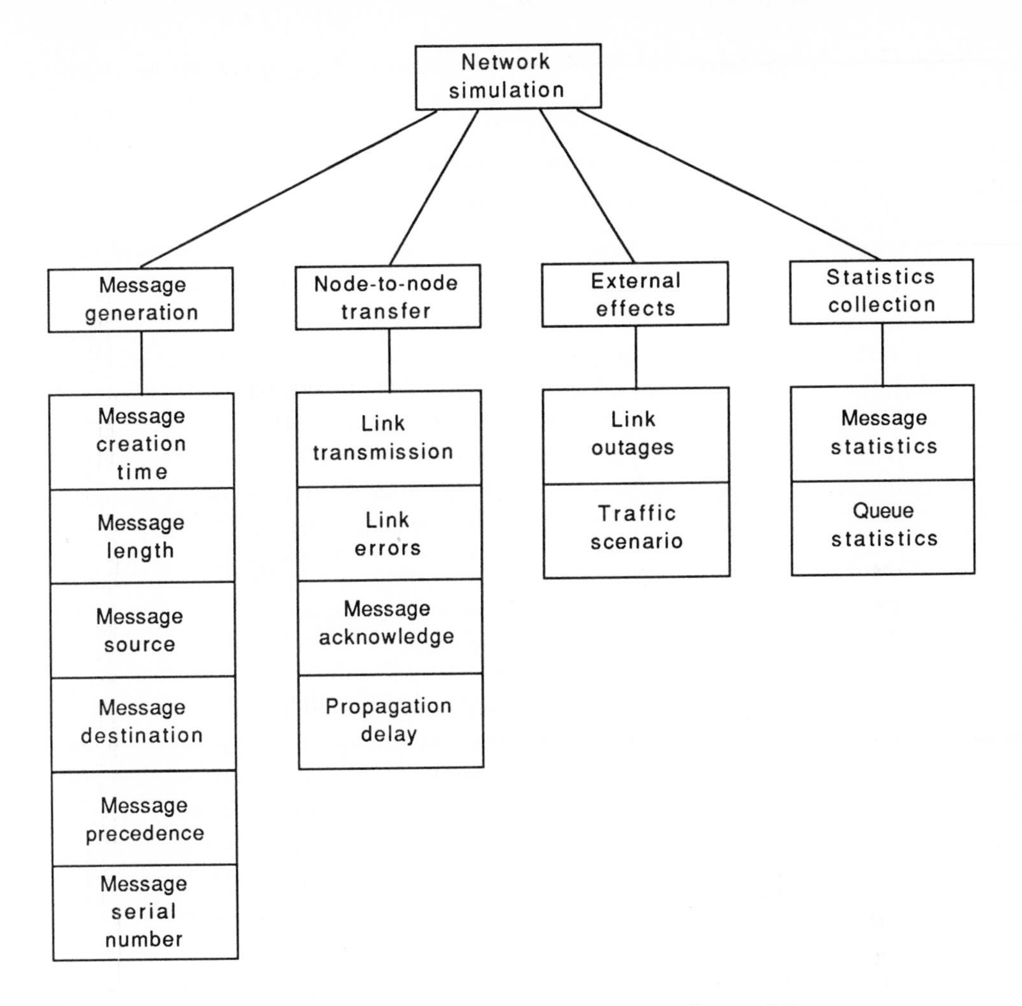

Figure 21-2 Modular structure of the simulation program. Source [4]

21.5.1 Message Generation Module

The SLAM II network segment for the message generation module is given in Figure 21-3. Messages are created at CREATE node C1 with an exponential time between creations. The mean interval times for different modules are obtained from the global variables XX(11) to XX(20). Attribute 1 is the time the message entity is generated. Attribute 2 defines a destination location for the message, attribute 3 is set to zero to indicate that the entity is an original message, and attribute 5 defines the originator of the message. The global variable XX(31) counts the number of messages created and it is placed in attribute 6 to provide a serial number for the message. The length of the message

Table 21-1 Definitions of ATRIB and XX Variables

Variable	Definition
ATRIB(1)	Creation time in seconds
ATRIB(2)	Destination node
ATRIB(3)	Type
ATRIB(4)	Precedence
ATRIB(5)	Origin node
ATRIB(6)	Serial number
ATRIB(7)	Length in bits
XX(1)	Link transmission rate in bits/second
XX(3)	Mean message length in bits
XX(4)	Minimum message length in bits
XX(5)	Maximum message length in bits
XX(8)	Acknowledge message length in bits/second
XX(9)	Node processing overhead in seconds
XX(11-20)	Message interarrival times in seconds
XX(31-35)	Message serial number counters
XX(38-49)	Link error probabilities
XX(51-54)	Message destination
XX(61-64)	Response probabilities by priority
XX(71-82)	Number of lines for each link
XX(83-87)	Message originator

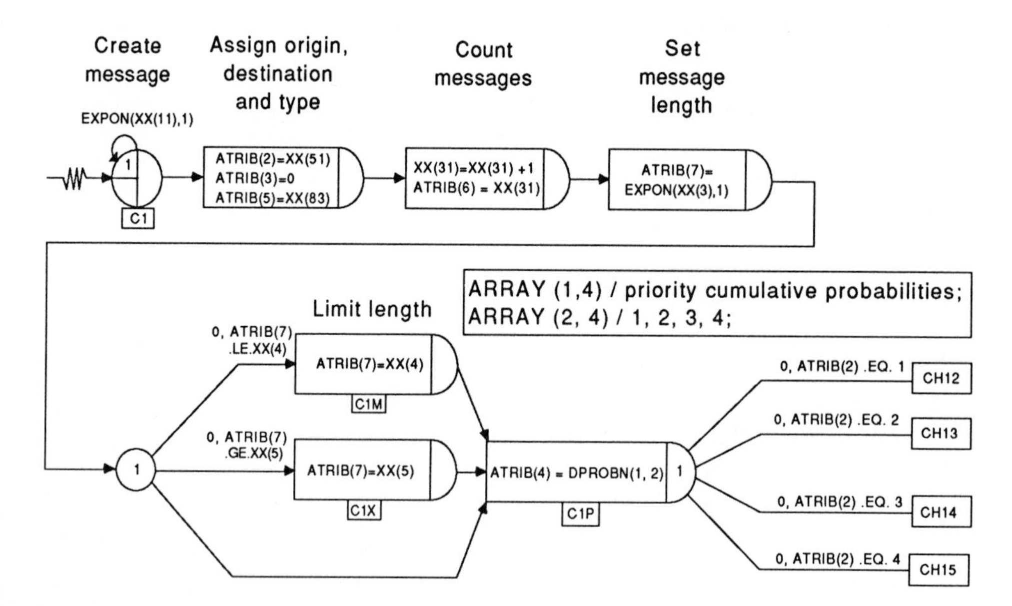

Figure 21-3 Message generation module segment. Source [4]

is then determined and assigned as the value of attribute 7. The length, in bits, of the message is adjusted to keep it in an allowable range for message transmission.

The priority for each message is determined at ASSIGN node C1P. Outgoing messages are assigned a priority ranging from 1 to 4. Routing messages are assigned a priority of 4 and others a higher priority according to their importance. ARRAY(1,4) contains the cumulative probabilities for an outgoing message and ARRAY(2,4) contains the different priority levels. The priority for the outgoing message is randomly determined using DPROBN(1,2). After the priority is assigned to attribute 4, the entity is routed over a channel to reach its destination.

21.5.2 Node-to-node Transfer Module

Each channel in the communications system is represented by a node-to-node transfer module. Since all of the channels are identical, only one subnetwork segment representing a single channel is illustrated in Figure 21-4. A RESOURCE block and a PRIORITY statement are required for each channel modeled.

Messages arrive at AWAIT node CH12 to acquire a communications channel. The single communication link is represented by the resource LINK12. If the channel is not in use, LINK12 is allocated to the arriving message, and the

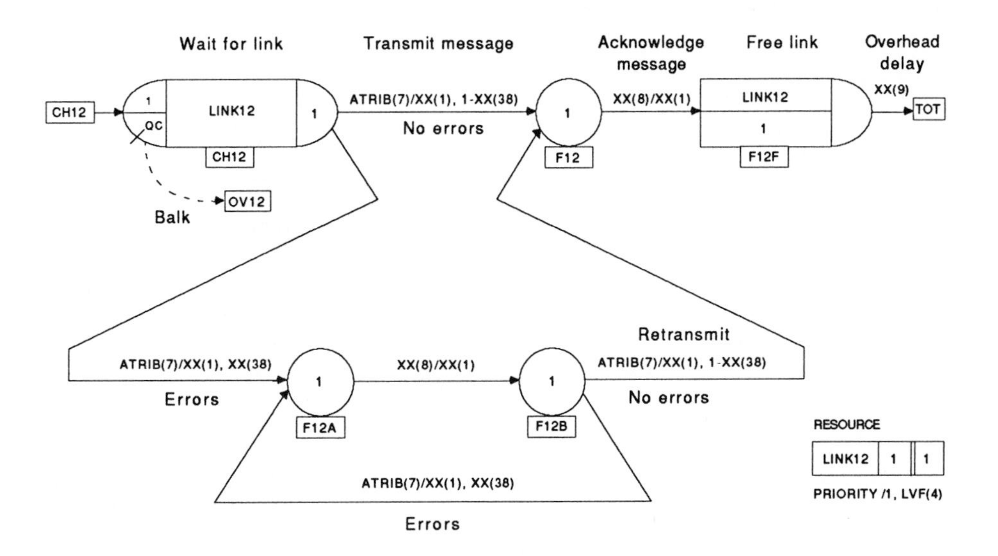

Figure 21-4 Node-to-node transfer module segment. Source [4]

message is transmitted. If the resource LINK12 is busy, then the arriving message waits according to its assigned priority value stored in ATRIB(4). If a message arrives when file 1 of AWAIT node CH12 is full, the message balks to COLCT node OV12. Node OV12 collects observations on message overflow and is described in the statistics collection module.

When a message is transmitted, there is a probability of a message transmission error that is defined by the global variable XX(38). If no transmission error is detected, the message transmission delay time is determined as a function of the message length, ATRIB(7), and the transmission bit rate of the channel, XX(1). If a transmission error is detected, the message is routed to GOON node F12A. After an acknowledge delay, the message is retransmitted from node F12B until an error-free transmission occurs. All successfully transmitted messages are then routed through node F12 and acknowledged by the receiver. The resource LINK12 is then freed at FREE node F12F for the next transmission. An overhead delay, XX(9), is incurred before the message entity is routed to node TOT of the statistics collection module.

21.5.3 Statistics Collection Module

Interval statistics for the time to transmit are collected at node TOT shown in the statistics collection module given in Figure 21-5. Transmission times are

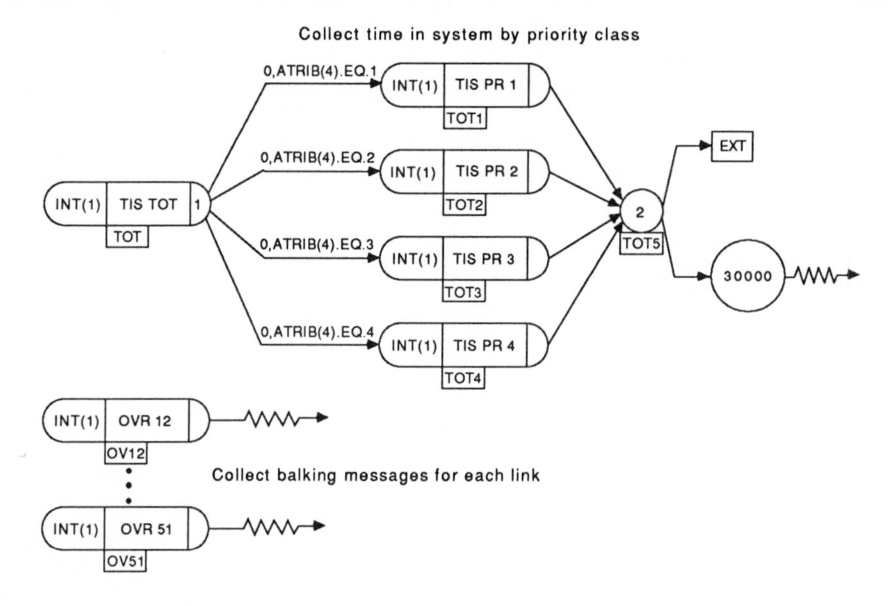

Figure 21-5 Statistics collection module segment. Source [4]

also collected by priority level and on message overflows from each communication link. Each entity is then routed to GOON node TOT5 which has two emanating activities. One activity routes an entity to the external effects module to determine if reply messages are to be generated in response to the messages received. The other activity leads to a TERMINATE node and defines the ending condition for the simulation run to be the transmission of 30,000 messages.

21.5.4 External Effects Module

The external effects module is presented in Figure 21-6. Entities enter at GOON node EXT. Only an original message is considered as a candidate for reply. The value of ATRIB(3) is zero for an original message, and is set to one for a reply message. If ATRIB(3) is one, the entity is terminated.

If ATRIB(3) is zero, the entity is routed to ASSIGN node EXR1, where the value of ATRIB(1) is reset to TNOW to provide a reference time for an observation of a reply message transmission time. Next, attribute 2, the originator of the reply message, is set to ATRIB(5) which was the destination of the original message. ATRIB(3) is set to 1 to designate the message as a reply. The variable II is set equal to the priority level of the message, ATRIB(4), in order that it may be used as a subscript in the probability specification of the activities emanating from node EXR1.

Entities are routed from node EXR1 based on the probability of a reply. This probability depends on the priority level of the original message. The values of XX(61) through XX(64) are the probabilities of replying to an originating

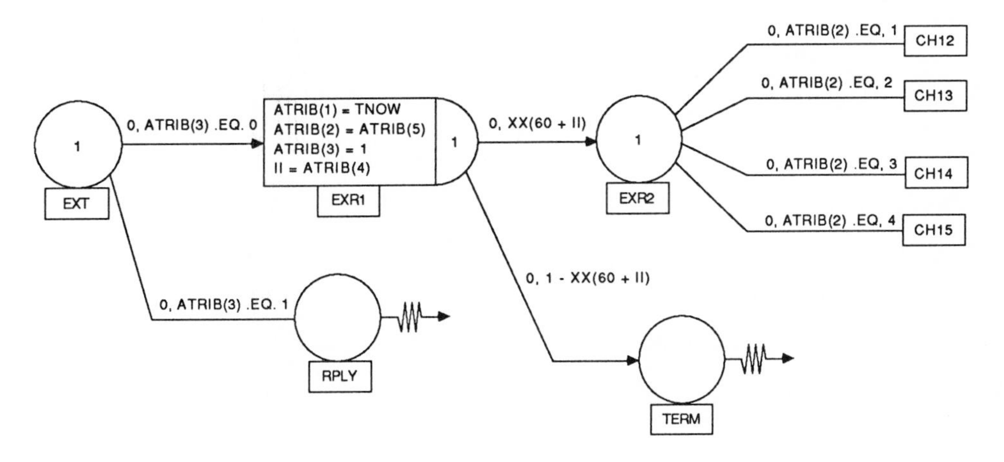

Figure 21-6 External effects module segment. Source [4]

message with priority levels 1 through 4, respectively. Therefore, entities will branch from node EXR1 to GOON node EXR2 with a probability of XX(60 + II) and depart the system with the probability of 1-XX(60 + II). The entities arriving at node EXR2 represent reply messages and are routed to the transmission links. When a reply message arrives at node EXT and ATRIB(3) is 1, it departs the model at node RPLY.

21.5.5 Computer Communication System Model Outputs

The model was used to analyze message delay time performance relating to a computer communication network's topology, message precedence, and link (transmission) error rates. The four topologies are shown in Figure 21-7. The SLAM II model was used to obtain statistics on message delay time as a function of traffic input varying one factor at a time. Traffic input was increased until system saturation was observed. The resulting plots from the simulations provide performance measure curves for expected message delay times relative to the message input rate.

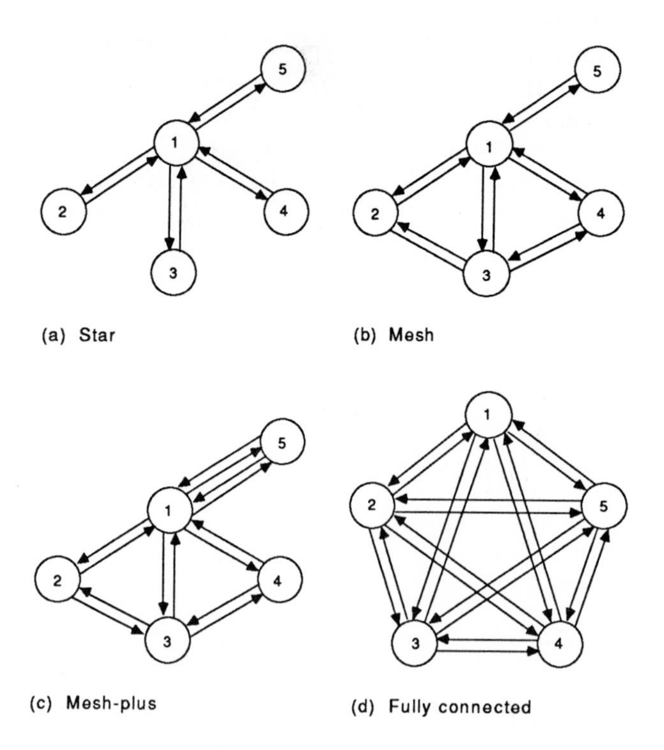

Figure 21-7 Network topologies. Source [4]

In Figure 21-8, the performance of a fully connected network topology yields the smallest message delay times for a given input rate. From Figure 21-9, it is seen that message priority levels cause few delays until the message traffic level exceeds approximately 12O messages per minute. The effect of the quality of the transmission medium on message delay time is evident from Figure 21-10. A communication system designer can use this type of data in conjunction with system cost to perform a cost versus system performance trade-off analysis when either designing a system or establishing precedence levels for working stations. Alternatively, given minimum requirements for system performance, a minimum cost design could be developed.

21.6 MODEL OF AN ETHERNET BUS

Erdbruegger and Bonin [2] developed a SLAM II simulation model of a proposed Ethernet communication network linking a VAX computer with work stations in a bus topology. The CSMA/CD network accessing method is employed in the system. The performance measures of primary interest for the study were estimates of network delay time, network response time, the probability of message collisions, and the message collision rate for the system. The controlling factor for the estimation of these performance measures is the message arrival rate per unit of time to the system. The description of the model is based on [2].

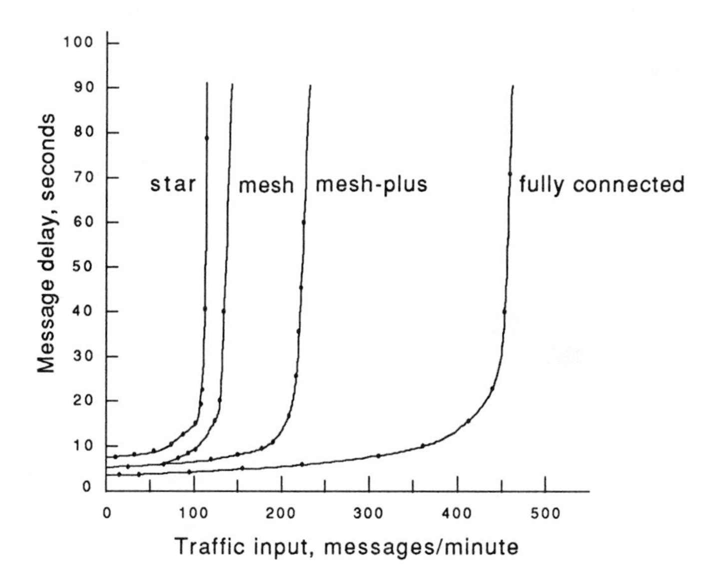

Figure 21-8 Message delay for network topologies. Source [4]

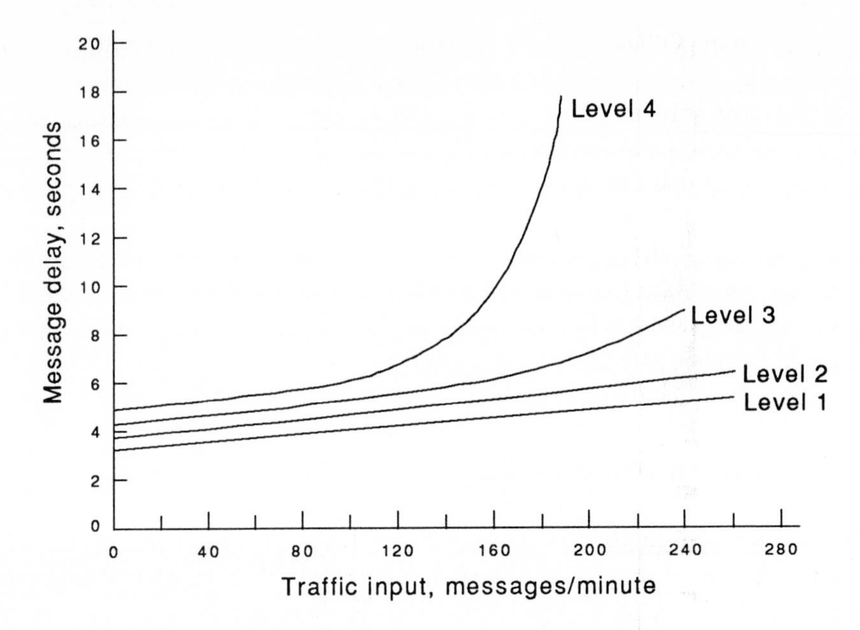

Figure 21-9 Message delay for precedence levels. Source [4]

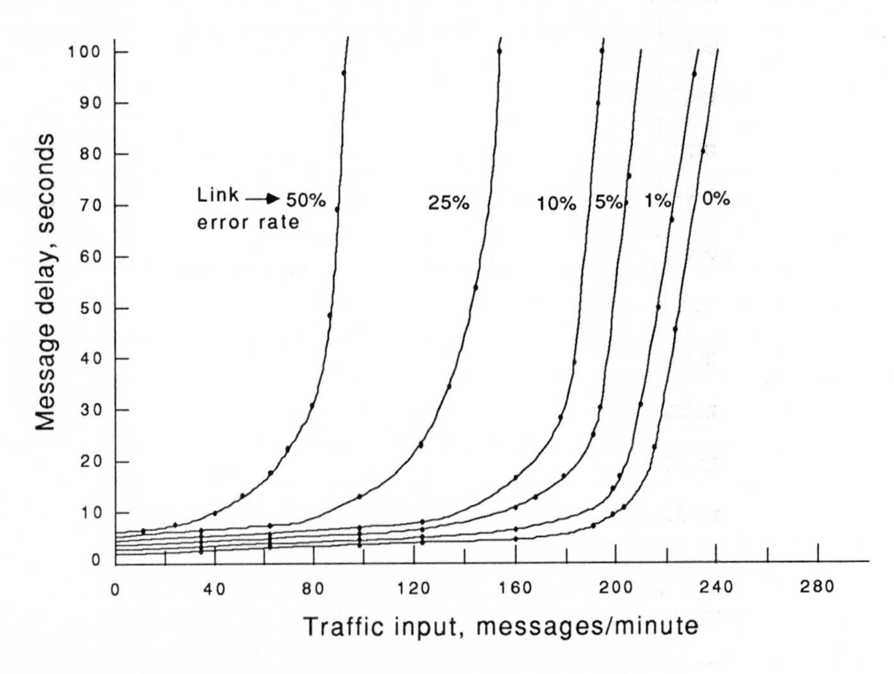

Figure 21-10 Message delay for link error rates. Source [4]

21.6.1 Ethernet Communication Methodology

The Ethernet network electronically transmits messages on a main transmission line which links work stations (nodes) together. The transmitted messages are data packets that consist of a stream of bytes. The data packet (frame) can be represented in SLAM II by two entities. Conceptually these entities represent the leading and trailing bits of the message and the length of the message determines the separation of the two entities. Therefore, a frame's leading and trailing bits are separated by the time required to transmit the entire message. When a node has a message to transmit and it does not detect another message on the line, it transmits a frame in each direction from the node as shown in Figure 21-11.

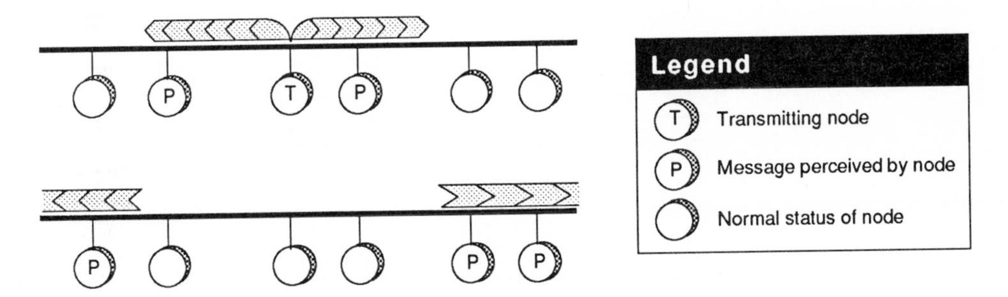

Figure 21-11 Message transmission and node states.

Other nodes on the network detect a frame on the line only after the leading bit of the frame reaches it. Should another node want to transmit a message and it has detected the leading bit of a frame on the line, it delays transmission until the trailing bit is detected. After detecting the trailing bit of a frame the node waits an additional 9.6 microseconds before transmitting, thus providing spacing between messages. The spacing of messages on the transmission line is illustrated in Figure 21-12.

Frame collisions occur on the line because a node can start the transmission of a message before it senses the signal of a transmission from another node. The collision is detected by the transmitting node when the leading bit of the other message arrives at the node. This situation is diagrammed in Figure 21-13.

When a transmitting node detects that a collision has occurred, the node sends a short jam signal and then terminates its transmission. The jam signal is required to assure that other concurrently transmitting nodes also detect the collision and respond appropriately. After a randomly determined delay time,

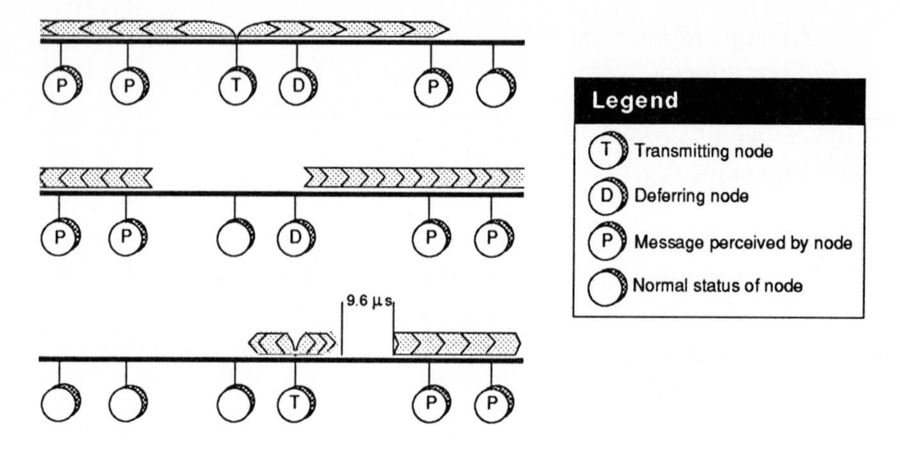

Figure 21-12 Message transmission with node deferring illustration.

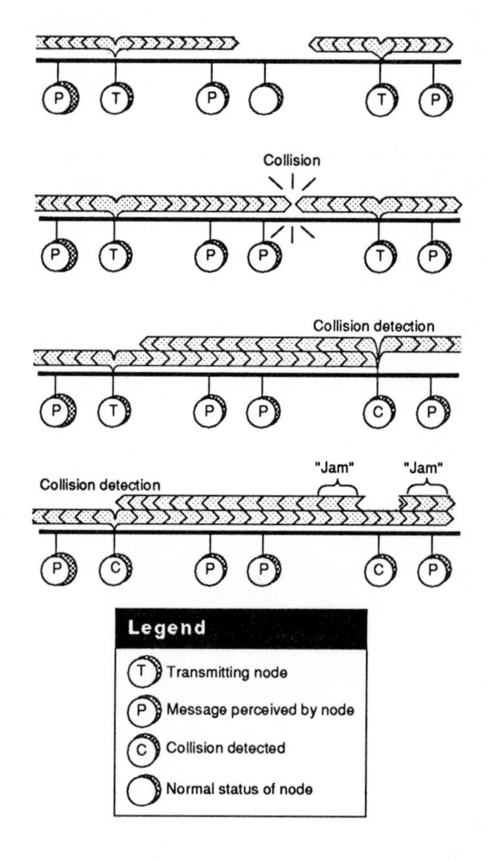

Figure 21-13 Message transmission with collision and jam illustrations.

the node attempts to retransmit its message. Should another collision occur on the subsequent attempt, the delay time is increased before retransmission. If the message cannot be transmitted after 15 attempts, it is terminated and an error is recorded. This procedure for message retransmission is employed to minimize the probability of subsequent collisions on the network.

21.6.2 SLAM II Model for an Ethernet Computer Communications Network

A SLAM II model was developed to simulate the above Ethernet operation for a basic system configuration and physical layout. In this model, resources are used to limit message access to a transmission line. The data describing the nodes, computers, message creation rates, and first transmission times are initialized in subroutine INTLC. The corresponding message collision logic is written in subroutine EVENT. Collisions on the transmission line are detected on the line at EVENT nodes, and the time between message arrivals, frame size, collision back-off delay times, and message transmit times are calculated in user-coded inserts to the SLAM II network. The Ethernet LAN network model is given in Figure 21-14.

The generation of messages to the system are created by the CREATE nodes VAX and MICR and ASSIGN nodes AVAX and AMIC. The CREATE nodes generate the first arrival of each type of message. These entities are routed to ASSIGN nodes AVAX and AMIC where attribute values are assigned to differentiate between messages and to record the length of each message in bits. Two entities then emanate from each of these nodes. The delay time for the looping entity around each node is used to enter the next message arrival into the system. The second entity departing node AMIC represents a message generated by microcomputers. This message is limited in frame length to 1K bytes and is routed to ASSIGN node ENID. The second entity departing node AVAX represents a transmission from a minicomputer. These transmissions are large files that may vary in size between 40K and 100K bytes. The maximum allowed length for a frame is 1K bytes. Therefore, the message is split into 1K byte frames at unbatch node BLKS and the resulting entities are routed to node ENID. The first attribute of each message frame is assigned an identification number at node ENID and then it attempts transmission over the network at AWAIT node ANET.

A node on the Ethernet network can only transmit (or attempt to transmit) one message at a time. AWAIT node ANET controls the access of messages at each originating node to the transmit logic of the model. A resource with capacity of one is defined for each work station. A message originating at a work station seizes its resource and holds it until the message transmission is completed.

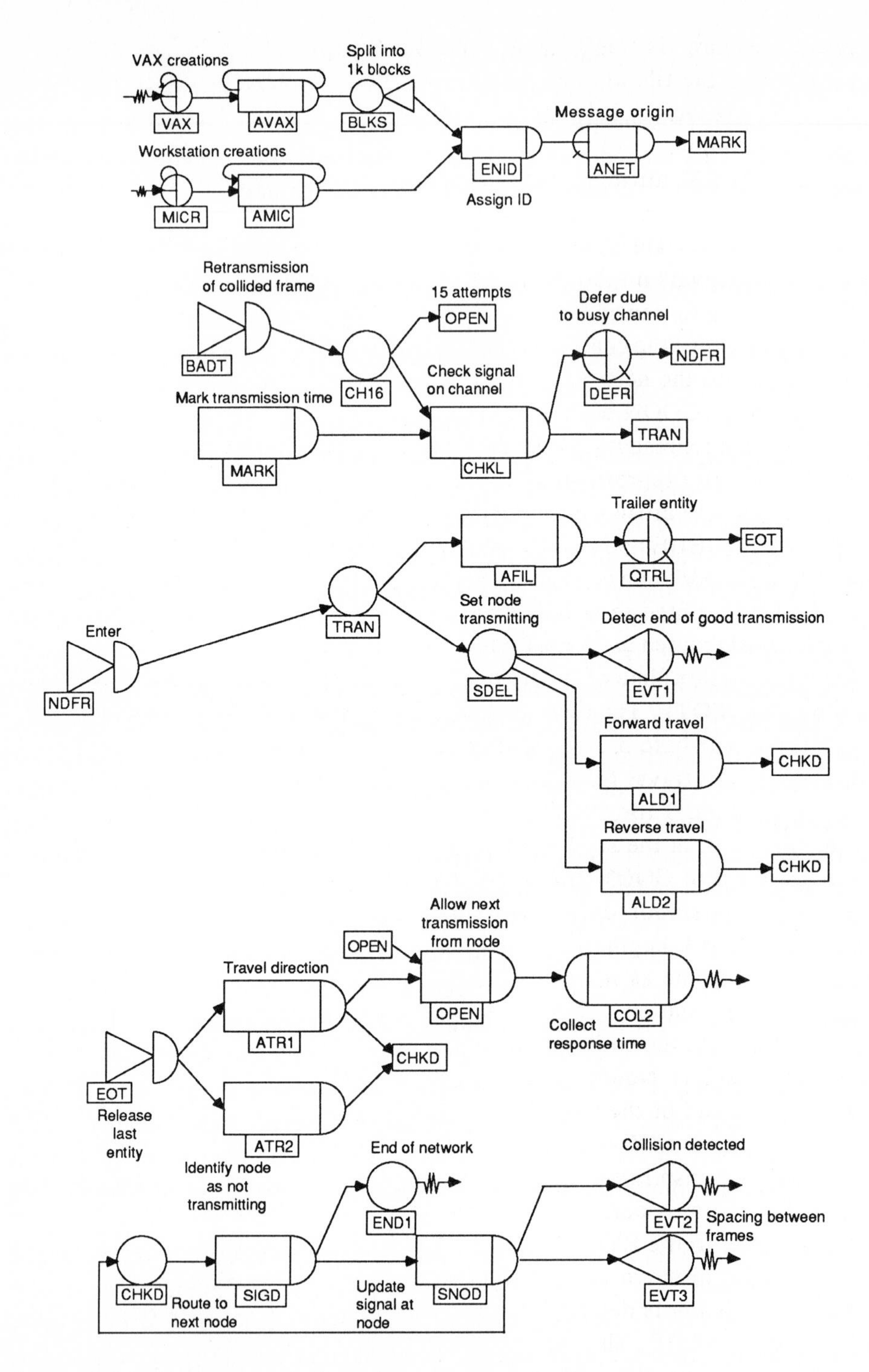

Figure 21-14 SLAM II model for Ethernet communications network.

When the resource is freed, the next message originating at the work station is removed from the file at node ANET. Once the message entity has seized the work station resource, it is routed to ASSIGN node MARK which begins the transmission logic.

At node MARK, attribute 7 is used to mark the time a frame entity begins its first attempt to transmit. The frame entity passes to ASSIGN node CHKL where network status variables, modeled by XX variables, are checked to determine the status of signals on the transmitting channel. If a signal is detected at the node attempting to transmit, the frame entity is branched to QUEUE node DEFR where it waits until the channel clears. When the channel is free, the frame entity is reentered into the network at ENTER node NDFR and message transmission is reinitiated at GOON node TRAN.

Two entities representing a leading bit and a trailing bit emanate from node TRAN. An entity representing the message's trailing bit is routed to ASSIGN node AFIL, where attribute 4 is given a value of +1. The trailing entity is then entered into QUEUE node QTRL and waits until the frame completes transmission on the network or a collision is detected. The second entity emanating from node TRAN is routed to GOON node SDEL, where three identical entities are created. One is routed to EVENT node EVT1 over an activity with duration equal to the transmission time of the frame. This entity represents a successful end of transmission (EOT). The other two entities are routed to ASSIGN nodes ALD1 and ALD2. Attribute 3 of the entity at node ALD1 is set to a value of +1 to indicate forward travel from the sending station on the transmission line. The value of attribute 3 for the entity at node ALD2 is set to equal -1 to indicate reverse travel from the sending station on the transmission line. Both entities are then routed to GOON node CHKD to simulate traversing the transmission line to the next station on the network.

The subnetwork beginning with node CHKD in Figure 21-14 simulates the progress of a frame as it travels in both directions simultaneously from work station to work station. If the leading bit entity has reached the end of the Ethernet network, the entity is routed from ASSIGN node SIGD to the TERMINATE node END1. Nodes SIGD and SNOD model the movement from node to node until the end of the Ethernet network is reached. As the leading bit entity reaches each node, XX variables are set to reflect the presence of the message. If the node is transmitting when the leading bit entity arrives, the node detects that a collision has occurred and begins to process the collision logic by sending a cloned entity to the EVENT node EVT2. The entity representing the leading byte continues to move along the transmission line until it reaches the end.

When a collision is detected, a trailing bit entity is entered into the network at ENTER node EOT. This entity is cloned and routed to ASSIGN nodes ATR1 and ATR2 where attribute 3 is assigned to the direction of travel on the transmission line. Both entities are then routed to GOON node CHKD. The trailing bit entities move through the subnetwork to set XX variables at each node to indicate that the message signal is no longer present. As the signal passes a node that is defering, a cloned entity is routed from ASSIGN node SNOD to EVENT node EVT3 over an activity with duration equal to the spacing time between frames. The event logic in EVENT node EVT3 tries to release the leading bit as the deferring message. The collision EVENT logic also enters an entity at ENTER node BADT. If the message has failed transmission after 15 attempts, the entity is routed from GOON node CHI6 to FREE node OPEN which frees the work station resource for a new message. If the message has not yet failed 15 times, this entity is routed to the transmission logic to represent the leading bit entity as the new transmission.

The end of a successful transmission is processed when the entity routed from GOON node SDEL arrives at EVENT node EVT1. If a collision has not occurred the trailing bit entity is removed from QUEUE node QTRL and entered back into the network at ENTER node EOT. The entity is cloned to represent the two trailing bit entities (one in each direction from the node). These entities are routed to node CHKD and moved down the transmission line as described previously. The control resource at the transmitting node is freed at FREE node OPEN after delaying on an activity for the 9.6 microsecond spacing time. After freeing up the resource, the entity is routed to COLCT node COL2 to collect an observation on response time.

21.6.3 Ethernet Communication Model Output

The performance measures estimated with the SLAM II simulation model for a 29-node Ethernet communications network are shown in Figure 21-15. The output presented here represents the results of a study for one specific Ethernet configuration. From these graphs, it can be determined that network response time, network wait time, and the message collision rate begin to deteriorate as the messages per second approach 800. At this point, the probability of message collision is approximately 0.40. Thus, an increase in message traffic above 800 messages per second would cause serious delays in the transfer of data. Furthermore, system saturation is expected to occur when the message arrival rate approaches 1200 messages per second. In this study, system saturation was achieved when the probability of a message collision exceeded 0.55.

Given the results of a simulation for a computer communication system, the system designer can estimate the limitations of a proposed or current communication network. If the analysis indicates that the system meets or exceeds stipulated minimum requirements for the network, then methods for reducing

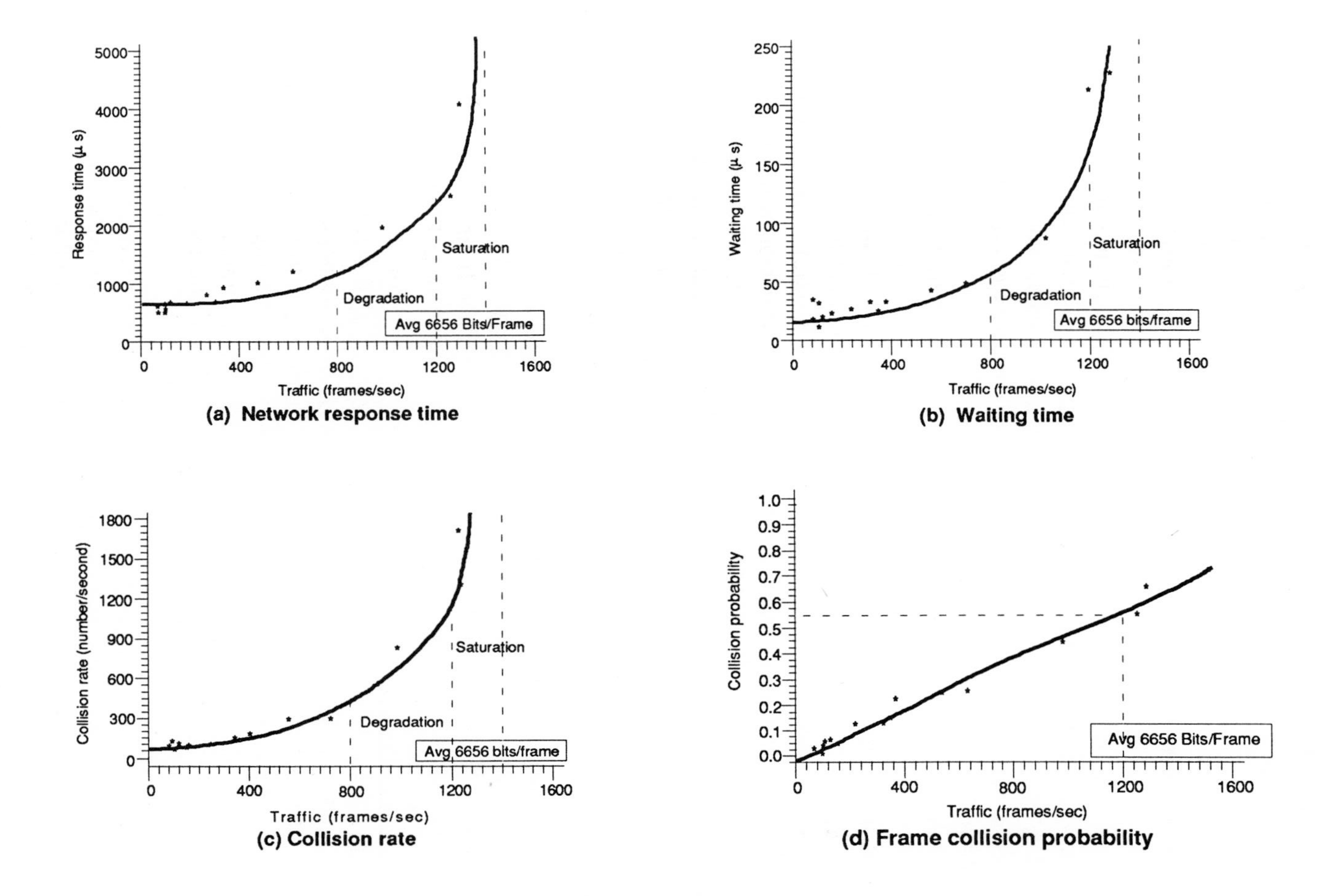

Figure 21-15 Performance measure plots the Ethernet communication network. Source [2]

the cost of the system are investigated. If the analysis indicates that the system does not meet requirements, then additional network configuration alternatives are considered.

21.7 SUMMARY

The characteristics of computer communication networks are described in this chapter. Performance measures for communication systems are defined. Two SLAM II network models are presented to illustrate techniques that have been successfully applied to the analysis of computer communication systems. Typical outputs required by system designers are presented.

21.8 EXERCISES

21-1. A microcomputer laboratory at a major university provides 24 microcomputers for the use of students. One of these microcomputers is connected to a 132 column line printer to provide students with hard copy output. Thus, a student's only communications link to the line printer consists of a dedicated microcomputer connected to the printer. If the printer is busy when a student wants to print output, the student waits until all previous printing jobs are completed, loads the file to be printed on the computer, then waits until the printing job is finished. Assuming that all twenty-four of the student microcomputers are in constant use, develop a SLAM II network that can be used to estimate student waiting time at the microcomputer/printer communication network and the average printer utilization per hour. Simulate the network given that the time between student arrivals at the printer is exponentially distributed with a mean of five minutes and the average printing time is normally distributed with a mean of four minutes and a standard deviation of one minute.

Embellishments

(a) Modify the SLAM II model to provide for a printer dedicated to every two microcomputers. Each of the 12 printers is connected to two microcomputers through an electronic server which automatically selects a microcomputer using a preferred order. Simulate this situation to estimate the average waiting time of jobs for printers and the average printer utilization.

(b) Prepare a table of student waiting time and printer utilization versus the number of microcomputers connected to a single printer through an electronic server-selector device. Discuss how you would decide on the number of printers to install in the microcomputer laboratory.

(c) Assume the microcomputer laboratory is open to students for an eight-hour period during which students arrive, on the average, every 10 minutes. Students arrive individually and the time between arrivals is exponentially distributed. If a microcomputer is available when a student arrives, the student immediately starts using it. If more than five students are waiting for a microcomputer when a student arrives, the arriving student exits the system. Student time on a microcomputer is normally distributed with a mean time of 1.5 hours and a standard deviation of 0.3 hours. Each student uses a printer from 1 to 5 times during a computing session with the time between uses being uniformly distributed. Build a SLAM II model to simulate the microcomputer laboratory. Would you change your recommendation for the number of printers required for the laboratory given this situation?

21-2. For a star topology communications system which connects six work stations to a main frame computer as illustrated in Figure 21-7, answer the following questions.
   (a) What are the major advantages of this topology?
   (b) What are the primary disadvantages of this topology?
   (c) Develop a SLAM II network to provide two-way communications between each working station and the mainframe computer.
   (d) What changes are required for the SLAM II network to provide communications between each work station via the mainframe computer?
   (e) What data would be required to simulate this network?
   (f) What modifications would be required to the SLAM II network model if each work station is assumed to be a microcomputer and the central computer is also a microcomputer that is designated as the communications server for the network?

21-3. Answer the following questions for a system consisting of seven microcomputers arranged in a ring topology as illustrated in Figure 21-7.
   (a) What are the primary advantages and disadvantages of this topology?
   (b) Develop a SLAM II network model for this topology, assuming that a token-passing message access method is employed on the network to establish priorities for the message traffic between the stations.
   (c) What changes would be required in the network if one of the microcomputers is designated as the server for the system?
   (d) What data would be required to simulate this network?

21-4. Develop a SLAM II network model for 10 microcomputers connected in the bus topology illustrated in Figure 21-7, assuming that CSMA/CD is used as the communication access method for the network.
   (a) What changes would be required in the network if a mainframe or if a minicomputer is incorporated into the system?
   (b) How could the model be modified to provide for communication links to parallel communication networks?
   (c) What changes would be required in the network if the 10 microcomputers were replaced by sensors that communicate system status to a controlling computer in an automated manufacturing environment?

21-5. In Section 21.4.1, Garcia's model employs a modular structured approach for the simulation of complex communication networks. Discuss how this approach might be used to model and simulate the telecommunications network illustrated in Figure 21-1. Include in your discussion, definitions of performance measures, and the variables in the model that provide estimates for them.

21.9 REFERENCES

1. Chlamtac, I., and W. R. Franta, "A Generalized Simulator for Computer Networks," *Simulation*, October 1982, pp. 123-132.
2. Erdbruegger, M. E., and R. Bonin, "Ethernet Simulation Report," Personal Communication, 1984.
3. Foster, S. J., "Design and Implementation of a Generic Computer Network Simulation System," unpublished MS Thesis, Air Force Institute of Technology, Wright-Patterson Air Force Base, Ohio, 1983.
4. Garcia, A. B., "Estimating Computer Communication Network Performance Using Network Simulations," unpublished dissertation, University of Dayton, April, 1985.
5. Green, R., and M. Fox, "AN-TTC-39 Circuit Switch Simulation," *Proceedings, Winter Simulation Conference*, 1975, pp. 211-216.
6. Kleinrock, L., *Queueing Systems Volume II: Computer Applications*, Wiley, New York, 1976.
7. Kuo, F. F., ed., *Protocols & Techniques for Data Communication*, Prentice-Hall, Englewood Cliffs, NJ, 1981.
8. Logendran, R., and M. P. Terrell, "Program Planning and Development of a National University Teleconference Network Using Simulation," *Proceedings, Winter Simulation Conference*, 1986, pp. 776-785.
9. Maira, A., "Local Area Networks - The Factory of the Future," *Manufacturing Engineering*, Vol. 96, March 1986, pp. 77-79.
10. Musselman, K. J. and R. J. Hannan, "A Network Simulation Model of a Computer Communications System," *Proceedings, Spring Conference*, 1984, pp. 425-433.
11. Pooch, V. W., C. Neblock, and R. Chattergy, "A Simulation Model for Network Routing," *Proceedings, Winter Simulation Conference*, 1979, pp. 435-445.
12. Schwetman, H., "CSIM: A C-Based, Process Oriented Simulation Language," *Proceedings, Winter Simulation Conference*, 1986, pp. 387-396.
13. Tanenbaum, A. S., *Computer Networks*, Prentice-Hall, Englewood Cliffs, NJ, 1981.
14. Telecommunications: A Survey, *The Economist*, November 29, 1985, pp. 5-40.
15. Warfield, J. N., *Societal Systems*, Wiley, New York, 1976.

AUTHOR INDEX

Q

R

S

T

SUBJECT INDEX

E

F

G

H

Q

R